**NASA
Reference
Publication
1178 Revised**

1987

Fifty Year Canon
of Solar Eclipses:
1986–2035

Fred Espenak

*Goddard Space Flight Center
Greenbelt, Maryland*

National Aeronautics
and Space Administration

**Scientific and Technical
Information Office**

FIFTY YEAR CANON OF SOLAR ECLIPSES: 1986 - 2035

Table of Contents

SECTION 1 - SOLAR ECLIPSE CATALOG: 1901 - 2100

SECTION 2 - WORLD MAPS OF ECLIPSE PATHS: 1901 - 2100

SECTION 3 - CENTRAL PATH CATALOG: 1986 - 2035

SECTION 4 - GLOBAL MAPS OF SOLAR ECLIPSES: 1986 - 2035

APPENDIX A - Solar Eclipses

APPENDIX B - Program SUNECL

INTRODUCTION

The primary purpose of the <u>Fifty Year Canon of Solar Eclipses:</u> <u>1986 - 2035</u> is to provide a reference of moderately detailed eclipse predictions and maps for use by the astronomical community. Such a work should be useful in identifying the most favorable eclipse opportunities, taking into account the celestial mechanics and geographic locations which are paramount in addressing the scientific goals, the issue of funding and the logistical problems of organizing an expedition to remote destinations. Special circulars containing detailed eclipse predictions are regularly issued by the Nautical Almanac Office of the U. S. Naval Observatory. Unfortunately, these circulars only become available 6 to 12 months before each event and cannot be of assistance to long range planning. <u>Canon of Solar Eclipses</u> [Meeus, Grosjean and Vanderleen, 1966] covers eclipses over a broad 612 year period from 1898 through 2510. Because of the sheer number of eclipses covered in this work (1448), the details for any one event must be rather brief. For instance, Meeus et al. give only the center line coordinates of each eclipse using a fairly coarse time interval of every 12 minutes. Center line coordinates over shorter time intervals as well as coordinates for the northern and southern limits and detailed maps of each path would be of great value to planners. The <u>Fifty Year Canon</u> contains this information (Sections 3 & 4), thereby filling a void in the published literature on eclipses.

The secondary purpose is to provide a general reference on future eclipses for teachers, students, amateur astronomers and interested laymen. The solar eclipse is unquestionably the most spectacular celestial phenomenon visible to the naked eye. As such, eclipses generate a great deal of interest among the general public and news media. Naturally, questions arise as to where a particular eclipse will be visible from, and when the next eclipse occurs. Unfortunately, there is very little information in print about the visiblity of future eclipses and what appears in the popular literature is often short on accuracy or even incorrect. Furthermore, most references are obscure, not easily accessible and/or out of print. The orthographic maps in the <u>Fifty Year Canon</u> show regions of visibility for both umbral and penumbral eclipses. From these, the time and maximum magnitude of partial phases or totality/annularity can be estimated for any locale. Furthermore, the 200 year eclipse catalog (Section 1) and world maps of every umbral eclipse during this period (Section 2) provide a convenient and useful reference.

1

ORGANIZATION OF THE CANON

The <u>Fifty Year Canon of Solar Eclipses: 1986 - 2035</u> is composed of four major sections and two appendices. Section 1 is a catalog which lists the general characteristics of every solar eclipse from 1901 through 2100. Section 2 presents a detailed set of cylindrical projection world maps which show the umbral paths of every solar eclipse from 1901 through 2100. Section 3 gives geodetic path coordinates and local circumstances on the center line for every central eclipse from 1986 through 2035. Finally, section 4 consists of a series of orthographic projection maps which show the regions of visibility of both partial and central phases for every solar eclipse from 1986 through 2035.

Appendix A provides some general background on solar eclipses and covers eclipse geometry, eclipse frequency and recurrence, modern eclipse prediction, geometry of the umbral shadow and time determination. Appendix B is a listing of a very simple Fortran program which can be used to predict the occurrence and general characteristics of solar eclipses. It makes use of many approximations while maintaining a reasonable level of accuracy and reliability. The program is based on algorithms devised by Meeus [1982] and the ample comments should make the program self-explanatory.

A detailed description of each section of the <u>Fifty Year Canon of Solar Eclipses: 1986 - 2035</u> follows.

SECTION 1 - SOLAR ECLIPSE CATALOG: 1901 - 2100

Section 1 consists of a catalog of the general characteristics at the instant of greatest eclipse for every solar eclipse during the two hundred year interval 1901 to 2100. During the first century, there are 228 eclipses of which 78 are penumbral and 150 are umbral. The second century contains 224 eclipses of which 77 are penumbral and 147 are umbral. However, to appreciate a realistic frequency and type distribution of present eclipses, it's necessary to sample a period commensurate with the 18 year 11 day Saros cycle. The period from 1986 to 2003 contains 39 eclipses of which 13 (33.3%) are penumbral and 26 (66.6%) are umbral. Of these, 12 (30.8%) are annular, 12 (30.8%) are total and the remaining 2 (5.1%) are annular/total. Since the Saros cycle is not static, these figures will change. For example, eclipses in Saros series 124 will change from annular/total to partial in 2004.

Column 1 of the catalog lists the Gregorian Date of each eclipse while column 2 gives the Julian Date. The Julian Date is the number of days elapsed since Greenwich Mean Noon on 1 January 4713 BC. Column 3 gives the Universal Time (hours:minutes:seconds) of greatest eclipse. Greatest eclipse is defined as the instant when the axis of the Moon's shadow passes closest to the Earth's center. Column 4 lists the value for delta T (in seconds) which was used in the calculations. Delta T is the difference between Terrestrial Dynamical Time and Universal Time. For the period 1901 - 1985, the values for delta T were determined from observations. Beyond 1985, the values for delta T were extrapolated and are only approximate since fluctuations in the Earth's rotation rate are unpredictable (See: Appendix A - Time Determination). Column 5 characterizes the nature of the eclipse as follows:

T = Total Eclipse.
A = Annular Eclipse.
A/T = Annular/Total Eclipse.
P = Partial Eclipse.

Additional information about the eclipse is defined as follows:

B = Begin (i.e. - the first eclipse of a new Saros series).
E = End (i.e. - the last eclipse of an old Saros series).
+ = Central Eclipse with no Northern Limit.
- = Central Eclipse with no Southern Limit.
* = Non-Central Umbral Eclipse
  (i.e. - edge of umbral cone grazes Earth but shadow axis misses).

5

The next column gives the Saros series to which the eclipse belongs. The Saros series numbers are consistent with those introduced by van den Bergh [1955]. Eclipses belonging to an odd numbered Saros take place at the ascending node of the Moon's orbit (Gamma decreases with each succeeding eclipse), while eclipses of an even numbered Saros take place at the descending node (Gamma increases with each succeeding eclipse). Column 7 lists the value of Gamma, which is defined as the minimum distance of the shadow cone axis from the center of the Earth in units of equatorial radii. This corresponds to the instant of greatest eclipse. The sign of Gamma indicates whether the shadow cone axis passes north (+) or south (-) of the Earth's center. Column 8 gives the maximum magnitude of the eclipse from any point on the Earth's surface at the instant of greatest eclipse. Eclipse magnitude is defined as the fraction of the Sun's diameter obscured by the Moon. For partial eclipses, the magnitude is always less than 1.000. For annular eclipses, the magnitude is also less than 1.000 and is equal to the topocentric ratio of the diameters of the Moon and Sun. The magnitude for total eclipses is always equal to 1.000. Therefore, the magnitude has been replaced with the topocentric ratio of the apparent diameters of the Moon and Sun. The apparent diameter ratio is always greater than or equal to 1.000 for total eclipses.

The geodetic coordinates of the shadow cone axis at the instant of greatest eclipse are listed in the columns 'LATITUDE' and 'LONGITUDE'. These coordinates are given in degrees and minutes to the nearest tenth of a minute. Negative latitudes are South of the Equator and negative longitudes are East of the Greenwich Meridian. Although greatest eclipse differs slightly from the instants of greatest magnitude and greatest duration (for total eclipses), the differences are usually negligible. For central eclipses, these coordinates represent the point where the Sun reaches its greatest obscuration from any location on Earth. For partial eclipses, the shadow axis misses the Earth entirely. Therefore, the point of greatest eclipse lies on the day/night terminator and the Sun appears in the horizon. The next two columns, 'SUN ALT' and 'SUN AZ', give the Sun's altitude and azimuth as seen from the previous coordinates at the instant of greatest eclipse. The width of the path of totality or annularity at that time follows (kilometers). Finally, the duration of the central phase (total or annular) at the instant of greatest eclipse is given in minutes and seconds.

SECTION 2 - WORLD MAPS OF ECLIPSE PATHS: 1901 - 2100

Section 2 presents maps of the path of every umbral eclipse during the two hundred year period 1901 through 2100 (as tabulated in Section 1). The eclipses are broken up into twenty year intervals and every decade is plotted on a set of three cylindrical projection maps of the Earth. Each map is centered on the equator and covers 160° in latitude (80° North to 80° South) and 130° in longitude. The first map of each trio runs from longitude 160° West to 30° West, the second map covers 40° West to 90° East and the third map encompasses 80° East to 150° West (= 210° East). Thus, the entire planet (excluding the extreme polar regions) is covered with a 10° of longitude overlap between each map.

The northern and southern limits of each eclipse path during a twenty year interval are plotted, along with the outline of the umbral shadow at every half hour Universal Time. Furthermore, the point of greatest eclipse is indicated with an '*', usually near the center of each path. Total eclipses are readily distinguished from annular eclipses because their paths are plotted with a heavier pen. The date of each eclipse appears directly above the point of greatest eclipse ('*'). In cases where the '*' appears within two overlapping eclipse paths, its identity can be established by noting that it usually lies midway between the ends of the path.

There are eight non-central eclipses during the two hundred year interval of Section 2. They occur in 1928, 1950, 1957 (two), 1967, 2014 and 2043 (two). These non-central eclipses are easily recognized by their distictive 'D' shaped paths and from the fact that the point of greatest eclipse lies on the terminator edge of the path. In addition, the annular eclipse of May 2003 has no northern limit, while the annular eclipse of Feb 2044 has no southern limit. These two events are also bounded by the terminator along their missing edges.

7

SECTION 3 - CENTRAL PATH CATALOG: 1986 - 2035

Section 3 consists of a series of tables which list the geodetic coordinates of the path of totality or annularity for every central eclipse over the fifty year interval 1986 to 2035. During this period, a total of 109 solar eclipses occur; 37 are penumbral or partial while the remaining 72 are umbral. Appearing at the top of each table is the type of eclipse (annular, total or annular/total) and date, followed by the Saros series and the extrapolated value of delta T which was used in the calculations. The first column of each table gives the Universal Time (UT) for the data which follows. Depending on the duration of the central eclipse, data are listed at one, two, four or six minute intervals. The next six columns define the geodetic coordinates of the northern and southern limits as well as the center line. The latitude and longitude of each point are given in degrees and minutes to the nearest tenth of a minute. Negative latitudes are South of the Equator and negative longitudes are East of the Greenwich Meridian.

The column identified as 'DIAMETER RATIO', is the ratio of the topocentric apparent diameters of the Moon and the Sun. For total eclipses, the ratio is always greater than or equal to 1.000. For annular eclipses, the ratio is less than 1.000 and is identical to the eclipse magnitude at maximum eclipse. Eclipse magnitude is defined as the fraction of the Sun's diameter obscured by the Moon. The next two columns, 'SUN ALT' and 'SUN AZ', give the Sun's altitude and azimuth at maximum eclipse as seen by an observer on the center line. The eleventh column lists the width of the path of totality or annularity in kilometers. Finally, the duration of the total or annular phase is given in minutes and seconds. For annular/total eclipses, the sections of the path which are total in nature are identified by a 'T' after the duration.

All path characteristics are calculated for sea level and the effects of refraction have been ignored. As we progress into the twenty-first century, the actual value for delta T will inevitably diverge from the extrapolated ones because of the unpredictable fluctuations in the Earth's rotation (See: Appendix A - Time Determination). The result of this divergence will be to induce a shift in the path coordinates east or west in longitude. Fortunately, corrections to the tabulated longitudes are quite straight forward, using the following equation:

$$\text{Shift (in degrees)} = 0.00417807 * (\Delta T1 - \Delta T2)$$

where: $\Delta T1$ = table value of delta T (in seconds)
$\Delta T2$ = true or observed delta T (in seconds)

This shift is added to the tabulated path longitudes to calculate the correct longitudes of the umbral path. Changes in delta T have no effect on the tabulated latitudes.

It should be pointed out that the path of the annular eclipse of 29 April 2014 is not included in this section because it has no central line. The shadow axis misses the Earth while the edge of the shadow cone barely grazes the planet. Furthermore, the annular eclipse of 31 May 2003 has no northern limit since the northern edge of the shadow cone misses the Earth. In this case, the northern boundary of the path is defined by the Earth's day/night terminator.

SECTION 4 - GLOBAL MAPS OF SOLAR ECLIPSES: 1986 - 2035

Section 4 consists of a series 109 maps, one for every solar eclipse during the fifty year interval 1986 to 2035. Each map is an orthographic projection of the Earth which shows the path of partial and central eclipse. North is to the top in all cases and the daylight terminator is plotted for the instant of greatest eclipse. The sub-solar point on the Earth is indicated by a star shaped character. The maps for partial eclipses are oriented with their origins (i.e. - disk center) at the sub-solar longitude at the instant of greatest eclipse and the latitude is equal to the Sun's declination plus or minus 45° (depending on the hemisphere). The maps for total or annular eclipses are oriented with the point of greatest eclipse at the origin.

The limits of the Moon's penumbral shadow delineate the region of visibility of the partial solar eclipse. This irregular or saddle shaped region often covers more than half of the daylight hemisphere of the Earth and consists of several distinct zones or limits. At the northern and/or southern boundaries lie the limits of the penumbra's path. Partial eclipses have only one of these limits, as do central eclipses when the shadow axis falls no closer than about 0.45 radii of the Earth's center. Great loops at the western and eastern extremes of the penumbra's path identify the areas where the eclipse begins/ends at sunrise and sunset, respectively. If the penumbra has both a northern and southern limit, the rising and setting curves form two separate, closed loops. Otherwise, the curves are connected in a distorted figure eight. Bisecting the 'eclipse begins/ends at sunrise and sunset' loops is the curve of maximum eclipse at sunrise (western loop) and sunset (eastern loop). The points 'P1' and 'P4' mark the coordinates where the penumbral shadow first contacts (partial eclipse begins) and last contacts (partial eclipse ends) the Earth's surface. If the penumbral path has both a northern and southern limit, then points 'P2' and 'P3' are also plotted. These correspond to the coordinates where the penumbral shadow cone becomes internally tangent to the Earth's disk.

A curve of maximum eclipse is the locus of all points where the eclipse is at maximum at a given time. Curves of maximum eclipse are plotted at each half hour Universal Time. They generally run from the northern to the southern penumbral limits, or from the maximum eclipse at sunrise and sunset curves to one of the limits. If the eclipse is central, the curves of maximum eclipse run through the half-hourly outlines of the umbral shadow, from which the Universal Time of each curve can be identified. The curves of constant eclipse magnitude delineate the locus of all points where the magnitude at maximum eclipse is constant. These curves run exclusively between the curves of maximum eclipse at sunrise and sunset. Furthermore, they're parallel to the northern/southern penumbral limits and the umbral paths of central eclipses. In fact the northern and southern limits of the penumbra can be thought of as curves of constant magnitude of 0.0. The adjacent curves are for magnitudes of 0.2, 0.4, 0.6 and 0.8. The northern and southern limits of the umbra which define the path of totality are curves of constant magnitude of 1.0.

Greatest eclipse is defined as the instant when the axis of the Moon's shadow passes closest to the Earth's center. Although greatest eclipse differs slightly from the instants of greatest magnitude and greatest duration (for total eclipses), the differences are usually negligible. The point on the Earth's surface which is at or is nearest to the axis at this time is marked by an '*'. For partial eclipses, the shadow axis misses the Earth entirely. Therefore, the point of greatest eclipse lies on the day/night terminator and the sun appears in the horizon. For central eclipses, the umbral path begins and ends along the curves of maximum eclipse at sunrise and sunset. The outline of the Moon's umbral shadow is plotted at every half hour Universal Time along the central path. In addition, the umbra's position is labeled at each integral hour Universal Time.

Data pertinent to the eclipse appear with each map. In the upper left corner are the Universal Times of greatest eclipse and conjunction of the Moon and Sun in right ascension, the minimum distance of the Moon's shadow axis from the Earth's center in Earth radii (Gamma) and the geocentric ratio of diameters of the Moon and the Sun. For partial eclipses, the geocentric ratio is replaced by the magnitude at greatest eclipse. To the upper right are exterior contact times of the Moon's shadow with the Earth. P1 and P4 are the first and last contacts of the penumbra; they mark the start and end of the partial eclipse. U1 and U4 are the first and last contacts of the umbra; they denote the start and end of the total or annular eclipse. Below each map are the geocentric coordinates of the Sun and Moon at the instant of greatest eclipse. They consist of the right ascension (RA), declination (DEC), apparent semi-diameter (SD) and horizontal parallax (HP). The Saros series for the eclipse is listed, followed by a pair of numbers in parentheses. The first number identifies the sequence order of the eclipse in the series, while the second number is the total number of eclipses in the series. The Julian Date (JD) at greatest eclipse is given, followed by the extrapolated value of delta T (ΔT) used in the calculations (delta

T is the difference between Terrestrial Dynamical Time and Universal Time). Finally, the geodetic coordinates of the point of greatest eclipse are given, as well as the local circumstances there. In particular, the Sun's altitude (ALT) and azimuth (AZ) are listed along with the duration of totality or annularity (minutes:seconds) and the width of the path (kilometers).

ACCURACY OF THE EPHEMERIDES

The solar ephemeris which was used for these predictions is based on the classic work of Newcomb [1895]. It includes all planetary perturbation terms in longitude and latitude with arguments greater than 0.01 arc-seconds. The lunar ephemeris was developed primarily from the the work of Brown [1919] with improvements from Eckert, Jones and Clark [1954]. All solar perturbation terms in longitude and latitude with coefficients greater than 0.025 arc-seconds have been included. The cut-off for planetary perturbations is 0.025 and 0.01 in longitude and latitude respectively. Perturbations in lunar parallax include all terms with coefficients greater than 0.0010. Finally, all terms additive to the Moon's fundamental arguments with coefficients greater than 0.025 arc-seconds have been retained.

In order to determine the accuracy of these ephemerides, they have been compared against the Jet Propulsion Laboratory's Developmental Ephemeris 200 (or JPL DE-200) for 260 new moon dates over the interval 1980 through 2000. The mean differences and standard deviations of the solar and lunar ephemerides with the JPL DE-200 are as follows:

Comparison of Solar/Lunar Ephemerides with JPL DE-200

	RA Mean (sec)	RA S. Dev. (sec)	Dec Mean (arc-sec)	Dec S. Dev. (arc-sec)
Sun	+0.036	0.044	-0.030	0.173
Moon	-0.002	0.031	-0.008	0.386

The solar and lunar deviations in declination correspond to an uncertainty in the umbral shadow's position of 1044 meters in the north/south direction in the Earth's fundamental plane. Similarly, the deviations in right ascension correspond to an uncertainty in the umbra's position of 2102 meters in the east/west direction. The mean angular velocity in right ascension of the Moon with respect to the Sun is 0.0343 seconds per second. Thus, the combined uncertainties in right ascension can also be thought of as an uncertainty of 2.2 seconds in time.

On the scale of the maps presented in Sections 2 and 4, path shifts of this magnitude are not visible. Thus, the uncertainties are well within the acceptable range for the purposes of predicting eclipse path coordinates and local circumstances for long range planning. In fact, the predictions presented here should be adaquate for most center line experiments. However, observers near the path limits will require predictions using higher accuracy ephemerides which also take into account a detailed analysis of the Moon's limb profile. Such information is available through special eclipse circulars published by: Nautical Almanac Office, U. S. Naval Observatory, Washington, D.C. 20392-5100, U.S.A.. These circulars are available 6 to 12 months before each major eclipse. For eclipses of low solar altitude, the observer must also make corrections for the effects of atmospheric refraction.

In the generation of eclipse predictions presented in the <u>Fifty Year Canon</u>, the author has applied a -0.6 arc-second correction to the Moon's ecliptic latitude. This takes into account the difference between the Moon's center of mass and center of figure. In addition, a correction of -1.34 seconds has been applied to the lunar ephemeris to reconcile it with the FK4 equinox.

In August 1982, the IAU General Assembly passed a resolution to adopt a value for the mean lunar radius of $k=0.2725076$, in units of the Earth's equatorial radius. This is the value currently used by the Nautical Almanac Office for solar eclipse predictions [Fiala and Lukac, 1983] and is believed to be the best mean radius, averaging mountain peaks and low valleys along the Moon's rugged limb. Previously, the Nautical Almanac Office used two separate values for 'k' in their eclipse predictions. The larger value ($k=0.2724880$) was used for annular eclipses and exterior contacts while a smaller value ($k=0.272281$) was reserved for second and third contact calculations of total eclipses [Explanatory Supplement, 1974]. In principle, the IAU's adoption of one single value for 'k' is commendable because it should eliminate confusion arising from the use of two different values. It also eliminates the discontinuity present since two different values of 'k' were used for total and annular eclipses. Nevertheless, a smaller value of 'k' is still necessary since it represents the lunar valley bottoms and hence the minimum solid disk. This is significant because a solar eclipse cannot be regarded as total as long as any photospheric rays reach the observer through valleys along the lunar limb [Meeus, Grosjean and Vanderleen, 1966].

Of primary interest to most observers are the times when central eclipse begins and ends (second and third contacts, respectively) and the duration of the central phase. When the IAU's mean value for 'k' is used to calculate these times, they must be corrected to accommodate low valleys (total) or high mountains (annular) along the Moon's limb. The calculation of these corrections is not trivial but must be performed, especially if one plans to observe near the path limits [Herald, 1983].

For observers near the center line of a total eclipse, the limb corrections can be closely approximated by using a smaller value of 'k' which accounts for the valleys along the profile.

For this reason, the author has chosen to adopt the smaller value for 'k' (k=0.272281). As a consequence, the current work predicts slightly shorter durations and narrower paths for total eclipses when compared with the Nautical Almanac Office's calculations which use the larger IAU value for 'k'. Furthermore, the smaller 'k' will produce longer durations and wider paths for annular eclipses.

The total solar eclipse of 3 October 1986 illustrates the effects of using different values of k in eclipse calculations. For example, the Fifty Year Canon predicts a maximum duration of totality of 0.2 seconds and a path width of 2.3 km (k=0.272281). However, the Astronomical Almanac for 1986 predicts a duration of 3.3 seconds and a path width of 31.1 km (k=0.2725076). Experienced observers in the path reported that the eclipse was never quite total. Instead, Baily's Beads surrounded the Moon for several seconds during maximum eclipse. The symmetry of the beading phenomena in photographs proves that the observers were near the center line and not the edge of the path. Thus, predictions using the smaller 'k' are in better agreement with actual observations made near the center line.

REFERENCES

Astronomical Almanac for 1986, 1985, Nautical Almanac Office, U. S. Naval Observatory, Washington, D.C..

Brown, E. W., 1919, Tables of the Motion of the Moon, Yale University Press, New Haven.

Chauvenet, W. A., 1960, A Manual of Spherical and Practical Astronomy, Dover (reprint), New York.

Eckert, W. J., Jones, R., and Clark, H. K., 1954, Improved Lunar Ephemeris 1952-1959, Nautical Almanac Office, U. S. Naval Observatory, Washington, D.C..

Espenak, F., 1980, "North American Solar Eclipses : the Next 50 Years", Astronomy, 8, 11.

Espenak, F., 1982, "Eclipse Chaser's Notebook", Astronomy, 10, 6.

Explanatory Supplement to the Astronomical Ephemeris and the American Ephemeris and Nautical Almanac, 1974, H. M. Nautical Almanac Office, London.

Fiala, A.D. and Lukac, M.R., 1983, "Annular Solar Eclipse of 30 May 1984", Circular No. 166, Nautical Almanac Office, U. S. Naval Observatory, Washington, D.C..

Herald, D., 1983, "Correcting Predictions of Solar Eclipse Contact Times ...", J. Brit. Ast. Assoc., 93, 6.

Meeus, J., Grosjean, C. C., and Vanderleen, W., 1966, Canon of Solar Eclipses, Pergamon Press, New York.

Meeus, J., 1982, Astronomical Formulae for Calculators, Willmann-Bell, Richmond.

Meeus, J., 1984, "Solar Eclipse Diary: 1985 - 1995", Sky and Telescope, 68, 4.

Mucke, H. and Meeus, J., 1983, Canon of Solar Eclipses, -2003 to +2526, Astronomisches Buro, Austria.

Newcomb, S., 1895, "Tables of the Motion of the Earth on its Axis Around the Sun", Astron. Papers Amer. Eph., Vol. 6, Part I.

Oppolzer, Th. von, 1887, Canon of Eclipses, (translated by O. Gingerich), Reprinted 1962, Dover, New York.

Smart, W. M., 1977, Text-Book on Spherical Astronomy, 6th Ed., Cambridge University Press, Cambridge.

van den Bergh, G., 1955, Periodicity and Variations of Solar (and Lunar) Eclipses, Tjeenk Willink, Haarlem, Netherlands.

FIFTY YEAR CANON OF SOLAR ECLIPSES : 1986 - 2035

SECTION 1 - SOLAR ECLIPSE CATALOG : 1901 - 2100

Table 1

CANON OF SOLAR ECLIPSES
LOCAL CIRCUMSTANCES AT GREATEST ECLIPSE

DATE	JULIAN DATE	GREATEST ECLIPSE	DELTA T	TYPE	SAROS	GAMMA	MAGNITUDE	LATITUDE	LONGITUDE	SUN ALT	SUN AZ	PATH WIDTH	CENTRAL DURATION
18 MAY 1901	2415522.73	5:33:48	-2.0	T	136	-0.3627	1.0680	-1-41.7	-98-25.4	68.7	352.7	237.7	6:28.6
11 NOV 1901	2415699.81	7:28:15	-0.9	A	141	0.4755	0.9218	10 46.4	-68-56.1	61.8	190.0	336.0	11: 1.2
8 APR 1902	2415848.09	14: 5: 2	-0.2	PE	108	1.5024	0.0638	71 41.1	142 27.8				
7 MAY 1902	2415877.44	22:34:14	-0.1	P	146	-1.0832	0.8588	-69-57.8	125 4.7				
31 OCT 1902	2416053.83	8: 0:11	0.5	P	151	1.1553	0.8961	70 47.0	-100-52.7				
29 MAR 1903	2416202.57	1:35:18	1.0	A	118	0.8412	0.9767	58 13.2	-130-20.1	32.4	147.3	153.0	1:52.7
21 SEP 1903	2416378.69	4:39:45	1.5	T	123	-0.8969	1.0315	-57-59.9	-77-14.0	25.9	35.1	240.9	2:11.9
17 MAR 1904	2416556.74	5:40:39	2.0	A	128	0.1298	0.9368	5 38.4	-94-44.3	82.5	162.4	237.1	8: 7.2
9 SEP 1904	2416733.36	20:44:16	2.4	T	133	-0.1627	1.0709	-3-43.1	134 31.2	80.6	17.1	233.6	6:19.8
6 MAR 1905	2416910.72	5:12:20	2.8	A	138	-0.5771	0.9289	-39-31.6	-117-25.0	54.6	338.0	334.1	7:57.8
30 AUG 1905	2417088.05	13: 7:22	3.6	T	143	0.5708	1.0477	42 27.8	4 18.5	55.0	201.6	192.4	3:46.1
23 FEB 1906	2417264.82	7:43:15	4.8	P	148	-1.2480	0.5379	-71-26.5	170 16.8				
21 JUL 1906	2417413.05	13:14:13	5.3	P	115	-1.3639	0.3351	-68-36.6	33 16.6				
20 AUG 1906	2417442.55	1:12:45	5.4	P	153	1.3729	0.3146	70 50.3	66 22.9				
14 JAN 1907	2417589.75	6: 5:38	5.9	T	120	0.8626	1.0281	38 16.8	-86-24.1	30.2	175.3	189.1	2:24.7
10 JUL 1907	2417767.14	15:24:28	6.6	A	125	-0.6315	0.9456	-16-55.3	50 53.6	50.8	1.5	258.5	7:22.9
3 JAN 1908	2417944.41	21:45:20	7.2	T	130	0.1933	1.0438	-11-49.6	145 8.3	78.8	180.6	148.8	4:13.8
28 JUN 1908	2418121.19	16:29:48	7.9	A	135	0.1387	0.9655	31 25.7	67 9.2	81.7	177.1	126.0	3:59.9
23 DEC 1908	2418298.99	11:44:24	8.6	AT	140	-0.4985	1.0024	-53-25.7	0 32.1	59.8	7.7	9.8	0:11.8
17 JUN 1909	2418475.47	23:18:34	9.2	AT	145	0.8964	1.0065	82 52.2	-123-53.1	26.0	110.3	51.1	0:23.6
12 DEC 1909	2418653.32	19:44:43	9.8	P	150	-1.2456	0.5422	-65-14.1	-86 -1.7				
9 MAY 1910	2418800.74	5:42: 9	10.3	T	117	-0.9441	1.0600	-48-14.3	-125-18.0	18.9	327.9	596.4	4:15.1
2 NOV 1910	2418977.59	2: 8:26	11.6	P	122	1.0603	0.8507	61 55.1	154 60.0				
28 APR 1911	2419155.44	22:27:19	12.2	T	127	-0.2297	1.0582	1 55.1	161 54.5	76.7	335.9	189.9	4:57.5
22 OCT 1911	2419331.68	4:12:58	12.4	A	132	0.3223	0.9650	8 14.9	-121-22.5	71.2	206.5	132.7	3:47.4
17 APR 1912	2419509.98	11:34:19	13.1	AT	137	0.5277	1.0003	38 20.9	11 14.0	58.0	146.0	1.3	0: 1.6
10 OCT 1912	2419686.07	13:36: 3	14.0	T	142	-0.4150	1.0229	-28 -6.0	40 3.6	65.4	31.8	84.9	1:55.0
6 APR 1913	2419864.23	17:33: 3	14.7	P	147	1.3145	0.4235	81 11.1	-176-38.2				
31 AUG 1913	2420011.37	20:52: 5	15.2	P	114	1.4510	0.1503	81 27.7	26 40.1				
30 SEP 1913	2420040.70	4:46:43	15.3	P	152	-1.1007	0.8234	-61 -2.8	-11-34.9				
25 FEB 1914	2420188.51	0:12:58	15.6	A	119	-0.9418	0.9248	-82 -6.1	113 10.4	19.2	286.5	840.9	5:35.1
21 AUG 1914	2420366.02	12:34:23	15.7	T	124	0.7653	1.0328	54 27.1	-27 -3.3	39.8	228.6	169.8	2:14.5
14 FEB 1915	2420542.89	4:33:15	15.8	A	129	-0.2028	0.9789	-24 -3.3	-120-45.2	78.2	333.2	76.5	2: 3.5
10 AUG 1915	2420720.45	22:52:20	16.2	A	134	0.0121	0.9853	16 25.8	161 24.3	88.5	204.7	52.1	1:33.2
3 FEB 1916	2420897.17	16: 0:19	17.7	T	139	0.4988	1.0280	11 6.9	67 39.0	60.1	158.5	108.4	2:36.2
30 JUL 1916	2421074.59	2: 6: 5	18.4	A	144	-0.7712	0.9447	-29 -3.4	-132-25.5	39.4	21.8	313.6	6:24.3
24 DEC 1916	2421222.37	20:46:17	19.0	P	111	-1.5323	0.0111	-85-42.6	-32 -9.1				
23 JAN 1917	2421251.81	7:28:28	19.0	P	149	1.1507	0.7250	63 10.3	-25-35.1				
19 JUN 1917	2421399.05	13:18:18	18.7	P	118	1.2855	0.4732	86 10.3	-150 -5.0				
19 JUL 1917	2421428.61	2:42:38	18.7	PB	154	-1.5104	0.0854	-63-40.3	-101-45.9				

23

Table 2

CANON OF SOLAR ECLIPSES
LOCAL CIRCUMSTANCES AT GREATEST ECLIPSE

DATE	JULIAN DATE	GREATEST ECLIPSE	DELTA T	TYPE	SAROS	GAMMA	MAGNITUDE	LATITUDE	LONGITUDE	SUN ALT	SUN AZ	PATH WIDTH	CENTRAL DURATION
14 DEC 1917	2421576.89	9:27:17	18.4	A	121	-0.9158	0.9791	-88 -2.9	-125-19.8	23.2	270.7	188.9	1:18.6
8 JUN 1918	2421753.42	22: 7:40	18.9	T	126	0.4658	1.0292	50 50.4	152 1.4	62.0	180.4	112.1	2:22.9
3 DEC 1918	2421931.14	15:21:58	19.5	A	131	-0.9382	0.9382	-36 -3.6	53 41.3	75.9	2.4	236.4	7: 5.8
29 MAY 1919	2422108.05	13: 8:53	19.8	T	136	-0.2968	1.0719	4 22.8	16 42.2	72.8	356.1	244.4	6:50.8
22 NOV 1919	2422285.13	15:14: 6	20.0	A	141	-0.4548	0.9197	6 55.4	48 51.7	62.9	186.5	340.9	11:36.7
18 MAY 1920	2422462.76	8:14:54	20.3	P	146	-1.0240	0.9731	-89 -3.9	-107-41.9				
10 NOV 1920	2422639.16	15:52:11	20.1	P	151	1.1285	0.7421	89 55.4	29 48.1				
8 APR 1921	2422787.89	9:14:57	20.5	A	118	0.8888	0.9753	84 28.5	-5-36.0	27.1	138.8	191.4	1:50.4
1 OCT 1921	2422964.02	12:35:54	21.4	T	123	-0.9384	1.0293	-66 -6.9	56 7.5	19.7	48.1	291.0	1:52.1
28 MAR 1922	2423142.05	13: 5:23	21.8	A	128	0.1708	0.9381	12 15.7	17 58.6	80.1	162.7	233.2	7:50.2
21 SEP 1922	2423318.69	4:40:28	21.8	T	133	-0.2131	1.0678	-10-45.0	-104-31.4	77.7	17.6	226.2	5:58.8
17 MAR 1923	2423496.03	12:44:54	21.9	A	138	-0.5441	0.9310	-33 -2.2	-2-22.1	56.9	338.8	305.3	7:51.3
10 SEP 1923	2423673.37	20:47:28	22.2	T	143	-0.5148	1.0430	34 40.0	121 46.6	58.9	201.1	166.8	3:36.8
5 MAR 1924	2423850.18	15:44:16	22.4	P	148	-1.2234	0.5811	-71-55.6	-55-39.2				
31 JUL 1924	2423998.33	19:58:16	22.5	P	115	-1.4461	0.1915	-89-33.3	146 32.0				
30 AUG 1924	2424027.85	8:22:57	22.6	P	153	1.3121	0.4244	71 27.2	-172-54.3				
24 JAN 1925	2424175.12	14:53:59	22.7	T	120	0.8660	1.0305	40 28.6	49 34.4	29.8	170.1	206.4	2:32.3
20 JUL 1925	2424352.41	21:48:38	22.7	A	125	-0.7196	0.9436	-25-21.0	150 1.6	43.9	5.7	299.9	7:14.9
14 JAN 1926	2424529.78	6:38:55	22.9	T	130	-0.1973	1.0431	-10 -8.6	-82-16.6	78.6	176.8	148.7	4:10.8
9 JUL 1926	2424706.46	23: 6: 0	22.9	A	135	0.0535	0.9680	25 35.3	165 4.2	86.3	183.0	115.5	3:51.0
3 JAN 1927	2424884.35	20:22:50	22.9	A	140	-0.4955	0.9995	-52-49.3	124 45.4	60.0	0.3	2.1	0: 2.6
29 JUN 1927	2425060.77	6:23:24	22.8	T	145	0.8181	1.0128	78 3.9	-73-49.8	34.9	167.6	76.7	0:50.0
24 DEC 1927	2425238.67	3:59:37	22.7	P	150	-1.2417	0.5489	-66 -4.1	47 42.5				
19 MAY 1928	2425386.06	13:24:17	22.8	T*	117	-1.0051	1.0592	-63-19.2	-22-21.9				
17 JUN 1928	2425415.35	20:27:28	22.8	PB	155	1.5105	0.0379	85 38.1	-70-36.9				
9 NOV 1928	2425562.91	9:48:20	22.9	P	122	1.0881	0.8072	62 34.3	-81 -7.9				
9 MAY 1929	2425740.76	8:10:34	23.0	T	127	-0.2890	1.0562	1 34.8	-92-42.2	73.2	338.9	193.1	5: 6.7
1 NOV 1929	2425917.00	12: 5: 5	23.1	A	132	0.3513	0.9649	4 28.2	-3 -6.9	89.4	203.9	134.4	3:53.9
28 APR 1930	2426095.29	19: 3:33	23.2	AT	137	0.4728	1.0003	39 23.8	121 9.0	61.6	149.0	1.1	0: 1.4
21 OCT 1930	2426271.41	21:43:50	23.3	T	142	-0.3806	1.0230	-30-33.5	161 7.2	67.5	30.3	84.3	1:55.2
18 APR 1931	2426449.53	0:45:32	23.4	P	147	1.2641	0.5099	61 29.4	-58-50.2				
12 SEP 1931	2426596.70	4:41:21	23.5	PE	114	1.5058	0.0461	61 14.1	152 43.3				
11 OCT 1931	2426828.04	12:55:37	23.6	P	152	-1.0609	0.8987	-61-13.0	119 33.0				
7 MAR 1932	2426773.83	7:55:47	23.6	A	119	-0.9875	0.9277	-60-45.0	-134-31.5	14.0	284.5	1086.2	5:19.0
31 AUG 1932	2426951.34	20: 3:39	23.5	T	124	0.8304	1.0267	54 28.7	79 31.6	33.6	231.6	164.6	1:44.8
24 FEB 1933	2427128.03	12:46:35	23.6	A	129	-0.2192	0.9841	-20-49.7	2 2.7	77.2	331.0	57.5	1:31.6
21 AUG 1933	2427305.74	5:49: 7	23.6	A	134	0.0868	0.9801	16 50.6	-95-52.3	84.9	206.9	70.9	2: 3.5
14 FEB 1934	2427482.53	0:38:40	23.7	T	139	0.0321	1.0321	13 14.8	-161-45.0	80.8	155.1	122.8	2:52.4
10 AUG 1934	2427659.88	8:37:42	23.7	A	144	-0.6893	0.9436	-24-33.1	-34-34.1	46.4	24.5	280.5	6:33.2
5 JAN 1935	2427807.73	5:35:43	23.6	PE	111	-1.5382	0.0009	-64-42.8	110 10.3				

Table 3

CANON OF SOLAR ECLIPSES
LOCAL CIRCUMSTANCES AT GREATEST ECLIPSE

DATE	JULIAN DATE	GREATEST ECLIPSE	DELTA T	TYPE	SAROS	GAMMA	MAGNITUDE	LATITUDE	LONGITUDE	SUN ALT	SUN AZ	PATH WIDTH	CENTRAL DURATION
3 FEB 1935	2427837.18	16:16:19	23.6	P	149	1.1438	0.7382	62 27.3	115 29.3				
30 JUN 1935	2427984.33	19:59:44	23.6	P	118	1.3621	0.3378	65 13.1	-39-11.0				
30 JUL 1935	2428013.89	9:18:24	23.6	P	154	-1.4261	0.2304	-62-53.5	5 59.1				
25 DEC 1935	2428162.25	17:59:49	23.5	A	121	-0.9229	0.9753	-83-29.9	-9-30.9	22.1	257.8	234.4	1:30.3
19 JUN 1936	2428338.72	5:20:28	23.6	T	126	0.5387	1.0329	58 7.2	-104-42.7	57.1	187.7	132.1	2:31.4
13 DEC 1936	2428518.48	23:28: 8	23.6	A	131	-0.2494	0.9349	-37-49.8	172 36.3	75.3	357.1	250.7	7:25.1
8 JUN 1937	2428693.36	20:41: 1	23.8	T	138	-0.2255	1.0751	9 54.1	130 30.6	78.9	359.9	249.8	7: 4.0
2 DEC 1937	2428870.46	23: 5:40	24.0	A	141	0.4387	0.9184	4 0.8	167 49.2	64.0	182.5	344.4	12: 0.4
29 MAY 1938	2429048.08	13:50:19	23.9	T	146	-0.9609	1.0552	-52-45.7	22 0.6	15.6	354.0	676.2	4: 4.4
21 NOV 1938	2429224.49	23:52:22	23.9	P	151	1.1075	0.7783	88 56.0	161 58.4				
19 APR 1939	2429373.20	16:45:50	23.8	A	118	-0.9386	0.9731	73 7.7	129 -0.2	19.6	118.6	284.2	1:49.2
12 OCT 1939	2429549.36	20:40:20	23.9	T	123	-0.9738	1.0285	-72-47.7	-155 -0.3	12.3	74.4	418.4	1:32.3
7 APR 1940	2429727.35	20:21:24	24.1	A	128	0.2188	0.9394	19 12.9	128 31.2	77.3	163.1	229.9	7:30.3
1 OCT 1940	2429904.03	12:44: 4	24.6	T	133	-0.2574	1.0644	-17-33.1	18 10.4	75.0	17.9	217.9	5:35.3
27 MAR 1941	2430081.34	20: 8: 5	25.0	A	138	-0.5027	0.9354	-26-10.0	110 51.4	59.7	340.3	275.9	7:40.9
21 SEP 1941	2430258.69	4:34: 1	25.0	T	143	-0.4649	1.0379	27 18.6	-119 -6.5	62.2	200.2	143.0	3:21.8
16 MAR 1942	2430435.48	23:37: 4	25.1	PE	148	-1.1910	0.6385	-72 -9.8	76 44.0				
12 AUG 1942	2430583.61	2:45: 8	25.4	P	115	-1.5247	0.0555	-70-24.2	-99-51.9				
10 SEP 1942	2430613.15	15:39:29	25.5	P	153	1.2570	0.5228	71 52.5	-50 -3.7				
4 FEB 1943	2430760.48	23:38: 8	25.8	T	120	0.8733	1.0331	43 34.9	-175 -6.9	28.9	164.8	229.4	2:39.2
1 AUG 1943	2430937.68	4:16:11	26.0	A	125	-0.8043	0.9409	-34-51.1	-108-33.4	36.3	10.6	367.3	6:58.8
25 JAN 1944	2431115.14	15:26:40	26.2	T	130	-0.2023	1.0428	-7-37.1	50 9.1	78.3	172.7	146.0	4: 8.9
20 JUL 1944	2431291.74	5:43:12	26.3	A	135	-0.0318	0.9700	19 0.8	-95-39.9	87.6	2.4	107.7	3:42.0
14 JAN 1945	2431469.71	5: 1:41	26.3	A	140	-0.4938	0.9970	-51 -6.0	-110-15.2	60.1	353.8	11.9	0:15.0
9 JUL 1945	2431646.06	13:27:44	26.4	T	145	0.7354	1.0180	69 58.7	17 14.4	42.3	184.3	91.6	1:15.4
3 JAN 1946	2431824.01	12:16: 7	26.9	P	150	-1.2394	0.5526	-87 -8.8	-177-39.2				
30 MAY 1946	2431971.38	21: 0:24	27.2	P	117	-1.0714	0.8856	-64 -4.6	100 56.6				
29 JUN 1946	2432000.66	3:51:58	27.3	P	155	1.4360	0.1804	66 36.3	50 47.2				
23 NOV 1946	2432148.23	17:37:10	27.6	P	122	1.1469	0.7756	63 21.7	45 12.5				
20 MAY 1947	2432328.07	13:47:48	27.9	T	127	-0.3630	1.0557	0 10.6	21 21.4	69.3	342.5	195.6	5:13.5
12 NOV 1947	2432502.34	20: 5:34	28.1	A	132	0.3742	0.9651	2 56.6	117 23.2	68.0	200.6	135.1	3:59.3
9 MAY 1948	2432680.60	2:26: 3	28.4	A	137	0.4131	0.9999	39 48.5	-131-13.8	65.4	153.1	0.2	0: 0.3
1 NOV 1948	2432856.75	5:59:18	28.8	T	142	-0.3519	1.0231	-33 -8.9	-76-13.9	69.2	27.9	83.8	1:55.7
28 APR 1949	2433034.83	7:48:51	29.0	P	147	1.2066	0.8086	61 58.5	55 47.0				
21 OCT 1949	2433211.38	21:13: 0	29.2	A	152	-1.0271	0.9622	-81-32.9	-107-23.2				
18 MAR 1950	2433359.15	15:31:59	29.3	A*	119	-0.8900	0.9294	-81 -5.1	-41 -1.8				
12 SEP 1950	2433536.65	3:38:44	29.5	T	124		1.0182	54 47.0	-172-12.6	26.8	235.7	134.3	1:13.6
7 MAR 1951	2433713.37	20:53:38	29.8	A	129	-0.2421	0.9896	-17-41.6	123 30.8	75.9	329.8	37.5	0:59.1
1 SEP 1951	2433891.04	12:51:48	30.1	A	134	-0.1554	0.9747	16 27.9	8 28.0	81.0	209.0	91.3	2:35.7
25 FEB 1952	2434067.88	9:11:35	30.4	T	139	0.4696	1.0366	15 37.2	-32-40.1	62.0	152.4	137.8	3: 9.4

Table 4

CANON OF SOLAR ECLIPSES
LOCAL CIRCUMSTANCES AT GREATEST ECLIPSE

DATE	JULIAN DATE	GREATEST ECLIPSE	DELTA T	TYPE	SAROS	GAMMA	MAGNITUDE	LATITUDE	LONGITUDE	SUN ALT	SUN AZ	PATH WIDTH	CENTRAL DURATION
20 AUG 1952	2434245.13	15:13:31	30.7	A	144	-0.6105	0.9420	-21-43.1	64 3.3	52.3	27.0	264.4	6:40.0
14 FEB 1953	2434422.54	0:59:29	31.0	P	149	1.1329	0.7589	61 52.1	-104-48.6				
11 JUL 1953	2434569.61	2:44:13	31.0	P	116	1.4387	0.2015	64 20.0	71 38.1				
9 AUG 1953	2434599.16	15:54:59	31.0	P	154	-1.3442	0.3717	-62-13.3	114 42.4				
5 JAN 1954	2434747.61	2:31:58	31.1	A	121	-0.9298	0.9720	-79 -3.9	120 39.7	21.1	259.6	278.3	1:42.3
30 JUN 1954	2434924.02	12:32:37	31.3	T	126	0.6133	1.0358	80 27.6	-4-10.6	51.9	197.2	152.7	2:35.1
25 DEC 1954	2435101.82	7:36:40	31.4	A	131	-0.2578	0.9323	-38-25.5	-68-10.9	74.8	351.4	262.0	7:39.0
20 JUN 1955	2435278.67	4:10:41	31.6	T	136	-0.1530	1.0776	14 45.2	-117 -0.1	81.1	3.8	253.7	7: 7.7
14 DEC 1955	2435455.79	7: 2:22	31.5	A	141	0.4265	0.9176	2 4.2	-72-13.0	64.8	178.3	345.8	12: 9.3
8 JUN 1956	2435633.39	21:20:39	31.7	T	146	-0.8936	1.0581	-40-46.5	140 44.5	26.5	0.3	428.9	4:44.7
2 DEC 1956	2435809.83	8: 0:32	31.9	P	151	1.0922	0.8048	67 52.0	-64-34.5				
30 APR 1957	2435958.50	0: 5:26	32.1	A*	118	0.9990	0.9684	70 39.7	-40 -4.6				
23 OCT 1957	2436134.70	4:54: 0	32.3	T*	123	-1.0023	1.0238	-71-15.3	23 20.5				
19 APR 1958	2436312.64	3:27:15	32.6	A	128	0.2748	0.9408	28 29.1	-123-33.9	73.9	184.1	227.8	7: 6.8
12 OCT 1958	2436489.37	20:55:24	32.8	T	133	-0.2952	1.0608	-24 -0.1	142 23.2	72.7	17.5	208.8	5:10.7
8 APR 1959	2436666.64	3:24: 7	33.0	A	138	-0.4549	0.9401	-19 -8.0	-137-35.8	62.9	342.2	247.5	7:25.5
2 OCT 1959	2436844.02	12:26:57	33.2	T	143	0.4207	1.0325	20 24.9	1 25.8	65.1	199.0	120.4	3: 1.7
27 MAR 1960	2437020.81	7:25: 5	33.3	P	148	-1.1540	0.7048	-72 -8.2	-152 -0.5				
20 SEP 1960	2437198.48	22:59:53	33.6	P	153	1.2055	0.6136	72 4.1	74 2.2				
15 FEB 1961	2437385.85	8:19:44	33.9	T	120	0.8828	1.0361	47 21.5	-40 -1.7	27.8	159.4	257.8	2:45.2
11 AUG 1961	2437522.95	10:46:44	34.1	A	125	-0.8861	0.9375	-45-50.8	-4 -2.5	27.4	16.9	499.8	8:35.3
5 FEB 1962	2437700.51	0:12:35	34.3	T	130	0.2105	1.0430	-4-14.5	-178 -4.3	77.8	169.5	146.7	4: 8.0
31 JUL 1962	2437877.02	12:25:32	34.5	A	135	-0.1131	0.9716	11 59.0	5 43.1	83.4	7.8	102.5	3:32.6
25 JAN 1963	2438055.07	13:37: 9	34.8	A	140	-0.4900	0.9951	-48-13.6	15 1.1	60.4	348.1	19.8	0:25.2
20 JUL 1963	2438231.36	20:36:11	35.1	T	145	0.6570	1.0223	61 41.1	119 33.3	48.6	191.3	101.4	1:39.6
14 JAN 1964	2438409.35	20:30: 5	35.4	P	150	-1.2356	0.5588	-68-13.7	-43 -7.7				
10 JUN 1964	2438556.69	4:34: 0	35.6	P	117	-1.1395	0.7539	-64-58.8	-135-54.1				
9 JUL 1964	2438585.97	11:17:53	35.7	P	155	1.3622	0.3223	87 36.0	172 58.8				
4 DEC 1964	2438733.58	1:31:51	35.9	P	122	1.1192	0.7518	64 16.1	173 16.2				
30 MAY 1965	2438911.39	21:17:32	36.1	T	127	-0.4228	1.0544	-2-28.8	133 50.3	65.0	346.3	197.7	5:15.5
23 NOV 1965	2439087.68	4:14:48	36.8	A	132	0.3905	0.9656	1 42.4	-119-50.7	67.0	196.9	134.1	4: 2.0
20 MAY 1966	2439265.90	9:39: 1	37.3	A	137	0.3465	0.9992	39 11.4	-26-26.8	69.5	158.3	3.1	0: 4.5
12 NOV 1966	2439442.10	14:23:26	37.8	T	142	-0.3302	1.0233	-35-37.9	48 9.4	70.5	24.5	84.0	1:57.1
2 MAY 1967	2439620.11	14:42:47	38.2	P	147	1.1420	0.7196	62 31.7	168 10.8				
2 NOV 1967	2439796.74	5:38:58	38.6	T*	152	-1.0009	1.0392	-62 -9.1	28 3.4				
28 MAR 1968	2439944.46	23: 0:29	39.0	P	119	-1.0372	0.8974	-60-59.1	79 42.5				
22 SEP 1968	2440121.97	11:18:45	39.4	T	124	0.9448	1.0099	56 9.4	-63-57.2	18.7	240.4	103.7	0:39.7
18 MAR 1969	2440298.70	4:54:55	39.8	A	129	-0.2706	0.9954	-14-47.4	-116-18.6	74.3	329.5	16.5	0:25.8
11 SEP 1969	2440476.33	19:58:56	40.2	A	134	0.2199	0.9691	15 35.1	114 7.7	77.2	210.3	113.6	3:11.0
7 MAR 1970	2440653.24	17:38:30	40.5	T	139	0.4471	1.0414	18 8.9	94 40.2	63.4	150.4	153.3	3:27.5

Table 5

CANON OF SOLAR ECLIPSES
LOCAL CIRCUMSTANCES AT GREATEST ECLIPSE

DATE	JULIAN DATE	GREATEST ECLIPSE	DELTA T	TYPE	SAROS	GAMMA	MAGNITUDE	LATITUDE	LONGITUDE	SUN ALT	SUN AZ	PATH WIDTH	CENTRAL DURATION
31 AUG 1970	2440830.41	21:55:28	41.0	A	144	-0.5367	0.9400	-20-17.9	164 1.0	57.5	28.9	258.3	6:47.5
25 FEB 1971	2441007.90	9:38: 4	41.9	P	149	1.1187	0.7860	61 26.0	33 35.8				
22 JUL 1971	2441154.90	9:31:53	42.3	PE	116	1.5128	0.0689	63 32.2	-177 -2.3				
20 AUG 1971	2441184.44	22:39:29	42.4	P	154	-1.2661	0.5065	-61-40.4	-135-17.7				
16 JAN 1972	2441332.96	11: 3:20	42.8	A	121	-0.9366	0.9693	-74-53.4	-107-44.1	20.0	263.1	321.5	1:53.2
10 JUL 1972	2441509.32	19:46:35	43.1	T	126	0.8869	1.0379	63 30.1	94 11.0	46.3	208.9	175.3	2:35.5
4 JAN 1973	2441687.16	15:46:18	43.4	A	131	-0.2645	0.9303	-37-51.3	51 11.0	74.4	345.7	270.7	7:48.4
30 JUN 1973	2441863.99	11:38:39	43.9	T	136	-0.0788	1.0792	18 49.3	-5-38.1	85.2	7.1	256.5	7: 3.5
24 DEC 1973	2442041.13	15: 2:39	44.5	A	141	0.4171	0.9174	1 4.8	48 27.1	65.3	174.0	344.6	12: 2.5
20 JUN 1974	2442218.70	4:48: 3	45.0	T	146	-0.8243	1.0592	-32 -8.8	-103-45.1	34.4	5.1	344.3	5: 8.8
13 DEC 1974	2442395.18	16:13:10	45.4	P	151	1.0797	0.8267	66 46.1	69 24.5				
11 MAY 1975	2442543.80	7:17:31	45.8	P	118	1.0644	0.8640	69 44.9	80 12.3				
3 NOV 1975	2442720.05	13:15:50	46.3	P	123	-1.0248	0.9584	-70-26.9	161 42.0				
29 APR 1976	2442897.93	10:24:16	46.8	A	128	0.3376	0.9421	33 56.1	-18-17.1	70.1	165.7	227.4	6:40.9
23 OCT 1976	2443074.72	5:13:43	47.3	T	133	-0.3271	1.0572	-30 -0.1	-92-16.6	70.8	16.2	199.1	4:46.4
18 APR 1977	2443251.94	10:31:28	47.8	A	138	-0.3993	0.9449	-11-57.4	-28-17.3	66.4	344.4	220.5	7: 4.2
12 OCT 1977	2443429.35	20:27:25	48.3	T	143	0.3837	1.0269	14 7.7	123 33.4	67.4	197.4	98.6	2:37.3
7 APR 1978	2443606.13	15: 3:44	48.8	P	148	-1.1084	0.7874	-71-51.4	-23-18.6				
2 OCT 1978	2443783.77	6:28:41	49.3	P	153	1.1616	0.6900	72 0.8	-159-41.1				
26 FEB 1979	2443931.20	16:55: 3	49.7	T	120	0.8979	1.0391	52 6.6	94 25.6	25.8	153.6	297.4	2:49.0
22 AUG 1979	2444108.22	17:22:35	50.2	A	125	-0.9633	0.9329	-59-35.9	108 32.4	15.0	29.0	954.3	6: 2.7
16 FEB 1980	2444285.87	8:53:59	50.6	T	130	0.2222	1.0434	0 -6.8	-47 -6.7	77.2	167.0	148.6	4: 8.1
10 AUG 1980	2444462.30	19:12:19	51.1	A	135	-0.1917	0.9727	4 34.3	108 54.1	78.9	10.8	99.6	3:23.2
4 FEB 1981	2444640.42	22: 9:21	51.5	A	140	-0.4840	0.9937	-44-26.7	140 45.5	80.8	343.9	25.1	0:33.2
31 JUL 1981	2444816.66	3:46:36	51.8	T	145	0.5790	1.0258	53 16.3	-134 -3.6	54.4	195.0	107.8	2: 2.4
25 JAN 1982	2444994.70	4:42:50	52.2	P	150	-1.2313	0.5659	-69-18.1	91 39.8				
21 JUN 1982	2445142.00	12: 4:31	52.5	P	117	-1.2104	0.6184	-65-56.5	-13-15.5				
20 JUL 1982	2445171.28	18:44:43	52.6	P	155	1.2885	0.4644	68 35.2	-64-12.4				
15 DEC 1982	2445318.90	9:32: 6	52.9	P	122	1.1292	0.7352	65 16.2	-56-58.4				
11 JUN 1983	2445496.70	4:43:33	53.3	T	127	-0.4949	1.0524	-6-15.4	-114-10.4	60.3	350.3	199.1	5:10.7
4 DEC 1983	2445673.02	12:31:12	53.7	A	132	0.4013	0.9866	0 53.1	4 43.9	66.3	192.9	131.1	4: 0.5
30 MAY 1984	2445851.20	16:45:41	54.0	A	137	0.2754	0.9980	37 29.0	76 44.4	73.7	164.1	7.3	0:11.5
22 NOV 1984	2446027.45	22:54:16	54.3	T	142	-0.3134	1.0237	-37-47.7	173 38.9	71.5	20.2	84.8	1:59.8
19 MAY 1985	2446205.40	21:29:36	54.8	P	147	1.0718	0.8404	63 14.5	-81 -3.4				
12 NOV 1985	2446382.09	14:11:26	54.8	T	152	-0.9797	1.0388	-88-31.8	142 46.5	10.7	111.2	693.1	1:58.5
9 APR 1986	2446529.78	6:21:21	55.0	P	119	-1.0823	0.8221	-81-12.7	-161-28.7				
3 OCT 1986	2446707.30	19: 6:13	55.3	AT	124	0.9928	1.0001	59 54.2	37 25.2	5.5	252.2	2.1	0: 0.2
29 MAR 1987	2446884.03	12:49:45	55.5	AT	129	-0.3055	1.0014	-12-20.4	2 17.3	72.2	329.9	5.0	0: 7.7
23 SEP 1987	2447061.63	3:12:19	55.8	A	134	0.2784	0.9634	14 20.0	-138-21.8	73.8	210.8	137.2	3:49.2
18 MAR 1988	2447238.58	1:58:55	56.1	T	139	0.4187	1.0464	20 42.0	-139-58.4	65.2	149.4	168.6	3:46.5

Table 6

CANON OF SOLAR ECLIPSES
LOCAL CIRCUMSTANCES AT GREATEST ECLIPSE

DATE	JULIAN DATE	GREATEST ECLIPSE	DELTA T	TYPE	SAROS	GAMMA	MAGNITUDE	LATITUDE	LONGITUDE	SUN ALT	SUN AZ	PATH WIDTH	CENTRAL DURATION
11 SEP 1988	2447415.70	4:44:24	56.4	A	144	-0.4684	0.9377	-20 -1.4	-94-24.5	62.0	30.1	257.8	8:58.9
7 MAR 1989	2447593.28	18: 8:40	56.7	P	149	1.0980	0.8255	61 9.6	169 54.0				
31 AUG 1989	2447769.73	5:31:45	57.0	P	154	-1.1931	0.6328	-81-15.9	-23-30.1				
26 JAN 1990	2447918.31	19:31:22	57.3	A	121	-0.9458	0.9670	-71 -1.0	-22 9.6	18.4	266.1	373.1	2: 2.8
22 JUL 1990	2448094.63	3: 3: 5	57.6	T	126	-0.7593	1.0391	65 9.4	-168-48.8	40.3	222.0	201.4	2:32.8
15 JAN 1991	2448272.50	23:53:49	58.0	A	131	-0.2728	0.9290	-36-22.5	170 23.9	73.9	340.5	276.7	7:52.8
11 JUL 1991	2448449.30	19: 7: 0	58.3	T	136	-0.0044	1.0800	21 58.9	105 12.6	88.2	304.7	258.0	6:53.2
4 JAN 1992	2448628.46	23: 5:34	58.7	A	141	0.4091	0.9179	0 59.0	169 40.8	65.9	169.8	340.5	11:41.0
30 JUN 1992	2448804.01	12:11:22	59.0	T	146	-0.7515	1.0592	-25-10.4	9 28.9	41.2	9.2	294.1	5:20.7
24 DEC 1992	2448980.52	0:31:39	59.4	P	151	1.0710	0.8422	65 41.0	-155-40.1				
21 MAY 1993	2449129.10	14:20:12	59.7	P	118	1.1369	0.7358	88 47.5	-162-18.8				
13 NOV 1993	2449305.41	21:45:48	60.1	P	123	-1.0412	0.9278	-69-33.7	-58-19.9				
10 MAY 1994	2449483.22	17:12:26	60.4	A	128	0.4075	0.9431	41 31.3	84 8.8	65.7	188.1	230.1	6:13.5
3 NOV 1994	2449660.07	13:40: 4	60.8	T	133	-0.3523	1.0535	-35-21.7	34 12.8	69.2	14.0	189.0	4:23.3
29 APR 1995	2449837.23	17:33:20	61.2	A	138	-0.3385	0.9497	-4-52.1	79 23.5	70.2	346.8	195.6	6:36.9
24 OCT 1995	2450014.69	4:33:29	61.5	T	143	0.3517	1.0214	8 24.3	-113-57.8	69.4	195.4	77.6	2: 9.6
17 APR 1996	2450191.44	22:38:11	61.9	P	148	-1.0583	0.8790	-71-20.8	103 57.8				
12 OCT 1996	2450369.09	14: 3: 2	62.3	P	153	1.1226	0.7672	71 42.5	-32 -9.1				
9 MAR 1997	2450516.56	1:24:49	62.6	T	120	0.9181	1.0420	57 45.6	-130-42.9	23.0	146.4	355.7	2:50.1
2 SEP 1997	2450693.50	0: 4:46	63.0	P	125	-1.0354	0.8980	-71-48.0	-114-12.9				
26 FEB 1998	2450871.23	17:29:26	63.3	T	130	0.2390	1.0441	4 42.9	82 42.7	76.2	165.0	151.4	4: 8.7
22 AUG 1998	2451047.59	2: 7: 9	63.7	A	135	-0.2848	0.9733	-2-59.3	-145-23.4	74.7	13.4	98.7	3:14.0
16 FEB 1999	2451225.77	6:34:37	64.1	A	140	-0.4727	0.9928	-39-49.6	-93-53.8	61.6	341.3	28.7	0:39.5
11 AUG 1999	2451401.96	11: 4: 9	64.5	T	145	0.5062	1.0286	45 4.3	-24-17.7	59.4	197.2	112.2	2:22.8
5 FEB 2000	2451580.04	12:50:24	64.8	P	150	-1.2235	0.5788	-70-13.1	-134-10.5				
1 JUL 2000	2451727.31	19:33:34	65.2	P	117	-1.2823	0.4765	-66-55.9	109 27.5				
31 JUL 2000	2451756.59	2:14: 9	65.2	P	155	1.2168	0.6033	69 31.7	59 50.8				
25 DEC 2000	2451904.23	17:35:54	65.5	P	122	1.1365	0.7231	66 20.2	74 6.3				

Table 7

CANON OF SOLAR ECLIPSES
LOCAL CIRCUMSTANCES AT GREATEST ECLIPSE

DATE	JULIAN DATE	GREATEST ECLIPSE	DELTA T	TYPE	SAROS	GAMMA	MAGNITUDE	LATITUDE	LONGITUDE	SUN ALT	SUN AZ	PATH WIDTH	CENTRAL DURATION
21 JUN 2001	2452082.00	12: 4:48	65.9	T	127	-0.5703	1.0498	-11-15.8	-2-44.6	55.2	354.4	200.0	4:56.6
14 DEC 2001	2452258.37	20:52:58	66.3	A	132	0.4088	0.9681	0 37.3	130 41.4	65.9	188.6	125.6	3:53.0
10 JUN 2002	2452438.49	23:45:21	66.7	A	137	0.1993	0.9982	34 32.7	178 36.7	78.2	170.2	13.4	0:22.7
4 DEC 2002	2452812.81	7:32:14	67.1	T	142	-0.3023	1.0244	-39-28.2	-59-33.6	72.1	15.0	86.9	2: 3.7
31 MAY 2003	2452790.87	4: 9:23	67.5	A+	147	0.9959	0.9384	66 37.7	24 21.4	2.9	35.4	4497.6	3:36.9
23 NOV 2003	2452987.45	22:50:23	67.9	T	152	-0.9639	1.0379	-72-39.3	-88-18.3	14.7	111.0	496.1	1:57.2
19 APR 2004	2453115.07	13:35: 6	68.2	P	119	-1.1336	0.7354	-81-35.6	-44-20.6				
14 OCT 2004	2453292.63	3: 0:22	68.6	P	124	1.0345	0.9275	61 15.0	153 36.6				
8 APR 2005	2453469.36	20:36:50	69.0	AT	129	-0.3475	1.0074	-10-34.6	118 57.7	89.6	331.2	27.0	0:42.0
3 OCT 2005	2453648.94	10:32:45	69.4	A	134	0.3303	0.9576	12 52.4	-28-45.0	70.7	210.2	162.1	4:31.4
29 MAR 2006	2453823.93	10:12:24	69.8	T	139	0.3842	1.0515	23 8.6	-16-45.9	67.3	149.3	183.5	4: 6.7
22 SEP 2006	2454000.99	11:41:13	70.2	A	144	-0.4066	0.9352	-20-39.9	9 3.4	65.9	30.6	261.1	7: 9.4
19 MAR 2007	2454178.61	2:32:59	70.6	P	149	1.0727	0.8742	61 2.9	-55-24.7				
11 SEP 2007	2454355.02	12:32:24	71.0	P	154	-1.1258	0.7489	-61 -0.3	90 17.4				
7 FEB 2008	2454503.66	3:56: 8	71.3	A	121	-0.9570	0.9650	-87-34.6	150 28.9	16.3	268.5	444.6	2:11.7
1 AUG 2008	2454679.93	10:22:11	71.7	T	126	0.8304	1.0394	65 38.2	-72-16.3	33.5	235.5	236.8	2:27.2
26 JAN 2009	2454857.83	7:59:42	72.1	A	131	-0.2820	0.9283	-34 -4.7	-70-16.6	73.4	336.0	280.2	7:53.5
22 JUL 2009	2455034.61	2:36:24	72.5	T	136	0.0694	1.0799	24 12.0	-144 -8.4	85.6	201.3	258.5	6:38.9
15 JAN 2010	2455211.80	7: 7:35	72.9	A	141	0.4001	0.9190	1 37.3	-89-20.2	66.4	165.8	333.2	11: 7.9
11 JUL 2010	2455389.32	19:34:38	73.4	T	146	-0.6791	1.0580	-19-46.5	121 51.0	47.2	13.1	258.7	5:20.2
4 JAN 2011	2455565.87	8:51:39	73.8	P	151	1.0626	0.8575	64 39.1	-20-48.8				
1 JUN 2011	2455714.39	21:17:16	74.1	P	118	1.2127	0.6016	67 47.0	-46-49.4				
1 JUL 2011	2455743.86	8:39:30	74.2	PB	156	-1.4921	0.0963	-65 -9.5	-28-38.9				
25 NOV 2011	2455890.76	8:21:22	74.5	P	123	-1.0537	0.9044	-68-34.1	82 24.0				
20 MAY 2012	2456068.50	23:53:53	74.9	A	128	0.4825	0.9439	49 4.6	-176-19.1	60.9	171.6	236.9	5:46.4
13 NOV 2012	2456245.43	22:12:55	75.4	T	133	-0.3719	1.0500	-39-57.6	161 17.9	67.9	10.6	179.0	4: 2.2
10 MAY 2013	2456422.52	0:26:20	75.8	A	138	-0.2696	0.9544	2 12.3	-175-30.5	74.4	349.4	172.6	6: 3.5
3 NOV 2013	2456600.03	12:47:38	76.2	AT	143	0.3271	1.0159	3 29.4	11 39.6	70.9	192.9	57.5	1:39.6
29 APR 2014	2456776.75	6: 4:32	76.6	A*	148	-1.0002	0.9855	-70-41.8	-131 -9.5				
23 OCT 2014	2456954.41	21:45:37	77.1	P	153	1.0907	0.8112	71 10.1	97 4.6				
20 MAR 2015	2457101.91	9:46:46	77.4	T	120	0.9452	1.0446	84 25.0	6 34.1	18.5	135.4	462.0	2:48.9
13 SEP 2015	2457278.79	6:55:18	77.8	P	125	-1.1005	0.7868	-72 -6.4	2 18.4				
9 MAR 2016	2457456.58	1:58:20	78.3	T	130	0.2607	1.0450	10 6.5	-148-50.2	74.9	163.7	155.0	4: 9.5
1 SEP 2016	2457632.88	9: 8: 2	78.7	A	135	-0.3332	0.9736	-10-41.4	-37-48.1	70.5	15.5	99.7	3: 5.5
26 FEB 2017	2457811.12	14:54:32	79.1	A	140	-0.4580	0.9922	-34-41.8	31 8.4	62.6	339.8	30.6	0:44.0
21 AUG 2017	2457987.27	18:26:40	79.6	T	145	0.4367	1.0306	36 58.0	87 37.7	63.9	198.6	114.8	2:40.1
15 FEB 2018	2458165.37	20:52:30	80.0	P	150	-1.2119	0.5982	-71 -1.8	0-44.5				
13 JUL 2018	2458312.63	3: 2:15	80.4	P	117	-1.3543	0.3384	-87-55.4	-127-28.2				
11 AUG 2018	2458341.91	9:47:27	80.5	P	155	1.1477	0.7363	70 23.3	-174-32.9				
6 JAN 2019	2458489.57	1:42:34	80.8	P	122	1.1415	0.7150	87 26.1	-153-37.0				

Table 8

CANON OF SOLAR ECLIPSES
LOCAL CIRCUMSTANCES AT GREATEST ECLIPSE

DATE	JULIAN DATE	GREATEST ECLIPSE	DELTA T	TYPE	SAROS	GAMMA	MAGNITUDE	LATITUDE	LONGITUDE	SUN ALT	SUN AZ	PATH WIDTH	CENTRAL DURATION
2 JUL 2019	2458867.31	19:24: 8	81.3	T	127	-0.6486	1.0459	-17-23.5	108 57.0	49.7	358.5	200.5	4:32.7
26 DEC 2019	2458843.72	5:18:49	81.7	A	132	0.4133	0.9701	0 59.7	-102-18.4	65.8	184.3	117.7	3:39.3
21 JUN 2020	2459021.78	6:41:14	82.2	A	137	0.1209	0.9940	30 31.0	-79-43.2	82.7	176.8	21.2	0:38.3
14 DEC 2020	2459198.18	16:14:37	82.6	T	142	-0.2942	1.0254	-40-21.1	67 54.4	72.6	9.3	90.3	2: 9.7
10 JUN 2021	2459375.95	10:43: 6	83.1	A	147	0.9150	0.9435	80 48.9	66 36.6	23.3	90.4	528.6	3:51.3
4 DEC 2021	2459552.82	7:34:38	83.5	T	152	-0.9528	1.0367	-76-45.6	46 17.8	17.0	114.6	419.1	1:54.3
30 APR 2022	2459700.36	20:42:36	83.9	P	119	-1.1902	0.6385	-62 -7.0	71 21.6				
25 OCT 2022	2459877.98	11: 1:17	84.4	P	124	1.0698	0.8613	61 38.9	-77-28.8				
20 APR 2023	2460054.68	4:17:55	84.8	AT	129	-0.3953	1.0132	-9-36.0	-125-50.5	68.7	333.2	49.1	1:18.3
14 OCT 2023	2460232.25	18: 0:38	85.3	A	134	0.3751	0.9520	11 21.2	83 2.8	68.0	208.8	187.4	5:17.2
8 APR 2024	2460409.26	18:18:31	85.7	T	139	0.3430	1.0566	25 16.9	104 5.1	69.8	150.3	197.5	4:28.1
2 OCT 2024	2460586.28	18:46:10	86.2	A	144	-0.3512	0.9326	-21-58.1	114 26.8	69.4	30.2	266.6	7:25.2
29 MAR 2025	2460763.95	10:48:38	86.7	P	149	1.0404	0.9365	61 6.0	77 6.2				
21 SEP 2025	2460940.32	19:43: 3	87.1	P	154	-1.0654	0.8531	-60-54.1	-153-29.1				
17 FEB 2026	2461089.01	12:13: 4	87.5	A	121	-0.9744	0.9630	-84-43.0	-86-53.1	12.3	267.9	618.2	2:19.7
12 AUG 2026	2461265.24	17:47: 4	88.0	T	128	-0.8974	1.0387	65 12.4	25 15.2	25.8	248.6	293.7	2:18.4
6 FEB 2027	2461443.17	16: 0:45	88.5	A	131	-0.2952	0.9281	-31-18.3	48 23.3	72.6	332.4	281.6	7:50.9
2 AUG 2027	2461619.92	10: 7:50	88.9	T	136	0.1417	1.0790	25 29.0	-33-14.7	81.6	204.3	257.7	6:22.7
26 JAN 2028	2461797.13	15: 8:58	89.4	A	141	0.3902	0.9208	2 57.7	51 28.6	67.0	161.8	323.1	10:27.2
22 JUL 2028	2461974.62	2:56:40	89.9	T	146	-0.6059	1.0560	-15-36.0	-126-46.2	52.7	16.7	230.2	5: 9.7
14 JAN 2029	2462151.22	17:13:45	90.4	P	151	1.0553	0.8710	83 42.1	114 11.0				
12 JUN 2029	2462299.67	4: 6:11	90.8	P	118	1.2940	0.4581	66 45.9	66 5.8				
11 JUL 2029	2462329.15	15:37:18	90.9	P	156	-1.4194	0.2295	-64-15.8	85 34.9				
5 DEC 2029	2462476.13	15: 3:56	91.3	P	123	-1.0609	0.8910	-87-30.8	-135-43.4				
1 JUN 2030	2462653.77	6:29:13	91.8	A	128	0.5623	0.9443	56 30.2	-80 -8.5	55.5	176.7	249.6	5:20.9
25 NOV 2030	2462830.79	6:51:36	92.2	T	133	-0.3868	1.0469	-43-36.9	-71-18.5	67.0	6.1	169.3	3:43.6
21 MAY 2031	2463007.80	7:16: 0	92.7	A	138	-0.1974	0.9589	8 54.3	-71-48.1	78.6	352.0	152.3	5:25.6
14 NOV 2031	2463185.38	21: 7:29	93.2	AT	143	0.3077	1.0106	0-38.1	137 33.7	72.1	189.9	38.4	1: 8.4
9 MAY 2032	2463362.06	13:26:42	93.7	A	148	-0.9378	0.9957	-51-18.9	6 58.2	20.0	344.2	44.2	0:22.2
3 NOV 2032	2463539.73	5:34:10	94.2	P	153	1.0642	0.8553	70 25.7	-132-44.2				
30 MAR 2033	2463687.25	18: 2:35	94.6	T	120	0.9776	1.0462	71 17.8	155 35.9	11.3	111.6	777.7	2:37.1
23 SEP 2033	2463864.08	13:54:29	95.1	P	125	-1.1584	0.6883	-72-12.2	121 13.1				
20 MAR 2034	2464041.93	10:18:46	95.7	T	130	-0.2892	1.0458	16 2.0	-22-18.0	73.1	162.9	159.2	4: 9.5
12 SEP 2034	2464218.18	16:19:27	96.2	A	135	-0.3936	0.9736	-18-14.7	72 32.4	66.8	17.3	102.1	2:57.7
9 MAR 2035	2464398.46	23: 5:52	96.7	A	140	-0.4371	0.9919	-29 -3.6	154 52.3	84.0	339.3	31.4	0:47.4
2 SEP 2035	2464572.58	1:56:45	97.2	T	145	0.3728	1.0320	29 5.8	-158 -6.2	68.0	199.5	116.2	2:54.1
27 FEB 2036	2464750.70	4:46:47	97.7	P	150	-1.1946	0.6275	-71-38.9	131 18.6				
23 JUL 2036	2464897.94	10:32: 0	98.1	P	117	-1.4250	0.1990	-68-53.0	-3-36.4				
21 AUG 2036	2464927.23	17:25:46	98.2	P	155	1.0826	0.8616	71 7.2	-47 -6.5				
16 JAN 2037	2465074.91	9:48:52	98.6	P	122	1.1474	0.7054	68 31.3	-20-52.6				

Table 9

CANON OF SOLAR ECLIPSES
LOCAL CIRCUMSTANCES AT GREATEST ECLIPSE

DATE	JULIAN DATE	GREATEST ECLIPSE	DELTA T	TYPE	SAROS	GAMMA	MAGNITUDE	LATITUDE	LONGITUDE	SUN ALT	SUN AZ	PATH WIDTH	CENTRAL DURATION
13 JUL 2037	2465252.61	2:40:37	99.2	T	127	-0.7246	1.0413	-24-46.1	-139 -8.0	43.5	2.8	200.7	3:58.5
5 JAN 2038	2465429.07	13:47: 8	99.7	A	132	0.4166	0.9728	2 4.7	25 22.2	65.4	180.1	107.2	3:18.5
2 JUL 2038	2465607.06	13:32:55	100.2	A	137	0.0398	0.9911	25 25.4	21 49.5	87.0	188.0	31.2	0:59.6
26 DEC 2038	2465783.54	1: 0:10	100.7	T	142	-0.2884	1.0268	-40-17.5	-164 -0.8	73.0	3.2	95.2	2:18.1
21 JUN 2039	2465961.22	17:12:54	101.3	A	147	0.8311	0.9454	78 52.3	102 2.2	33.4	153.1	365.3	4: 5.0
15 DEC 2039	2466138.18	16:13:46	101.8	T	152	-0.9460	1.0355	-80-50.0	-172-37.2	18.3	123.2	380.3	1:51.3
11 MAY 2040	2466285.65	3:43: 2	102.2	P	119	-1.2530	0.5297	-82-48.1	-174-32.3				
4 NOV 2040	2466463.30	19: 9: 0	102.8	P	124	1.0990	0.8071	62 12.1	53 14.4				
30 APR 2041	2466639.99	11:52:20	103.3	T	129	-0.4494	1.0189	-9-37.8	-12-16.0	63.3	335.8	71.7	1:50.0
25 OCT 2041	2466817.57	1:36:19	103.9	A	134	0.4130	0.9467	9 55.0	-162-55.8	65.6	208.6	213.1	6: 7.0
20 APR 2042	2466994.80	2:17:32	104.4	T	139	0.2954	1.0614	28 56.5	-137-22.8	72.7	152.6	210.4	4:51.0
14 OCT 2042	2467171.58	2: 0:38	105.0	A	144	-0.3033	0.9300	-23-45.7	-137-53.7	72.2	28.8	273.4	7:44.2
9 APR 2043	2467349.29	18:57:50	105.5	T*	149	1.0031	0.9429	61 27.6	-151-52.3				
3 OCT 2043	2467525.63	3: 1:47	106.1	A*	154	-1.0105	0.9459	-61 -7.2	-35-12.7				
28 FEB 2044	2467674.35	20:24:37	106.5	A-	121	-0.9954	0.9600	-82-10.4	25 25.5	3.6	259.6	2268.3	2:27.2
23 AUG 2044	2467850.55	1:17: 0	107.1	T	126	0.9810	1.0365	64 19.7	120 31.0	15.5	264.2	451.1	2: 4.0
16 FEB 2045	2468028.50	23:56: 3	107.6	A	131	-0.3125	0.9285	-28-15.3	166 6.3	71.6	329.8	281.2	7:46.5
12 AUG 2045	2468205.24	17:42:40	108.2	T	136	0.2112	1.0774	25 53.3	78 28.2	77.6	207.6	255.6	6: 5.8
5 FEB 2046	2468382.46	23: 6:22	108.8	A	141	0.3766	0.9232	4 47.1	171 17.7	67.9	158.5	310.1	9:42.4
2 AUG 2046	2468559.93	10:21:14	109.3	T	146	-0.5355	1.0531	-12-45.7	-15-14.9	57.6	19.9	208.0	4:51.2
26 JAN 2047	2468736.56	1:33:15	109.9	P	151	1.0449	0.8902	82 52.0	-111-46.8				
23 JUN 2047	2468884.95	10:52:29	110.4	P	118	1.3762	0.3135	65 46.0	177 53.8				
22 JUL 2047	2468914.44	22:36:17	110.5	P	156	-1.3481	0.3592	-83-26.5	-160-12.8				
16 DEC 2047	2469061.49	23:50:10	110.9	P	123	-1.0661	0.8816	-66-28.8	6 30.0				
11 JUN 2048	2469239.04	12:58:52	111.5	A	128	0.6465	0.9441	63 39.8	11 25.3	49.4	184.7	271.4	4:58.4
5 DEC 2048	2469416.15	15:35:27	112.1	T	133	-0.3973	1.0440	-46 -8.1	56 17.2	66.3	0.3	180.3	3:27.6
31 MAY 2049	2469593.08	13:59:57	112.7	A	138	-0.1190	0.9631	15 17.4	29 45.5	83.1	353.9	134.3	4:45.2
25 NOV 2049	2469770.73	5:33:46	113.3	AT	143	0.2943	1.0057	-3-48.2	-95-20.6	72.9	186.5	20.6	0:37.5
20 MAY 2050	2469947.36	20:42:49	113.9	AT	148	-0.8690	1.0038	-40 -7.4	123 37.0	29.5	351.5	28.7	0:21.3
14 NOV 2050	2470125.06	13:30:50	114.5	P	153	1.0447	0.8873	69 31.7	-1 -9.0				
11 APR 2051	2470272.59	2:10:37	115.0	P	120	1.0167	0.9848	71 37.6	-32-14.0				
4 OCT 2051	2470449.38	21: 2:12	115.5	P	125	-1.2095	0.8017	-72 -2.7	-117-46.1				
30 MAR 2052	2470627.27	18:31:52	116.1	T	130	0.3236	1.0466	22 21.9	102 26.8	71.0	162.8	163.7	4: 8.2
22 SEP 2052	2470803.49	23:39: 8	116.7	A	135	-0.4481	0.9734	-25-41.5	-175 -4.0	63.3	18.6	105.9	2:50.8
20 MAR 2053	2470981.80	7: 8:18	117.4	A	140	-0.4093	0.9919	-23 -2.1	-83 -2.6	65.8	339.6	31.1	0:49.6
12 SEP 2053	2471157.90	9:34: 8	118.0	T	145	0.3140	1.0329	21 28.3	-41-48.0	71.6	200.1	118.5	3: 4.1
9 MAR 2054	2471336.02	12:33:37	118.6	P	150	-1.1715	0.6668	-72 -2.6	-98 -3.8				
3 AUG 2054	2471483.25	18: 3:59	119.1	PE	117	-1.4941	0.0654	-69-46.4	121 17.7				
2 SEP 2054	2471512.55	1: 9:32	119.2	P	155	1.0216	0.9786	71 40.9	82 13.6				
27 JAN 2055	2471660.25	17:54: 2	119.7	P	122	1.1547	0.6936	69 32.9	112 12.0				

Table 10

CANON OF SOLAR ECLIPSES
LOCAL CIRCUMSTANCES AT GREATEST ECLIPSE

DATE	JULIAN DATE	GREATEST ECLIPSE	DELTA T	TYPE	SAROS	GAMMA	MAGNITUDE	LATITUDE	LONGITUDE	SUN ALT	SUN AZ	PATH WIDTH	CENTRAL DURATION
24 JUL 2055	2471837.92	9:57:49	120.3	T	127	-0.8012	1.0359	-33-16.9	-25-51.6	36.8	7.4	201.7	3:16.8
16 JAN 2056	2472014.43	22:16:42	120.9	A	132	-0.4196	0.9760	3 53.8	153 27.2	65.2	176.2	94.4	2:52.3
12 JUL 2056	2472192.35	20:21:58	121.6	A	137	-0.0428	0.9878	19 26.3	123 40.0	87.1	351.7	43.2	1:26.9
6 JAN 2057	2472368.91	9:47:50	122.2	T	142	-0.2840	1.0287	-39-14.0	-35-15.4	73.2	357.3	101.6	2:28.6
1 JUL 2057	2472546.49	23:40:13	122.8	A	147	0.7454	0.9464	71 29.3	176 8.1	41.5	177.4	298.1	4:22.5
26 DEC 2057	2472723.55	1:14:33	123.5	T	152	-0.9407	1.0347	-84-49.4	-21-36.2	19.2	141.2	355.0	1:49.7
22 MAY 2058	2472870.94	10:39:25	124.0	P	119	-1.3195	0.4135	-63-31.8	-61-14.5				
21 JUN 2058	2472900.51	0:19:34	124.1	PB	157	1.4869	0.1260	65 57.0	-9-55.4				
16 NOV 2058	2473048.64	3:23: 2	124.6	P	124	1.1221	0.7643	62 54.0	-174-18.4				
11 MAY 2059	2473225.31	19:22:14	125.3	T	129	-0.5080	1.0242	-10-43.3	100 20.2	59.5	339.0	94.6	2:23.5
5 NOV 2059	2473402.89	9:18:10	125.9	A	134	0.4452	0.9417	8 43.8	-47 -8.6	63.6	203.7	238.3	6:59.6
30 APR 2060	2473579.92	10:10: 1	126.6	T	139	0.2420	1.0660	27 58.8	-20-57.7	75.8	158.1	221.9	5:15.2
24 OCT 2060	2473756.89	9:24: 4	127.2	A	144	-0.2828	0.9276	-25-48.1	-28-11.0	74.6	26.4	280.6	8: 5.8
20 APR 2061	2473934.82	2:56:51	127.9	T	149	-0.9576	1.0475	64 31.9	-59-18.6	18.2	97.4	557.6	2:37.4
13 OCT 2061	2474110.94	10:32: 6	128.5	A	154	-0.9643	0.9469	-82 -9.0	54 28.5	14.8	78.6	746.6	3:41.0
11 MAR 2062	2474259.68	4:26:12	129.1	P	121	-1.0239	0.9316	-60-58.3	146 59.6				
3 SEP 2062	2474435.87	8:54:24	129.8	P	128	1.0187	0.9744	61 18.3	-150-27.0				
28 FEB 2063	2474613.82	7:43:27	130.4	A	131	-0.3360	0.9293	-25-13.8	-77-48.0	70.3	328.0	279.6	7:41.2
24 AUG 2063	2474790.58	1:22: 9	131.1	T	136	0.2767	1.0760	25 32.3	-168-28.3	73.8	210.3	252.1	5:49.1
17 FEB 2064	2474967.79	7: 0:18	131.8	A	141	0.3598	0.9262	7 2.3	-69-48.1	68.9	155.8	294.6	8:56.5
12 AUG 2064	2475145.24	17:46: 5	132.5	T	146	-0.4656	1.0495	-10-57.2	95 56.5	62.2	22.7	183.8	4:27.7
5 FEB 2065	2475321.91	9:52:22	133.1	P	151	1.0336	0.9114	62 9.7	21 54.4				
3 JUL 2065	2475470.23	17:33:49	133.7	P	118	1.4615	0.1644	64 49.3	-71-56.7				
2 AUG 2065	2475499.73	5:34:14	133.7	P	158	-1.2762	0.4889	-62-42.7	-46-31.9				
27 DEC 2065	2475646.86	8:39:52	134.4	P	123	-1.0688	0.8789	-65-23.8	149 5.3				
22 JUN 2066	2475824.31	19:25:47	135.1	A	128	-0.7326	0.9435	70 6.0	96 21.2	42.6	198.2	308.3	4:39.9
17 DEC 2066	2476001.52	0:23:35	135.8	T	133	-0.4043	1.0416	-47-22.2	-175-49.8	65.9	353.8	152.3	3:14.5
11 JUN 2067	2476178.36	20:42:25	136.5	A	138	-0.0391	0.9670	21 1.0	130 8.4	87.2	347.9	119.0	4: 5.4
6 DEC 2067	2476356.09	14: 3:38	137.2	AT	143	0.2844	1.0011	-6 -2.9	32 19.5	73.5	182.7	4.2	0: 7.7
31 MAY 2068	2476533.66	3:56:37	137.9	T	148	-0.7974	1.0110	-31 -3.1	-123-16.3	37.0	356.6	62.7	1: 5.6
24 NOV 2068	2476710.40	21:32:24	138.6	P	153	1.0299	0.9109	68 31.0	131 0.5				
21 APR 2069	2476857.92	10:11: 6	139.2	P	120	1.0622	0.8992	71 2.7	101 16.3				
20 MAY 2069	2476887.25	17:53:17	139.3	PB	168	-1.4865	0.0873	-88-45.6	89 52.0				
15 OCT 2069	2477034.68	4:19:52	139.9	P	125	-1.2525	0.5293	-71-38.6	5 28.2				
11 APR 2070	2477212.61	2:36: 8	140.6	T	130	0.3650	1.0472	29 2.4	-135 -6.3	88.5	163.2	168.3	4: 4.4
4 OCT 2070	2477388.80	7: 8:53	141.4	A	135	-0.4950	0.9731	-32-51.1	-60-26.1	60.2	19.4	110.3	2:44.2
31 MAR 2071	2477587.13	15: 1: 4	142.1	A	140	-0.3743	0.9919	-18-43.0	36 57.7	68.0	340.5	30.6	0:51.6
23 SEP 2071	2477743.22	17:20:25	142.8	T	145	0.2621	1.0333	14 13.9	76 43.0	74.8	200.3	118.1	3:10.8
19 MAR 2072	2477921.34	20:10:28	143.6	P	150	-1.1409	0.7187	-72-11.0	30 18.5				
12 SEP 2072	2478097.87	8:59:19	144.3	T	155	0.9658	1.0558	69 46.9	-102 -4.2	14.4	240.7	733.2	3:12.8

Table 11

CANON OF SOLAR ECLIPSES
LOCAL CIRCUMSTANCES AT GREATEST ECLIPSE

DATE	JULIAN DATE	GREATEST ECLIPSE	DELTA T	TYPE	SAROS	GAMMA	MAGNITUDE	LATITUDE	LONGITUDE	SUN ALT	SUN AZ	PATH WIDTH	CENTRAL DURATION
7 FEB 2073	2478245.58	1:55:54	144.9	P	122	1.1648	0.8771	70 28.0	-114-54.5				
3 AUG 2073	2478423.22	17:15:22	145.7	T	127	-0.8762	1.0294	-43-14.3	89 22.8	28.6	13.2	206.5	2:29.3
27 JAN 2074	2478599.78	6:44:11	146.4	A	132	-0.4249	0.9798	6 32.9	-78-48.9	64.9	172.6	79.3	2:21.2
24 JUL 2074	2478777.63	3:10:30	147.2	A	137	-0.1242	0.9838	12 47.1	-133-44.6	82.8	2.2	57.7	1:57.2
16 JAN 2075	2478954.28	18:36: 1	147.9	T	142	-0.2802	1.0311	-37-11.5	94 4.7	73.5	351.8	109.7	2:41.9
13 JUL 2075	2479131.75	6: 5:41	148.7	A	147	0.6583	0.9467	63 7.8	-95-13.0	48.5	187.3	262.0	4:44.6
6 JAN 2076	2479308.92	10: 7:25	149.5	T	152	-0.9376	1.0342	-87 -7.6	173 16.9	19.7	202.4	340.9	1:49.3
1 JUN 2076	2479456.23	17:31:20	150.1	P	119	-1.3897	0.2893	-64-23.0	51 11.4				
1 JUL 2076	2479485.79	6:50:40	150.2	P	157	1.4008	0.2745	68 57.0	98 6.6				
26 NOV 2076	2479633.99	11:42:57	150.9	P	124	1.1398	0.7317	63 43.8	-40-10.4				
22 MAY 2077	2479810.62	2:46: 4	151.7	T	129	-0.5726	1.0290	-13 -6.7	-148-19.6	55.0	342.5	118.9	2:53.7
15 NOV 2077	2479988.21	17: 7:52	152.5	A	134	0.4701	0.9371	7 46.4	70 48.4	61.9	200.3	262.4	7:53.8
11 MAY 2078	2480165.25	17:56:54	153.2	T	139	0.1837	1.0701	28 8.2	93 42.8	79.2	161.0	232.1	5:40.5
4 NOV 2078	2480342.21	16:55:38	154.0	A	144	-0.2289	0.9255	-27-50.8	83 18.2	76.8	22.9	287.5	8:29.1
1 MAY 2079	2480519.96	10:50:12	154.8	T	149	-0.9081	1.0511	66 10.0	46 19.9	24.3	108.9	405.6	2:54.9
24 OCT 2079	2480696.26	12:11:17	155.6	A	154	-0.9246	0.9484	-83-25.6	160 38.5	22.0	71.6	495.7	3:38.8
21 MAR 2080	2480845.01	12:20:11	156.3	P	121	-1.0577	0.8721	-60-55.4	-86 -1.3				
13 SEP 2080	2481021.19	16:38: 6	157.1	P	126	1.0720	0.8736	61 7.4	-25-54.9				
10 MAR 2081	2481199.14	15:23:26	157.9	A	131	-0.3652	0.9304	-22-23.5	36 42.0	68.5	327.2	277.3	7:36.3
3 SEP 2081	2481375.88	9: 7:26	158.7	T	136	0.3373	1.0720	24 36.0	-53-38.2	70.2	212.2	247.2	5:33.2
27 FEB 2082	2481553.12	14:46:55	159.6	A	141	0.3362	0.9298	9 25.8	47 6.2	70.3	153.9	276.9	8:11.8
24 AUG 2082	2481730.55	1:16:18	160.4	T	146	-0.4008	1.0451	-10-18.4	-151-45.4	66.4	24.8	162.9	4: 1.0
16 FEB 2083	2481907.25	18: 6:32	161.2	P	151	1.0170	0.9422	61 36.4	154 9.6				
15 JUL 2083	2482055.51	0:14:19	161.9	PE	118	1.5460	0.0173	83 57.1	37 39.9				
14 AUG 2083	2482085.02	12:34:38	162.1	P	156	-1.2068	0.8130	-82 -5.6	67 31.9				
7 JAN 2084	2482232.23	17:30:19	162.8	P	123	-1.0715	0.8720	-84-24.6	-68-34.4				
3 JUL 2084	2482409.58	1:50:22	163.6	A	128	0.8204	0.9421	74 59.0	169 8.6	34.5	222.6	377.0	4:25.1
27 DEC 2084	2482586.88	9:13:44	164.5	T	133	-0.4094	1.0459	-47-19.4	-47-41.7	65.8	347.1	145.7	3: 4.3
22 JUN 2085	2482763.64	3:21:14	165.3	A	138	-0.0449	0.9704	26 7.7	-131-15.6	86.7	198.9	106.3	3:28.9
16 DEC 2085	2482941.44	22:37:44	166.2	A	143	0.2787	0.9971	-7-15.3	160 48.2	73.8	178.8	10.5	0:19.3
11 JUN 2086	2483117.96	11: 7:11	167.0	T	148	-0.7219	1.0174	-23-14.2	-12-28.4	43.7	0.9	85.9	1:48.0
6 DEC 2086	2483295.74	5:38:51	167.9	P	153	0.0194	0.9270	67 26.7	-96-15.1				
2 MAY 2087	2483443.25	18: 4:40	168.6	P	120	1.1137	0.8012	70 17.6	-127-33.2				
1 JUN 2087	2483472.58	1:27:14	168.8	P	158	-1.4189	0.2139	-87-47.3	-165-22.4				
26 OCT 2087	2483619.99	11:46:51	169.5	P	125	-1.2883	0.4692	-71 -1.3	130 34.4				
21 APR 2088	2483797.94	10:31:49	170.4	A	130	0.4132	0.0474	-35 58.8	-15 -5.2	65.4	184.3	173.0	3:57.8
14 OCT 2088	2483974.12	14:48: 2	171.3	A	135	-0.5349	0.9727	-39-39.5	56 10.4	57.5	19.5	115.1	2:38.1
10 APR 2089	2484152.45	22:44:38	172.2	A	140	-0.3322	0.9919	-10-13.2	154 48.6	70.8	341.7	30.0	0:53.3
4 OCT 2089	2484328.55	1:15:18	173.1	T	145	0.2168	1.0333	7 25.9	-162-47.2	77.5	200.5	115.0	3:14.2
31 MAR 2090	2484506.65	3:38: 3	174.0	P	150	-1.1032	0.7830	-72 -3.9	156 16.7				

Table 12

CANON OF SOLAR ECLIPSES
LOCAL CIRCUMSTANCES AT GREATEST ECLIPSE

DATE	JULIAN DATE	GREATEST ECLIPSE	DELTA T	TYPE	SAROS	GAMMA	MAGNITUDE	LATITUDE	LONGITUDE	SUN ALT	SUN AZ	PATH WIDTH	CENTRAL DURATION
23 SEP 2090	2484683.21	16:56:33	174.9	T	155	0.9158	1.0561	60 41.9	40 28.2	23.3	219.3	463.0	3:35.8
18 FEB 2091	2484830.91	9:54:35	175.6	P	122	1.1775	0.6561	71 13.5	17 48.6				
15 AUG 2091	2485008.52	0:34:39	176.6	T	127	-0.9489	1.0215	-55-34.2	-150-28.6	18.0	22.5	236.3	1:38.1
7 FEB 2092	2485185.13	15:10:14	177.5	A	132	0.4318	0.9841	9 54.8	48 43.7	64.4	169.5	62.4	1:47.6
3 AUG 2092	2485362.92	9:59:28	178.4	T	137	-0.2044	0.9794	5 35.2	-30-17.8	78.2	6.8	74.7	2:31.4
27 JAN 2093	2485539.64	3:22:12	179.4	T	142	-0.2741	1.0340	-34 -8.6	-136-23.6	73.9	6.6	119.3	2:58.0
23 JUL 2093	2485717.02	12:31:59	180.3	A	147	0.5717	0.9463	54 33.9	-1-19.5	54.9	347.0	241.3	5:11.3
16 JAN 2094	2485894.29	18:59: 0	181.2	T	152	-0.9337	1.0342	-84-46.3	9 60.0	20.4	265.7	329.9	1:51.3
13 JUN 2094	2486041.52	0:22:10	182.0	P	119	-1.4613	0.1617	-65-18.3	163 38.4				
12 JUL 2094	2486071.06	13:24:32	182.2	P	157	1.3150	0.4223	67 58.1	-152-44.1				
7 DEC 2094	2486219.34	20: 5:51	183.0	P	124	1.1544	0.7049	64 40.1	94 58.4				
2 JUN 2095	2486395.92	10: 7:38	184.0	T	129	-0.6396	1.0332	-16-42.3	-37 -7.4	50.2	346.3	144.8	3:18.5
27 NOV 2095	2486573.54	1: 2:51	184.9	A	134	0.4900	0.9330	7 13.3	-169-47.3	60.7	196.5	284.7	8:46.7
22 MAY 2096	2486750.57	1:37:14	185.9	T	139	0.1195	1.0737	27 15.7	-153-24.0	82.8	168.2	240.9	6: 6.5
15 NOV 2096	2486927.53	0:36: 7	186.9	A	144	-0.2021	0.9237	-29-44.5	-163-17.1	78.1	18.3	293.7	8:52.5
11 MAY 2097	2487105.27	18:34:31	187.9	T	149	0.8514	1.0538	67 24.3	149 31.8	31.3	121.4	339.4	3:10.0
4 NOV 2097	2487281.58	2: 1:20	188.9	A	154	-0.8930	0.9494	-85-47.2	-86-39.7	28.3	67.4	411.6	3:35.6
1 APR 2098	2487430.34	20: 2:26	189.7	P	121	-1.1005	0.7970	-61 -2.5	38 1.4				
25 SEP 2098	2487606.52	0:31:11	190.7	P	126	1.1180	0.7865	61 5.6	100 54.6				
24 OCT 2098	2487635.94	10:36: 6	190.9	PB	164	-1.5412	0.0037	-61-45.7	95 33.0				
21 MAR 2099	2487784.45	22:54:27	191.7	A	131	-0.4016	0.9318	-20 -1.1	149 4.2	66.3	327.2	275.1	7:32.4
14 SEP 2099	2487961.21	16:57:48	192.7	T	136	0.3938	1.0684	23 20.4	62 50.4	66.7	213.2	241.0	5:17.9
10 MAR 2100	2488138.44	22:28: 6	193.7	A	141	0.3077	0.9338	11 57.4	162 27.0	72.0	152.9	257.5	7:29.3
4 SEP 2100	2488315.87	8:49:16	194.8	T	146	-0.3388	1.0402	-10-29.4	-38-57.7	70.2	26.1	142.2	3:32.6

34

FIFTY YEAR CANON OF SOLAR ECLIPSES : 1986 - 2035

SECTION 2 - <u>WORLD MAPS OF ECLIPSE PATHS : 1901 - 2100</u>

SOLAR ECLIPSES : 1901 - 1920

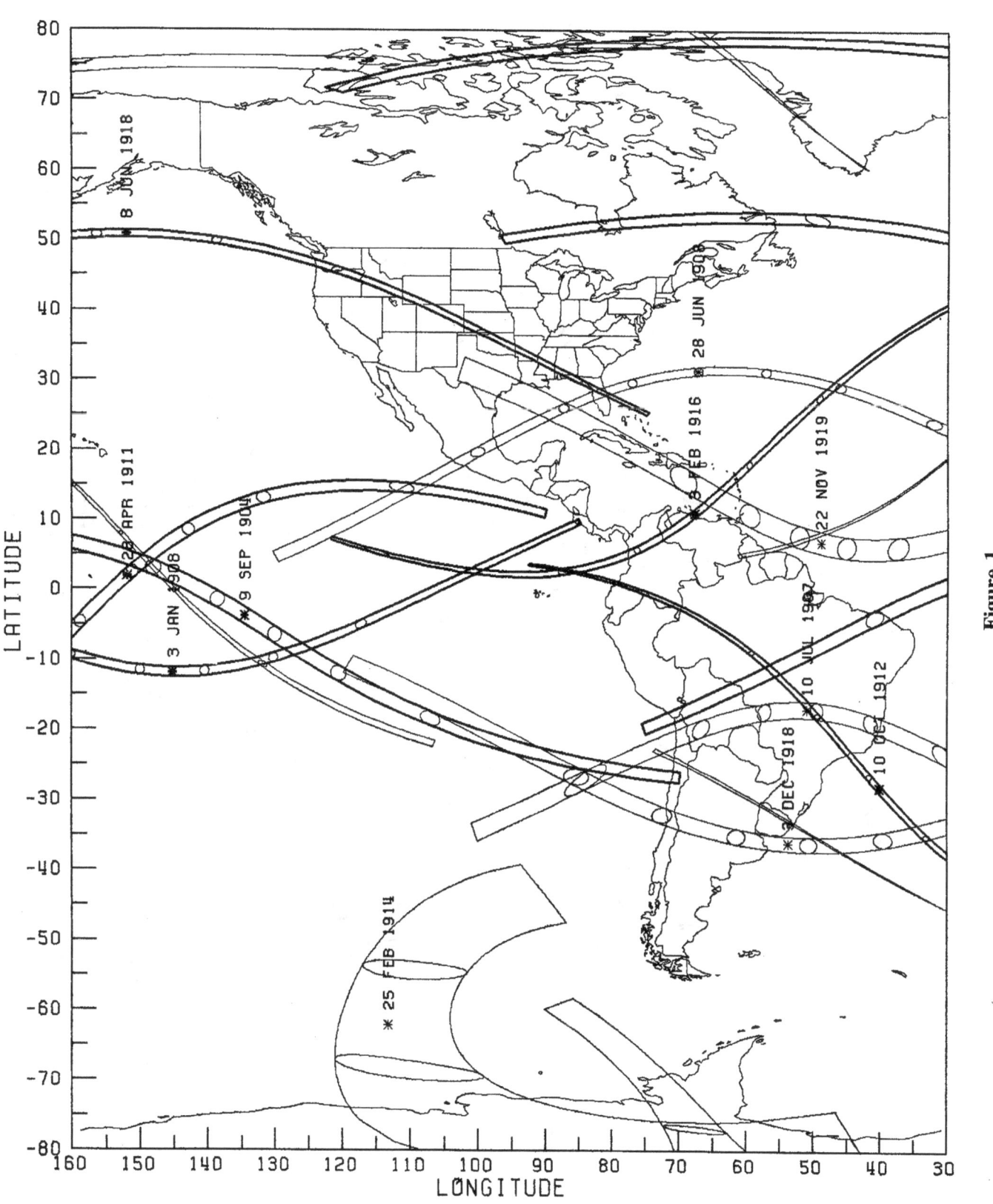

Figure 1

SOLAR ECLIPSES :   1901 - 1920

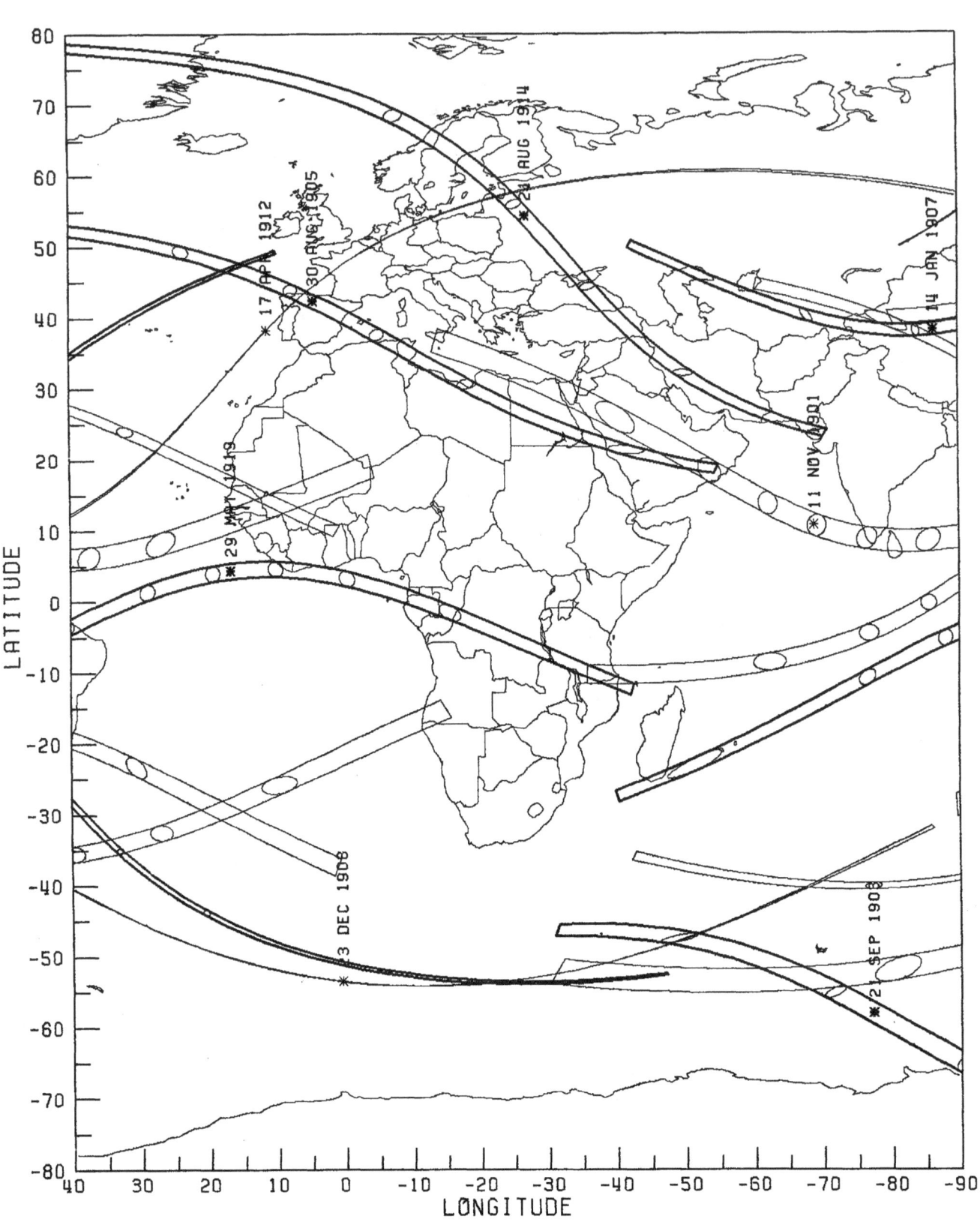

SOLAR ECLIPSES : 1901 - 1920

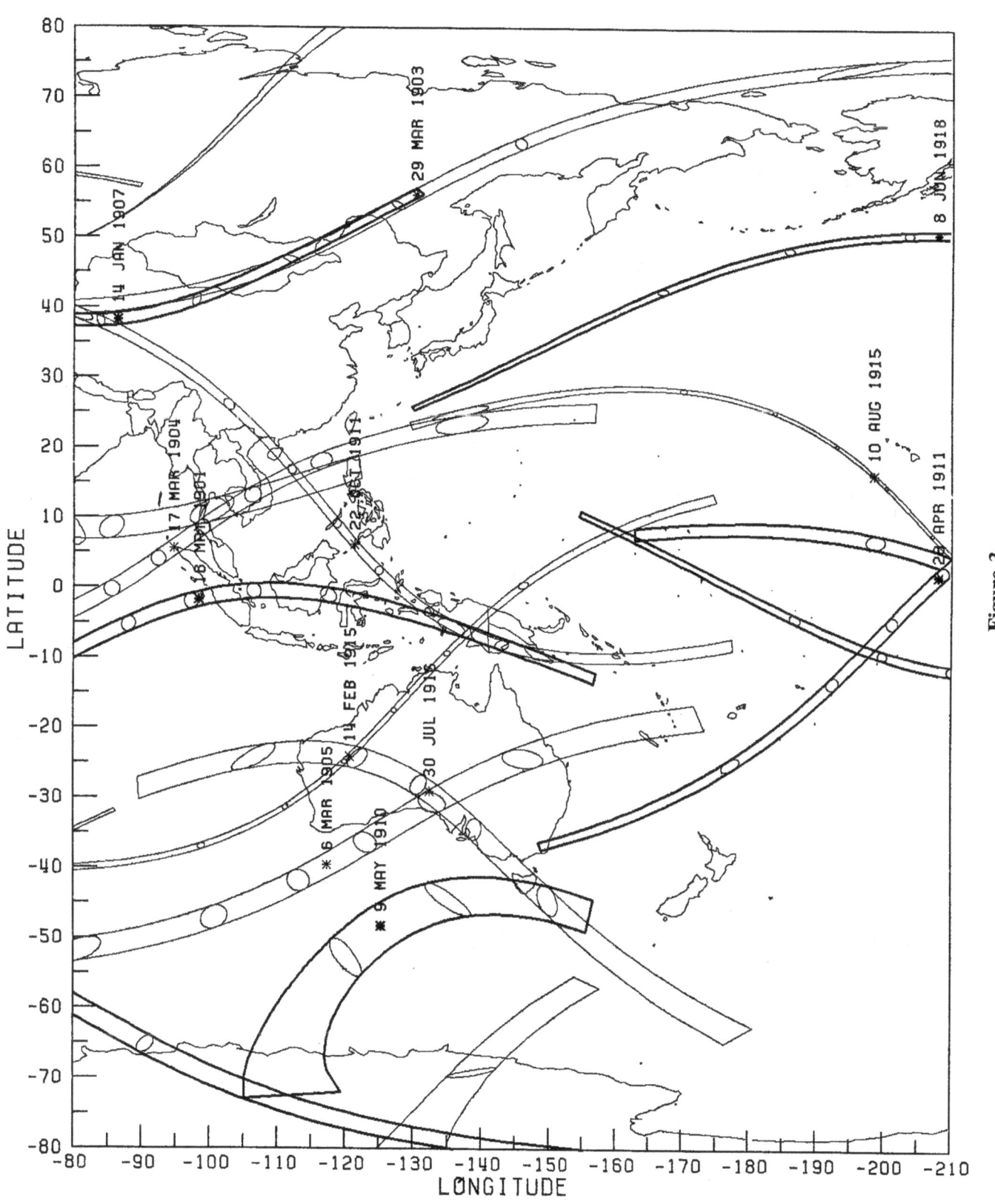

Figure 3

SOLAR ECLIPSES :   1921 - 1940

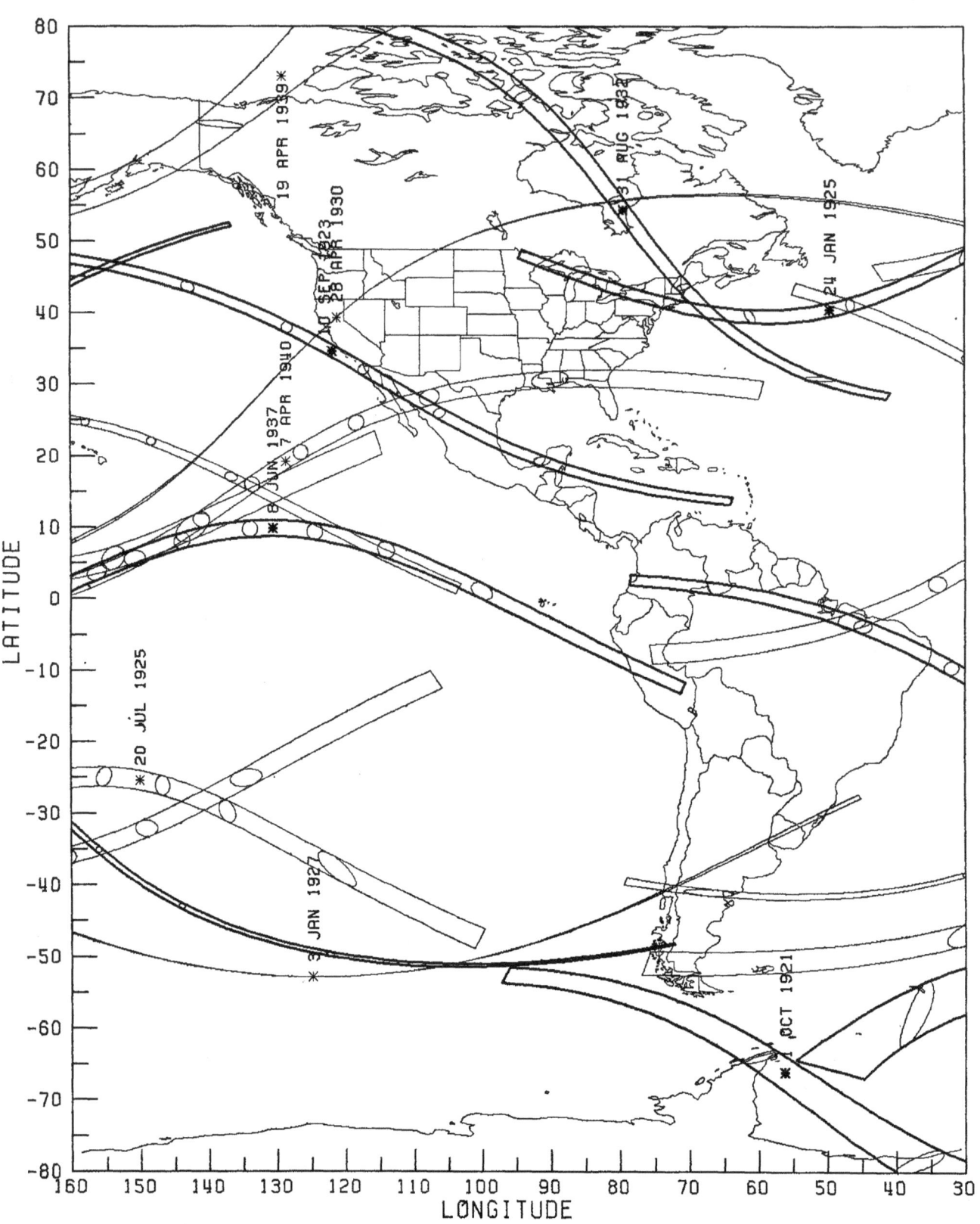

SØLAR ECLIPSES : 1921 - 1940

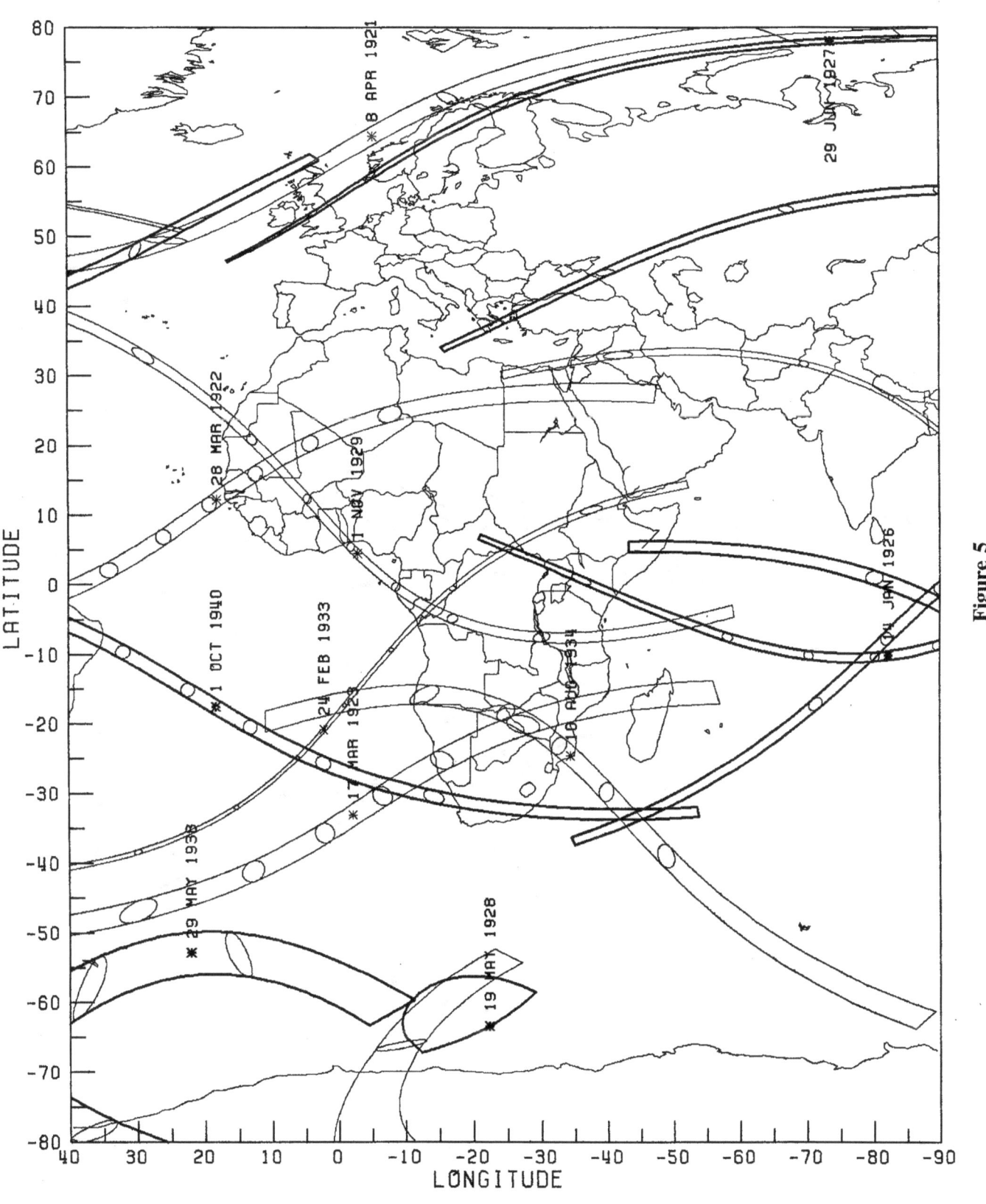

Figure 5

41

SOLAR ECLIPSES :   1921 - 1940

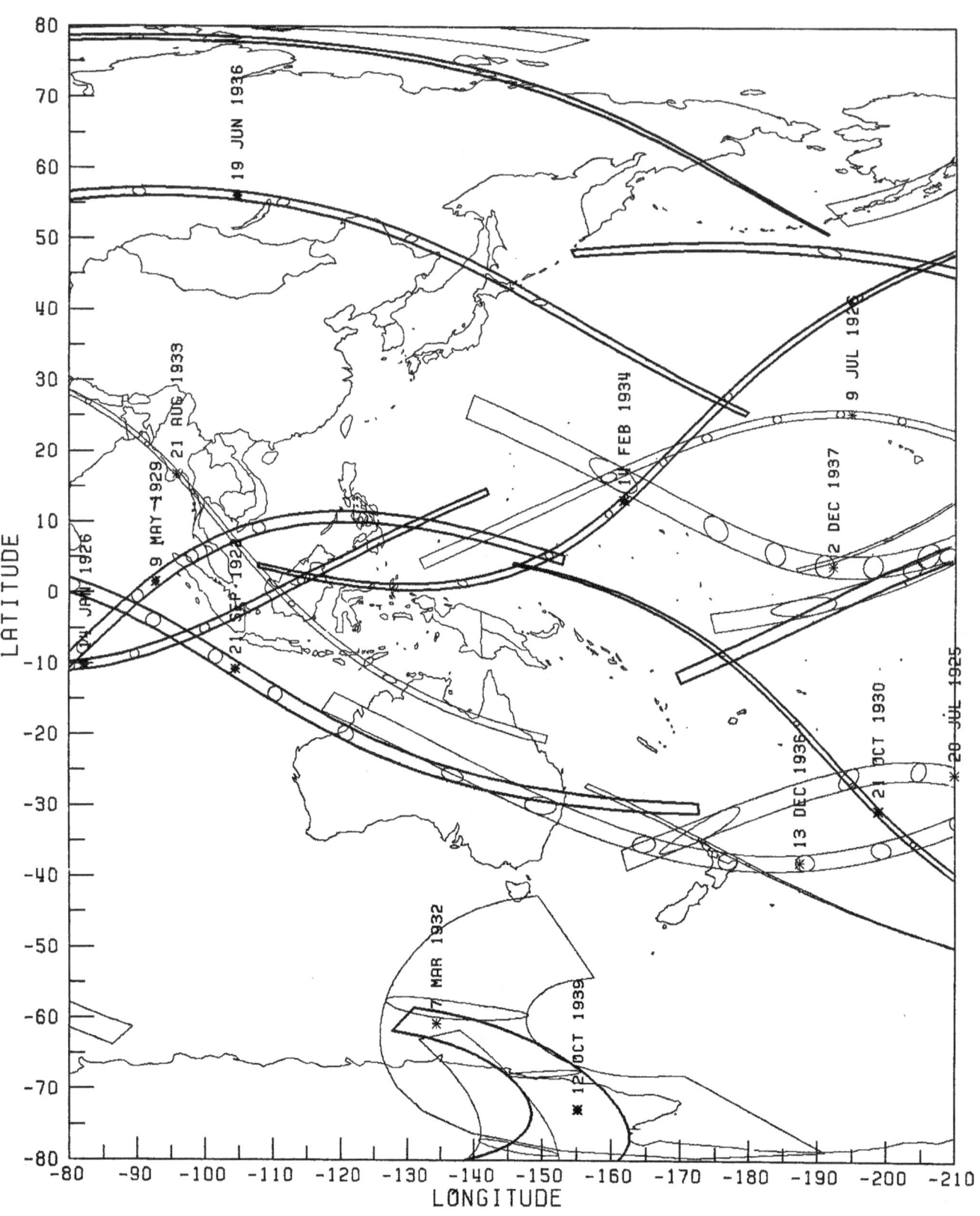

SOLAR ECLIPSES : 1941 - 1960

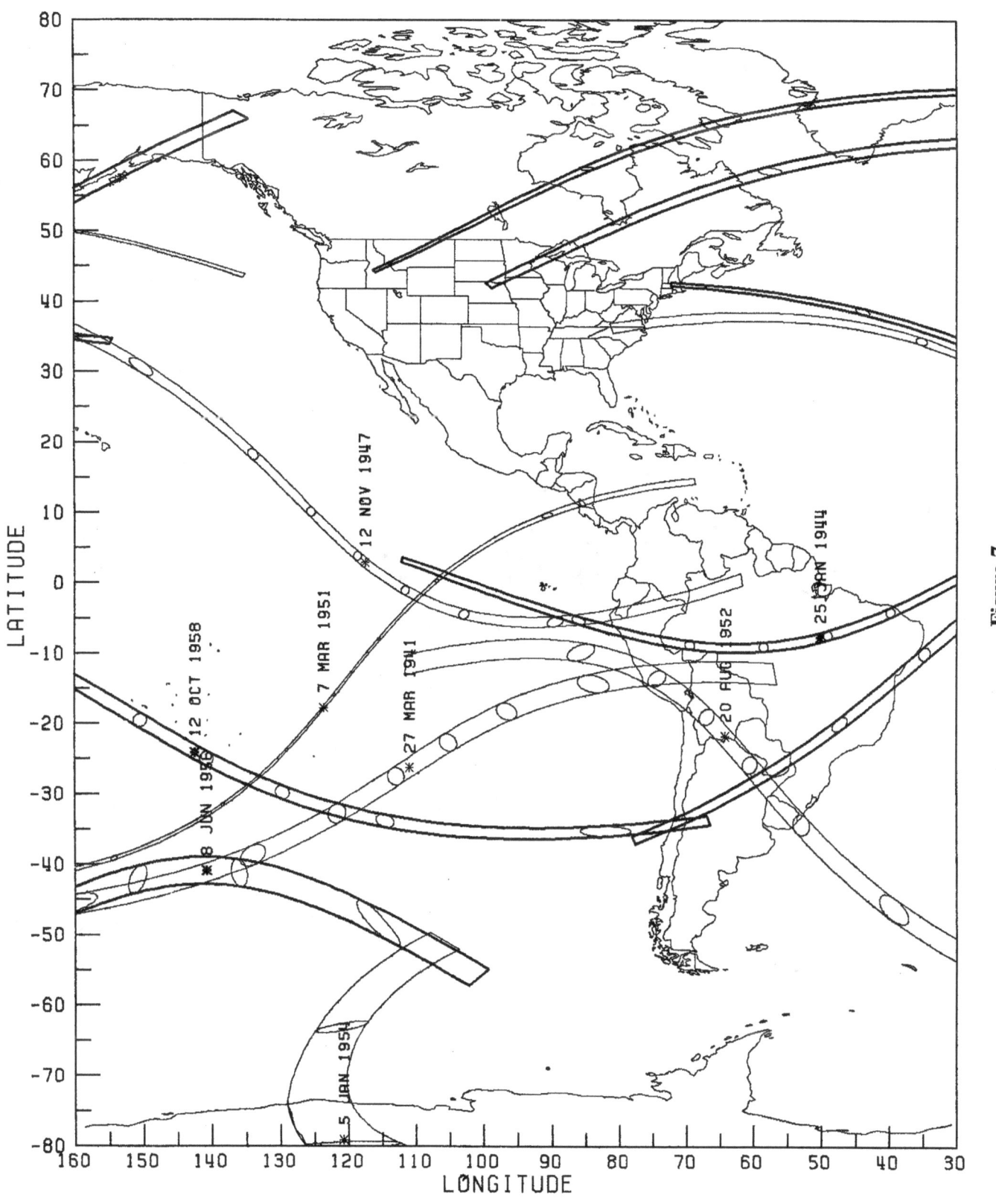

Figure 7

43

SOLAR ECLIPSES :   1941 - 1960

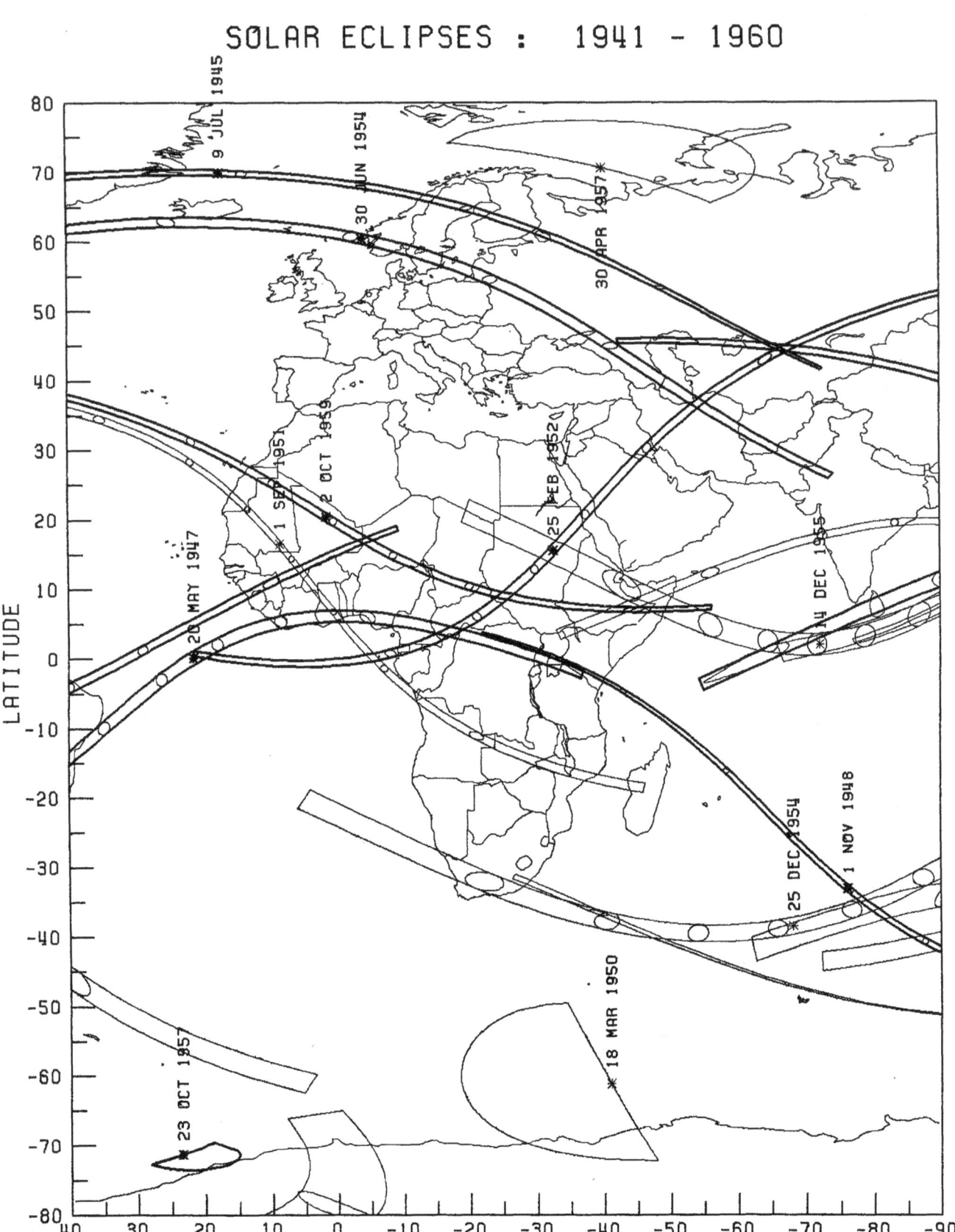

SOLAR ECLIPSES : 1941 - 1960

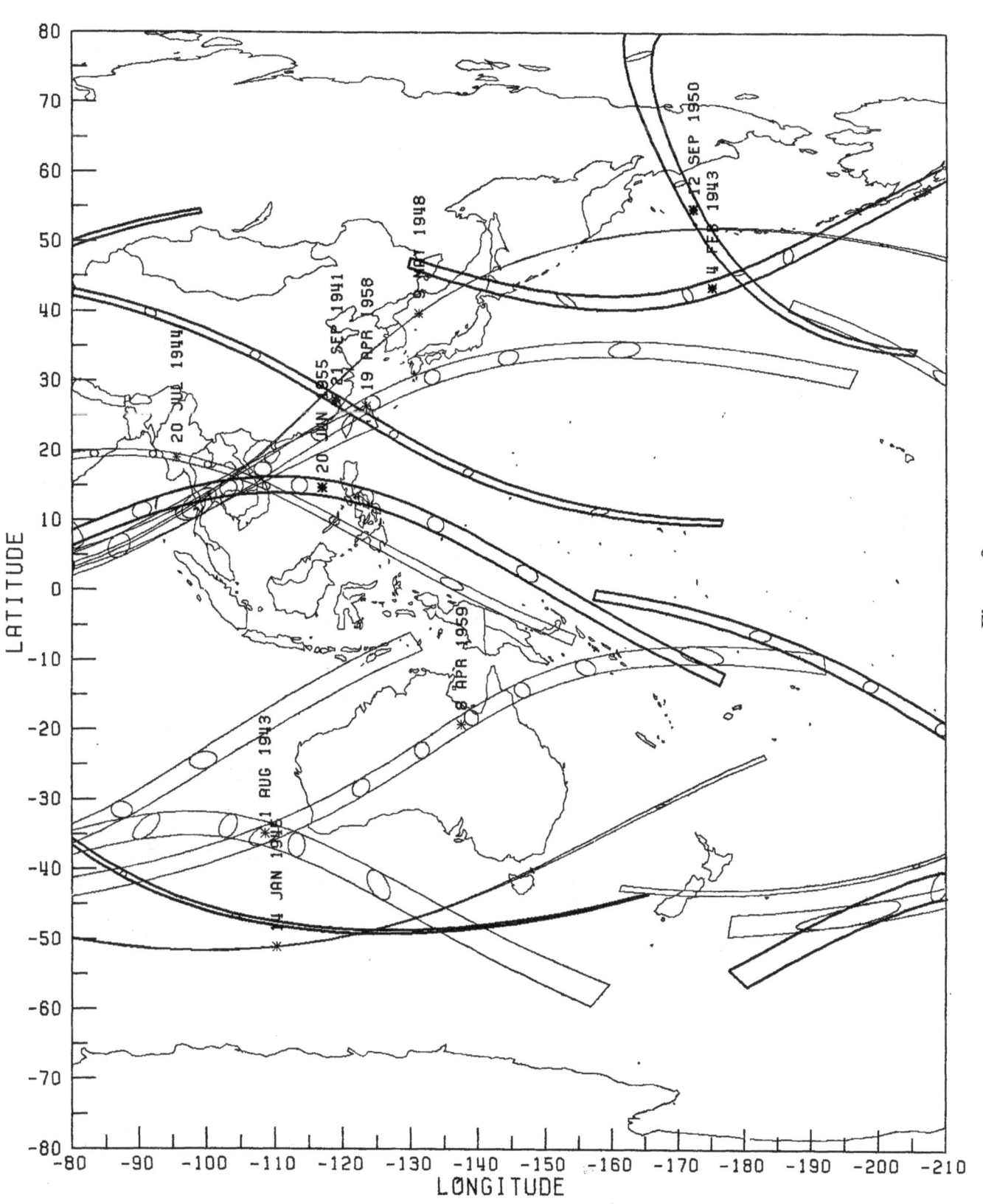

Figure 9

SOLAR ECLIPSES : 1961 – 1980

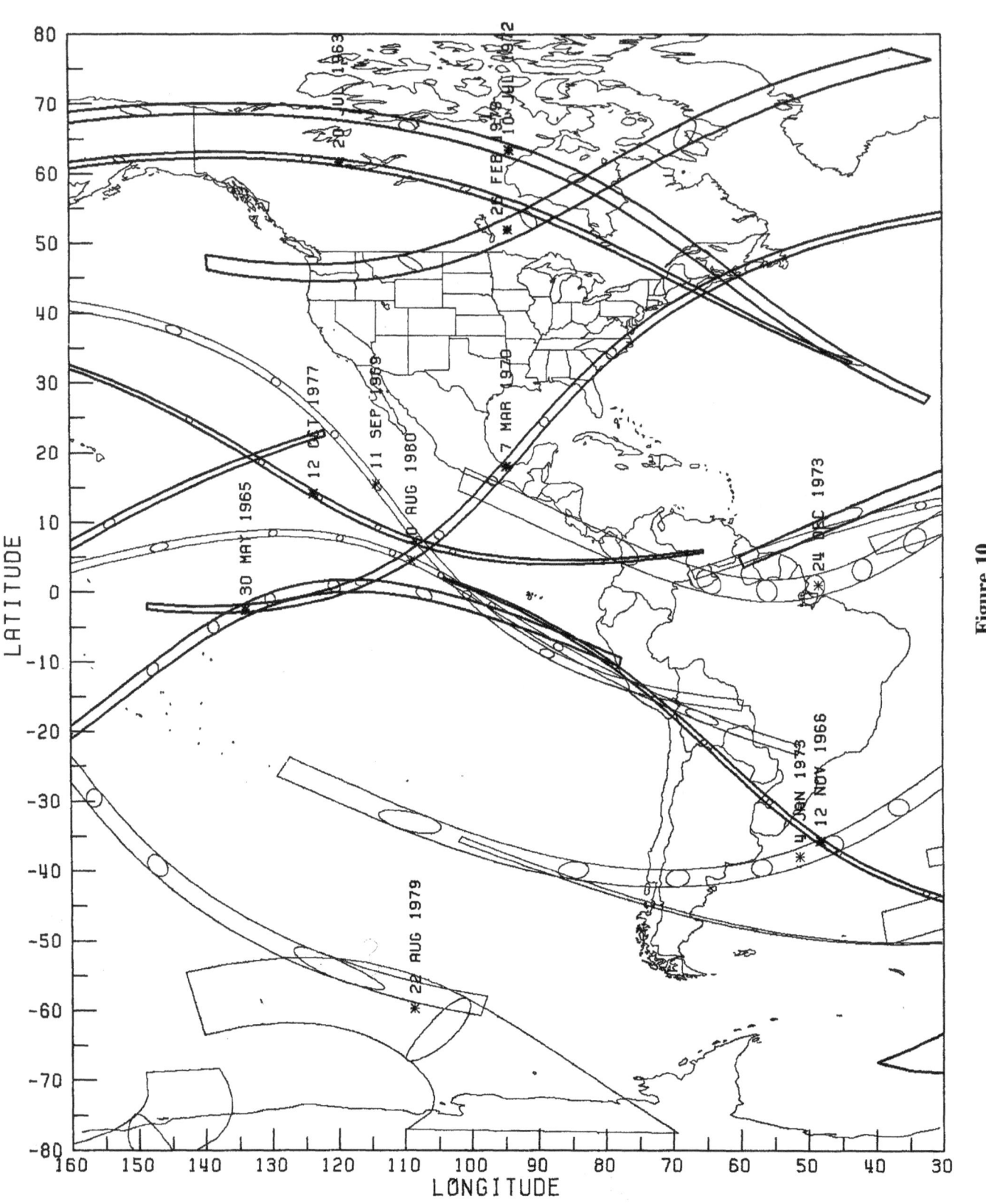

Figure 10

SOLAR ECLIPSES : 1961 - 1980

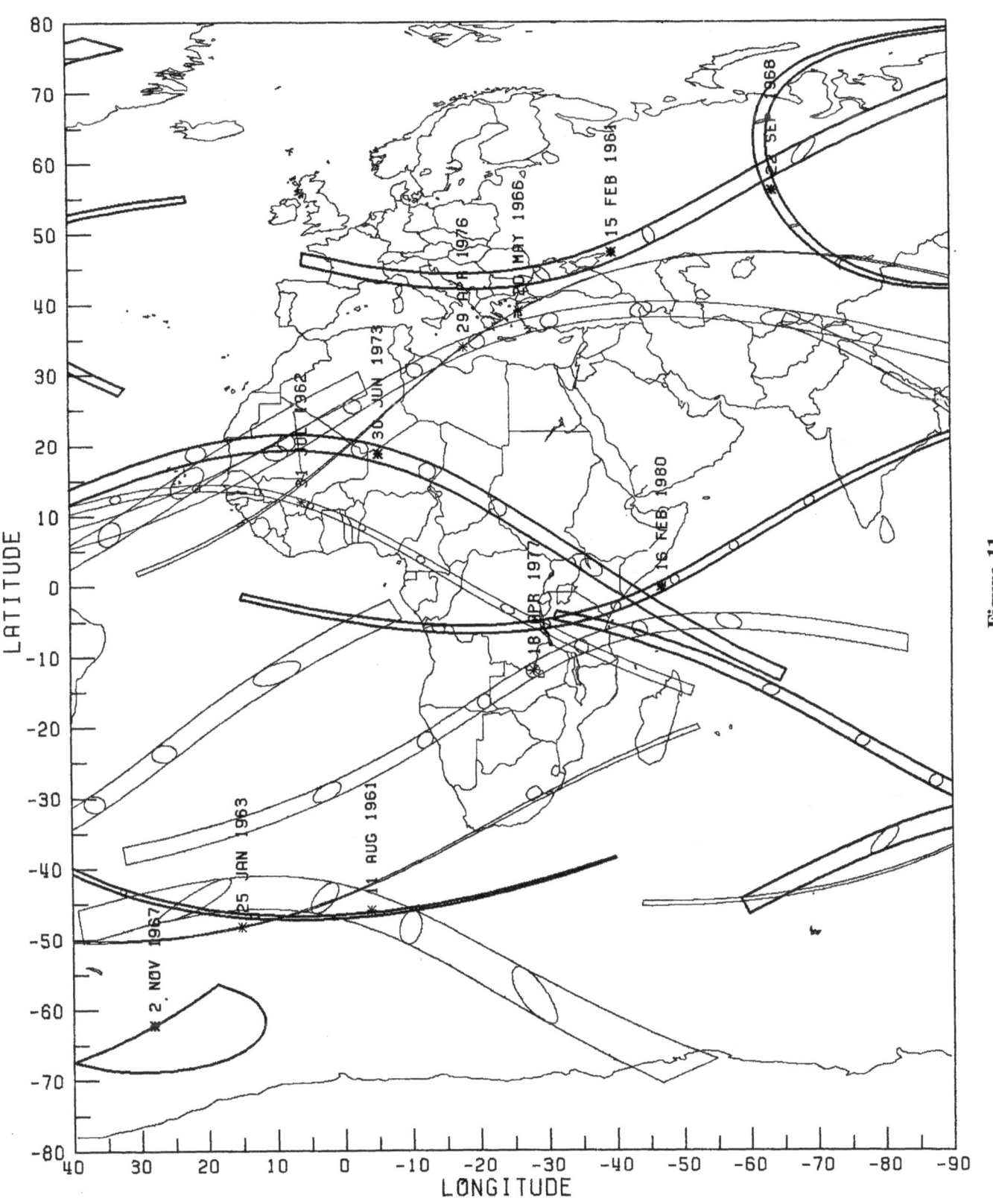

Figure 11

SOLAR ECLIPSES : 1961 - 1980

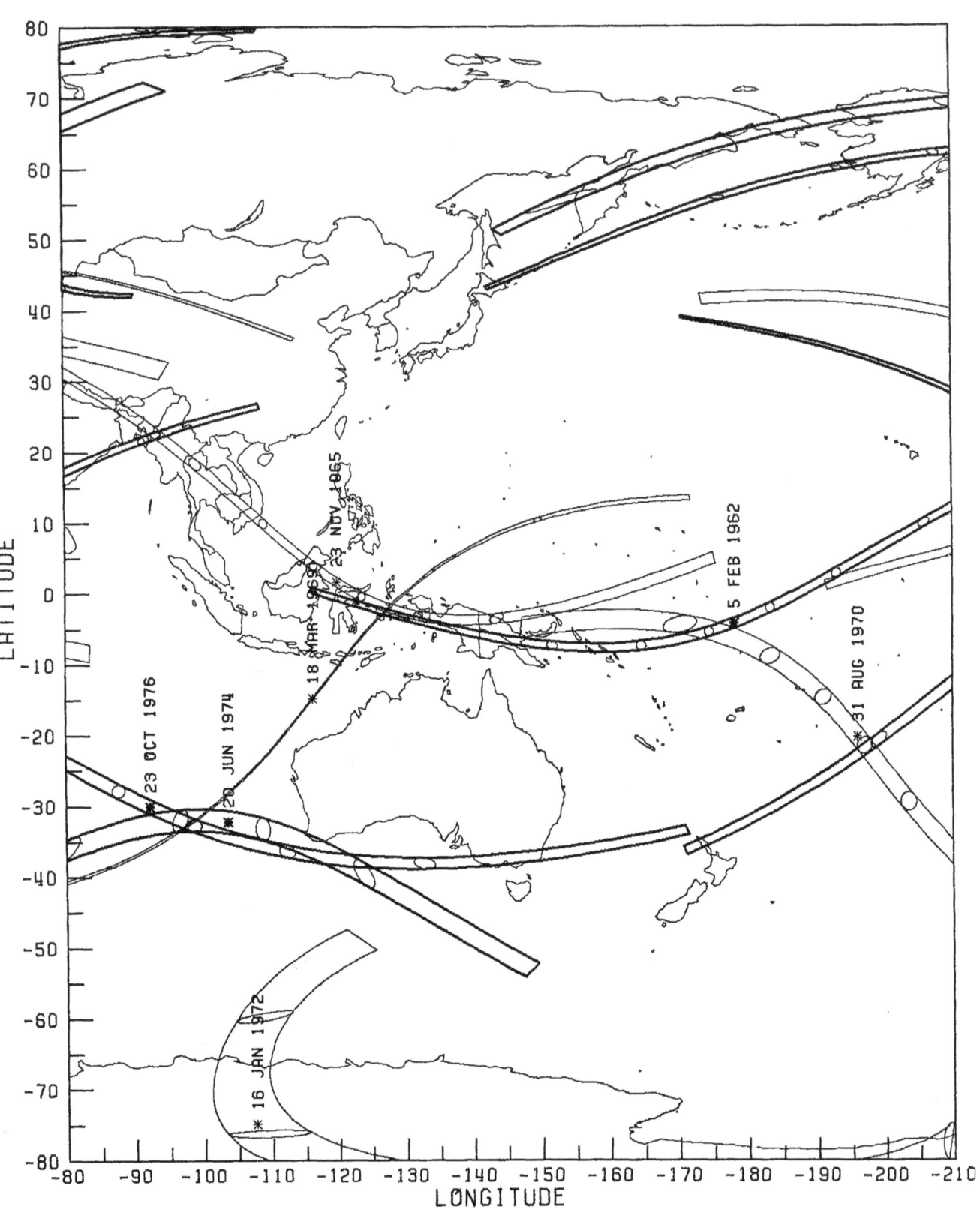

SOLAR ECLIPSES : 1981 - 2000

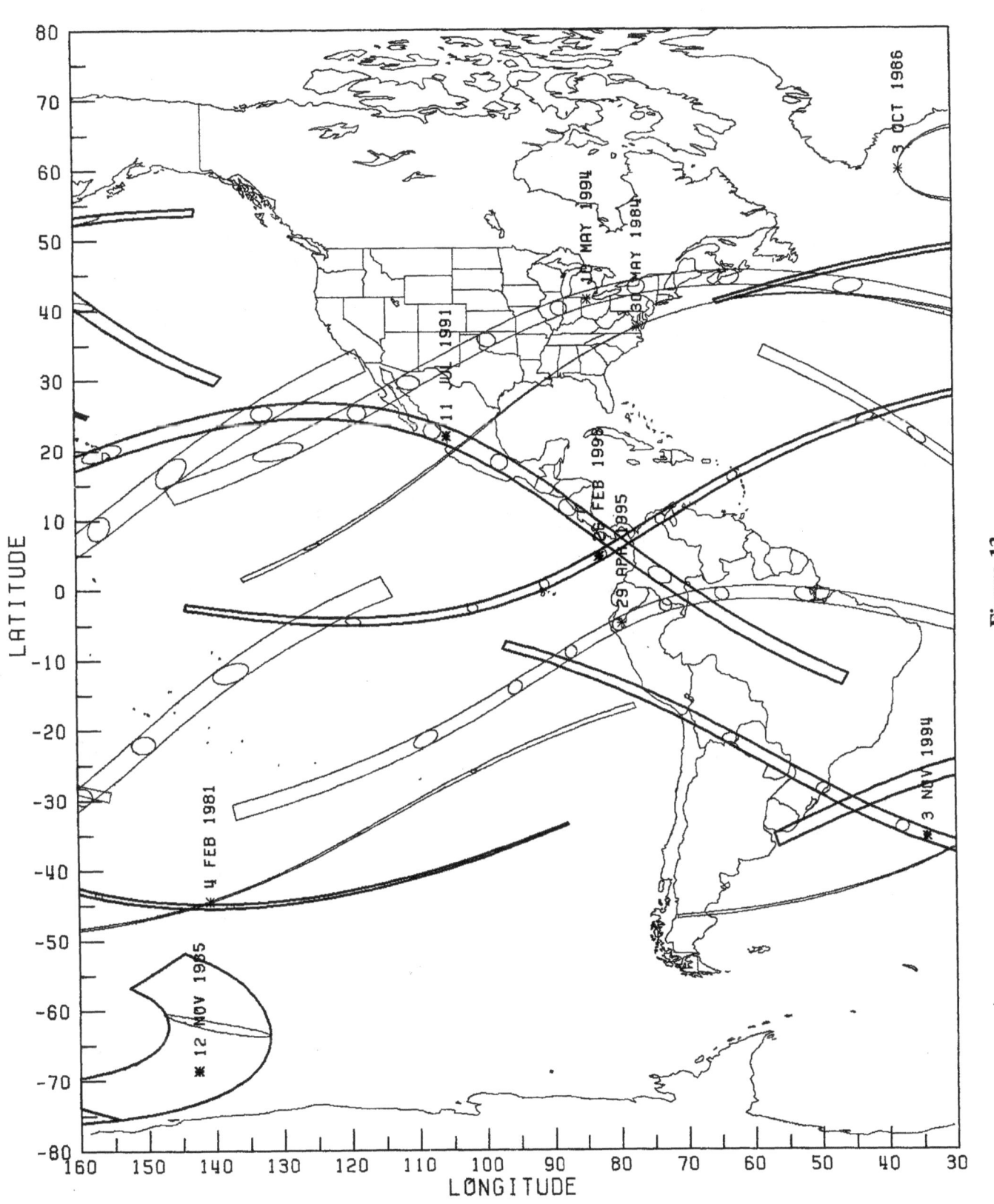

Figure 13

SOLAR ECLIPSES : 1981 - 2000

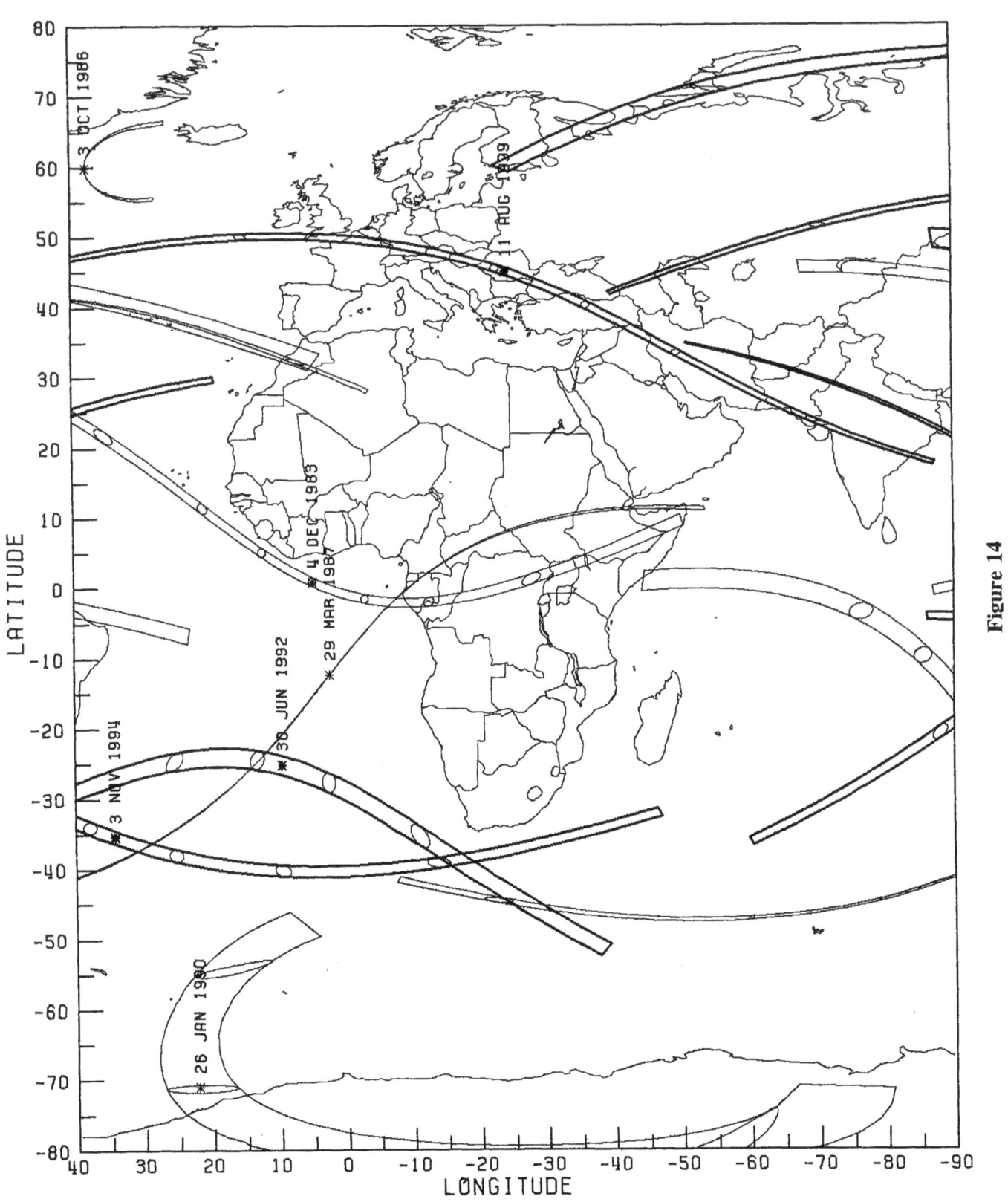

Figure 14

SOLAR ECLIPSES : 1981 - 2000

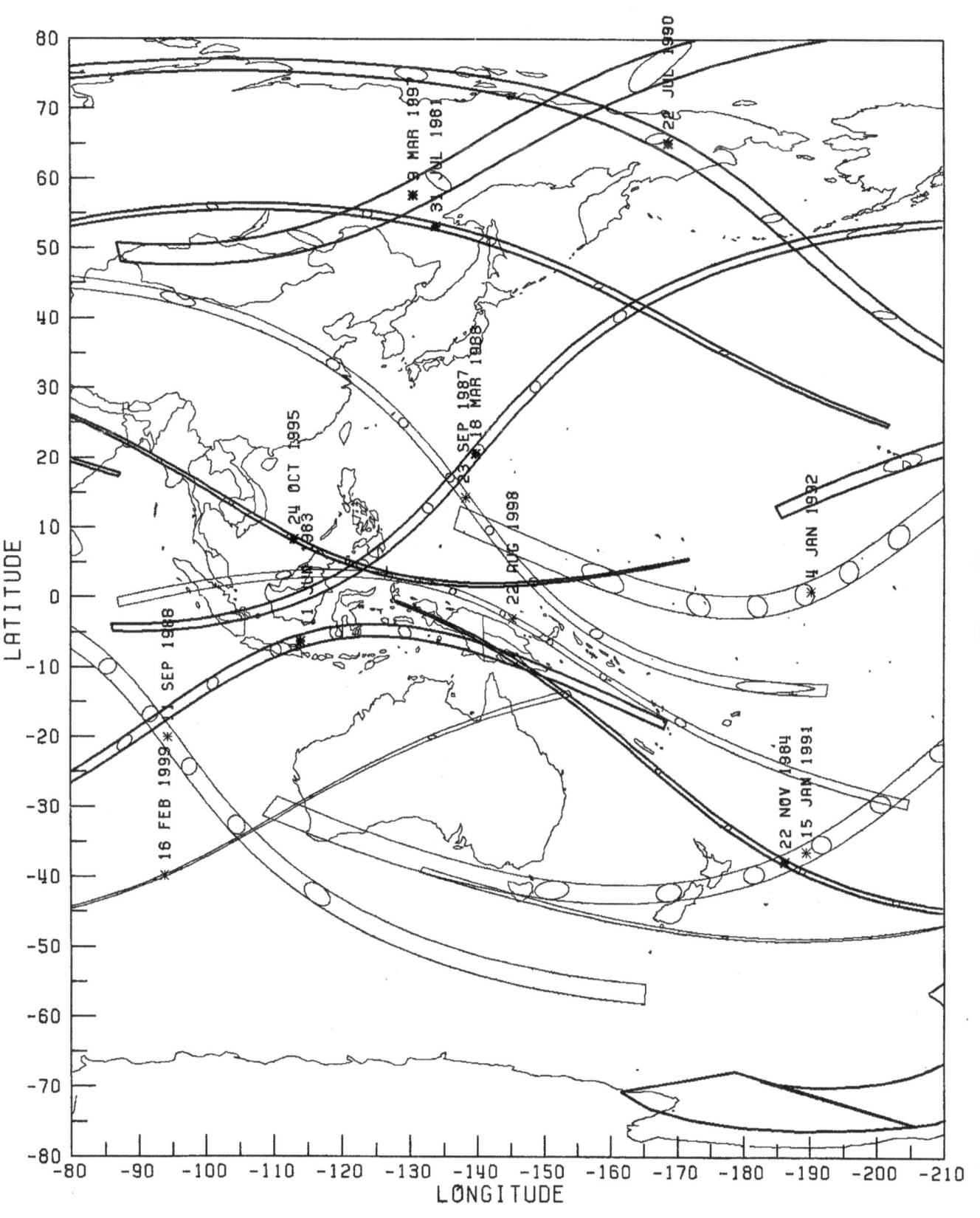

Figure 15

SOLAR ECLIPSES :   2001 - 2020

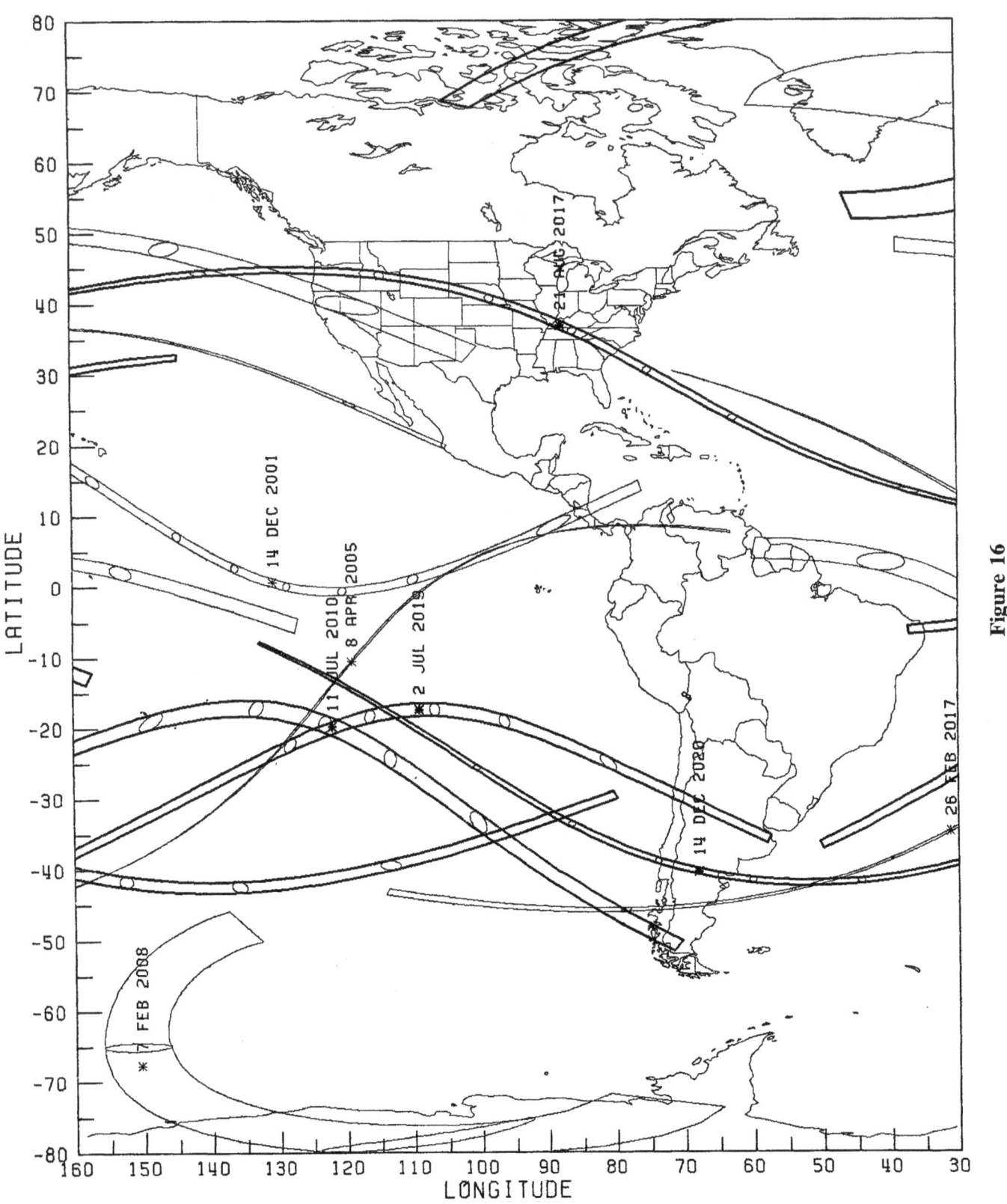

Figure 16

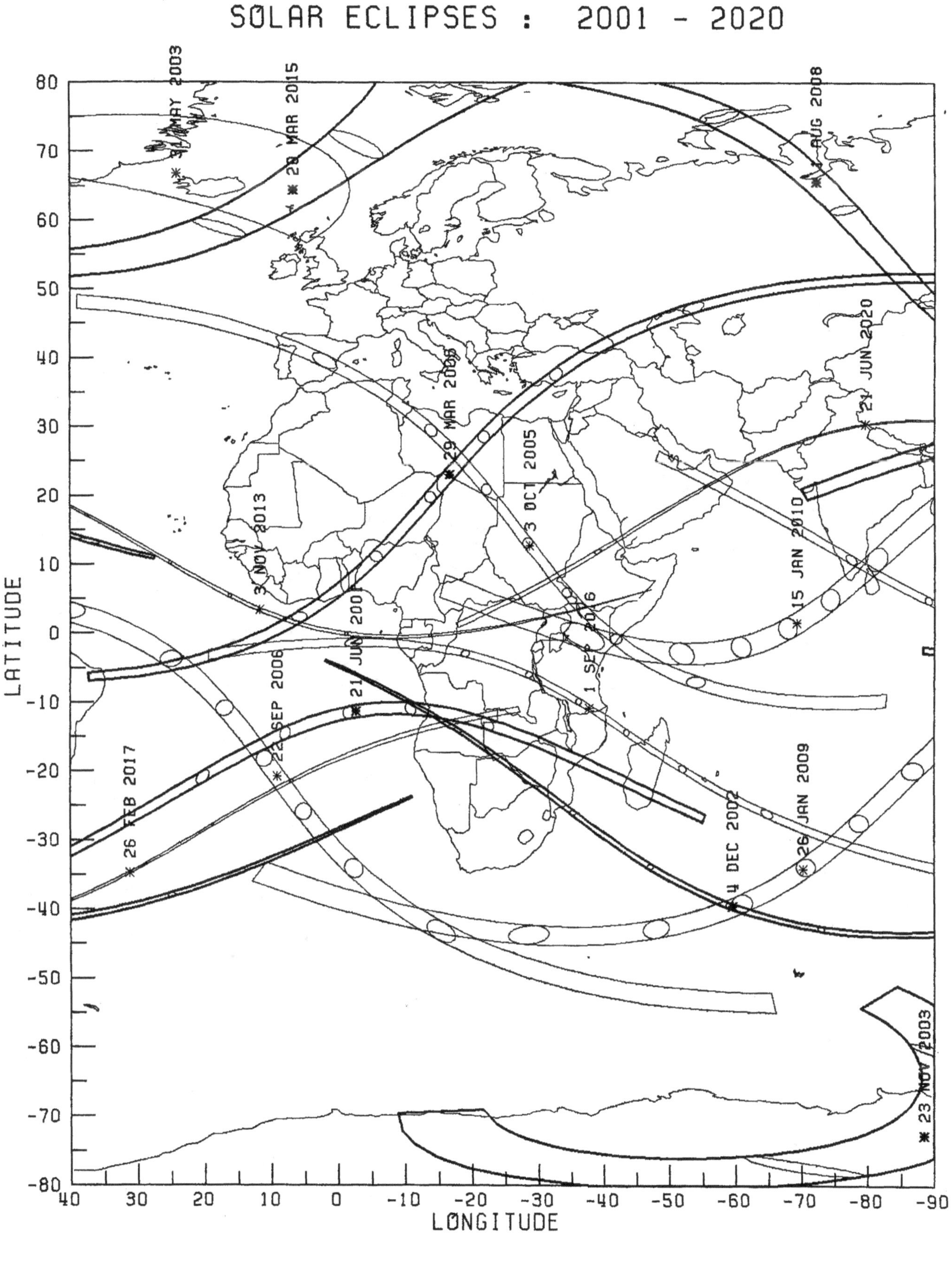
SOLAR ECLIPSES : 2001 - 2020
LATITUDE
LONGITUDE
31 MAY 2003
20 MAR 2015
1 AUG 2008
21 JUN 2020
29 MAR 2006
3 OCT 2005
3 NOV 2013
15 JAN 2010
1 SEP 2016
21 JUN 2001
22 SEP 2006
26 FEB 2017
4 DEC 2002
26 JAN 2009
23 NOV 2003
80
70
60
50
40
30
20
10
0
-10
-20
-30
-40
-50
-60
-70
-80
40
30
20
10
0
-10
-20
-30
-40
-50
-60
-70
-80
-90

Figure 17

SOLAR ECLIPSES :   2001 - 2020

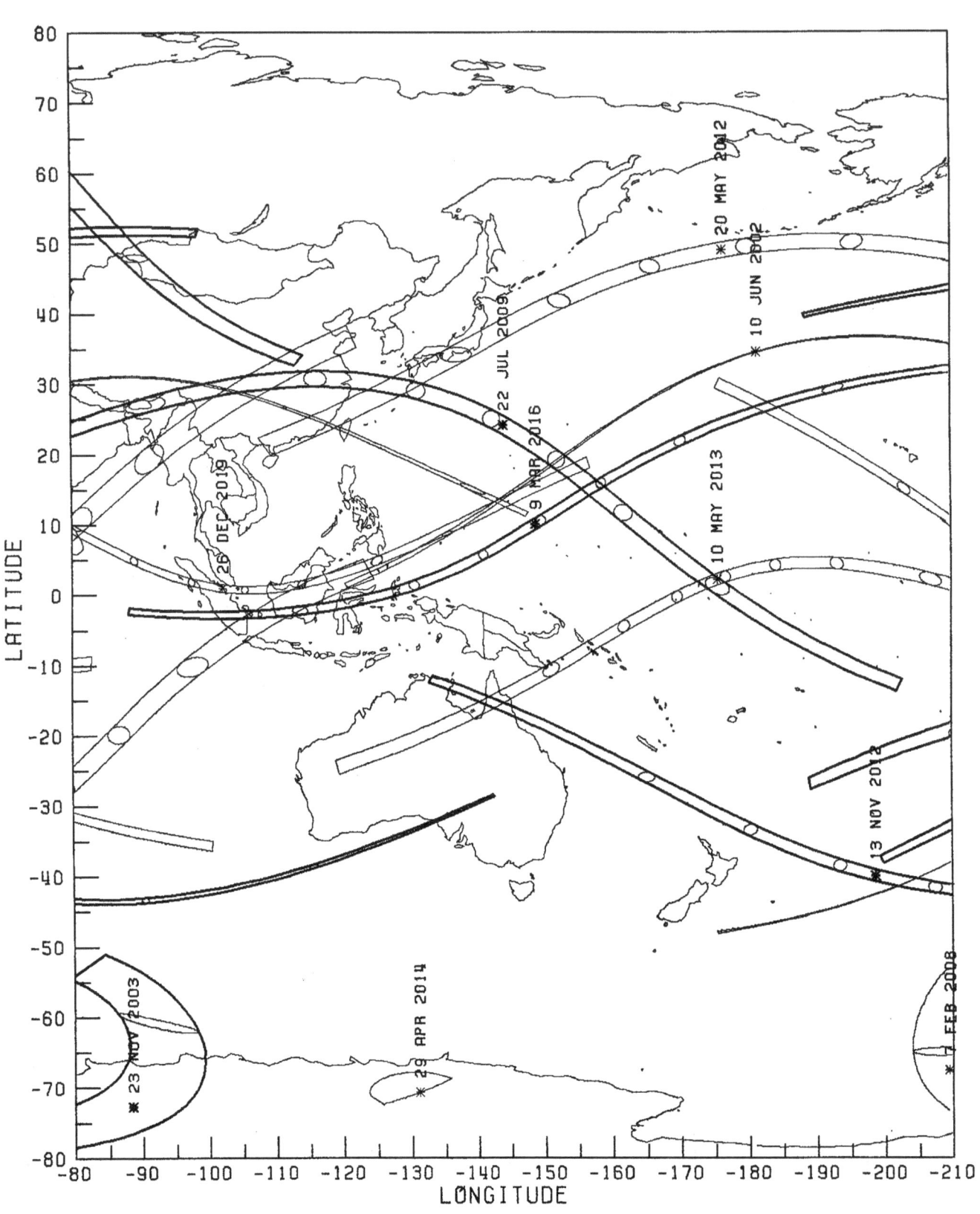

SOLAR ECLIPSES : 2021 - 2040

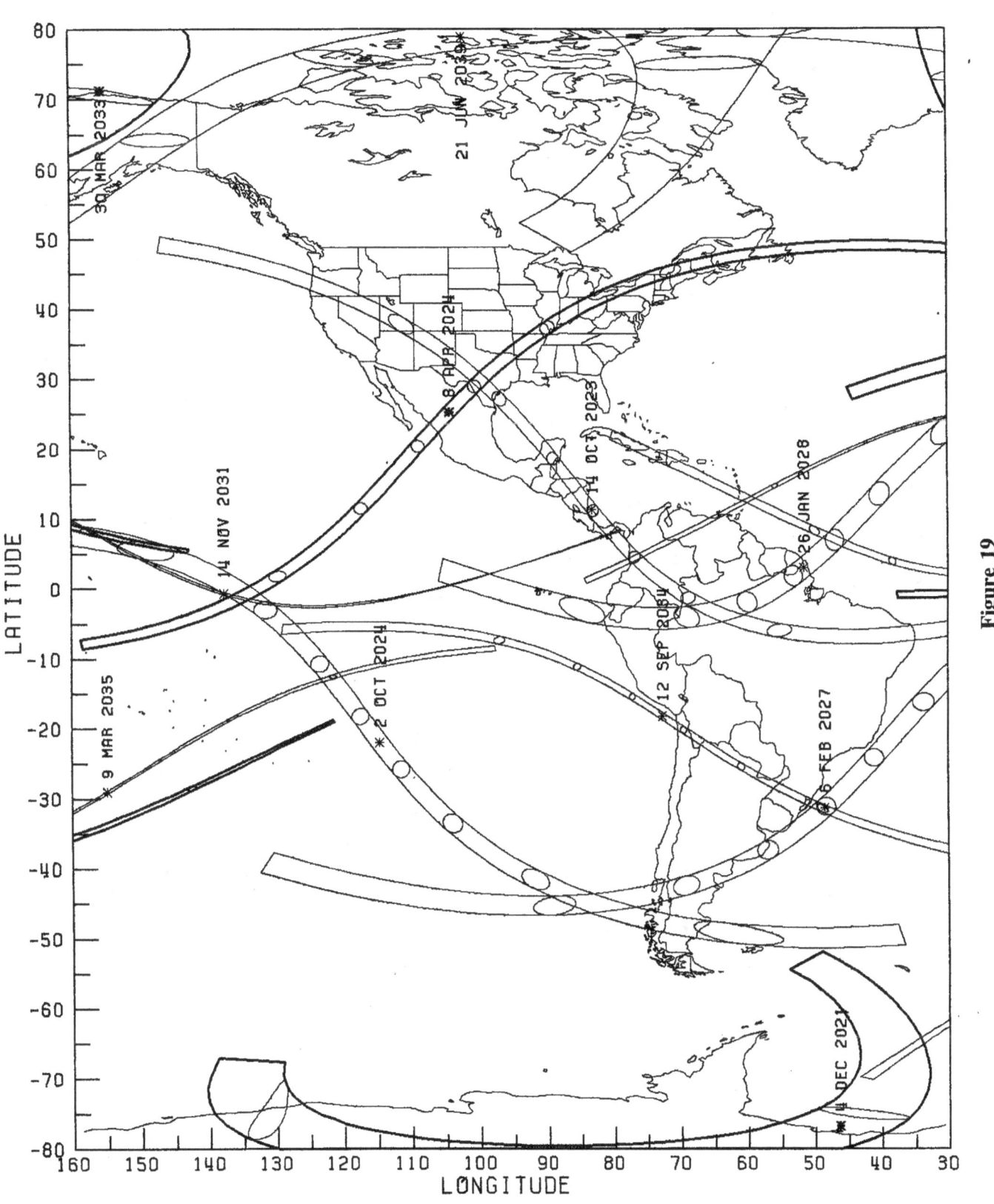

Figure 19

SOLAR ECLIPSES : 2021 - 2040

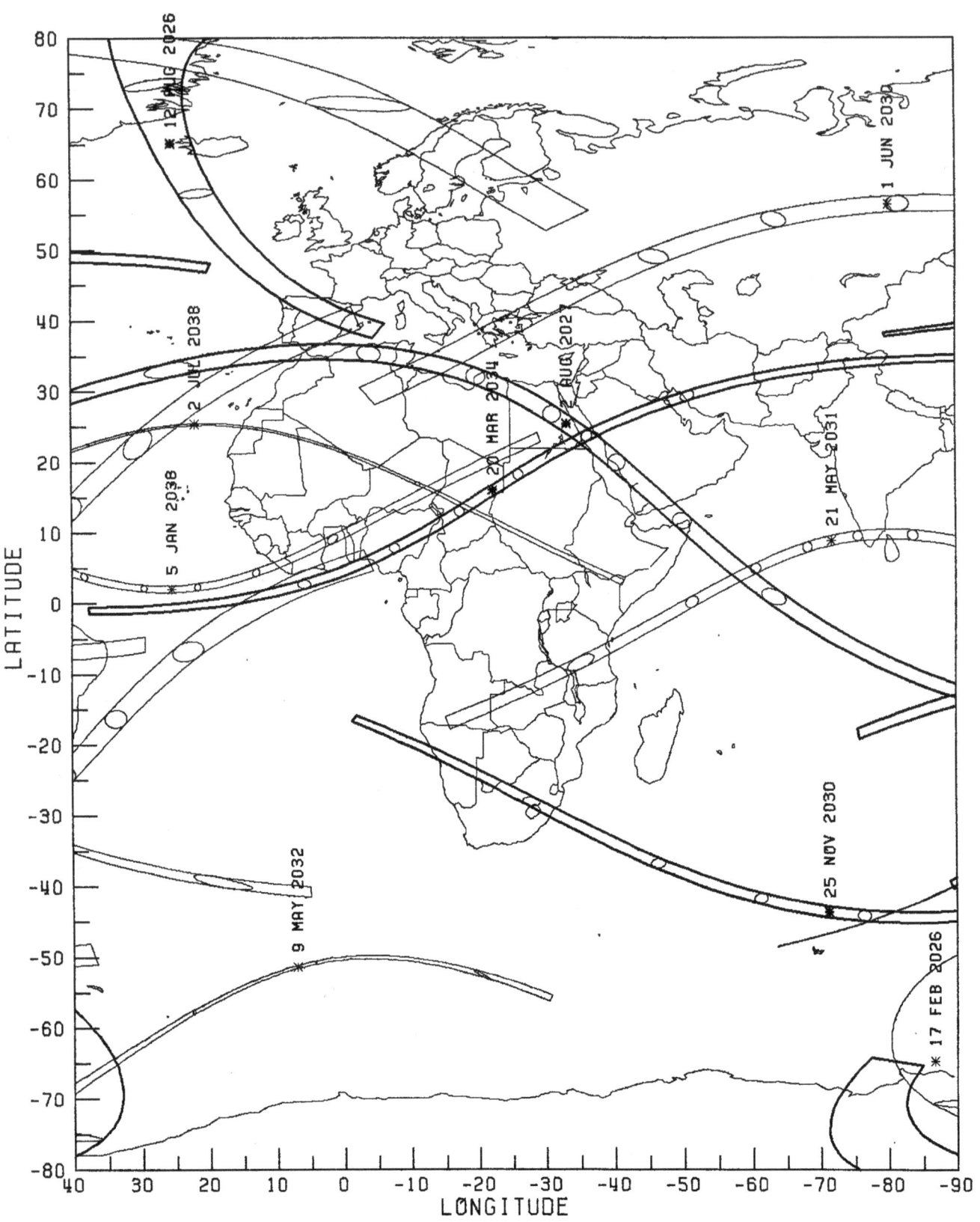

SOLAR ECLIPSES : 2021 - 2040

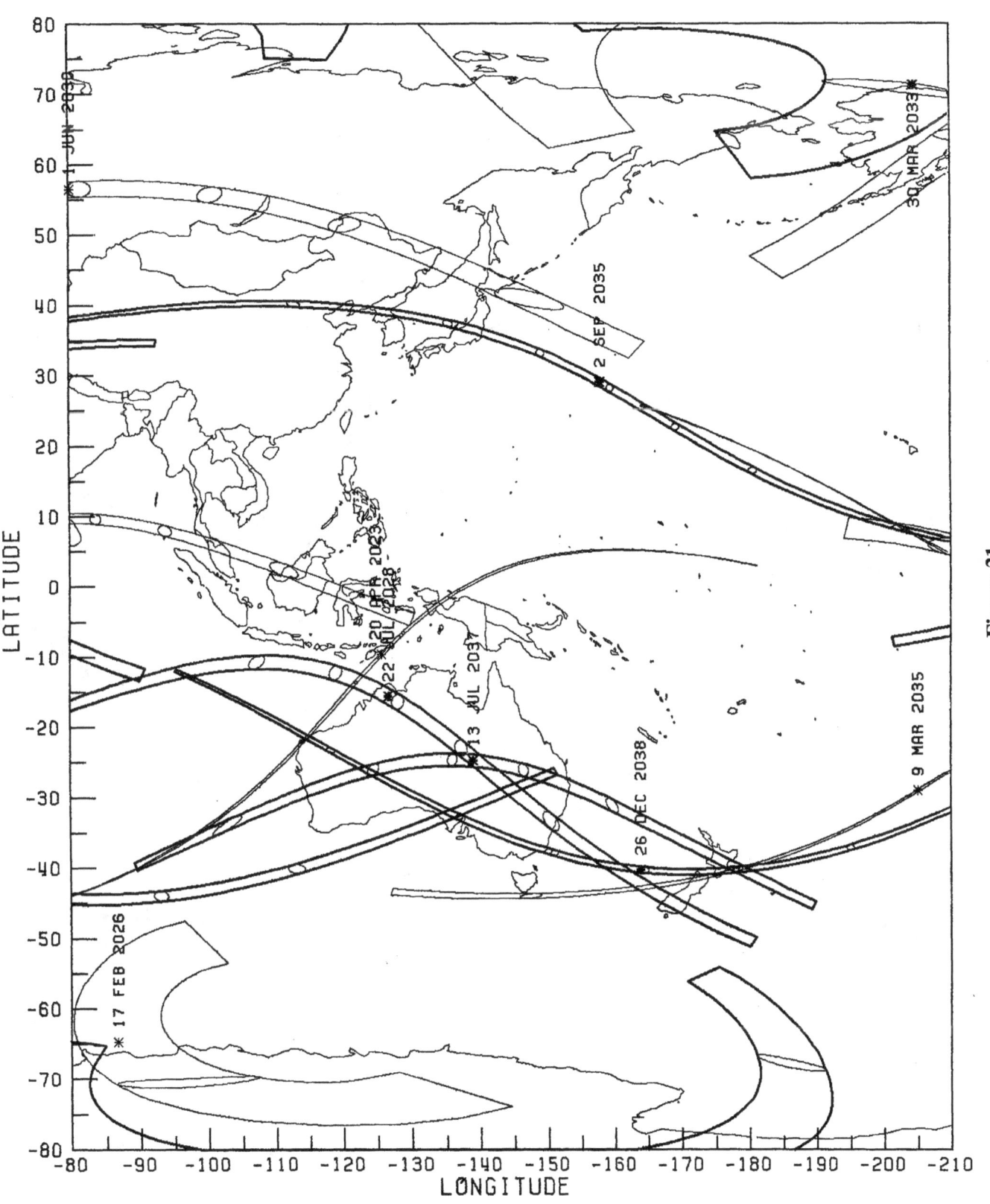

Figure 21

SOLAR ECLIPSES :   2041 - 2060

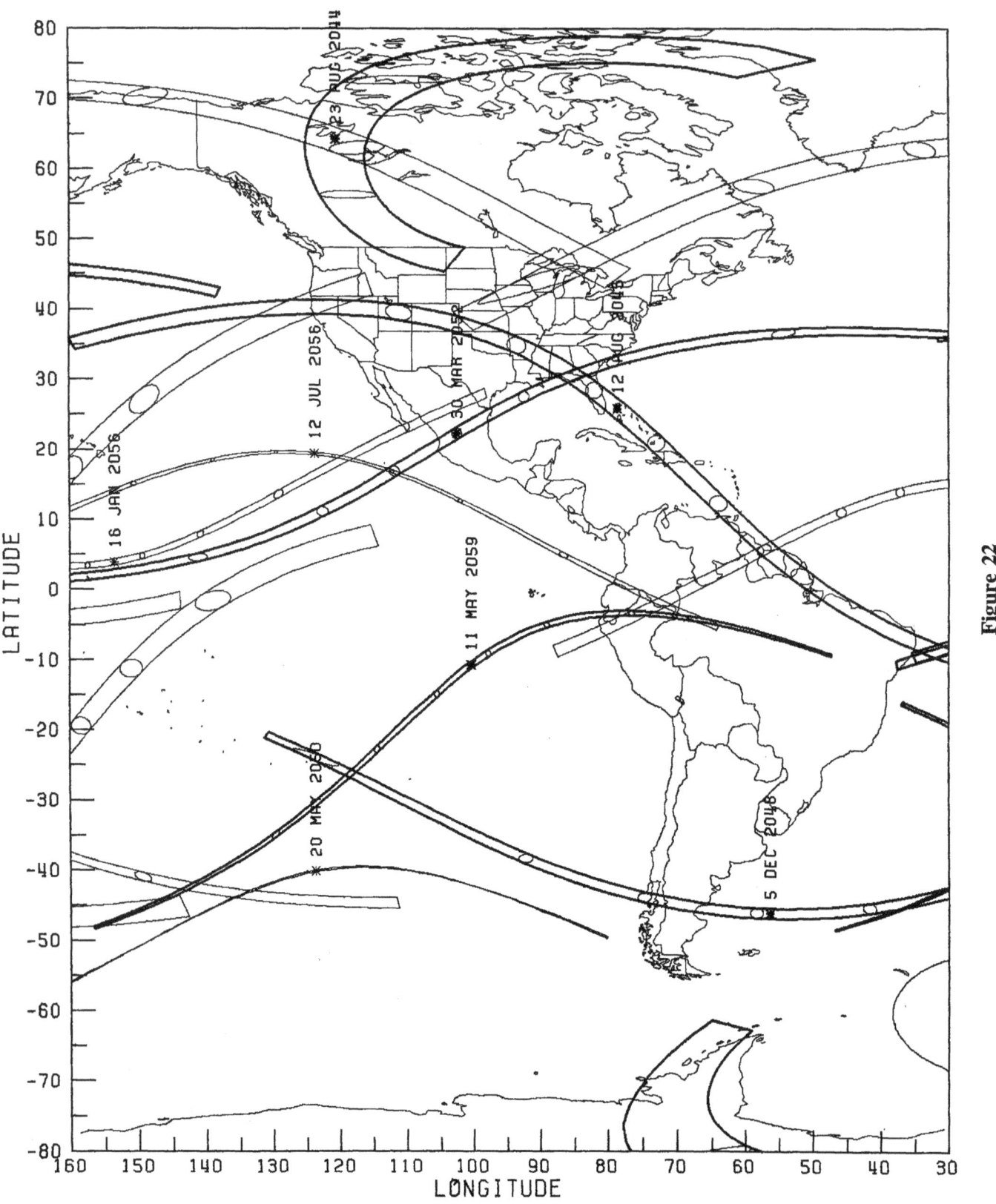

Figure 22

SOLAR ECLIPSES :   2041 - 2060

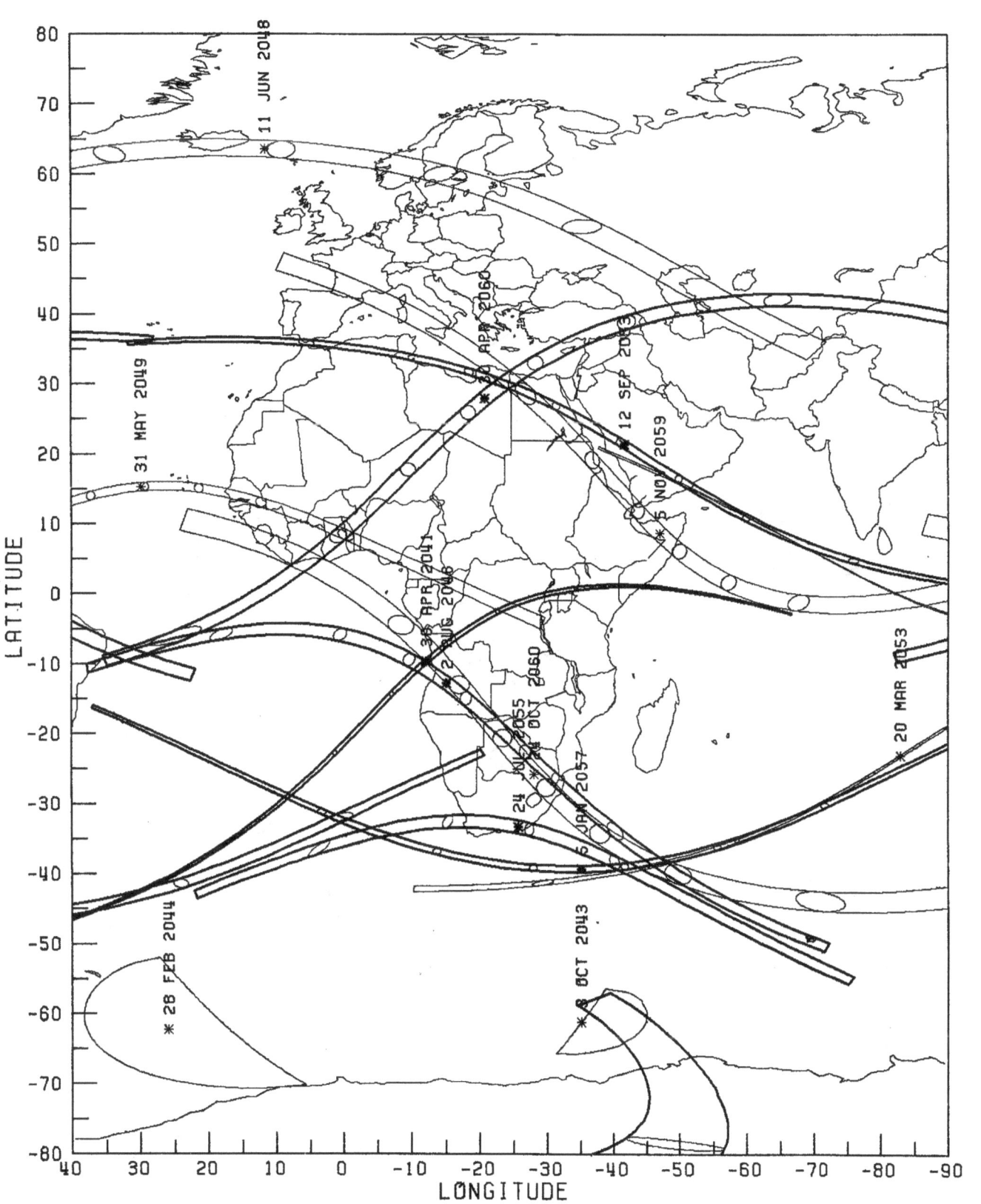

Figure 23

SOLAR ECLIPSES : 2041 - 2060

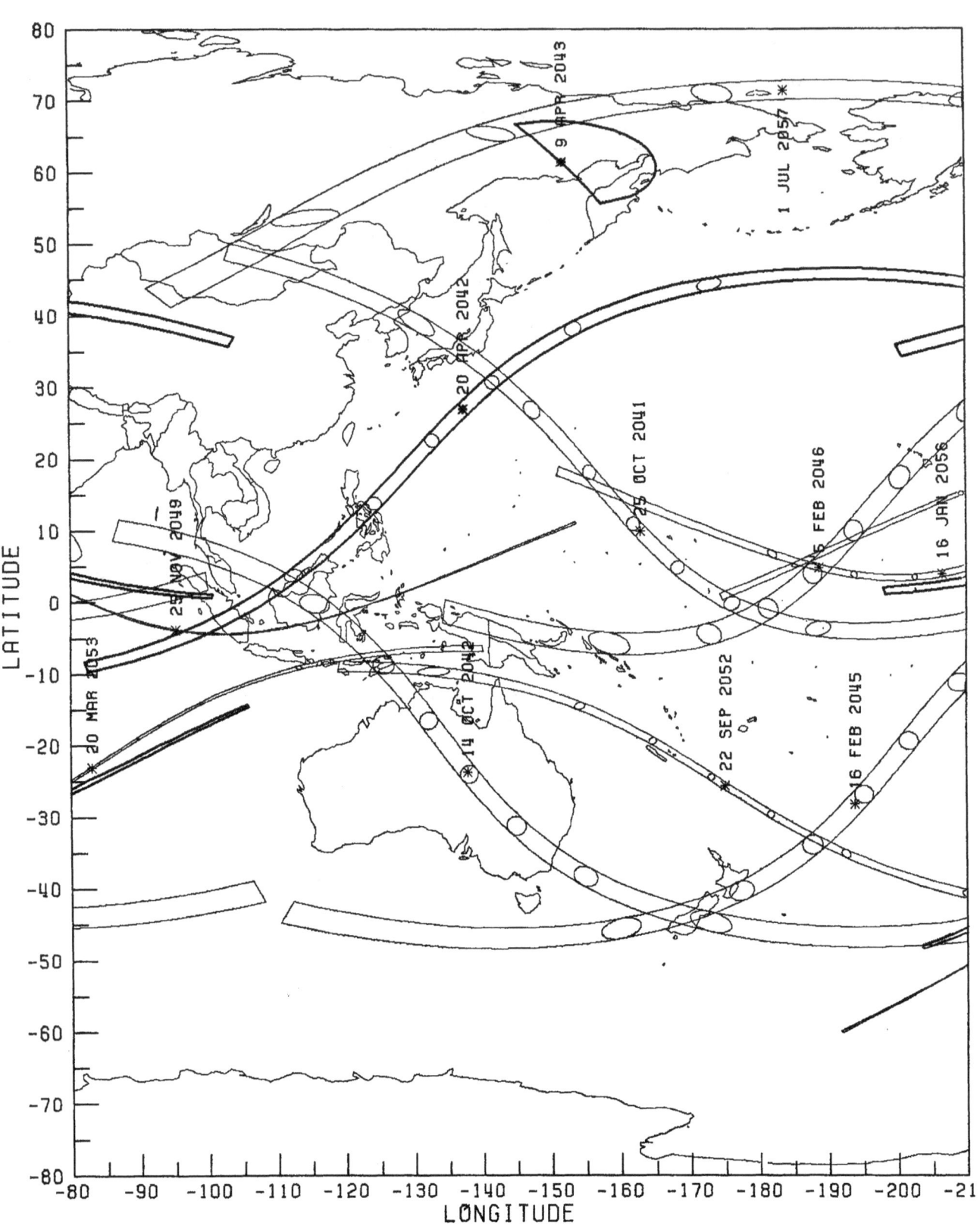

SOLAR ECLIPSES : 2061 - 2080

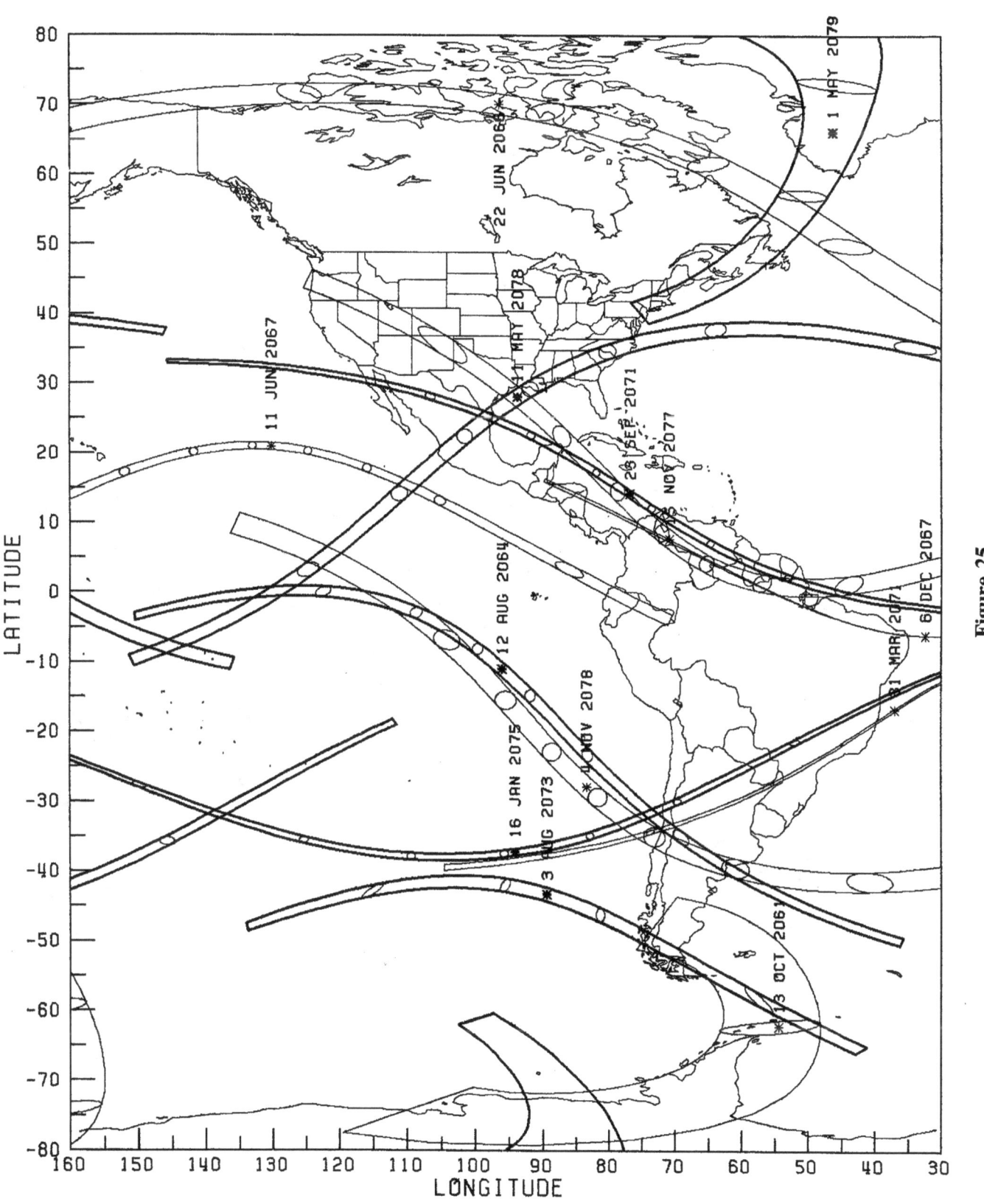

Figure 25

SOLAR ECLIPSES :   2061 - 2080

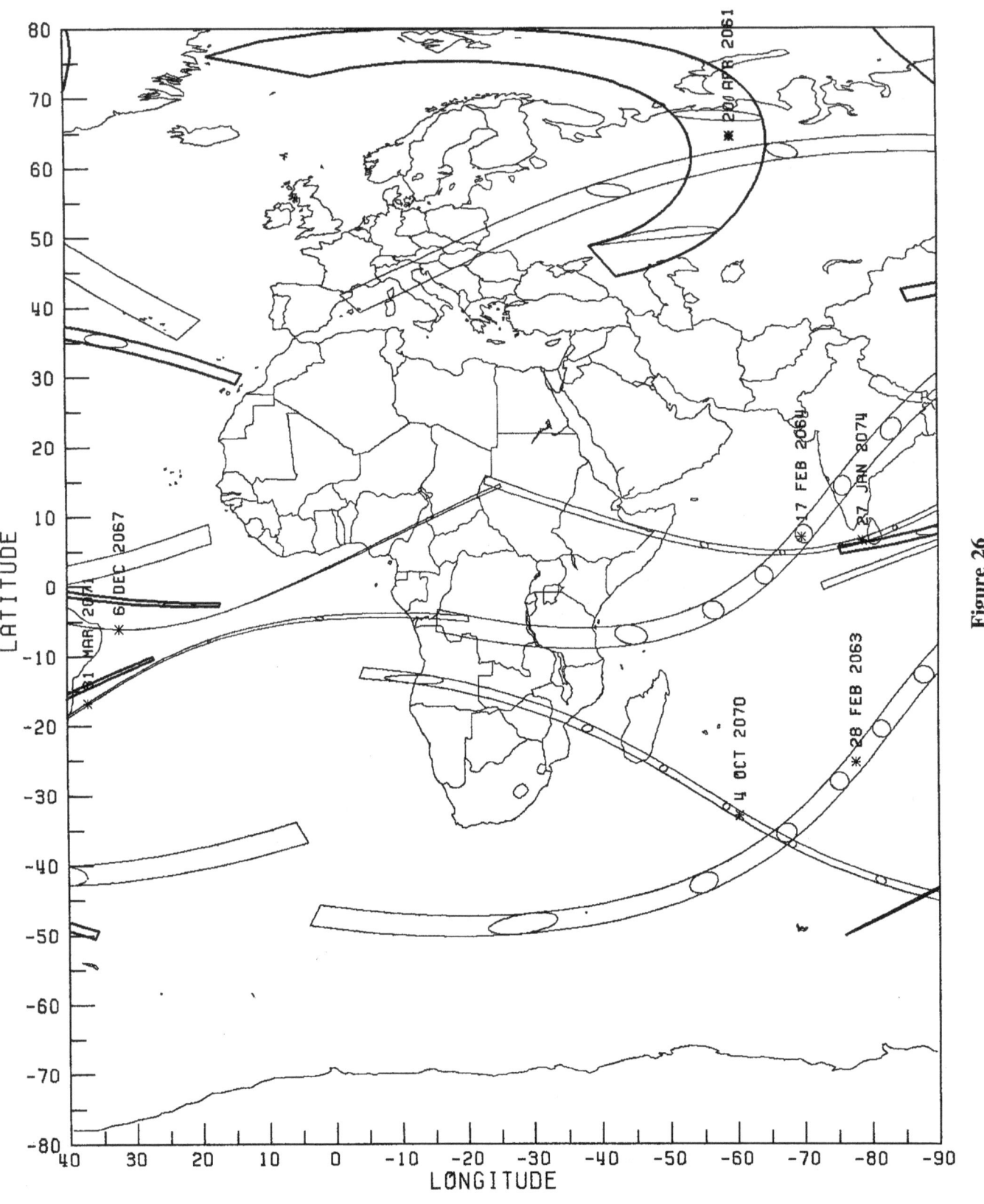

Figure 26

SØLAR ECLIPSES : 2061 - 2080

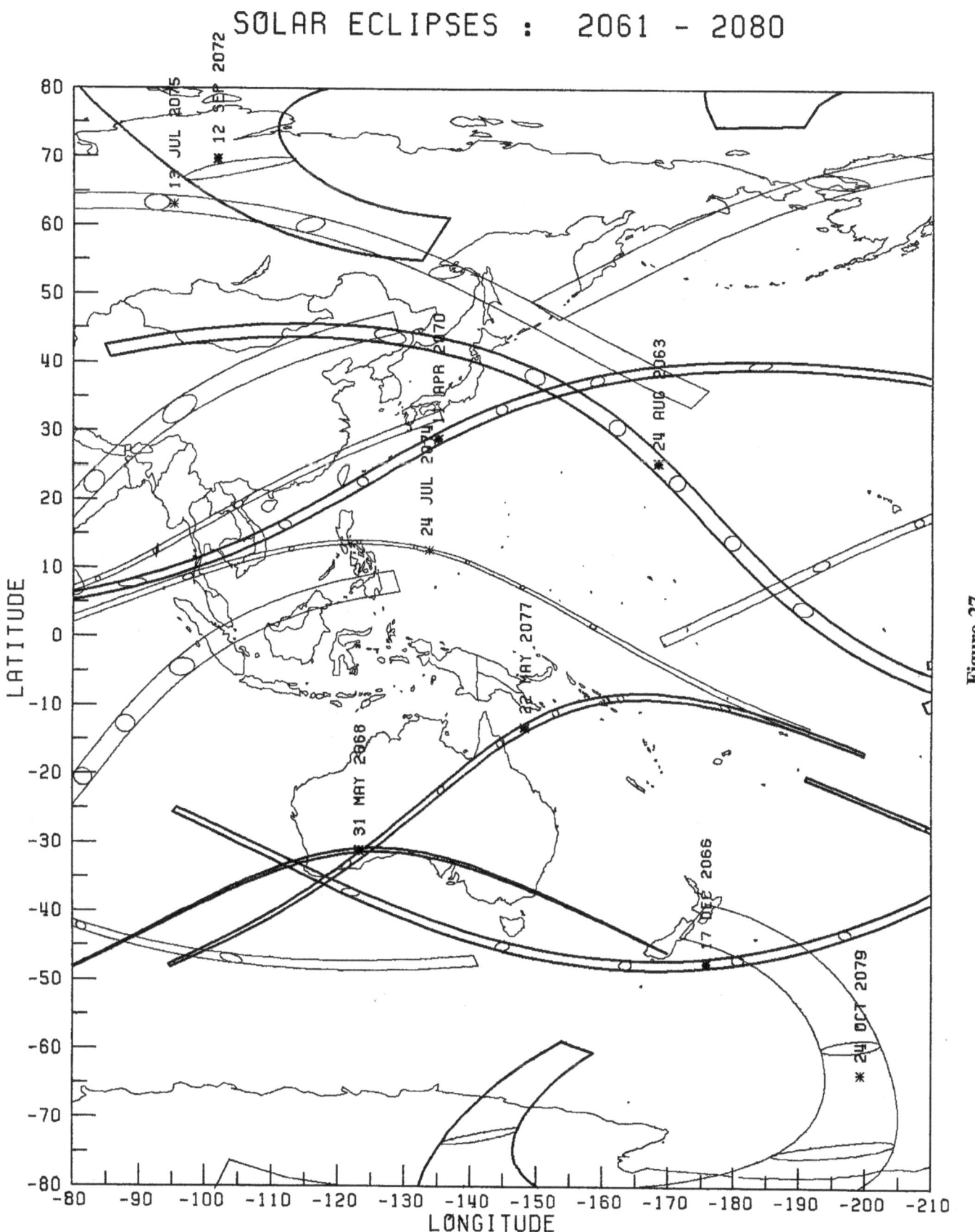

Figure 27

SØLAR ECLIPSES : 2081 - 2100

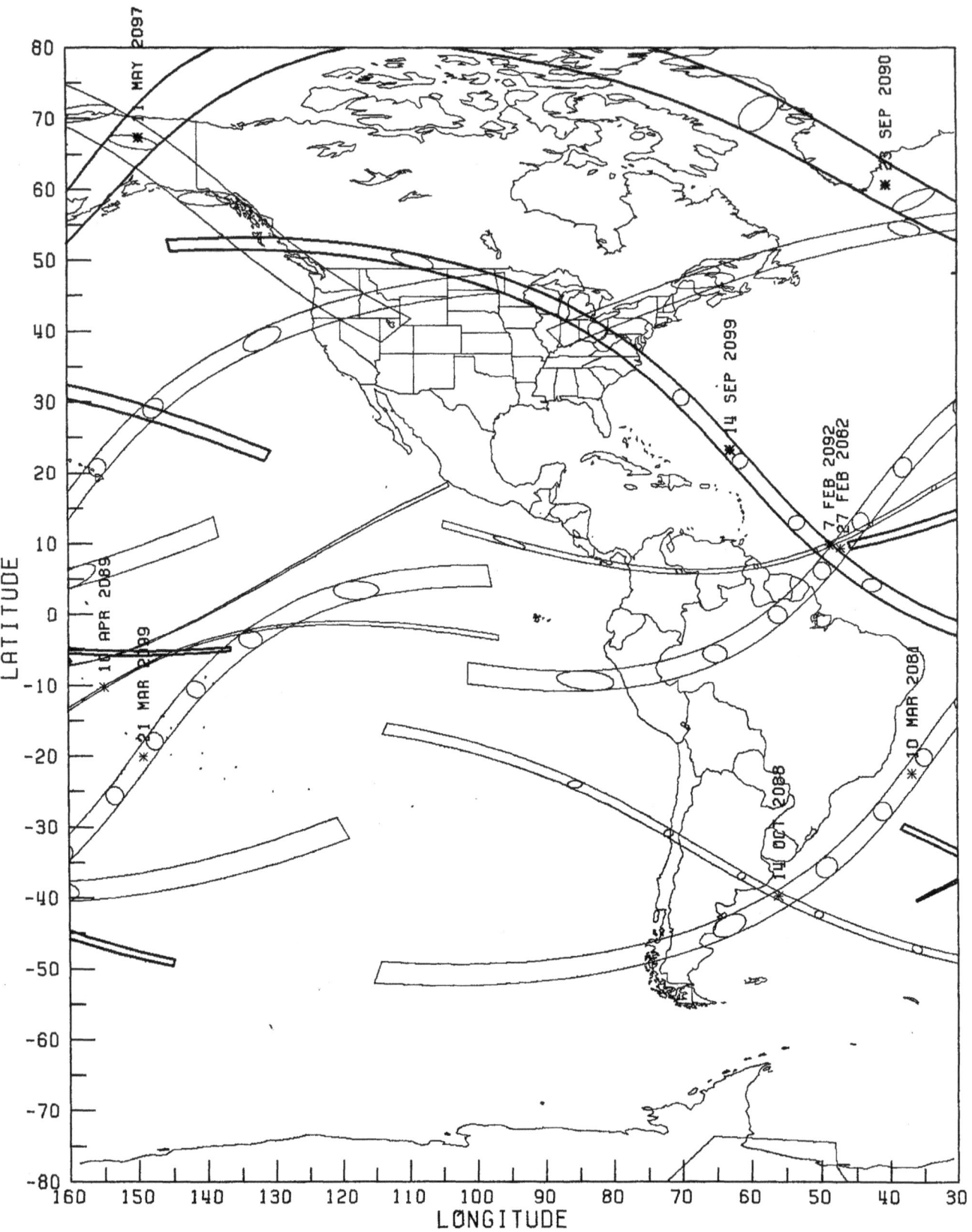

SOLAR ECLIPSES :   2081 - 2100

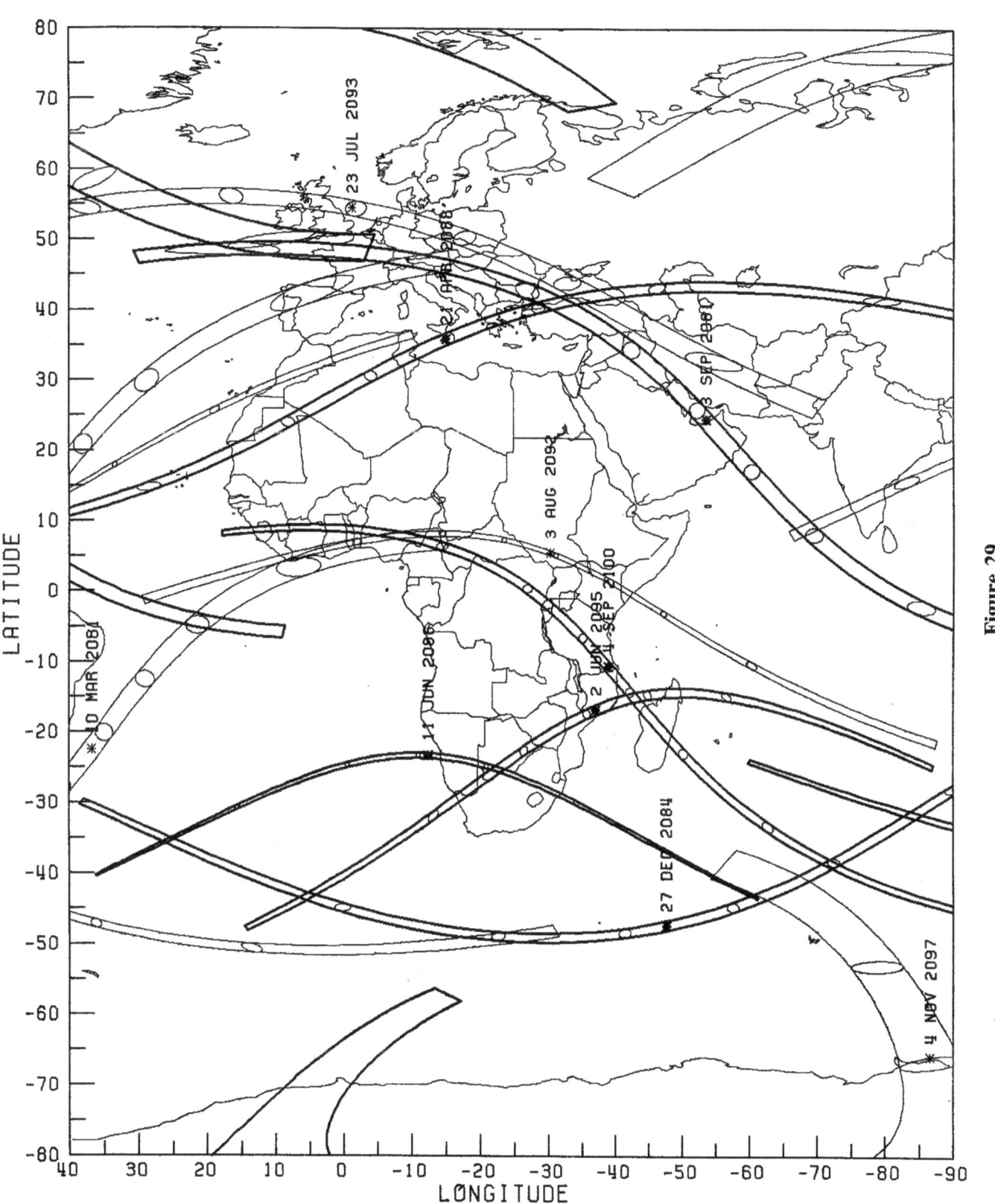

Figure 29

SOLAR ECLIPSES : 2081 - 2100

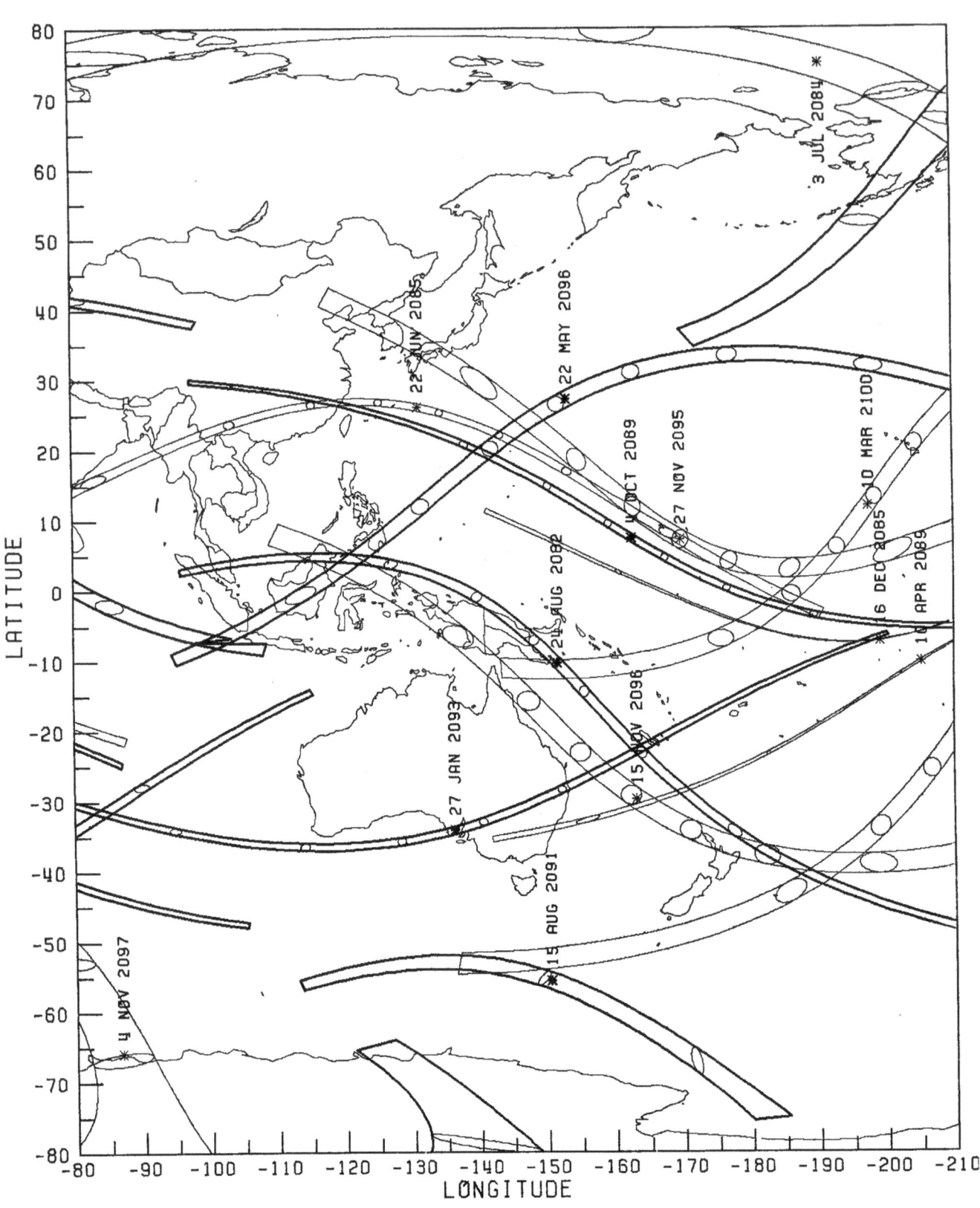

SECTION 3 - <u>CENTRAL PATH CATALOG : 1986 - 2035</u>

Table 13

ANN/TOT SOLAR ECLIPSE OF 3 OCT 1986

SAROS 124   Delta T = 55.3 Sec

UNIVERSAL TIME	NORTHERN LIMIT LATITUDE	LONGITUDE	SOUTHERN LIMIT LATITUDE	LONGITUDE	CENTER LINE LATITUDE	LONGITUDE	DIAMETER RATIO	SUN ALT	SUN AZ	PATH WIDTH	DURATION ANNULARITY
LIMITS	66 10.9	25 57.1	66 42.7	25 56.2	66 24.0	26 34.4	0.9984	0.0	259.4	55.9	0: 5.8
18:56	65 43.7	29 37.5	65 29.9	31 48.7	65 35.3	30 59.2	0.9990	2.1	255.6	33.7	0: 3.8
18:57	64 53.0	32 37.5	64 46.0	33 34.3	64 49.1	33 9.3	0.9993	3.1	253.8	23.2	0: 2.7
18:58	64 9.6	34 14.1	64 5.3	34 46.2	64 7.2	34 31.6	0.9995	3.8	252.9	16.4	0: 1.9
18:59	63 29.4	35 19.1	63 26.7	35 38.3	63 28.0	35 29.4	0.9996	4.3	252.3	11.4	0: 1.3
19: 0	62 51.5	36 5.3	62 49.9	36 16.6	62 50.7	36 11.3	0.9998	4.7	251.9	7.5	0: 0.9
19: 1	62 15.4	36 38.5	62 14.4	36 44.6	62 14.9	36 41.7	0.9999	5.0	251.7	4.5	0: 0.5
19: 2	61 40.7	37 1.6	61 40.2	37 4.2	61 40.4	37 3.0	0.9999	5.2	251.7	2.1	0: 0.2
19: 3	61 7.2	37 16.7	61 7.1	37 16.9	61 7.2	37 16.8	1.0000	5.4	251.7	0.2	0: 0.0
19: 4	60 34.8	37 24.8	60 35.0	37 23.6	60 35.0	37 24.2	1.0000	5.5	251.9	1.1	0: 0.1T
19: 5	60 3.5	37 26.9	60 3.9	37 24.8	60 3.7	37 25.8	1.0001	5.5	252.1	2.0	0: 0.2T
19: 6	59 33.2	37 23.4	59 33.6	37 21.0	59 33.4	37 22.1	1.0001	5.5	252.5	2.3	0: 0.2T
19: 7	59 3.8	37 14.4	59 4.1	37 12.3	59 3.9	37 13.3	1.0001	5.5	252.8	2.1	0: 0.2T
19: 8	58 35.2	37 0.1	58 35.4	36 58.8	58 35.3	36 59.4	1.0000	5.4	253.3	1.3	0: 0.1T
19: 9	58 7.5	36 40.3	58 7.5	36 40.4	58 7.5	36 40.3	1.0000	5.2	253.8	0.1	0: 0.0
19:10	57 40.8	36 14.4	57 40.4	36 16.7	57 40.6	36 15.6	0.9999	5.0	254.4	2.3	0: 0.2
19:11	57 14.9	35 41.7	57 14.1	35 47.2	57 14.5	35 44.6	0.9999	4.7	255.1	5.3	0: 0.5
19:12	56 50.1	35 0.4	56 48.7	35 10.9	56 49.4	35 6.0	0.9998	4.4	255.9	9.3	0: 0.9
19:13	56 26.6	34 7.8	56 24.2	34 25.9	56 25.3	34 17.6	0.9996	3.9	256.8	14.5	0: 1.4
19:14	56 4.8	32 57.2	56 0.9	33 29.0	56 2.7	33 14.8	0.9995	3.3	258.0	21.2	0: 2.1
19:15	55 46.3	31 7.2	55 39.2	32 13.0	55 42.2	31 46.0	0.9992	2.4	259.4	30.1	0: 3.0
LIMITS	55 47.6	27 39.1	55 18.3	27 46.4	55 31.0	28 9.7	0.9986	0.0	262.6	49.1	0: 5.2

Table 14

ANN/TOT SOLAR ECLIPSE OF 29 MAR 1987

SAROS 129   Delta T = 55.5 Sec

UNIVERSAL TIME	NORTHERN LIMIT LATITUDE	LONGITUDE	SOUTHERN LIMIT LATITUDE	LONGITUDE	CENTER LINE LATITUDE	LONGITUDE	DIAMETER RATIO	SUN ALT	SUN AZ	PATH WIDTH	DURATION ANNULARITY
LIMITS	-46-34.3	71 27.3	-47 -1.9	71 27.8	-46-47.6	71 17.7	0.9862	0.0	84.9	50.8	0:50.6
11: 6	-45-23.8	59 12.8	-45-52.8	59 58.6	-45-38.1	59 33.8	0.9886	8.6	76.2	41.4	0:44.3
11:12	-42-27.7	44 54.9	-42-45.1	44 57.6	-42-36.4	44 56.2	0.9919	20.5	64.7	28.9	0:34.2
11:18	-39-58.7	37 17.8	-40-11.0	37 15.3	-40 -4.8	37 16.5	0.9938	27.7	58.1	22.0	0:27.6
11:24	-37-41.3	31 52.1	-37-50.2	31 48.5	-37-45.8	31 50.3	0.9952	33.5	53.1	16.9	0:22.1
11:30	-35-31.7	27 37.9	-35-38.1	27 34.4	-35-34.9	27 36.1	0.9964	38.4	48.8	12.8	0:17.4
11:36	-33-27.9	24 9.5	-33-32.5	24 6.5	-33-30.2	24 8.0	0.9973	42.9	44.8	9.5	0:13.3
11:42	-31-28.9	21 13.1	-31-31.9	21 10.8	-31-30.4	21 11.9	0.9981	46.9	41.1	6.7	0: 9.6
11:48	-29-33.7	18 40.2	-29-35.6	18 38.7	-29-34.6	18 39.5	0.9988	50.6	37.3	4.3	0: 6.3
11:54	-27-41.9	16 25.3	-27-42.8	16 24.4	-27-42.3	16 24.9	0.9993	54.0	33.5	2.3	0: 3.4
12: 0	-25-52.9	14 24.4	-25-53.2	14 24.1	-25-53.0	14 24.2	0.9998	57.2	29.4	0.6	0: 0.9
12: 6	-24 -6.5	12 34.4	-24 -6.2	12 34.7	-24 -6.4	12 34.6	1.0006	60.2	24.9	0.9	0: 1.3T
12:12	-22-22.4	10 53.2	-22-21.6	10 54.0	-22-22.0	10 53.6	1.0006	62.9	19.8	2.1	0: 3.2T
12:18	-20-40.3	9 18.9	-20-39.2	9 20.2	-20-39.7	9 19.6	1.0009	65.4	14.1	3.1	0: 4.8T
12:24	-19 -0.1	7 50.1	-18-58.6	7 51.7	-18-59.3	7 50.9	1.0011	67.6	7.5	3.8	0: 6.0T
12:30	-17-21.4	6 25.6	-17-19.8	6 27.4	-17-20.6	6 26.5	1.0012	69.4	359.8	4.4	0: 6.9T
12:36	-15-44.3	5 4.1	-15-42.6	5 6.1	-15-43.5	5 5.1	1.0013	70.9	351.1	4.8	0: 7.5T
12:42	-14 -8.6	3 44.9	-14 -6.8	3 46.9	-14 -7.7	3 45.9	1.0014	71.8	341.5	5.0	0: 7.8T
12:48	-12-34.2	2 26.9	-12-32.4	2 29.0	-12-33.3	2 27.9	1.0014	72.2	331.3	5.0	0: 7.7T
12:54	-11 -0.9	1 9.4	-10-59.2	1 11.4	-11 -0.1	1 10.4	1.0013	72.0	321.1	4.8	0: 7.4T
13: 0	-9-28.7	0 -8.4	-9-27.2	0 -6.6	-9-27.9	0 -7.5	1.0012	71.2	311.5	4.4	0: 6.7T
13: 6	-7-57.5	-1-27.4	-7-56.2	-1-25.8	-7-56.9	-1-26.6	1.0010	69.9	303.0	3.8	0: 5.8T
13:12	-6-27.3	-2-48.4	-6-26.3	-2-47.2	-6-26.8	-2-47.8	1.0008	68.2	295.7	3.0	0: 4.5T
13:18	-4-58.0	-4-12.3	-4-57.3	-4-11.5	-4-57.7	-4-11.9	1.0005	66.1	289.6	1.9	0: 2.9T
13:24	-3-29.6	-5-40.0	-3-29.7	-5-39.7	-3-29.5	-5-39.9	1.0002	63.7	284.6	0.7	0: 1.0T
13:30	-2 -2.0	-7-12.8	-2 -2.3	-7-13.1	-2 -2.1	-7-12.9	0.9998	61.0	280.5	0.9	0: 1.2
13:36	0-35.2	-8-51.8	0-36.2	-8-52.9	0-35.7	-8-52.3	0.9993	58.1	277.2	2.7	0: 3.8
13:42	0 50.7	-10-38.7	0 48.9	-10-40.7	0 49.8	-10-39.7	0.9988	55.0	274.6	4.8	0: 6.7
13:48	2 15.7	-12-35.5	2 13.0	-12-38.5	2 14.3	-12-37.0	0.9981	51.6	272.5	7.3	0: 9.9
13:54	3 39.7	-14-44.7	3 35.9	-14-48.8	3 37.8	-14-46.8	0.9974	48.0	270.9	10.1	0:13.4
14: 0	5 2.7	-17 -9.8	4 57.4	-17-15.4	5 0.1	-17-12.6	0.9966	44.1	269.8	13.3	0:17.3
14: 6	6 24.3	-19-56.1	6 17.4	-20 -3.4	6 20.8	-19-59.7	0.9956	39.8	269.1	17.0	0:21.6
14:12	7 44.1	-23-11.4	7 35.2	-23-20.9	7 39.7	-23-16.2	0.9944	35.1	268.7	21.4	0:26.4
14:18	9 1.5	-27-10.1	8 50.0	-27-22.6	8 55.8	-27-16.3	0.9931	29.6	268.8	26.6	0:31.8
14:24	10 14.8	-32-22.3	9 59.8	-32-39.9	10 7.3	-32-31.0	0.9913	22.9	269.4	33.1	0:38.0
14:30	11 18.5	-40-27.8	10 57.5	-41 -0.2	11 8.1	-40-43.9	0.9886	13.4	270.8	42.6	0:46.3
LIMITS	11 24.7	-53-34.8	10 54.0	-53-36.8	11 9.5	-53-33.2	0.9848	0.0	273.4	56.2	0:56.1

70

Table 15

ANNULAR SOLAR ECLIPSE OF 23 SEP 1987

SAROS 134   Delta T = 55.8 Sec

UNIVERSAL TIME	NORTHERN LIMIT LATITUDE	LONGITUDE	SOUTHERN LIMIT LATITUDE	LONGITUDE	CENTER LINE LATITUDE	LONGITUDE	DIAMETER RATIO	SUN ALT	SUN AZ	PATH WIDTH	DURATION ANNULARITY
LIMITS	46 27.4	-67-31.1	44 39.9	-67-50.6	45 33.6	-68 -0.3	0.9481	0.0	90.1	199.1	3:26.8
1:24	45 36.0	-83-36.0	43 42.3	-85-31.8	44 38.3	-84-40.3	0.9513	12.5	102.5	183.3	3:32.6
1:30	43 30.8	-96 -4.2	41 49.9	-96-20.5	42 39.9	-96-13.7	0.9538	22.1	111.8	171.8	3:37.2
1:36	41 30.3	-103-24.8	39 58.6	-103-10.1	40 44.1	-103-18.0	0.9554	28.8	118.0	164.7	3:40.2
1:42	39 33.6	-108-46.7	38 9.2	-108-16.1	38 51.2	-108-31.6	0.9567	34.2	122.9	159.3	3:42.5
1:48	37 40.3	-113 -0.9	36 21.9	-112-20.8	37 0.9	-112-40.9	0.9578	38.9	127.2	155.0	3:44.2
1:54	35 49.8	-116-30.4	34 36.6	-115-44.2	35 13.1	-116 -7.3	0.9587	43.1	131.1	151.4	3:45.6
2: 0	34 1.9	-119-28.1	32 53.1	-118-37.9	33 27.4	-119 -2.9	0.9594	47.0	134.8	148.4	3:46.8
2: 6	32 16.3	-122 -1.8	31 11.3	-121 -8.9	31 43.7	-121-35.3	0.9601	50.6	138.5	145.8	3:47.6
2:12	30 32.7	-124-17.2	29 31.0	-123-22.4	30 1.8	-123-49.7	0.9607	54.0	142.2	143.6	3:48.3
2:18	28 51.0	-126-18.0	27 52.2	-125-22.1	28 21.5	-125-49.8	0.9612	57.1	146.1	141.8	3:48.8
2:24	27 10.9	-128 -7.1	26 14.8	-127-10.1	26 42.7	-127-38.5	0.9617	60.1	150.3	140.2	3:49.2
2:30	25 32.4	-129-46.8	24 38.2	-128-49.1	25 5.2	-129-17.9	0.9620	62.8	154.9	138.9	3:49.5
2:36	23 55.2	-131-18.8	23 2.9	-130-20.7	23 29.0	-130-49.7	0.9624	65.4	160.2	137.9	3:49.6
2:42	22 19.2	-132-44.6	21 28.5	-131-46.2	21 53.8	-132-15.3	0.9627	67.7	166.2	137.2	3:49.6
2:48	20 44.4	-134 -5.5	19 55.0	-133 -6.8	20 19.6	-133-36.1	0.9629	69.8	173.3	136.7	3:49.7
2:54	19 10.5	-135-22.5	18 22.3	-134-23.5	18 46.3	-134-52.9	0.9631	71.5	181.4	136.5	3:49.8
3: 0	17 37.6	-136-36.5	16 50.2	-135-37.3	17 13.8	-136 -6.8	0.9632	72.8	190.7	136.5	3:49.5
3: 6	16 5.5	-137-48.3	15 18.8	-136-48.9	15 42.0	-137-18.5	0.9633	73.5	201.0	136.7	3:49.4
3:12	14 34.2	-138-58.8	13 47.9	-137-59.0	14 10.9	-138-28.8	0.9634	73.8	211.8	137.2	3:49.2
3:18	13 3.5	-140 -8.6	12 17.4	-139 -8.5	12 40.3	-139-38.5	0.9634	73.5	222.5	138.0	3:48.9
3:24	11 33.4	-141-18.5	10 47.4	-140-18.0	11 10.3	-140-48.2	0.9633	72.6	232.3	139.0	3:48.6
3:30	10 3.8	-142-29.2	9 17.6	-141-28.2	9 40.6	-141-58.6	0.9632	71.2	240.9	140.2	3:48.3
3:36	8 34.7	-143-41.4	7 48.2	-142-39.9	8 11.3	-143-10.6	0.9631	69.4	248.2	141.7	3:47.9
3:42	7 6.0	-144-56.0	6 18.9	-143-53.8	6 42.3	-144-24.8	0.9629	67.3	254.3	143.4	3:47.5
3:48	5 37.6	-146-13.7	4 49.8	-145-10.7	5 13.5	-145-42.1	0.9627	65.0	259.2	145.3	3:47.0
3:54	4 9.4	-147-35.5	3 20.8	-146-31.7	3 45.0	-147 -3.5	0.9624	62.4	263.2	147.4	3:46.4
4: 0	2 41.5	-149 -2.6	1 51.8	-147-57.7	2 16.5	-148-30.1	0.9621	59.6	266.5	149.7	3:45.8
4: 6	1 13.8	-150-36.2	0 22.7	-149-30.2	0 48.1	-150 -3.1	0.9617	56.6	269.1	152.1	3:45.0
4:12	0-13.8	-152-18.1	-1 -6.5	-151-10.6	0-40.4	-151-44.2	0.9612	53.4	271.2	154.7	3:44.1
4:18	-1-41.4	-154-10.1	-2-35.9	-153 -1.0	-2 -8.8	-153-35.4	0.9607	50.0	272.8	157.5	3:43.1
4:24	-3 -8.9	-156-15.3	-4 -5.6	-155 -4.0	-3-37.5	-155-39.5	0.9601	46.4	274.0	160.3	3:41.9
4:30	-4-36.5	-158-37.3	-5-35.7	-157-23.4	-5 -6.3	-158 -0.1	0.9594	42.4	274.8	163.3	3:40.5
4:36	-6 -4.1	-161-21.8	-7 -6.2	-160 -4.5	-6-35.4	-160-43.0	0.9585	38.1	275.5	166.4	3:38.8
4:42	-7-31.8	-164-38.3	-8-37.3	-163-16.0	-8 -4.8	-163-58.0	0.9575	33.3	275.5	169.8	3:36.8
4:48	-8-59.6	-168-44.0	-10 -9.2	-167-13.6	-9-34.7	-167-58.3	0.9562	27.7	275.2	173.7	3:34.4
4:54	-10-27.5	-174-19.5	-11-42.2	-172-32.2	-11 -5.2	-173-24.9	0.9546	20.8	274.4	178.5	3:31.2
5: 0	-11-54.8	175 26.6	-13-17.6	178 35.4	-12-36.7	177 7.7	0.9518	10.0	272.4	186.4	3:26.2
LIMITS	-12-12.8	167 10.6	-13-58.2	167 28.6	-13 -5.6	167 37.9	0.9492	0.0	270.2	194.2	3:21.9

Table 16

TOTAL SOLAR ECLIPSE OF 18 MAR 1988

SAROS 139   Delta T = 56.1 Sec

UNIVERSAL TIME	NORTHERN LIMIT LATITUDE	NORTHERN LIMIT LONGITUDE	SOUTHERN LIMIT LATITUDE	SOUTHERN LIMIT LONGITUDE	CENTER LINE LATITUDE	CENTER LINE LONGITUDE	DIAMETER RATIO	SUN ALT	SUN AZ	PATH WIDTH	DURATION TOTALITY
LIMITS	-3 51.5	-85 58.7	-4 52.5	-86 10.5	-4 22.3	-86 16.8	1.0302	0.0	90.9	112.3	1:40.5
0:24	-3 51.6	-88 16.8	-4 47.1	-93 15.3	-4 19.4	-91 18.0	1.0319	5.3	90.5	119.2	1:49.6
0:30	-2 27.5	-102 58.7	-3 24.2	-104 40.1	-2 55.7	-103 50.3	1.0362	19.3	90.0	138.9	2:17.1
0:36	0 56.5	-109 -7.4	-1 54.8	-110 34.2	-1 25.5	-109 51.2	1.0384	26.8	90.3	149.2	2:33.0
0:42	0 35.9	-113 31.5	0 22.9	-114 52.8	0 6.6	-114 12.4	1.0400	32.7	91.2	156.6	2:45.6
0:48	2 9.3	-117 -1.8	1 10.0	-118 20.4	1 39.7	-117 41.4	1.0412	37.6	92.5	162.2	2:56.2
0:54	3 43.4	-119 58.5	2 43.7	-121 15.5	3 13.6	-120 37.2	1.0423	41.9	94.2	166.4	3: 5.4
1: 0	5 18.0	-122 31.7	4 18.2	-123 47.9	4 48.2	-123 10.0	1.0431	45.8	96.4	169.5	3:13.4
1: 6	6 53.3	-124 47.9	5 53.4	-126 -3.5	6 23.4	-125 25.8	1.0439	49.3	99.0	171.6	3:20.4
1:12	8 29.3	-126 51.3	7 29.2	-128 -6.4	7 59.2	-127 29.0	1.0445	52.5	102.2	173.0	3:26.4
1:18	10 5.9	-128 44.8	9 5.7	-129 59.5	9 35.8	-129 22.3	1.0450	55.4	106.0	173.7	3:31.6
1:24	11 43.2	-130 30.8	10 43.0	-131 45.1	11 13.1	-131 -8.0	1.0454	57.9	110.5	173.8	3:36.0
1:30	13 21.4	-132 11.0	12 21.0	-133 25.0	12 51.2	-132 48.1	1.0458	60.1	115.7	173.5	3:39.5
1:36	15 0.5	-133 47.0	13 59.9	-135 -0.5	14 30.2	-134 23.8	1.0460	62.0	121.7	172.9	3:42.4
1:42	16 40.6	-135 20.0	15 39.7	-136 33.1	16 10.1	-135 56.6	1.0462	63.5	128.5	172.0	3:44.4
1:48	18 21.8	-136 51.3	17 20.5	-138 -3.9	17 51.1	-137 27.7	1.0464	64.5	135.9	170.8	3:45.8
1:54	20 4.2	-138 22.1	19 2.5	-139 34.0	19 33.3	-138 58.1	1.0464	65.1	143.9	169.5	3:46.4
2: 0	21 48.0	-139 53.3	20 45.6	-141 -4.4	21 16.7	-140 28.9	1.0464	65.2	152.1	168.1	3:46.4
2: 6	23 33.2	-141 28.1	22 30.2	-142 36.3	23 1.6	-142 -1.3	1.0463	64.7	160.3	166.5	3:45.6
2:12	25 20.1	-143 -1.8	24 16.2	-144 10.9	24 48.1	-143 36.4	1.0462	63.8	168.2	164.8	3:44.2
2:18	27 8.8	-144 41.6	26 4.0	-145 49.4	26 36.3	-145 15.5	1.0459	62.5	175.6	163.1	3:42.0
2:24	28 59.5	-146 27.1	27 53.6	-147 33.1	28 26.4	-147 -0.1	1.0456	60.7	182.5	161.2	3:39.1
2:30	30 52.5	-148 19.9	29 45.3	-149 24.0	30 18.8	-148 51.9	1.0453	58.6	188.8	159.2	3:35.4
2:36	32 47.9	-150 22.3	31 39.4	-151 23.8	32 13.5	-150 53.0	1.0448	56.1	194.6	157.1	3:31.1
2:42	34 46.2	-152 36.8	33 36.2	-153 35.3	34 11.1	-153 -6.0	1.0443	53.4	200.0	154.8	3:25.9
2:48	36 47.7	-155 -7.2	35 36.0	-156 -1.6	36 11.7	-155 34.3	1.0436	50.3	205.1	152.3	3:19.9
2:54	38 52.8	-157 57.9	37 39.2	-158 47.2	38 15.8	-158 22.4	1.0429	46.9	210.0	149.7	3:13.0
3: 0	41 2.1	-161 15.8	39 48.5	-161 58.3	40 24.1	-161 36.9	1.0420	43.1	215.0	146.7	3: 5.2
3: 6	43 16.3	-165 11.0	41 58.4	-165 44.1	42 37.1	-165 27.2	1.0410	38.9	220.1	143.4	2:56.2
3:12	45 36.2	-169 59.7	44 16.0	-170 19.5	44 55.9	-170 -9.1	1.0397	34.2	225.6	139.6	2:46.6
3:18	48 3.2	-176 12.2	46 40.6	-176 11.4	47 21.6	-176 10.9	1.0382	28.7	232.0	135.0	2:34.0
3:24	50 39.3	175 3.0	49 14.3	175 41.5	49 56.5	175 23.9	1.0362	21.8	240.2	129.2	2:19.4
3:30	53 29.6	159 11.5	52 2.8	161 50.3	52 45.9	160 37.6	1.0330	11.3	253.4	120.0	1:58.3
LIMITS	54 33.5	142 11.5	53 34.4	142 20.4	54 3.5	142 37.8	1.0295	0.0	268.4	109.7	1:38.0

Table 17

ANNULAR SOLAR ECLIPSE OF 11 SEP 1988

SAROS 144   Delta T = 56.4 Sec

UNIVERSAL TIME	NORTHERN LIMIT LATITUDE	LONGITUDE	SOUTHERN LIMIT LATITUDE	LONGITUDE	CENTER LINE LATITUDE	LONGITUDE	DIAMETER RATIO	SUN ALT	SUN AZ	PATH WIDTH	DURATION ANNULARITY
LIMITS	2 17.2	-44-40.7	0-36.3	-44-15.2	0 54.1	-45 -9.4	0.9249	0.0	85.5	314.6	5:23.9
3: 0	2 29.7	-55-54.5	0-24.8	-46-54.0	1 10.8	-52-40.8	0.9269	8.6	85.8	311.2	5:36.2
3: 6	1 43.4	-63-38.0	0-35.6	-59-50.4	0 35.4	-61-50.0	0.9295	19.2	85.4	308.2	5:54.3
3:12	0 43.6	-68-26.7	-1-25.4	-65-24.4	0-19.9	-66-58.3	0.9310	25.7	84.8	306.8	6: 5.9
3:18	0-21.9	-72 -5.0	-2-24.3	-69-23.6	-1-22.2	-70-46.0	0.9322	30.9	83.9	305.3	6:15.1
3:24	-1-31.0	-75 -2.8	-3-28.0	-72-33.5	-2-28.8	-73-49.4	0.9332	35.3	82.7	303.8	6:22.8
3:30	-2-42.5	-77-33.5	-4-35.2	-75-12.3	-3-38.0	-76-23.8	0.9339	39.1	81.2	301.3	6:29.4
3:36	-3-58.1	-79-44.6	-5-44.9	-77-29.2	-4-49.7	-78-37.7	0.9346	42.6	79.3	298.6	6:35.0
3:42	-5-11.3	-81-41.0	-6-56.9	-79-30.1	-6 -3.3	-80-36.2	0.9352	45.7	77.1	295.4	6:39.8
3:48	-6-28.0	-83-25.9	-8-10.6	-81-18.5	-7-18.6	-82-22.8	0.9357	48.6	74.5	291.8	6:43.9
3:54	-7-46.0	-85 -1.7	-9-26.1	-82-57.1	-8-35.3	-83-59.9	0.9361	51.2	71.6	287.9	6:47.4
4: 0	-9 -5.3	-86-30.2	-10-43.2	-84-28.0	-9-53.5	-85-29.6	0.9365	53.5	68.2	283.8	6:50.2
4: 6	-10-25.7	-87-52.9	-12 -1.8	-85-52.7	-11-13.1	-86-53.2	0.9368	55.6	64.3	279.7	6:52.5
4:12	-11-47.4	-89-10.9	-13-21.3	-87-12.5	-12-34.0	-88-12.1	0.9371	57.4	59.9	275.6	6:54.4
4:18	-13-10.2	-90-25.3	-14-43.6	-88-28.4	-13-56.2	-89-27.2	0.9373	59.0	55.1	271.7	6:55.7
4:24	-14-34.3	-91-37.0	-16 -6.7	-89-41.3	-15-19.9	-90-39.6	0.9374	60.2	49.8	267.9	6:56.6
4:30	-15-59.7	-92-46.8	-17-31.5	-90-52.3	-16-44.9	-91-49.9	0.9375	61.2	44.0	264.4	6:57.1
4:36	-17-28.3	-93-55.4	-18-57.9	-92 -1.9	-18-11.4	-92-59.0	0.9376	61.8	38.0	261.3	6:57.2
4:42	-18-54.3	-95 -3.5	-20-25.9	-93-11.0	-19-39.5	-94 -7.6	0.9377	62.0	31.7	258.4	6:57.0
4:48	-20-23.8	-96-12.0	-21-55.9	-94-20.3	-21 -9.2	-95-16.5	0.9378	61.9	25.3	256.0	6:58.4
4:54	-21-54.9	-97-21.5	-23-27.7	-95-30.7	-22-40.7	-96-26.4	0.9378	61.5	19.0	253.9	6:55.4
5: 0	-23-27.6	-98-32.9	-25 -1.6	-96-42.9	-24-14.0	-97-38.2	0.9375	60.7	12.9	252.2	6:54.1
5: 6	-25 -2.2	-99-47.0	-26-37.8	-97-57.9	-25-49.4	-98-52.7	0.9374	59.6	7.0	250.9	6:52.5
5:12	-26-38.8	-101 -4.8	-28-16.4	-99-16.6	-27-27.0	-100-11.0	0.9372	58.1	1.4	250.1	6:50.5
5:18	-28-17.6	-102-27.3	-29-57.8	-100-40.3	-29 -7.1	-101-34.1	0.9369	56.5	356.1	249.6	6:48.2
5:24	-29-58.8	-103-55.9	-31-42.1	-102-10.4	-30-49.8	-103 -3.4	0.9367	54.5	351.2	249.6	6:45.5
5:30	-31-42.7	-105-32.1	-33-29.7	-103-48.7	-32-35.8	-104-40.6	0.9363	52.3	346.5	250.0	6:42.4
5:36	-33-29.7	-107-18.0	-35-21.2	-105-37.2	-34-24.7	-106-27.8	0.9359	49.8	342.1	250.9	6:38.9
5:42	-35-20.2	-109-16.0	-37-17.0	-107-39.0	-36-17.8	-108-27.6	0.9355	47.0	337.8	252.3	6:35.0
5:48	-37-14.7	-111-29.6	-39-17.8	-109-57.8	-38-15.3	-110-43.8	0.9349	44.0	333.5	254.3	6:30.5
5:54	-39-13.9	-114 -3.4	-41-24.6	-112-39.4	-40-18.2	-113-21.3	0.9343	40.7	329.3	256.9	6:25.4
6: 0	-41-18.7	-117 -4.2	-43-38.7	-115-52.0	-42-27.5	-118-27.7	0.9336	37.0	324.8	260.2	6:19.7
6: 8	-43-30.5	-120-42.7	-46 -2.2	-119-49.4	-44-44.8	-120-14.9	0.9327	32.9	320.0	264.5	6:13.1
6:12	-45-51.3	-125-17.2	-48-38.1	-124-56.3	-47-12.7	-125 -4.2	0.9317	28.1	314.6	270.2	6: 5.4
6:18	-48-24.7	-131-23.5	-51-33.1	-132 -8.4	-49-56.1	-131-39.8	0.9303	22.3	307.8	278.0	5:56.0
6:24	-51-19.4	-140-39.6	-55 -6.9	-144-42.3	-53 -7.4	-142-15.0	0.9284	14.2	297.9	290.3	5:43.3
LIMITS	-55-16.4	-165-20.2	-58 -7.7	-165-15.0	-56-35.4	-164 -3.6	0.9250	0.0	279.0	314.2	5:23.4

73

Table 18

ANNULAR SOLAR ECLIPSE OF 26 JAN 1990

SAROS 121 Delta T = 57.3 Sec

UNIVERSAL TIME	NORTHERN LIMIT		SOUTHERN LIMIT		CENTER LINE		DIAMETER RATIO	SUN ALT	SUN AZ	PATH WIDTH	DURATION ANNULARITY
	LATITUDE	LONGITUDE	LATITUDE	LONGITUDE	LATITUDE	LONGITUDE					
LIMITS	-71 22.0	-80 45.3	-71 0.3	-66 41.8	-73 8.4	-73 6.9	0.9622	0.0	185.6	491.9	2: 6.2
19: 0	-82 25.0	-48 35.7	-74 34.5	-62 29.9	-79 38.3	-55 47.6	0.9642	8.9	201.2	429.9	2: 5.2
19: 2	-82 53.7	-36 9.4	-77 14.5	-53 59.2	-80 40.4	-45 54.2	0.9646	10.6	210.3	416.0	2: 4.9
19: 4	-82 55.9	-23 45.1	-78 33.0	-45 2.5	-81 11.0	-35 22.0	0.9650	11.9	220.0	405.1	2: 4.7
19: 6	-82 36.5	-12 39.1	-79 12.2	-35 51.5	-81 16.1	-25 2.0	0.9653	13.0	229.5	396.4	2: 4.4
19: 8	-82 1.4	-3 28.9	-79 23.9	-26 57.3	-81 1.2	-15 39.4	0.9656	14.0	238.0	389.3	2: 4.3
19:10	-81 15.8	3 45.4	-79 15.1	-18 47.6	-80 31.2	-7 38.2	0.9659	14.8	245.2	383.5	2: 4.1
19:12	-80 23.5	9 21.9	-78 51.1	-11 38.9	-79 50.7	-1 2.1	0.9661	15.5	250.9	378.9	2: 3.9
19:14	-79 27.1	13 41.8	-78 16.0	-5 35.3	-79 2.9	4 17.6	0.9662	16.1	255.4	375.3	2: 3.7
19:16	-78 28.2	17 3.0	-77 33.1	0 -33.0	-78 10.3	8 33.3	0.9664	16.7	258.7	372.5	2: 3.6
19:18	-77 27.8	19 40.1	-76 44.5	3 35.7	-77 14.5	11 57.7	0.9665	17.1	261.3	370.5	2: 3.5
19:20	-76 26.7	21 43.2	-75 52.0	6 59.1	-76 16.7	14 41.1	0.9667	17.5	263.1	369.3	2: 3.3
19:22	-75 25.1	23 20.1	-74 56.7	9 45.1	-75 17.4	16 52.1	0.9668	17.8	264.4	368.7	2: 3.2
19:24	-74 23.3	24 36.4	-73 59.4	12 0.4	-74 17.2	18 37.0	0.9668	18.0	265.3	368.8	2: 3.1
19:26	-73 21.5	25 36.1	-73 0.8	13 50.4	-73 16.5	20 1.0	0.9669	18.2	265.8	369.5	2: 3.0
19:28	-72 19.7	26 22.4	-72 1.0	15 19.3	-72 15.3	21 7.8	0.9669	18.3	266.1	370.8	2: 2.9
19:30	-71 18.0	26 57.6	-71 0.4	16 30.6	-71 13.8	22 0.3	0.9670	18.4	266.1	372.7	2: 2.8
19:32	-70 16.3	27 23.3	-69 59.2	17 26.9	-70 12.1	22 40.7	0.9670	18.4	265.9	375.1	2: 2.8
19:34	-69 14.6	27 41.1	-68 57.3	18 10.3	-69 10.2	23 10.8	0.9670	18.3	265.6	378.2	2: 2.7
19:36	-68 12.9	27 51.8	-67 54.8	18 42.4	-68 8.0	23 31.8	0.9669	18.2	265.1	381.8	2: 2.7
19:38	-67 11.0	27 56.2	-66 51.6	19 4.3	-67 5.4	23 44.8	0.9669	18.0	264.5	386.1	2: 2.6
19:40	-66 9.0	27 54.9	-65 47.8	19 17.0	-66 2.5	23 50.5	0.9669	17.7	263.8	390.9	2: 2.6
19:42	-65 6.7	27 48.2	-64 43.1	19 21.0	-64 59.1	23 49.3	0.9668	17.4	263.0	396.4	2: 2.6
19:44	-64 4.0	27 36.4	-63 37.4	19 16.5	-63 55.1	23 41.5	0.9667	17.0	262.1	402.5	2: 2.6
19:46	-63 0.9	27 19.6	-62 30.6	19 3.8	-62 50.4	23 27.4	0.9666	16.5	261.1	409.2	2: 2.6
19:48	-61 57.0	26 57.8	-61 22.1	18 42.3	-61 44.6	23 6.7	0.9664	15.9	260.1	416.5	2: 2.6
19:50	-60 52.3	26 30.8	-60 11.8	18 11.5	-60 37.6	22 39.2	0.9663	15.3	259.0	424.5	2: 2.7
19:52	-59 46.6	25 58.3	-58 58.8	17 30.0	-59 29.0	22 4.4	0.9661	14.5	257.8	433.0	2: 2.7
19:54	-58 39.4	25 20.7	-57 42.4	16 35.6	-58 18.4	21 21.2	0.9658	13.6	256.5	442.1	2: 2.8
19:56	-57 30.5	24 34.4	-56 20.9	15 24.4	-57 5.1	20 28.1	0.9656	12.6	255.1	451.6	2: 2.9
19:58	-56 19.3	23 40.9	-54 51.2	13 48.1	-55 48.1	19 22.8	0.9653	11.4	253.5	461.5	2: 3.1
20: 0	-55 5.1	22 37.5	-53 5.6	11 25.5	-54 25.6	18 0.3	0.9649	10.0	251.8	471.5	2: 3.3
20: 2	-53 46.6	21 21.1	-49 60.0	5 19.8	-52 54.4	16 12.1	0.9644	8.2	249.8	481.2	2: 3.5
LIMITS	-46 6.3	8 40.4	-49 32.4	4 4.9	-48 41.2	8 55.3	0.9626	0.0	243.3	493.9	2: 4.4

Table 19

TOTAL SOLAR ECLIPSE OF 22 JUL 1990

SAROS 126   Delta T = 57.6 Sec

UNIVERSAL TIME	NORTHERN LIMIT LATITUDE	NORTHERN LIMIT LONGITUDE	SOUTHERN LIMIT LATITUDE	SOUTHERN LIMIT LONGITUDE	CENTER LINE LATITUDE	CENTER LINE LONGITUDE	DIAMETER RATIO	SUN ALT	SUN AZ	PATH WIDTH	DURATION TOTALITY
LIMITS	60 13.8	-22-35.4	59 16.8	-24-53.6	60 5.4	-24-27.6	1.0282	0.0	47.1	166.3	1:24.9
1:56	65 46.2	-34-42.3	67 24.4	-45 -6.9	66 46.6	-40-10.4	1.0312	10.2	61.9	178.2	1:40.1
2: 0	70 20.4	-47-42.5	70 36.5	-58-43.8	70 33.0	-52-20.2	1.0329	16.0	74.6	183.3	1:50.0
2: 4	73 6.0	-58-54.3	72 42.1	-67-30.6	72 57.1	-63-21.0	1.0341	20.2	86.6	186.1	1:57.2
2: 8	74 59.4	-70-10.4	74 6.2	-78-13.1	74 35.0	-74-22.8	1.0350	23.4	98.9	188.1	2: 3.1
2:12	76 14.0	-81-50.5	74 57.9	-88-55.2	75 37.4	-85-35.5	1.0358	26.2	111.5	189.6	2: 8.2
2:16	76 55.5	-93-43.9	75 22.5	-99-24.1	76 9.5	-96-46.5	1.0364	28.6	124.3	190.8	2:12.5
2:20	77 7.7	-105-22.8	75 33.3	-109-24.1	76 15.3	-107-33.1	1.0370	30.7	136.9	191.8	2:16.3
2:24	76 54.9	-116-16.6	75 3.3	-118-37.1	75 58.9	-117-33.0	1.0374	32.5	148.7	192.8	2:19.7
2:28	76 21.8	-126 -3.6	74 28.2	-126-54.6	75 24.7	-126-31.8	1.0378	34.1	159.7	193.7	2:22.6
2:32	75 33.0	-134-35.8	73 40.8	-134-14.4	74 36.5	-134-25.1	1.0381	35.5	169.7	194.6	2:25.1
2:36	74 32.6	-141-55.1	72 43.4	-140-39.7	73 37.7	-141-15.8	1.0384	36.7	178.7	195.5	2:27.2
2:40	73 23.8	-148-10.4	71 39.0	-146-18.1	72 31.2	-147-10.8	1.0386	37.7	186.9	196.3	2:29.0
2:44	72 8.8	-153-31.8	70 29.0	-151-10.5	71 18.8	-152-18.2	1.0388	38.6	194.4	197.2	2:30.4
2:48	70 49.4	-158 -8.8	69 14.9	-155-29.3	70 2.2	-156-46.0	1.0390	39.2	201.2	198.1	2:31.5
2:52	69 26.7	-162 -9.8	67 57.8	-159-18.5	68 42.3	-160-41.2	1.0390	39.7	207.6	199.0	2:32.3
2:56	68 1.6	-165-41.9	66 37.8	-162-43.2	67 19.8	-164 -9.8	1.0391	40.1	213.5	200.0	2:32.7
3: 0	66 34.6	-168-50.7	65 15.6	-165-47.7	65 55.3	-167-16.6	1.0391	40.2	219.1	200.9	2:32.8
3: 4	65 5.9	-171-40.7	63 51.9	-168-35.8	64 29.2	-170 -5.9	1.0391	40.3	224.4	201.9	2:32.7
3: 8	63 35.9	-174-15.8	62 26.6	-171-10.0	63 1.5	-172-40.8	1.0390	40.1	229.4	202.9	2:32.2
3:12	62 4.8	-176-39.0	60 59.8	-173-33.4	61 32.5	-175 -4.3	1.0389	39.8	234.2	203.9	2:31.4
3:16	60 32.0	-178-52.8	59 31.6	-175-48.0	60 2.2	-177-18.7	1.0388	39.3	238.8	204.9	2:30.2
3:20	58 58.0	179 0.5	58 2.0	-177-55.8	58 30.4	-179-26.1	1.0386	38.7	243.2	205.8	2:28.8
3:24	57 22.6	176 59.0	56 30.8	-179-58.5	56 57.2	178 31.7	1.0384	37.8	247.4	206.7	2:27.1
3:28	55 45.6	175 1.0	54 58.0	178 2.3	55 22.2	176 33.0	1.0381	36.9	251.4	207.6	2:25.1
3:32	54 6.6	173 4.8	53 23.3	176 4.9	53 45.4	174 36.0	1.0378	35.7	255.3	208.3	2:22.7
3:36	52 25.3	171 8.2	51 46.3	174 8.0	52 6.3	172 39.3	1.0374	34.3	259.1	208.8	2:20.0
3:40	50 41.3	169 9.4	50 6.7	172 9.6	50 24.6	170 40.9	1.0370	32.8	262.7	209.0	2:16.9
3:44	48 53.8	167 7.4	48 24.4	170 7.9	48 39.5	168 38.8	1.0365	31.0	266.2	208.9	2:13.5
3:48	47 2.1	164 58.2	46 37.4	168 0.5	46 50.3	166 30.5	1.0359	28.9	269.6	208.4	2: 9.6
3:52	45 4.6	162 38.3	44 45.7	165 44.2	44 55.9	164 12.5	1.0352	26.6	273.0	207.1	2: 5.3
3:56	42 59.2	160 2.2	42 47.4	163 14.5	42 54.1	161 39.7	1.0344	23.9	276.3	204.9	2: 0.3
4: 0	40 42.2	157 0.0	40 39.6	160 24.0	40 42.0	158 43.8	1.0335	20.7	279.6	201.2	1:54.6
4: 4	38 5.5	153 11.2	38 17.0	156 58.6	38 12.8	155 7.9	1.0323	16.7	283.1	195.3	1:47.7
4: 8	34 42.4	147 28.1	35 26.0	152 23.1	35 8.4	150 4.6	1.0308	11.2	287.2	185.1	1:38.6
LIMITS	30 30.1	138 20.8	29 20.0	139 25.3	30 6.9	139 24.8	1.0273	0.0	293.5	160.8	1:22.3

75

Table 20

ANNULAR SOLAR ECLIPSE OF 15 JAN 1991

SAROS 131   Delta T = 58.0 Sec

UNIVERSAL TIME	NORTHERN LIMIT LATITUDE	LONGITUDE	SOUTHERN LIMIT LATITUDE	LONGITUDE	CENTER LINE LATITUDE	LONGITUDE	DIAMETER RATIO	SUN ALT	SUN AZ	PATH WIDTH	DURATION ANNULARITY
LIMITS	-28-29.7	-110-38.6	-31-14.1	-108-31.6	-30 -1.8	-110 -1.9	0.9146	0.0	114.2	360.4	5:38.4
22: 6	-35-48.2	-130-43.9	-37-23.0	-125-28.7	-36-38.9	-128-18.8	0.9191	18.1	103.0	334.5	6:10.1
22:12	-37-46.8	-137-40.2	-39-53.0	-133-58.4	-38-50.9	-135-55.8	0.9209	25.6	97.1	323.0	6:25.8
22:18	-39 -4.2	-143-11.5	-41-24.2	-140-19.9	-40-14.6	-141-50.2	0.9222	31.4	91.8	314.4	6:38.8
22:24	-39-57.1	-147-55.9	-42-24.9	-145-40.5	-41-11.1	-146-51.7	0.9233	36.3	86.9	307.4	6:50.2
22:30	-40-32.8	-152 -9.2	-43 -5.1	-150-23.3	-41-48.8	-151-19.0	0.9242	40.7	82.1	301.5	7: 0.4
22:36	-40-55.1	-155-59.8	-43-30.0	-154-39.0	-42-12.3	-155-21.5	0.9249	44.7	77.6	296.8	7: 9.6
22:42	-41 -8.4	-159-31.9	-43-42.8	-158-33.9	-42-24.2	-159 -4.7	0.9256	48.4	72.6	292.4	7:17.9
22:48	-41 -8.7	-162-49.5	-43-45.0	-162-11.8	-42-26.5	-162-31.9	0.9262	51.8	67.7	288.9	7:25.3
22:54	-41 -2.9	-165-54.3	-43-38.7	-165-35.1	-42-20.4	-165-45.7	0.9267	55.0	62.6	285.9	7:31.9
23: 0	-40-50.2	-168-48.2	-43-24.6	-168-45.8	-42 -7.0	-168-47.6	0.9271	58.1	57.3	283.3	7:37.6
23: 6	-40-31.2	-171-32.3	-43 -3.8	-171-45.2	-41-47.1	-171-39.1	0.9275	60.9	51.7	281.2	7:42.5
23:12	-40 -6.5	-174 -7.8	-42-36.9	-174-34.5	-41-21.3	-174-21.2	0.9279	63.6	45.5	279.5	7:46.6
23:18	-39-36.7	-176-35.0	-42 -4.4	-177-14.7	-40-50.2	-176-54.8	0.9282	66.0	38.7	278.2	7:49.8
23:24	-39 -2.1	-178-55.2	-41-27.0	-179-48.5	-40-14.2	-179-20.7	0.9284	68.2	31.1	277.1	7:52.2
23:30	-38-23.1	178 51.1	-40-44.9	177 49.2	-39-33.7	178 20.5	0.9286	70.2	22.6	276.4	7:53.8
23:36	-37-39.9	176 43.2	-39-58.4	175 31.8	-38-48.9	176 7.9	0.9288	71.8	12.9	276.1	7:54.6
23:42	-36-52.9	174 40.6	-39 -8.0	173 20.5	-38 -0.2	174 1.0	0.9289	73.0	2.1	276.0	7:54.6
23:48	-36 -2.0	172 42.8	-38-13.8	171 14.7	-37 -7.7	171 59.3	0.9290	73.7	350.3	276.2	7:53.9
23:54	-35 -7.6	170 48.9	-37-15.9	169 13.8	-36-11.6	170 2.0	0.9290	73.9	338.2	276.8	7:52.4
0: 0	-34 -9.7	168 58.7	-36-14.6	167 17.1	-35-12.0	168 8.6	0.9290	73.5	326.3	277.7	7:50.4
0: 6	-33 -8.4	167 11.5	-35 -9.9	165 24.0	-34 -9.0	166 18.5	0.9289	72.6	315.3	278.9	7:47.6
0:12	-32 -3.8	165 26.8	-34 -2.0	163 33.8	-33 -2.7	164 31.1	0.9289	71.3	305.4	280.4	7:44.3
0:18	-30-55.7	163 44.0	-32-50.7	161 45.8	-31-53.0	162 45.7	0.9287	69.6	296.9	282.3	7:40.5
0:24	-29-44.3	162 2.4	-31-36.0	159 59.2	-30-40.0	161 1.6	0.9286	67.5	289.7	284.6	7:36.1
0:30	-28-29.3	160 21.3	-30-18.0	158 13.3	-29-23.5	159 18.2	0.9284	65.2	283.5	287.2	7:31.2
0:36	-27-10.7	158 39.9	-28-56.4	156 27.3	-28 -3.4	157 34.5	0.9281	62.7	278.2	290.1	7:25.9
0:42	-25-48.2	156 57.4	-27-31.0	154 40.0	-26-39.5	155 49.6	0.9278	60.0	273.6	293.5	7:20.2
0:48	-24-21.6	155 12.7	-26 -1.5	152 50.5	-25-11.4	154 2.6	0.9274	57.1	269.7	297.2	7:14.2
0:54	-22-50.4	153 24.6	-24-27.5	150 57.2	-23-38.9	152 12.0	0.9270	54.0	266.2	301.2	7: 7.7
1: 0	-21-14.2	151 31.7	-22-48.5	148 58.4	-22 -1.3	150 16.2	0.9265	50.7	263.2	305.7	7: 0.8
1: 6	-19-32.3	149 31.9	-21 -3.5	146 51.9	-20-17.8	148 13.1	0.9259	47.2	260.5	310.5	6:53.6
1:12	-17-43.4	147 22.5	-19-11.5	144 34.6	-18-27.5	145 60.0	0.9253	43.4	258.1	315.6	6:45.5
1:18	-15-46.3	144 59.9	-17-10.6	142 1.9	-16-28.5	143 32.5	0.9245	39.2	256.0	321.0	6:37.8
1:24	-13-38.4	142 18.2	-14-58.1	139 8.6	-14-18.4	140 44.4	0.9236	34.7	254.1	326.8	6:29.0
1:30	-11-15.6	139 7.5	-12-28.6	135 35.5	-11-52.5	137 24.3	0.9225	29.5	252.4	332.9	6:19.4
1:36	-8-29.9	135 8.2	-9-30.8	130 59.5	-9 -1.4	133 8.5	0.9211	23.2	250.9	339.4	6: 8.5
1:42	-4-59.2	129 26.0	-5-22.9	123 31.7	-5-15.7	126 43.8	0.9190	14.6	249.6	346.9	5:54.5
LIMITS	1 50.8	115 8.4	0-56.2	113 24.7	0 17.8	114 42.4	0.9155	0.0	248.9	356.4	5:34.8

Table 21

TOTAL SOLAR ECLIPSE OF 11 JUL 1991

SAROS 136   Delta T = 58.3 Sec

UNIVERSAL TIME	NORTHERN LIMIT LATITUDE	NORTHERN LIMIT LONGITUDE	SOUTHERN LIMIT LATITUDE	SOUTHERN LIMIT LONGITUDE	CENTER LINE LATITUDE	CENTER LINE LONGITUDE	DIAMETER RATIO	SUN ALT	SUN AZ	PATH WIDTH	DURATION TOTALITY
LIMITS	13 17.2	175 1.4	11 38.4	174 11.1	12 30.4	174 29.6	1.0611	0.0	67.4	203.3	3:11.5
17:24	15 22.2	169 57.5	14 34.1	188 49.2	15 3.0	168 12.2	1.0633	6.9	69.0	210.1	3:28.2
17:30	20 27.9	158 36.3	18 48.2	155 12.1	19 38.8	155 52.1	1.0678	20.8	73.1	223.1	4: 7.5
17:36	22 42.3	149 43.7	20 50.7	148 39.3	21 46.9	149 10.0	1.0702	28.8	76.0	229.8	4:32.4
17:42	24 9.9	144 27.1	22 11.0	143 35.7	23 10.8	144 0.1	1.0720	35.1	78.6	234.8	4:53.2
17:48	25 11.1	140 0.2	23 6.9	139 19.9	24 9.2	139 38.9	1.0734	40.5	80.9	238.7	5:11.4
17:54	25 54.0	136 5.2	23 45.8	135 34.9	24 50.0	135 49.1	1.0746	45.4	83.1	242.1	5:27.8
18: 0	26 23.0	132 32.7	24 11.7	132 11.8	25 17.4	132 21.4	1.0756	50.0	85.2	245.0	5:42.6
18: 6	26 40.7	129 17.3	24 27.1	129 5.4	25 34.0	129 10.7	1.0764	54.2	87.3	247.5	5:55.9
18:12	26 48.8	126 15.4	24 33.7	126 12.2	25 41.3	126 13.2	1.0772	58.2	89.2	249.7	6: 7.9
18:18	26 48.6	123 24.7	24 32.5	123 29.8	25 40.5	123 26.7	1.0778	62.1	91.0	251.6	6:18.5
18:24	26 41.0	120 43.2	24 24.4	120 56.3	25 32.7	120 49.4	1.0784	65.8	92.8	253.3	6:27.6
18:30	26 26.6	118 9.6	24 10.1	118 30.4	25 18.4	118 19.8	1.0788	69.4	94.5	254.6	6:35.5
18:36	26 6.1	115 42.9	23 50.1	116 11.0	24 58.1	115 56.8	1.0792	72.9	96.1	255.8	6:41.9
18:42	25 39.9	113 22.1	23 24.7	113 57.1	24 32.3	113 39.6	1.0795	76.3	97.6	256.7	6:46.9
18:48	25 8.3	111 6.5	22 54.3	111 48.0	24 1.3	111 27.3	1.0797	79.7	99.1	257.3	6:50.5
18:54	24 31.5	108 55.3	22 19.1	109 43.0	23 25.3	109 19.3	1.0799	83.0	100.3	257.7	6:52.8
19: 0	23 49.9	106 48.0	21 39.2	107 41.3	22 44.6	107 14.8	1.0800	86.2	101.1	258.0	6:53.6
19: 6	23 3.5	104 43.8	20 54.9	105 42.4	21 59.3	105 13.3	1.0800	88.2	307.2	258.0	6:53.2
19:12	22 12.5	102 42.2	20 6.1	103 45.7	21 9.4	103 16.9	1.0799	86.3	286.2	257.9	6:51.4
19:18	21 16.9	100 42.6	19 12.9	101 50.5	20 15.0	101 16.9	1.0798	83.1	286.8	257.6	6:48.4
19:24	20 16.8	98 44.2	18 15.3	99 56.2	19 16.2	99 20.6	1.0797	79.8	287.8	257.2	6:44.1
19:30	19 12.0	96 46.5	17 13.2	98 2.1	18 12.8	97 24.7	1.0794	76.4	288.8	256.8	6:38.7
19:36	18 2.5	94 48.5	16 6.5	96 7.5	17 4.7	95 28.4	1.0791	73.0	289.8	255.8	6:32.1
19:42	16 48.1	92 49.4	14 54.9	94 11.5	15 51.7	93 30.9	1.0787	69.5	290.8	255.0	6:24.5
19:48	15 28.5	90 48.3	13 38.2	92 13.0	14 33.6	91 31.1	1.0782	65.9	291.7	253.9	6:15.8
19:54	14 3.4	88 43.8	12 15.9	90 11.0	13 9.9	89 27.9	1.0777	62.2	292.5	252.7	6: 6.1
20: 0	12 32.1	86 34.4	10 47.6	88 3.9	11 40.1	87 19.7	1.0770	58.3	293.3	251.3	5:55.4
20: 6	10 53.9	84 18.2	9 12.5	85 49.7	10 3.5	85 4.5	1.0763	54.3	294.0	249.5	5:43.7
20:12	9 7.7	81 52.5	7 29.4	83 26.0	8 18.9	82 39.8	1.0754	50.1	294.6	247.5	5:30.9
20:18	7 11.9	79 13.5	5 37.0	80 49.1	6 24.8	80 1.9	1.0744	45.6	295.1	244.9	5:17.1
20:24	5 4.1	76 15.7	3 32.7	77 53.5	4 18.8	77 5.3	1.0732	40.7	295.6	241.8	5: 1.9
20:30	2 40.0	72 49.5	1 12.7	74 30.2	1 56.8	73 40.6	1.0718	35.2	295.8	237.8	4:45.2
20:36	0 -8.2	68 36.9	-1 -30.4	70 22.0	0 -48.7	69 30.4	1.0700	28.9	295.9	232.6	4:28.3
20:42	-3 -40.2	62 52.5	-4 -54.4	64 47.5	-4 -16.4	63 51.4	1.0676	21.0	295.6	225.1	4: 3.4
20:48	-9 -52.4	51 0.9	-10 -26.3	54 17.6	-10 -5.4	52 47.9	1.0632	7.3	295.0	210.4	3:27.4
LIMITS	-12 -10.0	45 47.5	-13 -48.4	46 37.5	-12 -57.8	46 16.1	1.0608	0.0	292.7	202.4	3:10.7

Table 22

ANNULAR SOLAR ECLIPSE OF 4 JAN 1992

SAROS 141   Delta T = 58.7 Sec

UNIVERSAL TIME	NORTHERN LIMIT LATITUDE	LONGITUDE	SOUTHERN LIMIT LATITUDE	LONGITUDE	CENTER LINE LATITUDE	LONGITUDE	DIAMETER RATIO	SUN ALT	SUN AZ	PATH WIDTH	DURATION ANNULARITY
LIMITS	13 11.6	-137-38.2	9 55.8	-136-37.5	11 15.1	-137-50.8	0.9052	0.0	113.4	374.2	7:19.6
21:18	10 18.6	-144-23.3	5 35.6	-147 -9.2	7 47.5	-148 -8.3	0.9075	10.1	114.9	364.9	7:47.0
21:24	6 28.6	-153-45.3	2 41.7	-154-37.6	4 32.7	-154-15.8	0.9099	20.1	116.3	356.5	8:22.6
21:30	4 33.8	-158-51.9	1 0.4	-159-24.3	2 45.5	-159-10.8	0.9114	26.6	117.3	352.3	8:49.1
21:36	3 16.3	-162-41.4	0-10.2	-163 -7.1	1 31.7	-162-56.0	0.9125	31.9	118.2	349.8	9:12.2
21:42	2 19.9	-165-49.7	-1 -2.1	-166-13.5	0 37.7	-166 -2.9	0.9134	36.3	119.3	348.2	9:33.2
21:48	1 38.0	-168-31.7	-1-40.8	-168-55.7	0 -2.4	-168-44.8	0.9142	40.3	120.5	347.4	9:52.6
21:54	1 7.0	-170-55.0	-2 -9.2	-171-20.7	0-32.0	-171 -8.7	0.9149	43.9	121.9	347.0	10:10.4
22: 0	0 44.8	-173 -4.5	-2-29.2	-173-32.6	0-53.1	-173-19.2	0.9154	47.2	123.6	346.9	10:26.9
22: 6	0 29.9	-175 -3.3	-2-42.1	-175-34.2	-1 -7.0	-175-19.3	0.9159	50.2	125.6	346.9	10:41.9
22:12	0 21.3	-176-53.5	-2-48.7	-177-27.6	-1-14.5	-177-11.0	0.9163	53.0	128.0	347.1	10:55.5
22:18	0 18.4	-178-36.7	-2-49.8	-179-14.2	-1-16.5	-178-55.8	0.9167	55.5	130.8	347.2	11: 7.4
22:24	0 20.5	179 45.7	-2-45.9	179 4.7	-1-13.5	179 25.0	0.9170	57.8	134.0	347.1	11:17.7
22:30	0 27.4	178 12.8	-2-37.2	177 28.3	-1 -5.7	177 50.4	0.9173	59.9	137.8	346.8	11:26.3
22:36	0 38.6	176 43.7	-2-24.1	175 55.7	0-53.6	176 19.6	0.9175	61.7	142.1	346.3	11:33.1
22:42	0 53.9	175 17.7	-2 -6.9	174 26.3	0-37.3	174 52.0	0.9176	63.2	147.0	345.6	11:38.1
22:48	1 13.2	173 54.3	-1-45.8	172 59.5	0-17.0	173 26.9	0.9178	64.4	152.5	344.5	11:41.3
22:54	1 36.4	172 32.9	-1-20.5	171 34.9	0 7.1	172 3.9	0.9178	65.2	158.5	343.2	11:42.8
23: 0	2 3.4	171 12.9	0-51.6	170 11.8	0 35.1	170 42.4	0.9179	65.7	164.8	341.7	11:42.4
23: 6	2 34.2	169 53.9	0-18.9	168 49.8	1 6.8	169 22.0	0.9179	65.8	171.3	340.1	11:40.4
23:12	3 8.7	168 35.3	0 17.4	167 28.5	1 42.2	168 2.1	0.9178	65.6	177.7	338.4	11:36.8
23:18	3 47.1	167 16.8	0 57.5	166 7.4	2 21.4	166 42.3	0.9178	64.9	183.9	336.7	11:31.6
23:24	4 29.4	165 57.7	1 41.4	164 46.0	3 4.5	165 22.1	0.9176	63.9	189.7	335.0	11:25.1
23:30	5 15.7	164 37.6	2 29.1	163 23.8	3 51.5	164 1.0	0.9175	62.5	195.1	333.6	11:17.2
23:36	6 6.2	163 15.8	3 20.8	162 0.1	4 42.6	162 38.2	0.9173	60.9	199.9	332.3	11: 8.1
23:42	7 1.3	161 51.7	4 16.8	160 34.3	5 38.1	161 13.3	0.9170	58.9	204.2	331.3	10:58.0
23:48	8 1.1	160 24.4	5 17.2	159 5.6	6 38.2	159 45.4	0.9167	56.8	208.0	330.6	10:46.9
23:54	-9 6.1	158 53.0	6 22.5	157 33.1	7 43.3	158 13.5	0.9164	54.3	211.4	330.3	10:34.8
0: 0	10 16.9	157 16.3	7 33.1	155 55.6	8 54.0	156 36.4	0.9160	51.7	214.4	330.4	10:22.0
0: 6	11 34.1	155 32.7	8 49.7	154 11.8	10 10.8	154 52.8	0.9155	48.8	217.1	331.0	10: 8.3
0:12	12 58.8	153 40.3	10 13.0	152 19.6	11 34.8	153 0.6	0.9150	45.7	219.6	332.0	9:54.0
0:18	14 32.4	151 36.2	11 44.3	150 16.6	13 7.2	150 57.1	0.9143	42.2	221.9	333.6	9:38.8
0:24	16 16.9	149 18.2	13 25.4	147 58.8	14 49.9	148 38.4	0.9136	38.5	224.2	335.8	9:22.8
0:30	18 15.5	146 33.8	15 19.0	145 20.7	16 45.8	145 58.4	0.9128	34.3	226.3	338.8	9: 5.8
0:36	20 33.9	143 17.3	17 29.5	142 12.3	18 59.9	142 46.5	0.9117	29.5	228.6	342.8	8:47.5
0:42	23 23.8	139 2.6	20 6.0	138 14.3	21 42.4	138 41.2	0.9104	23.7	231.2	348.2	8:27.1
0:48	27 20.0	132 35.4	23 31.2	132 35.5	25 20.9	132 42.3	0.9086	16.1	234.7	356.4	8: 2.4
LIMITS	34 28.8	118 30.7	31 19.2	116 59.2	32 32.8	118 31.6	0.9049	0.0	242.3	375.7	7:21.1

Table 23

TOTAL SOLAR ECLIPSE OF 30 JUN 1992

SAROS 146   Delta T = 59.0 Sec

UNIVERSAL TIME	NORTHERN LIMIT LATITUDE	NORTHERN LIMIT LONGITUDE	SOUTHERN LIMIT LATITUDE	SOUTHERN LIMIT LONGITUDE	CENTER LINE LATITUDE	CENTER LINE LONGITUDE	DIAMETER RATIO	SUN ALT	SUN AZ	PATH WIDTH	DURATION TOTALITY
LIMITS	-34-55.4	56 50.7	-36-46.0	56 13.2	-35-26.8	55 40.4	1.0469	0.0	60.3	207.2	2:56.7
11: 4	-29-31.6	44 34.9	-34 -1.5	49 58.3	-31-32.0	48 43.9	1.0497	9.7	55.0	221.0	3:21.8
11: 8	-27-29.0	39 26.4	-30-50.6	42 13.4	-29 -6.4	40 42.3	1.0517	16.0	51.1	232.9	3:42.1
11:12	-26 -9.0	35 44.2	-29-12.0	37 46.3	-27-38.5	36 41.1	1.0531	20.3	48.1	242.3	3:57.2
11:16	-25-10.7	32 43.6	-28 -4.7	34 22.5	-26-36.3	33 30.2	1.0541	23.7	45.5	250.4	4: 9.8
11:20	-24-28.6	30 8.5	-27-15.2	31 33.0	-25-49.8	30 48.6	1.0550	26.6	43.0	257.8	4:20.9
11:24	-23-52.8	27 50.7	-26-37.9	29 5.3	-25-14.4	28 26.3	1.0557	29.1	40.5	264.6	4:30.6
11:28	-23-27.1	25 45.7	-26 -9.8	26 52.9	-24-47.6	26 17.9	1.0563	31.2	38.1	270.8	4:39.4
11:32	-23 -8.0	23 50.3	-25-49.1	24 52.0	-24-27.8	24 19.9	1.0569	33.1	35.6	276.4	4:47.2
11:36	-22-54.7	22 2.4	-25-34.8	22 59.7	-24-13.9	22 30.1	1.0573	34.8	33.1	281.4	4:54.1
11:40	-22-46.4	20 20.5	-25-25.3	21 14.3	-24 -5.2	20 46.6	1.0577	36.2	30.5	285.7	5: 0.2
11:44	-22-42.7	18 43.5	-25-20.9	19 34.3	-24 -1.1	19 8.2	1.0581	37.5	27.9	289.4	5: 5.4
11:48	-22-43.1	17 10.3	-25-20.7	17 58.6	-24 -1.3	17 33.9	1.0584	38.5	25.2	292.2	5: 9.9
11:52	-22-47.4	15 40.3	-25-24.6	16 26.4	-24 -5.4	16 2.8	1.0586	39.4	22.5	294.4	5:13.6
11:56	-22-55.4	14 12.8	-25-32.3	14 56.9	-24-13.2	14 34.4	1.0588	40.1	19.7	295.7	5:16.5
12: 0	-23 -8.9	12 47.2	-25-43.6	13 29.5	-24-24.6	13 7.9	1.0590	40.6	16.8	296.2	5:18.7
12: 4	-23-21.9	11 23.0	-25-58.4	12 3.5	-24-39.5	11 43.0	1.0591	41.0	13.9	296.0	5:20.1
12: 8	-23-40.3	9 59.7	-26-16.6	10 38.6	-24-57.8	10 18.9	1.0591	41.2	11.0	295.0	5:20.7
12:12	-24 -2.1	8 36.9	-26-38.4	9 14.1	-25-19.6	8 55.4	1.0592	41.2	8.1	293.4	5:20.6
12:16	-24-27.3	7 14.2	-27 -3.6	7 49.6	-25-44.8	7 31.8	1.0591	41.0	5.1	291.2	5:19.7
12:20	-24-55.9	5 51.0	-27-32.4	6 24.5	-26-13.5	6 7.8	1.0591	40.7	2.2	288.4	5:18.1
12:24	-25-28.1	4 27.0	-28 -4.9	4 58.5	-26-45.8	4 42.8	1.0590	40.2	359.3	285.2	5:15.8
12:28	-26 -4.0	3 1.5	-28-41.3	3 30.8	-27-21.9	3 16.3	1.0588	39.5	356.4	281.5	5:12.8
12:32	-26-43.9	1 34.1	-29-21.8	2 0.8	-28 -2.0	1 47.7	1.0586	38.7	353.5	277.6	5: 9.0
12:36	-27-28.0	0 4.0	-30 -8.7	0 27.8	-28-46.5	0 16.2	1.0584	37.7	350.7	273.4	5: 4.6
12:40	-28-16.6	-1 -29.5	-30-56.5	-1 -9.2	-29-35.6	-1 -18.9	1.0581	36.5	347.8	269.0	4:59.4
12:44	-29-10.2	-3 -7.3	-31-51.7	-2 -51.2	-30-29.9	-2 -58.7	1.0578	35.0	345.1	264.4	4:53.5
12:48	-30 -9.5	-4 -50.8	-32-53.1	-4 -39.7	-31-30.2	-4 -44.5	1.0573	33.4	342.3	259.7	4:46.8
12:52	-31-15.4	-6 -41.4	-34 -1.7	-6 -36.6	-32-37.3	-6 -38.0	1.0569	31.6	339.5	254.8	4:39.4
12:56	-32-28.9	-8 -41.3	-35-19.1	-8 -44.6	-33-52.6	-8 -41.7	1.0563	29.5	336.7	249.9	4:31.1
13: 0	-33-51.8	-10-53.7	-36-47.5	-11 -7.8	-35-18.0	-10-59.1	1.0556	27.1	333.9	244.8	4:21.9
13: 4	-35-26.8	-13-23.2	-38-30.5	-13-52.8	-36-58.6	-13-35.7	1.0548	24.3	330.9	239.5	4:11.5
13: 8	-37-18.4	-16-17.9	-40-34.9	-17-11.8	-38-53.9	-16-41.2	1.0539	20.9	327.6	233.9	3:59.7
13:12	-39-35.1	-19-53.7	-43-15.8	-21-32.7	-41-20.9	-20-36.3	1.0526	16.8	323.9	227.8	3:45.8
13:16	-42-39.1	-24-51.3	-47-30.3	-28-43.7	-44-50.7	-28-22.7	1.0508	11.0	318.8	220.5	3:27.4
LIMITS	-50-45.5	-39-20.7	-52-24.6	-37-49.1	-51 -4.5	-37-37.1	1.0476	0.0	309.8	210.1	2:59.2

Table 24

ANNULAR SOLAR ECLIPSE OF 10 MAY 1994

SAROS 128

Delta T = 60.4 Sec

UNIVERSAL TIME	NORTHERN LIMIT LATITUDE	LONGITUDE	SOUTHERN LIMIT LATITUDE	LONGITUDE	CENTER LINE LATITUDE	LONGITUDE	DIAMETER RATIO	SUN ALT	SUN AZ	PATH WIDTH	DURATION ANNULARITY
LIMITS	14 47.1	146 54.3	12 19.4	145 21.8	13 42.9	145 36.3	0.9297	0.0	72.0	310.2	4:34.6
15:30	20 17.7	132 4.7	19 26.4	127 25.4	19 53.9	129 36.8	0.9342	18.4	77.4	294.3	4:55.8
15:36	23 6.0	125 54.8	21 58.0	122 19.8	22 31.5	124 3.3	0.9358	25.4	80.5	286.4	5: 5.5
15:42	25 17.8	121 27.9	23 59.3	118 23.7	24 38.8	119 53.2	0.9371	30.8	83.5	279.6	5:13.6
15:48	27 11.0	117 50.1	25 47.1	115 5.1	26 29.1	116 25.6	0.9381	35.5	86.6	273.5	5:20.8
15:54	28 51.9	114 41.4	27 24.1	112 10.2	28 8.0	113 24.2	0.9389	39.5	89.8	267.9	5:27.4
16: 0	30 23.9	111 51.7	28 52.9	109 31.3	29 38.3	110 40.1	0.9396	43.2	93.3	262.7	5:33.5
16: 6	31 48.8	109 15.1	30 15.1	107 3.8	31 1.8	108 8.2	0.9403	46.6	97.1	257.9	5:39.2
16:12	33 7.9	106 47.7	31 31.7	104 44.4	32 19.6	105 44.9	0.9408	49.6	101.1	253.5	5:44.4
16:18	34 21.9	104 26.9	32 43.4	102 30.8	33 32.4	103 27.8	0.9413	52.4	105.6	249.5	5:49.3
16:24	35 31.5	102 10.6	33 50.6	100 21.4	34 40.8	101 15.0	0.9417	55.0	110.5	245.9	5:53.8
16:30	36 36.9	99 57.2	34 53.9	98 14.6	35 45.1	99 5.0	0.9420	57.4	115.8	242.7	5:57.9
16:36	37 38.5	97 45.6	35 53.3	96 9.4	36 45.6	96 56.7	0.9423	59.5	121.7	239.8	6: 1.5
16:42	38 36.5	95 34.6	36 49.1	94 5.0	37 42.5	94 49.0	0.9426	61.3	128.3	237.3	6: 4.8
16:48	39 30.9	93 23.3	37 41.4	92 0.4	38 35.9	92 41.1	0.9428	62.9	135.4	235.2	6: 7.5
16:54	40 21.9	91 11.0	38 30.3	89 55.0	39 25.8	90 32.3	0.9429	64.1	143.1	233.4	6: 9.8
17: 0	41 9.5	88 56.9	39 15.7	87 48.1	40 12.3	88 21.9	0.9430	65.0	151.4	231.9	6:11.6
17: 6	41 53.5	86 40.4	39 57.8	85 39.2	40 55.3	86 9.3	0.9431	65.6	160.0	230.8	6:12.9
17:12	42 34.1	84 20.9	40 36.4	83 27.7	41 34.9	83 53.9	0.9431	65.7	168.9	230.0	6:13.6
17:18	43 11.0	81 57.8	41 11.4	81 13.1	42 10.8	81 35.2	0.9431	65.5	177.9	229.6	6:13.7
17:24	43 44.1	79 30.6	41 42.8	78 55.9	42 43.1	79 12.6	0.9431	64.9	186.6	229.4	6:13.3
17:30	44 13.4	76 58.8	42 10.4	76 32.6	43 11.5	76 45.6	0.9430	63.9	195.0	229.6	6:12.3
17:36	44 38.5	74 21.3	42 34.0	74 5.7	43 35.8	74 13.5	0.9428	62.6	203.0	230.2	6:10.8
17:42	44 59.2	71 38.0	42 53.5	71 33.5	43 55.9	71 36.0	0.9426	60.9	210.5	231.0	6: 8.6
17:48	45 15.2	68 48.2	43 8.4	68 55.4	44 11.4	68 52.2	0.9424	59.0	217.6	232.3	6: 5.8
17:54	45 26.2	65 50.9	43 18.7	66 10.9	44 22.0	66 1.5	0.9421	56.9	224.2	233.9	6: 2.5
18: 0	45 31.7	62 45.4	43 23.8	63 18.9	44 27.3	63 3.0	0.9418	54.5	230.4	236.0	5:58.5
18: 6	45 31.2	59 30.5	43 23.3	60 18.7	44 26.8	59 55.7	0.9414	51.9	236.3	238.5	5:54.0
18:12	45 23.9	56 4.8	43 16.5	57 8.7	44 19.9	56 38.1	0.9409	49.0	241.8	241.4	5:49.0
18:18	45 9.2	52 26.2	43 2.8	53 47.5	44 5.7	53 8.5	0.9404	45.9	247.2	244.9	5:43.4
18:24	44 45.6	48 32.1	42 41.1	50 12.6	43 43.1	49 24.4	0.9398	42.5	252.5	249.1	5:37.2
18:30	44 11.7	44 18.1	42 10.0	46 20.6	43 10.7	45 21.8	0.9391	38.7	257.6	253.9	5:30.4
18:36	43 24.6	39 37.5	41 27.4	42 8.1	42 26.0	40 55.0	0.9382	34.5	262.7	259.6	5:22.9
18:42	42 20.1	34 18.1	40 29.9	37 20.2	41 25.2	35 53.3	0.9372	29.8	268.0	266.4	5:14.6
18:48	40 48.7	27 53.7	39 11.0	31 45.3	40 0.7	29 55.7	0.9359	24.1	273.5	274.8	5: 5.2
18:54	38 23.2	19 3.1	37 15.4	24 38.1	37 52.5	22 4.4	0.9341	16.6	279.9	285.9	4:53.7
LIMITS	33 28.3	3 14.2	31 4.3	5 2.2	32 27.8	4 45.4	0.9301	0.0	290.9	308.3	4:32.9

Table 25

TOTAL SOLAR ECLIPSE OF   3 NOV 1994

SAROS 133   Delta T =   60.8 Sec

UNIVERSAL TIME	NORTHERN LIMIT LATITUDE	LONGITUDE	SOUTHERN LIMIT LATITUDE	LONGITUDE	CENTER LINE LATITUDE	LONGITUDE	DIAMETER RATIO	SUN ALT	SUN AZ	PATH WIDTH	DURATION TOTALITY
LIMITS	-7-20.2	96 37.4	-8-27.5	97 10.5	-7-57.0	96 42.4	1.0360	0.0	105.2	135.6	1:52.3
12: 6	-11-45.0	82 40.8	-12-15.1	85 21.0	-12 -1.1	83 58.1	1.0405	14.1	102.7	153.8	2:18.8
12:12	-14-45.9	75 9.0	-15-33.7	77 5.2	-15-10.2	76 6.1	1.0434	23.7	100.0	165.2	2:39.3
12:18	-17 -4.0	70 2.7	-17-59.7	71 45.2	-17-32.2	70 53.2	1.0454	30.4	97.5	172.3	2:54.8
12:24	-19 -2.4	65 58.6	-20 -3.4	67 33.7	-19-33.2	66 45.6	1.0469	35.9	94.7	177.3	3: 7.8
12:30	-20-48.2	62 30.2	-21-53.6	64 0.2	-21-21.1	63 14.7	1.0481	40.7	91.7	181.0	3:19.3
12:36	-22-25.2	59 24.7	-23-34.2	60 50.9	-22-59.9	60 7.3	1.0492	45.0	88.5	183.8	3:29.7
12:42	-23-55.2	56 35.0	-25 -7.5	57 57.9	-24-31.5	57 16.0	1.0500	48.9	84.8	185.8	3:39.0
12:48	-25-19.4	53 56.5	-26-34.8	55 18.2	-25-57.3	54 35.9	1.0508	52.5	80.7	187.4	3:47.3
12:54	-26-38.8	51 25.9	-27-57.0	52 42.5	-27-18.0	52 3.8	1.0514	55.7	76.0	188.5	3:54.8
13: 0	-27-53.8	49 0.8	-29-14.7	50 14.3	-28-34.3	49 37.2	1.0520	58.7	70.7	189.3	4: 1.5
13: 6	-29 -4.9	46 39.5	-30-28.3	47 49.7	-29-46.7	47 14.2	1.0524	61.4	64.6	189.7	4: 7.2
13:12	-30-12.4	44 20.4	-31-38.1	45 27.1	-30-55.3	44 53.4	1.0528	63.8	57.6	190.0	4:12.2
13:18	-31-16.3	42 2.2	-32-44.3	43 42.7	-32 -0.3	42 33.4	1.0531	65.8	49.6	190.0	4:16.2
13:24	-32-16.9	39 43.9	-33-47.0	40 42.7	-33 -1.9	40 13.0	1.0533	67.4	40.5	189.9	4:19.4
13:30	-33-14.2	37 24.4	-34-46.2	38 18.8	-34 -0.2	37 51.3	1.0534	68.5	30.4	189.7	4:21.6
13:36	-34 -8.1	35 2.8	-35-42.0	35 52.3	-34-55.1	35 27.4	1.0535	69.1	19.6	189.3	4:23.0
13:42	-34-58.8	32 38.3	-36-34.3	33 22.5	-35-46.5	33 0.2	1.0535	69.1	8.6	188.8	4:23.4
13:48	-35-46.0	30 9.9	-37-23.0	30 48.3	-36-34.4	30 29.0	1.0535	68.5	357.7	188.1	4:22.9
13:54	-36-29.6	27 36.8	-38 -7.9	28 8.9	-37-18.7	27 52.8	1.0533	67.4	347.4	187.4	4:21.4
14: 0	-37 -9.5	24 58.0	-38-48.7	25 23.2	-37-59.1	25 10.7	1.0531	65.8	338.0	186.5	4:18.9
14: 6	-37-45.5	22 12.5	-39-25.3	22 30.3	-38-35.3	22 21.5	1.0529	63.8	329.5	185.5	4:15.4
14:12	-38-17.1	19 19.2	-39-57.2	19 28.8	-39 -7.1	19 24.2	1.0525	61.4	321.7	184.3	4:11.0
14:18	-38-44.0	16 16.7	-40-24.1	16 17.6	-39-34.0	16 17.5	1.0521	58.7	314.7	183.0	4: 5.6
14:24	-39 -5.7	13 3.4	-40-45.3	12 55.0	-39-55.4	12 59.7	1.0516	55.7	308.3	181.5	3:59.2
14:30	-39-21.7	9 37.5	-41 -0.1	9 19.1	-40-10.8	9 28.8	1.0509	52.5	302.3	179.8	3:51.9
14:36	-39-30.9	5 58.5	-41 -7.6	5 27.1	-40-19.2	5 42.4	1.0502	48.9	296.6	177.9	3:43.5
14:42	-39-32.3	1 58.7	-41 -8.5	1 15.5	-40-19.4	1 36.9	1.0493	45.0	291.2	175.6	3:34.0
14:48	-39-24.2	-2-28.7	-40-54.9	-3-21.0	-40 -9.6	-2-52.9	1.0483	40.7	285.8	172.9	3:23.4
14:54	-39 -4.1	-7-21.7	-40-30.1	-8-31.0	-39-47.2	-7-55.2	1.0471	35.9	280.4	169.7	3:11.4
15: 0	-38-27.7	-13 -2.4	-39-46.9	-14-29.9	-39 -7.6	-13-44.8	1.0456	30.4	274.9	165.6	2:57.8
15: 6	-37-26.1	-19-58.6	-38-34.7	-21-52.4	-38 -0.9	-20-53.8	1.0437	23.7	268.8	160.2	2:41.8
15:12	-35-32.5	-29-44.5	-36-17.5	-32-40.9	-35-56.3	-31 -8.6	1.0408	14.1	261.3	151.4	2:20.5
LIMITS	-31-25.3	-46-26.1	-32-32.0	-47 -5.9	-32 -2.1	-46-33.1	1.0363	0.0	252.1	136.7	1:53.1

Table 26

ANNULAR SOLAR ECLIPSE OF 29 APR 1995

SAROS 138   Delta T = 61.2 Sec

UNIVERSAL TIME	NORTHERN LIMIT LATITUDE	LONGITUDE	SOUTHERN LIMIT LATITUDE	LONGITUDE	CENTER LINE LATITUDE	LONGITUDE	DIAMETER RATIO	SUN ALT	SUN AZ	PATH WIDTH	DURATION ANNULARITY
LIMITS	-30-37.5	137 16.6	-32-47.4	136 45.1	-31-34.8	136 32.6	0.9362	0.0	72.6	243.6	4:32.8
15:48	-24-53.4	119 16.0	-27-10.9	119 30.2	-26 -0.9	119 20.8	0.9405	17.7	63.3	223.8	4:55.1
15:54	-22-22.5	113 11.1	-24-25.7	112 55.6	-23-23.4	113 2.3	0.9423	25.2	59.6	216.2	5: 6.7
16: 0	-20-22.1	108 50.2	-22-17.4	108 22.6	-21-19.2	108 35.8	0.9436	30.9	56.7	210.9	5:16.3
16: 6	-18-38.3	105 22.5	-20-28.4	104 48.4	-19-33.0	105 5.0	0.9447	35.8	54.2	206.8	5:25.0
16:12	-17 -5.7	102 28.1	-18-52.0	101 50.0	-17-58.5	102 8.7	0.9455	40.1	51.8	203.5	5:33.0
16:18	-15-41.4	99 58.5	-17-24.7	99 15.8	-16-32.7	99 35.9	0.9463	44.0	49.3	200.8	5:40.5
16:24	-14-23.6	97 41.5	-16 -4.7	96 59.2	-15-13.8	97 20.2	0.9469	47.6	46.8	198.7	5:47.5
16:30	-13-11.2	95 39.2	-14-50.5	94 55.7	-14 -0.6	95 17.3	0.9474	50.9	44.1	196.9	5:54.2
16:36	-12 -3.5	93 46.6	-13-41.4	93 2.3	-12-52.1	93 24.3	0.9479	54.0	41.2	195.5	6: 0.4
16:42	-10-59.7	92 1.8	-12-36.6	91 17.0	-11-47.9	91 39.2	0.9483	56.8	37.9	194.5	6: 6.3
16:48	-9-59.5	90 22.8	-11-35.7	89 37.9	-10-47.3	90 0.2	0.9487	59.5	34.2	193.7	6:11.8
16:54	-9 -2.6	88 48.7	-10-38.3	88 3.8	-9-50.2	88 26.2	0.9490	61.9	30.0	193.3	6:16.8
17: 0	-8 -8.7	87 18.4	-9-44.1	86 33.7	-8-56.2	86 56.0	0.9492	64.1	25.1	193.1	6:21.4
17: 6	-7-17.6	85 51.0	-8-53.6	85 6.5	-8 -5.0	85 28.7	0.9494	66.0	19.6	193.1	6:25.5
17:12	-6-29.2	84 25.8	-8 -4.8	83 41.7	-7-16.7	84 3.7	0.9495	67.6	13.2	193.4	6:29.1
17:18	-5-43.3	83 2.1	-7-19.2	82 18.4	-6-31.0	82 40.2	0.9496	68.9	6.1	193.8	6:32.1
17:24	-4-59.9	81 39.2	-6-36.3	80 56.0	-5-47.9	81 17.6	0.9497	69.8	358.3	194.5	6:34.6
17:30	-4-19.0	80 16.7	-5-56.0	79 34.1	-5 -7.2	79 55.4	0.9497	70.2	350.0	195.2	6:36.4
17:36	-3-40.4	78 54.0	-5-18.2	78 12.1	-4-29.1	78 33.0	0.9497	70.1	341.7	196.2	6:37.5
17:42	-3 -4.3	77 30.5	-4-43.0	76 49.3	-3-53.4	77 9.9	0.9496	69.6	333.7	197.2	6:37.9
17:48	-2-30.5	76 5.8	-4-10.3	75 25.4	-3-20.1	75 45.6	0.9495	68.7	326.2	198.3	6:37.6
17:54	-1-59.2	74 39.3	-3-40.1	73 59.7	-2-49.4	74 19.5	0.9493	67.3	319.6	199.4	6:36.6
18: 0	-1-30.5	73 10.3	-3-12.6	72 31.6	-2-21.2	72 51.0	0.9492	65.6	313.8	200.7	6:34.8
18: 6	-1 -4.4	71 38.4	-2-47.7	71 0.6	-1-55.8	71 19.5	0.9489	63.6	308.8	201.9	6:32.3
18:12	0-41.0	70 2.7	-2-25.8	69 25.8	-1-33.1	69 44.3	0.9486	61.4	304.6	203.2	6:28.9
18:18	0-20.5	68 22.4	-2 -6.9	67 46.4	-1-13.4	68 4.5	0.9483	58.9	301.1	204.5	6:24.8
18:24	0 -3.3	66 36.6	-1-51.2	66 1.4	0-57.0	66 19.1	0.9479	56.2	298.2	205.8	6:20.0
18:30	0 10.6	64 44.0	-1-39.2	64 9.6	0-44.0	64 26.9	0.9474	53.3	295.7	207.3	6:14.4
18:36	0 20.6	62 43.2	-1-31.1	62 9.4	0-35.0	62 26.5	0.9469	50.2	293.7	208.8	6: 8.1
18:42	0 26.6	60 32.1	-1-27.5	59 58.8	0-30.4	60 15.7	0.9463	46.9	291.9	210.5	6: 1.1
18:48	0 26.8	58 8.1	-1-29.3	57 34.8	0-31.0	57 51.8	0.9456	43.2	290.5	212.4	5:53.3
18:54	0 21.2	55 27.3	-1-37.7	54 53.4	0-37.9	55 10.7	0.9448	39.2	289.3	214.7	5:44.8
19: 0	0 7.8	52 23.7	-1-54.3	51 48.1	0-52.9	52 6.4	0.9439	34.8	288.3	217.4	5:35.5
19: 6	0-16.1	48 47.0	-2-22.4	48 7.5	-1-18.8	48 27.9	0.9427	29.8	287.5	220.8	5:25.2
19:12	0-56.1	44 16.4	-3 -8.6	43 28.0	-2 -1.8	43 53.3	0.9413	23.8	286.8	225.3	5:13.5
19:18	-2 -7.3	37 54.8	-4-32.6	36 37.6	-3-18.8	37 18.9	0.9393	15.7	286.0	232.1	4:59.1
LIMITS	-5-37.9	22 54.5	-7-51.4	23 14.1	-6-38.6	23 26.8	0.9354	0.0	284.6	246.6	4:36.3

Table 27

TOTAL SOLAR ECLIPSE OF 24 OCT 1995

SAROS 143   Delta T = 61.5 Sec

UNIVERSAL TIME	NORTHERN LIMIT LATITUDE LONGITUDE	SOUTHERN LIMIT LATITUDE LONGITUDE	CENTER LINE LATITUDE LONGITUDE	DIAMETER RATIO	SUN ALT	SUN AZ	PATH WIDTH	DURATION TOTALITY
LIMITS	34 54.5 -51 -6.9	34 46.0 -51 -5.4	34 49.6 -51 -9.0	1.0044	0.0	104.3	15.9	0:15.9
2:54	32 37.5 -80-56.6	32 18.2 -81-15.1	32 27.9 -61 -6.1	1.0071	9.0	110.0	25.3	0:27.3
3: 0	29 3.7 -72-38.5	28 40.6 -72-42.1	28 52.1 -72-40.4	1.0106	20.7	117.1	37.4	0:44.6
3: 6	26 40.2 -78-59.8	26 14.5 -78-58.1	26 27.4 -78-59.1	1.0126	27.9	121.3	44.4	0:56.5
3:12	24 41.2 -83-39.6	24 13.4 -83-34.6	24 27.3 -83-37.2	1.0141	33.7	124.7	49.7	1: 6.4
3:18	22 56.3 -87-25.2	22 27.0 -87-17.6	22 41.6 -87-21.5	1.0154	38.6	127.8	54.0	1:15.2
3:24	21 21.2 -90-36.4	20 50.5 -90-26.8	21 5.8 -90-31.7	1.0164	43.0	130.9	57.7	1:23.1
3:30	19 53.5 -93-23.8	19 21.7 -93-12.4	19 37.5 -93-18.1	1.0173	46.9	134.0	60.9	1:30.4
3:36	18 31.7 -95-53.6	17 58.9 -95-40.8	18 15.3 -95-47.3	1.0181	50.5	137.3	63.6	1:37.0
3:42	17 14.9 -98-10.3	16 41.2 -97-56.3	16 58.0 -98 -3.4	1.0188	53.9	140.9	66.1	1:43.0
3:48	16 2.3 -100-18.8	15 27.8 -100 -1.8	15 45.0 -100 -9.4	1.0193	56.9	144.8	68.3	1:48.4
3:54	14 53.5 -102-15.4	14 18.3 -101-59.6	14 35.9 -102 -7.6	1.0198	59.7	149.3	70.2	1:53.2
4: 0	13 48.1 -104 -7.9	13 12.2 -103-51.4	13 30.1 -103-59.7	1.0203	62.3	154.4	71.9	1:57.4
4: 6	12 45.7 -105-55.5	12 9.3 -105-38.5	12 27.5 -105-47.1	1.0206	64.5	160.2	73.4	2: 1.1
4:12	11 46.3 -107-39.6	11 9.4 -107-22.2	11 27.8 -107-31.0	1.0209	66.4	166.9	74.7	2: 4.1
4:18	10 49.6 -109-21.2	10 12.3 -109 -3.4	10 30.9 -109-12.3	1.0211	67.8	174.5	75.8	2: 6.5
4:24	9 55.4 -111 -1.0	9 17.8 -110-43.1	9 36.6 -110-52.1	1.0212	68.9	182.8	76.7	2: 8.2
4:30	9 3.8 -112-40.0	8 25.8 -112-21.9	8 44.8 -112-31.0	1.0213	69.3	191.7	77.4	2: 9.3
4:36	8 14.7 -114-18.8	7 36.4 -114 -0.8	7 55.6 -114 -9.8	1.0214	69.3	200.6	77.9	2: 9.7
4:42	7 28.0 -115-58.4	6 49.6 -115-40.4	7 8.8 -115-49.4	1.0213	68.7	209.3	78.1	2: 9.5
4:48	6 43.7 -117-39.3	6 5.2 -117-21.5	6 24.4 -117-30.4	1.0213	67.6	217.2	78.1	2: 8.5
4:54	6 1.8 -119-22.4	5 23.4 -119 -4.8	5 42.6 -119-13.6	1.0211	66.1	224.2	77.8	2: 6.8
5: 0	5 22.6 -121 -8.5	4 44.2 -120-51.3	5 3.4 -120-59.9	1.0209	64.1	230.3	77.3	2: 4.5
5: 6	4 45.9 -122-58.6	4 7.8 -122-41.7	4 26.9 -122-50.2	1.0206	61.9	235.4	76.4	2: 1.4
5:12	4 12.1 -124-53.8	3 34.3 -124-37.3	3 53.2 -124-45.5	1.0203	59.3	239.7	75.2	1:57.6
5:18	3 41.2 -126-55.3	3 3.9 -126-39.3	3 22.6 -126-47.3	1.0199	56.4	243.3	73.7	1:53.1
5:24	3 13.6 -129 -4.7	2 36.9 -128-49.2	2 55.3 -128-56.9	1.0193	53.3	246.2	71.7	1:48.0
5:30	2 49.7 -131-24.1	2 13.9 -131 -9.1	2 31.8 -131-16.6	1.0187	50.0	248.7	69.4	1:42.1
5:36	2 30.0 -133-56.2	1 55.2 -133-41.7	2 12.7 -133-48.9	1.0180	46.3	250.8	66.6	1:35.5
5:42	2 15.4 -136-44.8	1 41.8 -136-30.7	1 58.7 -136-37.7	1.0172	42.3	252.4	63.3	1:28.2
5:48	2 7.0 -139-55.8	1 35.0 -139-42.0	1 51.1 -139-48.8	1.0162	37.8	253.8	59.4	1:20.0
5:54	2 6.9 -143-38.8	1 36.9 -143-25.2	1 52.0 -143-31.9	1.0150	32.8	255.0	54.7	1:10.9
6: 0	2 19.1 -148-12.4	1 51.6 -147-58.5	2 5.4 -148 -5.4	1.0134	26.9	255.9	48.9	1: 0.5
6: 6	2 53.4 -154-23.7	2 29.1 -154 -8.2	2 41.3 -154-15.8	1.0114	19.3	256.8	41.1	0:47.8
6:12	4 53.3 -167-22.4	4 31.7 -166-43.3	4 42.5 -167 -2.2	1.0072	4.9	257.9	25.7	0:26.7
LIMITS	5 44.7 -171-47.6	5 33.6 -171-48.5	5 38.7 -171-45.7	1.0057	0.0	258.4	20.4	0:20.4

83

Table 28

TOTAL SOLAR ECLIPSE OF 9 MAR 1997

SAROS 120   Delta T = 62.6 Sec

UNIVERSAL TIME	NORTHERN LIMIT LATITUDE LONGITUDE		SOUTHERN LIMIT LATITUDE LONGITUDE		CENTER LINE LATITUDE LONGITUDE		DIAMETER RATIO	SUN ALT	SUN AZ	PATH WIDTH	DURATION TOTALITY
LIMITS	50 48.9	-86-51.2	48 7.0	-87-27.9	49 17.5	-89 -3.0	1.0353	0.0	98.7	295.7	2: 0.5
0:46	50 36.4	-90-39.7	47 55.0	-103-27.5	49 7.9	-98-53.9	1.0375	8.0	106.7	325.9	2:14.7
0:48	50 35.6	-97-27.8	48 10.6	-106-13.7	49 20.5	-102-30.2	1.0383	10.5	110.0	337.5	2:20.2
0:50	50 49.2	-101-15.9	48 28.8	-108-34.9	49 37.4	-105-19.6	1.0389	12.4	112.7	346.2	2:24.6
0:52	51 7.3	-104-11.1	48 49.0	-110-39.5	49 57.0	-107-42.9	1.0393	14.0	115.2	353.1	2:28.3
0:54	51 28.2	-106-38.1	49 10.6	-112-32.2	50 18.4	-109-49.1	1.0398	15.3	117.5	358.5	2:31.5
0:56	51 50.9	-108-47.1	49 33.5	-114-16.1	50 41.4	-111-43.3	1.0401	16.5	119.6	362.7	2:34.3
0:58	52 15.3	-110-43.5	49 57.5	-115-53.0	51 5.6	-113-28.4	1.0404	17.5	121.7	365.9	2:36.7
1: 0	52 40.9	-112-30.4	50 22.5	-117-24.4	51 30.9	-115 -6.5	1.0407	18.4	123.7	368.2	2:39.0
1: 2	53 7.7	-114-10.2	50 48.5	-118-51.3	51 57.3	-116-39.1	1.0409	19.2	125.7	369.8	2:40.9
1: 4	53 35.6	-115-44.3	51 15.3	-120-14.5	52 24.7	-118 -7.1	1.0411	19.9	127.6	370.6	2:42.7
1: 6	54 4.6	-117-13.8	51 43.1	-121-34.7	52 53.0	-119-31.5	1.0413	20.5	129.6	370.8	2:44.2
1: 8	54 34.5	-118-39.6	52 11.6	-122-52.5	53 22.3	-120-52.9	1.0415	21.1	131.5	370.5	2:45.5
1:10	55 5.4	-120 -2.4	52 41.1	-124 -8.2	53 52.4	-122-11.9	1.0416	21.5	133.4	369.7	2:46.7
1:12	55 37.3	-121-22.9	53 11.3	-125-22.3	54 23.5	-123-29.0	1.0417	21.9	135.2	368.5	2:47.7
1:14	56 10.2	-122-41.4	53 42.5	-126-35.1	54 55.4	-124-44.5	1.0418	22.3	137.1	367.0	2:48.5
1:16	56 44.1	-123-58.6	54 14.5	-127-47.0	55 28.3	-125-58.9	1.0419	22.5	139.0	365.1	2:49.1
1:18	57 19.0	-125-14.6	54 47.4	-128-58.2	56 2.2	-127-12.4	1.0420	22.7	140.9	363.0	2:49.6
1:20	57 55.9	-128-29.9	55 21.2	-130 -9.1	56 37.0	-128-25.4	1.0420	22.9	142.8	360.6	2:49.9
1:22	58 32.1	-127-44.9	55 55.9	-131-19.8	57 12.9	-129-38.2	1.0420	22.9	144.7	358.1	2:50.1
1:24	59 10.4	-128-59.9	56 31.7	-132-30.7	57 49.8	-130-51.2	1.0420	23.0	146.6	355.4	2:50.1
1:26	59 50.0	-130-15.2	57 8.5	-133-42.0	58 27.9	-132 -4.5	1.0420	22.9	148.5	352.6	2:50.0
1:28	60 30.9	-131-31.2	57 46.4	-134-54.1	59 7.2	-133-18.5	1.0420	22.8	150.5	349.7	2:49.7
1:30	61 13.2	-132-48.2	58 25.5	-136 -7.2	59 47.8	-134-33.6	1.0419	22.7	152.5	346.7	2:49.2
1:32	61 57.1	-134 -6.7	59 5.8	-137-21.7	60 29.8	-135-50.1	1.0418	22.4	154.5	343.7	2:48.6
1:34	62 42.7	-135-27.0	59 47.5	-138-37.8	61 13.3	-137 -8.4	1.0417	22.1	156.5	340.6	2:47.9
1:36	63 30.1	-136-49.8	60 30.7	-139-56.1	61 58.4	-138-29.1	1.0416	21.8	158.6	337.5	2:46.9
1:38	64 19.6	-138-15.7	61 15.4	-141-17.0	62 45.3	-139-52.5	1.0415	21.4	160.7	334.5	2:45.8
1:40	65 11.5	-139-45.3	62 1.9	-142-41.1	63 34.2	-141-19.4	1.0413	20.9	162.8	331.4	2:44.6
1:42	66 5.9	-141-19.6	62 50.3	-144 -8.9	64 25.4	-142-50.5	1.0412	20.3	165.2	328.3	2:43.1
1:44	67 3.5	-142-59.8	63 40.9	-145-41.3	65 19.1	-144-26.8	1.0410	19.7	167.3	325.3	2:41.4
1:46	68 4.5	-144-47.4	64 33.9	-147-19.3	66 15.6	-146 -9.6	1.0407	18.9	169.7	322.3	2:39.6
1:48	69 9.8	-146-44.5	65 29.7	-149 -4.2	67 15.6	-148 -0.4	1.0405	18.1	172.2	319.3	2:37.5
1:50	70 20.2	-148-54.2	66 28.8	-150-57.7	68 19.6	-150 -1.5	1.0402	17.1	174.9	316.4	2:35.1
1:52	71 37.3	-151-21.2	67 31.7	-153 -2.1	69 28.4	-152-16.1	1.0398	16.1	177.8	313.5	2:32.5
1:54	73 3.0	-154-13.2	68 39.3	-155-20.6	70 43.3	-154-49.0	1.0394	14.8	180.9	310.5	2:29.5
1:56	74 41.1	-157-44.7	69 52.7	-157-58.4	72 6.3	-157-48.3	1.0390	13.4	184.5	307.6	2:26.1
1:58	76 39.2	-162-28.1	71 13.6	-161 -3.4	73 40.6	-161-28.2	1.0385	11.7	188.7	304.6	2:22.2
2: 0	79 18.3	-170-11.5	72 45.1	-164-50.2	75 32.3	-166-20.1	1.0378	9.6	194.1	301.5	2:17.4
LIMITS	83 56.5	166 4.0	81 54.1	152 41.5	82 6.6	165 14.8	1.0351	0.0	223.1	292.6	1:59.9

Table 29

TOTAL SOLAR ECLIPSE OF 26 FEB 1998

SAROS 130   Delta T = 63.3 Sec

UNIVERSAL TIME	NORTHERN LIMIT LATITUDE	LONGITUDE	SOUTHERN LIMIT LATITUDE	LONGITUDE	CENTER LINE LATITUDE	LONGITUDE	DIAMETER RATIO	SUN ALT	SUN AZ	PATH WIDTH	DURATION TOTALITY
LIMITS	-1 57.5	143 56.0	-2 46.1	144 3.4	-2 22.8	143 53.9	1.0260	0.0	98.6	89.5	1:27.5
15:48	-2 36.0	139 19.5	-3 39.6	137 51.9	-3 -8.2	138 31.6	1.0277	5.6	98.4	95.6	1:37.4
15:54	-3 49.4	125 51.9	-4 51.5	125 20.1	-4 20.4	125 35.5	1.0322	20.1	97.6	111.4	2: 7.1
16: 0	-3 57.8	119 24.7	-5 -2.7	118 56.4	-4 30.2	119 10.2	1.0345	27.9	97.4	119.7	2:25.3
16: 6	-3 49.4	114 39.2	-4 56.5	114 11.4	-4 22.9	114 24.9	1.0362	34.1	97.5	126.0	2:40.3
16:12	-3 31.7	110 46.7	-4 40.4	110 18.3	-4 -6.0	110 32.2	1.0376	39.4	97.8	131.1	2:53.6
16:18	-3 -7.6	107 27.8	-4 17.6	106 58.5	-3 42.6	107 12.9	1.0388	44.1	98.4	135.3	3: 5.4
16:24	-2 38.8	104 32.4	-3 49.8	104 2.0	-3 14.3	104 17.0	1.0397	48.4	99.4	138.9	3:16.0
16:30	-2 -6.1	101 54.4	-3 18.0	101 22.9	-2 42.0	101 38.5	1.0406	52.4	100.7	142.0	3:25.0
16:36	-1 30.3	99 29.7	-2 42.7	98 57.2	-2 -6.5	99 13.3	1.0413	56.1	102.4	144.5	3:34.3
16:42	0 51.7	97 15.6	-2 -4.5	96 42.1	-1 28.1	96 58.7	1.0419	59.6	104.7	146.6	3:41.9
16:48	0 10.6	95 9.9	-1 23.7	94 35.4	0 47.1	94 52.5	1.0425	62.9	107.7	148.3	3:48.6
16:54	0 32.8	93 10.9	0 40.5	92 35.5	0 -3.8	92 53.1	1.0429	65.9	111.6	149.6	3:54.3
17: 0	1 18.3	91 17.3	0 4.9	90 41.1	0 41.6	90 59.1	1.0433	68.7	116.6	150.6	3:59.1
17: 6	2 5.7	89 27.9	0 52.4	88 51.1	1 29.1	89 9.5	1.0436	71.2	123.2	151.2	4: 2.9
17:12	2 55.1	87 41.8	1 41.9	87 4.4	2 18.5	87 23.1	1.0438	73.4	131.7	151.8	4: 5.7
17:18	3 46.3	85 58.1	2 33.3	85 20.2	3 9.8	85 39.1	1.0440	75.0	142.4	151.7	4: 7.6
17:24	4 39.3	84 15.9	3 26.5	83 37.7	4 2.9	83 56.8	1.0441	76.0	155.1	151.6	4: 8.6
17:30	5 34.2	82 34.5	4 21.7	81 56.1	4 57.9	82 15.3	1.0441	76.1	168.7	151.3	4: 8.6
17:36	6 30.9	80 53.1	5 18.7	80 14.7	5 54.8	80 33.9	1.0441	75.5	181.9	150.7	4: 7.6
17:42	7 29.6	79 10.9	6 17.7	78 32.6	6 53.6	78 51.8	1.0440	74.2	193.4	150.0	4: 5.8
17:48	8 30.2	77 27.0	7 18.6	76 49.1	7 54.3	77 8.1	1.0439	72.3	202.8	149.1	4: 3.0
17:54	9 32.8	75 40.7	8 21.6	75 3.2	8 57.1	75 22.0	1.0436	69.9	210.3	148.0	3:59.4
18: 0	10 37.6	73 50.9	9 26.7	73 14.0	10 2.1	73 32.5	1.0434	67.2	216.3	146.7	3:54.8
18: 6	11 44.8	71 56.4	10 34.3	71 20.4	11 9.4	71 38.5	1.0430	64.3	221.1	145.3	3:49.5
18:12	12 54.4	69 55.9	11 44.3	69 20.9	12 19.3	69 38.5	1.0425	61.1	225.1	143.6	3:43.2
18:18	14 6.9	67 47.7	12 57.2	67 14.1	13 31.9	67 31.0	1.0420	57.7	228.5	141.7	3:36.2
18:24	15 22.5	65 29.6	14 13.2	64 57.6	14 47.8	65 13.8	1.0414	54.1	231.4	139.5	3:28.2
18:30	16 41.7	62 58.9	15 32.9	62 28.9	16 7.2	62 44.1	1.0407	50.2	234.1	137.0	3:19.4
18:36	18 5.1	60 11.6	16 57.0	59 44.1	17 30.9	59 58.1	1.0398	46.1	236.7	134.1	3: 9.7
18:42	19 33.8	57 1.9	18 26.3	56 37.6	18 59.9	56 50.0	1.0388	41.6	239.2	130.8	2:58.9
18:48	21 9.2	53 20.7	20 2.5	53 0.5	20 35.7	53 10.9	1.0378	36.6	241.7	126.8	2:46.9
18:54	22 54.1	48 51.6	21 48.2	48 37.2	22 21.0	48 44.8	1.0361	30.8	244.5	122.0	2:33.4
19: 0	24 54.0	42 58.8	23 49.2	42 53.6	24 21.5	42 56.7	1.0341	23.9	247.9	115.8	2:17.5
19: 6	27 27.8	33 40.1	26 23.8	33 57.6	26 55.6	33 49.9	1.0312	14.0	252.7	106.4	1:56.4
LIMITS	30 15.5	19 5.5	29 25.7	18 59.8	29 50.2	19 5.2	1.0268	0.0	260.2	92.4	1:30.2

Table 30

ANNULAR SOLAR ECLIPSE OF 22 AUG 1998

SAROS 135   Delta T = 63.7 Sec

UNIVERSAL TIME	NORTHERN LIMIT LATITUDE	LONGITUDE	SOUTHERN LIMIT LATITUDE	LONGITUDE	CENTER LINE LATITUDE	LONGITUDE	DIAMETER RATIO	SUN ALT	SUN AZ	PATH WIDTH	DURATION ANNULARITY
LIMITS	0 -1.5	-86-54.5	-1-23.1	-87 -5.2	0-39.3	-87-13.9	0.9590	0.0	78.1	150.9	2:44.4
0:18	2 4.7	-98 -8.5	0 38.3	-97-12.2	1 22.0	-97-40.7	0.9620	11.5	78.1	139.9	2:48.0
0:24	3 15.6	-106-31.1	2 0.8	-106 -3.0	2 38.4	-106-17.4	0.9646	21.6	78.2	130.7	2:52.1
0:30	3 45.2	-111-48.9	2 35.4	-111-26.1	3 10.4	-111-37.7	0.9663	28.4	78.2	125.0	2:55.2
0:36	3 57.5	-115-54.0	2 51.2	-115-54.8	3 24.5	-115-54.7	0.9675	33.9	77.9	120.8	2:57.8
0:42	3 59.4	-119-17.5	2 55.8	-118-57.4	3 27.7	-119 -7.5	0.9685	38.7	77.5	117.4	3: 0.2
0:48	3 54.0	-122-13.3	2 52.7	-121-53.5	3 23.4	-122 -3.4	0.9694	43.0	76.8	114.5	3: 2.3
0:54	3 42.9	-124-49.3	2 43.5	-124-29.4	3 13.3	-124-39.4	0.9701	46.9	75.9	112.1	3: 4.3
1: 0	3 27.2	-127-10.2	2 29.5	-126-50.1	2 58.4	-127 -0.2	0.9707	50.6	74.7	110.0	3: 6.1
1: 6	3 7.6	-129-19.3	2 11.5	-128-58.9	2 39.6	-129 -9.1	0.9713	54.0	73.2	108.2	3: 7.7
1:12	2 44.6	-131-18.9	1 49.9	-130-58.1	2 17.3	-131 -8.5	0.9717	57.2	71.3	106.5	3: 9.1
1:18	2 18.6	-133-10.8	1 25.2	-132-49.6	1 52.0	-133 -0.2	0.9721	60.2	68.8	105.1	3:10.4
1:24	1 49.9	-134-56.4	0 57.6	-134-34.8	1 23.8	-134-45.6	0.9725	63.1	65.7	103.8	3:11.4
1:30	1 18.7	-136-36.7	0 27.3	-136-14.8	0 53.1	-136-25.8	0.9727	65.7	61.9	102.6	3:12.3
1:36	0 45.1	-138-12.8	0 -5.4	-137-50.5	0 20.0	-138 -1.6	0.9730	68.1	57.1	101.6	3:13.0
1:42	0 9.3	-139-45.4	0-40.4	-139-22.7	0-15.5	-139-34.0	0.9731	70.2	51.0	100.8	3:13.5
1:48	0-28.6	-141-15.2	-1-17.7	-140-52.1	0-53.1	-141 -3.7	0.9733	72.1	43.6	100.0	3:13.9
1:54	-1 -8.7	-142-42.8	-1-57.2	-142-19.4	-1-32.9	-142-31.1	0.9733	73.4	34.7	99.4	3:14.1
2: 0	-1-50.7	-144 -9.0	-2-38.8	-143-45.2	-2-14.7	-143-57.1	0.9734	74.3	24.5	99.0	3:14.1
2: 6	-2-34.8	-145-34.1	-3-22.6	-145-10.0	-2-58.6	-145-22.1	0.9733	74.7	13.5	98.7	3:14.0
2:12	-3-20.9	-146-58.8	-4 -8.4	-146-34.4	-3-44.6	-146-46.6	0.9733	74.4	2.6	98.6	3:13.8
2:18	-4 -9.0	-148-23.7	-4-56.4	-147-58.9	-4-32.6	-148-11.3	0.9732	73.5	352.4	98.7	3:13.4
2:24	-4-59.2	-149-49.3	-5-46.8	-149-24.2	-5-22.8	-149-36.7	0.9730	72.1	343.6	98.9	3:12.9
2:30	-5-51.5	-151-16.2	-6-39.1	-150-50.8	-6-15.2	-151 -3.5	0.9728	70.3	336.1	99.3	3:12.3
2:36	-6-46.0	-152-45.1	-7-33.9	-152-19.4	-7 -9.9	-152-32.3	0.9726	68.2	329.9	99.9	3:11.6
2:42	-7-42.9	-154-16.7	-8-31.2	-153-50.9	-8 -7.0	-154 -3.8	0.9723	65.8	324.8	100.8	3:10.9
2:48	-8-42.2	-155-52.0	-9-31.2	-155-25.9	-9 -6.7	-155-39.0	0.9720	63.2	320.6	101.8	3:10.9
2:54	-9-44.3	-157-31.9	-10-34.0	-157 -5.7	-10 -9.1	-157-18.8	0.9716	60.3	317.0	103.1	3: 9.0
3: 0	-10-49.2	-159-17.6	-11-40.0	-158-51.3	-11-14.6	-159 -4.5	0.9711	57.3	313.9	104.7	3: 8.0
3: 6	-11-57.5	-161-10.8	-12-49.4	-160-44.4	-12-23.4	-160-57.5	0.9706	54.1	311.2	106.5	3: 6.9
3:12	-13 -9.4	-163-12.9	-14 -2.8	-162-46.8	-13-36.0	-162-59.9	0.9700	50.7	308.7	108.6	3: 5.7
3:18	-14-25.8	-165-27.1	-15-20.8	-165 -1.4	-14-53.1	-165-14.3	0.9693	47.0	306.4	111.1	3: 4.5
3:24	-15-47.0	-167-56.8	-16-44.2	-167-31.8	-16-15.6	-167-44.3	0.9685	43.1	304.3	114.0	3: 3.1
3:30	-17-14.7	-170-47.4	-18-14.4	-170-23.6	-17-44.6	-170-35.4	0.9676	38.8	302.1	117.4	3: 1.6
3:36	-18-50.7	-174 -7.4	-19-53.9	-173-45.7	-19-22.2	-173-56.5	0.9665	34.0	299.8	121.4	2:60.0
3:42	-20-38.5	-178-12.6	-21-46.2	-177-54.9	-21-12.2	-178 -3.6	0.9652	28.5	297.3	126.4	2:58.2
3:48	-22-45.8	176 22.6	-24 -0.1	176 31.5	-23-22.7	176 27.4	0.9635	21.8	294.3	133.0	2:56.0
3:54	-25-39.2	167 33.7	-27 -8.5	167 7.7	-26-23.2	167 22.3	0.9608	11.8	289.6	143.3	2:53.0
LIMITS	-28-45.6	155 3.2	-30 -9.2	155 19.3	-29-25.0	155 22.1	0.9577	0.0	283.6	156.0	2:49.8

Table 31

ANNULAR SOLAR ECLIPSE OF 16 FEB 1999

SAROS 140   Delta T = 64.1 Sec

UNIVERSAL TIME	NORTHERN LIMIT LATITUDE	LONGITUDE	SOUTHERN LIMIT LATITUDE	LONGITUDE	CENTER LINE LATITUDE	LONGITUDE	DIAMETER RATIO	SUN ALT	SUN AZ	PATH WIDTH	DURATION ANNULARITY
LIMITS	-41	-8.2 -8 -7.7	-41-55.2	-7-35.7	-41-34.7	-8 -5.2	0.9775	0.0	106.5	96.0	1:18.4
5: 0	-44-21.1	-23-52.0	-44-41.2	-21-29.5	-44-31.6	-22-44.4	0.9808	11.7	95.9	80.5	1:13.0
5: 6	-46 -2.1	-35-24.9	-46-31.8	-34-18.9	-46-16.9	-34-52.6	0.9835	21.3	85.7	68.0	1: 7.8
5:12	-46-46.7	-43-21.2	-47-16.4	-42-39.7	-47 -1.5	-43 -0.8	0.9853	27.7	77.9	60.1	1: 3.9
5:18	-47 -5.7	-49-46.4	-47-34.2	-49-18.6	-47-19.9	-49-32.7	0.9866	32.9	71.0	54.1	1: 0.5
5:24	-47 -8.7	-55-16.8	-47-35.5	-54-58.1	-47-22.1	-55 -7.5	0.9877	37.3	64.4	49.2	0:57.5
5:30	-47 -0.1	-60 -8.9	-47-25.2	-59-58.7	-47-12.8	-60 -2.8	0.9886	41.3	58.1	45.2	0:54.9
5:36	-46-42.6	-64-32.0	-47 -6.0	-64-24.6	-46-54.3	-64-28.3	0.9894	44.8	51.8	41.9	0:52.4
5:42	-46-17.8	-68-32.0	-46-39.7	-68-28.3	-46-28.7	-68-30.2	0.9901	47.9	45.5	39.1	0:50.1
5:48	-45-47.1	-72-13.1	-46 -7.5	-72-12.3	-45-57.3	-72-12.7	0.9907	50.7	39.0	36.7	0:48.1
5:54	-45-11.2	-75-38.3	-45-30.4	-75-39.7	-45-20.8	-75-39.0	0.9912	53.3	32.4	34.7	0:46.3
6: 0	-44-30.9	-78-50.0	-44-48.9	-78-53.1	-44-39.9	-78-51.5	0.9916	55.5	25.4	33.0	0:44.7
6: 6	-43-46.7	-81-50.0	-44 -3.6	-81-54.6	-43-55.1	-81-52.3	0.9919	57.4	18.1	31.6	0:43.3
6:12	-42-58.0	-84-40.6	-43-15.0	-84-45.8	-43 -6.9	-84-42.9	0.9922	59.0	10.5	30.5	0:42.1
6:18	-42 -8.0	-87-21.5	-42-23.3	-87-28.2	-42-15.6	-87-24.8	0.9925	60.2	2.6	29.7	0:41.1
6:24	-41-14.1	-89-55.7	-41-28.8	-90 -3.3	-41-21.4	-89-59.5	0.9926	61.1	354.5	29.1	0:40.3
6:30	-40-17.5	-92-23.7	-40-31.7	-92-32.1	-40-24.6	-92-27.9	0.9927	61.5	346.2	28.8	0:39.8
6:36	-39-18.3	-94-46.7	-39-32.1	-94-55.9	-39-25.2	-94-51.3	0.9928	61.6	337.9	28.7	0:39.4
6:42	-38-16.6	-97 -5.7	-38-32.1	-97-15.6	-38-23.3	-97-10.6	0.9928	61.2	329.8	28.9	0:39.3
6:48	-37-12.4	-99-21.6	-37-25.8	-99-32.4	-37-19.1	-99-27.0	0.9927	60.4	322.0	29.3	0:39.5
6:54	-36 -5.9	-101-35.7	-36-19.1	-101-47.3	-36-12.5	-101-41.5	0.9926	59.3	314.7	30.0	0:39.8
7: 0	-34-56.8	-103-48.9	-35-10.1	-104 -1.4	-35 -3.4	-103-55.1	0.9924	57.8	307.9	31.0	0:40.4
7: 6	-33-45.2	-106 -2.4	-33-58.6	-106-16.0	-33-51.9	-106 -9.2	0.9922	55.9	301.6	32.3	0:41.3
7:12	-32-30.8	-108-17.5	-32-44.5	-108-32.4	-32-37.6	-108-24.9	0.9918	53.8	295.9	33.9	0:42.3
7:18	-31-13.6	-110-35.8	-31-27.6	-110-52.2	-31-20.5	-110-43.9	0.9914	51.3	290.7	35.8	0:43.6
7:24	-29-53.1	-112-59.0	-30 -7.5	-113-17.2	-30 -0.2	-113 -8.1	0.9910	48.6	285.9	38.1	0:45.2
7:30	-28-28.8	-115-29.6	-28-43.7	-115-49.9	-28-36.3	-115-39.7	0.9904	45.5	281.6	40.8	0:47.0
7:36	-27 -0.3	-118-10.6	-27-15.7	-118-33.5	-27 -8.0	-118-22.0	0.9897	42.1	277.7	44.0	0:49.0
7:42	-25-26.3	-121 -6.4	-25-42.5	-121-32.8	-25-34.3	-121-19.5	0.9889	38.3	274.1	47.8	0:51.4
7:48	-23-45.5	-124-23.8	-24 -2.3	-124-55.0	-23-53.8	-124-33.3	0.9879	33.9	270.7	52.2	0:54.1
7:54	-21-54.9	-128-14.5	-22-12.3	-128-52.9	-22 -3.6	-128-33.5	0.9867	28.9	267.5	57.6	0:57.2
8: 0	-19-49.0	-133 -2.7	-20 -6.5	-133-53.7	-19-57.7	-133-27.8	0.9852	22.8	264.4	64.4	1: 1.0
8: 6	-17-11.0	-140 -0.3	-17-25.0	-141-25.8	-17-18.1	-140-41.7	0.9829	14.3	261.1	74.0	1: 6.0
LIMITS	-13-13.7	-153-55.4	-13-58.7	-154-18.3	-13-39.0	-153-53.8	0.9789	0.0	257.2	89.8	1:13.5

Table 32

TOTAL SOLAR ECLIPSE OF 11 AUG 1999

SAROS 145

Delta T = 64.5 Sec

UNIVERSAL TIME	NORTHERN LIMIT LATITUDE	LONGITUDE	SOUTHERN LIMIT LATITUDE	LONGITUDE	CENTER LINE LATITUDE	LONGITUDE	DIAMETER RATIO	SUN ALT	SUN AZ	PATH WIDTH	DURATION TOTALITY
LIMITS	41 16.1	65 16.8	40 47.4	84 53.9	41 3.7	64 58.1	1.0143	0.0	69.7	60.9	0:46.5
9:36	48 35.1	44 35.8	46 12.1	42 48.7	46 24.0	43 41.4	1.0193	17.1	85.8	80.5	1:11.6
9:42	48 27.3	34 52.2	47 51.8	33 29.8	48 9.8	34 10.4	1.0214	24.6	94.6	87.8	1:24.1
9:48	49 32.3	27 23.0	48 49.4	26 15.4	49 11.0	26 48.6	1.0229	30.3	102.2	92.7	1:34.0
9:54	50 11.4	21 1.0	49 23.2	20 5.5	49 47.4	20 32.8	1.0241	35.0	109.4	96.5	1:42.5
10: 0	50 32.6	15 22.2	49 40.6	14 37.6	50 6.7	14 59.5	1.0250	39.1	116.4	99.4	1:49.9
10: 6	50 40.2	10 14.9	49 45.3	9 40.5	50 12.8	9 57.3	1.0258	42.7	123.3	101.9	1:56.4
10:12	50 36.9	5 32.2	49 39.9	5 7.4	50 8.4	5 19.5	1.0265	45.9	130.2	103.9	2: 2.1
10:18	50 24.4	1 9.9	49 26.0	0 54.1	49 55.2	1 1.8	1.0270	48.7	137.3	105.6	2: 7.1
10:24	50 4.1	-2-55.1	49 4.8	-3 -2.4	49 34.4	-1 -8.9	1.0275	51.3	144.6	107.1	2:11.4
10:30	49 36.9	-6-45.6	48 37.2	-6-44.5	49 7.0	-6-44.9	1.0278	53.5	152.0	108.3	2:15.0
10:36	49 3.6	-10-21.8	48 3.9	-10-14.0	48 33.8	-10-18.0	1.0281	55.4	159.8	109.4	2:17.9
10:42	48 24.9	-13-46.9	47 25.6	-13-32.5	47 55.2	-13-39.7	1.0283	56.9	167.8	110.3	2:20.1
10:48	47 41.2	-17 -1.8	46 42.6	-16-41.3	47 11.9	-16-51.5	1.0285	58.1	176.0	111.0	2:21.7
10:54	46 53.0	-20 -7.6	45 55.3	-19-41.7	46 24.2	-19-54.6	1.0286	58.9	184.4	111.6	2:22.6
11: 0	46 0.6	-23 -5.7	45 4.1	-22-34.8	45 32.4	-22-50.1	1.0286	59.3	192.9	112.0	2:22.9
11: 6	45 4.3	-25-57.0	44 9.1	-25-21.8	44 36.7	-25-39.3	1.0286	59.3	201.3	112.3	2:22.6
11:12	44 4.2	-28-42.9	43 10.5	-28 -3.7	43 37.4	-28-23.2	1.0284	58.9	209.5	112.5	2:21.5
11:18	43 0.4	-31-24.3	42 8.4	-30-41.7	42 34.4	-31 -2.9	1.0283	58.2	217.5	112.4	2:19.9
11:24	41 53.0	-34 -2.6	41 2.8	-33-16.9	41 28.0	-33-39.6	1.0280	57.0	224.9	112.2	2:17.7
11:30	40 42.0	-36-38.9	39 53.8	-35-50.5	40 18.0	-36-14.5	1.0277	55.5	232.0	111.8	2:14.8
11:36	39 27.3	-39-14.6	38 41.1	-38-23.9	39 4.3	-38-49.0	1.0274	53.6	238.5	111.2	2:11.3
11:42	38 8.7	-41-51.3	37 24.7	-40-58.5	37 46.8	-41-24.7	1.0269	51.4	244.5	110.3	2: 7.3
11:48	36 45.8	-44-30.8	36 4.1	-43-36.1	36 25.1	-44 -3.3	1.0264	48.9	250.0	109.1	2: 2.6
11:54	35 18.3	-47-15.5	34 39.1	-46-19.0	34 58.8	-46-47.1	1.0257	46.0	255.1	107.4	1:57.3
12: 0	33 45.4	-50 -8.2	33 8.8	-49-10.1	33 27.2	-49-38.9	1.0250	42.9	259.7	105.3	1:51.4
12: 6	32 6.1	-53-13.1	31 32.4	-52-13.1	31 49.4	-52-42.9	1.0241	39.3	264.1	102.5	1:44.8
12:12	30 18.6	-56-36.1	29 48.2	-55-34.2	30 3.5	-58 -5.0	1.0231	35.2	268.1	99.0	1:37.4
12:18	28 20.1	-60-27.7	27 53.6	-59-22.9	28 7.0	-59-55.1	1.0219	30.6	271.9	94.3	1:29.1
12:24	26 4.8	-65 -7.5	25 43.2	-63-58.0	25 54.2	-64-32.5	1.0203	25.0	275.7	88.1	1:19.4
12:30	23 17.6	-71-27.8	23 4.0	-70 -6.6	23 11.1	-70-46.9	1.0182	17.7	279.6	79.0	1: 7.4
LIMITS	17 46.6	-87-24.9	17 19.9	-87 -9.0	17 35.4	-87 -8.9	1.0130	0.0	286.1	55.3	0:42.3

Table 33

TOTAL SOLAR ECLIPSE OF 21 JUN 2001

SAROS 127 Delta T = 65.9 Sec

UNIVERSAL TIME	NORTHERN LIMIT LATITUDE	LONGITUDE	SOUTHERN LIMIT LATITUDE	LONGITUDE	CENTER LINE LATITUDE	LONGITUDE	DIAMETER RATIO	SUN ALT	SUN AZ	PATH WIDTH	DURATION TOTALITY
LIMITS	-36 -5.3	50 17.8	-37-10.2	49 46.6	-36-26.6	49 38.2	1.0343	0.0	59.9	127.8	2: 6.9
10:42	-27-57.9	34 20.3	-29-58.3	35 13.9	-28-56.9	34 44.6	1.0391	15.9	50.9	143.9	2:43.7
10:48	-24-28.8	28 7.8	-28-18.3	28 26.2	-25-22.0	28 15.9	1.0414	23.5	46.5	152.4	3: 5.5
10:54	-21-58.2	23 45.0	-23-41.9	23 49.7	-22-49.6	23 46.6	1.0431	29.2	43.0	159.0	3:22.8
11: 0	-19-57.9	20 14.5	-21-40.2	20 11.6	-20-48.7	20 12.5	1.0443	33.8	39.8	164.7	3:37.9
11: 6	-18-17.5	17 15.3	-19-59.4	17 7.5	-19 -8.1	17 10.9	1.0454	37.8	36.8	169.8	3:51.3
11:12	-16-51.6	14 36.8	-18-33.8	14 25.8	-17-42.4	14 30.9	1.0463	41.2	33.2	174.5	4: 3.3
11:18	-15-37.3	12 12.9	-17-20.0	11 59.8	-16-28.4	12 6.0	1.0470	44.3	29.7	178.9	4:14.2
11:24	-14-32.8	9 59.8	-16-15.9	9 45.2	-15-24.0	9 52.2	1.0476	46.9	26.0	182.9	4:23.8
11:30	-13-36.2	7 54.6	-15-20.2	7 39.1	-14-28.0	7 46.6	1.0482	49.2	21.9	186.7	4:32.3
11:36	-12-47.2	5 55.4	-14-31.9	5 39.3	-13-39.4	5 47.1	1.0486	51.1	17.6	190.1	4:39.6
11:42	-12 -4.9	4 0.5	-13-50.3	3 44.2	-12-57.4	3 52.2	1.0489	52.7	13.0	193.1	4:45.7
11:48	-11-28.8	2 8.7	-13-15.0	1 52.4	-12-21.7	2 0.4	1.0492	53.9	8.1	195.8	4:50.4
11:54	-10-58.7	0 18.9	-12-45.5	0 2.8	-11-52.0	0 10.8	1.0494	54.7	2.9	197.9	4:53.9
12: 0	-10-34.4	-1-29.8	-12-21.7	-1-45.8	-11-27.9	-1-37.7	1.0495	55.1	357.6	199.4	4:56.0
12: 6	-10-15.7	-3-18.4	-12 -3.4	-3-33.7	-11 -9.4	-3-26.0	1.0496	55.2	352.3	200.3	4:56.7
12:12	-10 -2.6	-5 -7.6	-11-50.6	-5-22.4	-10-56.4	-5-14.9	1.0495	54.8	347.1	200.6	4:56.1
12:18	-9-55.1	-6-58.3	-11-43.3	-7-12.6	-10-49.1	-7 -5.4	1.0494	54.1	342.0	200.1	4:54.0
12:24	-9-53.6	-8-51.5	-11-41.7	-9 -5.3	-10-47.5	-8-58.3	1.0493	53.0	337.2	199.0	4:50.5
12:30	-9-58.0	-10-48.1	-11-46.2	-9 -1.6	-10-52.0	-10-54.7	1.0490	51.5	332.8	197.1	4:45.6
12:36	-10 -9.0	-12-49.3	-11-56.9	-11 -1.8	-11 -2.8	-12-55.8	1.0487	49.7	328.4	194.7	4:39.3
12:42	-10-26.9	-14-56.6	-12-14.5	-13 -2.7	-11-20.6	-15 -3.0	1.0483	47.5	324.6	191.5	4:31.7
12:48	-10-52.6	-17-11.7	-12-39.9	-15-10.0	-11-46.1	-17-18.3	1.0477	44.9	321.1	187.9	4:22.7
12:54	-11-27.1	-19-37.1	-13-14.0	-17-25.6	-12-20.4	-19-44.2	1.0471	42.0	317.8	183.6	4:12.4
13: 0	-12-12.0	-22-16.2	-13-58.6	-22-32.9	-13 -5.1	-22-24.1	1.0463	38.7	314.9	178.9	4: 0.7
13: 6	-13 -9.8	-25-14.1	-14-56.3	-25-33.9	-14 -2.8	-25-23.5	1.0454	34.8	312.1	173.5	3:47.6
13:12	-14-24.4	-28-39.5	-16-11.5	-29 -4.6	-15-17.7	-28-51.3	1.0443	30.4	309.5	167.5	3:32.7
13:18	-16 -3.8	-32-48.5	-17-52.9	-33-23.5	-16-58.0	-33 -5.0	1.0428	25.0	306.8	160.6	3:15.7
13:24	-18-26.9	-38-21.6	-20-23.2	-39-20.1	-19-24.3	-38-49.0	1.0408	18.0	303.8	151.9	2:54.4
LIMITS	-28-11.0	-55-24.8	-27-20.1	-55 -1.7	-26-38.1	-54-52.3	1.0354	0.0	296.5	131.6	2:10.7

Table 34

ANNULAR SOLAR ECLIPSE OF 14 DEC 2001

SAROS 132   Delta T = 66.3 Sec

UNIVERSAL TIME	NORTHERN LIMIT LATITUDE	LONGITUDE	SOUTHERN LIMIT LATITUDE	LONGITUDE	CENTER LINE LATITUDE	LONGITUDE	DIAMETER RATIO	SUN ALT	SUN AZ	PATH WIDTH	DURATION ANNULARITY
LIMITS	30 48.1	-176 -2.1	29 21.8	-175 -22.1	29 54.4	-176 -5.2	0.9544	0.0	117.5	171.0	3: 7.7
19:12	25 57.5	174 3.7	23 58.6	173 26.2	24 58.2	173 42.0	0.9574	11.2	122.5	158.6	3:11.3
19:18	21 6.5	165 25.1	19 36.9	165 33.9	20 21.1	165 28.9	0.9600	21.2	126.7	148.2	3:15.6
19:24	18 2.7	160 24.9	16 42.4	160 43.4	17 22.2	160 33.9	0.9616	27.8	129.4	142.0	3:19.0
19:30	15 39.0	156 40.3	14 24.0	157 2.5	15 1.2	156 51.2	0.9628	33.1	131.7	137.5	3:22.2
19:36	13 38.9	153 36.4	12 27.2	154 0.3	13 2.8	153 48.2	0.9638	37.7	133.9	134.1	3:25.2
19:42	11 54.9	150 58.4	10 45.7	151 23.0	11 20.1	151 10.6	0.9647	41.8	136.2	131.3	3:28.1
19:48	10 23.2	148 38.3	9 15.7	149 3.1	9 49.3	148 50.6	0.9654	45.5	138.5	129.1	3:30.9
19:54	9 1.4	146 31.1	7 55.0	146 55.7	8 28.0	146 43.4	0.9659	48.8	141.1	127.4	3:33.7
20: 0	7 47.7	144 33.7	6 42.3	144 57.9	7 14.8	144 45.7	0.9665	51.9	144.0	126.1	3:36.4
20: 6	6 41.2	142 43.6	5 36.4	143 7.3	6 8.6	142 55.4	0.9669	54.7	147.2	125.2	3:38.9
20:12	5 41.0	140 59.2	4 36.5	141 22.2	5 8.6	141 10.7	0.9672	57.2	150.9	124.5	3:41.4
20:18	4 46.5	139 19.2	3 42.2	139 41.4	4 14.2	139 30.3	0.9675	59.4	155.1	124.1	3:43.7
20:24	3 57.3	137 42.4	2 53.0	138 3.8	3 25.0	137 53.1	0.9677	61.4	159.8	124.0	3:45.9
20:30	3 13.1	136 8.1	2 8.8	136 28.6	2 40.7	136 18.3	0.9679	63.1	165.1	124.0	3:47.9
20:36	2 33.5	134 35.4	1 28.8	134 54.9	2 1.0	134 45.1	0.9680	64.4	171.1	124.3	3:49.6
20:42	1 58.5	133 3.8	0 53.4	133 22.1	1 25.8	133 12.8	0.9681	65.3	177.5	124.7	3:51.1
20:48	1 28.0	131 32.2	0 22.3	131 49.7	0 55.0	131 40.9	0.9681	65.8	184.2	125.2	3:52.3
20:54	1 1.7	130 0.6	0 -4.5	130 17.0	0 28.5	130 8.8	0.9681	65.9	191.1	125.8	3:53.3
21: 0	0 39.9	128 28.2	0-27.0	128 43.5	0 6.3	128 35.8	0.9680	65.5	197.8	126.5	3:53.9
21: 6	0 22.4	126 54.5	0-45.3	127 8.6	0-11.6	127 1.6	0.9679	64.7	204.2	127.2	3:54.1
21:12	0 9.3	125 19.0	0-59.1	125 31.9	0-25.4	125 25.4	0.9677	63.5	210.0	128.0	3:54.0
21:18	0 0.9	123 40.9	-1 -8.4	123 52.5	0-33.9	123 46.7	0.9674	61.9	215.3	128.9	3:53.6
21:24	0 -2.7	121 59.5	-1-13.0	122 10.0	0-38.0	122 4.8	0.9672	60.1	219.8	129.8	3:52.7
21:30	0 -1.3	120 14.2	-1-12.6	120 23.5	0-37.1	120 18.8	0.9668	57.9	223.8	130.8	3:51.6
21:36	0 5.6	118 23.8	-1 -6.8	118 32.0	0-30.8	118 27.9	0.9664	55.4	227.2	131.9	3:50.0
21:42	0 18.4	116 27.4	0-55.3	116 34.5	0-18.6	116 30.9	0.9659	52.7	230.1	133.1	3:48.2
21:48	0 37.8	114 23.3	0-37.3	114 29.4	0 -0.1	114 26.4	0.9653	49.7	232.5	134.6	3:46.0
21:54	1 4.7	112 9.6	0-12.1	112 14.9	0 26.1	112 12.3	0.9647	46.5	234.6	136.3	3:43.5
22: 0	1 40.4	109 43.6	0 21.6	109 48.4	1 0.8	109 46.1	0.9639	42.9	236.4	138.4	3:40.7
22: 6	2 26.6	107 1.3	1 5.5	107 6.0	1 45.9	107 3.7	0.9631	39.5	237.9	140.9	3:37.6
22:12	3 26.6	103 56.3	2 44.3	104 1.6	2 44.3	103 59.1	0.9621	34.6	239.2	143.9	3:34.2
22:18	4 45.4	100 17.4	3 17.1	100 24.9	4 1.0	100 21.4	0.9608	29.5	240.5	147.9	3:30.5
22:24	6 34.2	95 40.6	4 59.3	95 54.2	5 46.3	95 47.9	0.9593	23.4	241.7	153.1	3:26.3
22:30	9 27.5	88 50.1	7 37.1	89 26.1	8 31.4	89 9.5	0.9570	15.0	243.1	161.1	3:21.0
LIMITS	14 58.3	76 19.2	13 26.7	75 48.5	14 4.4	76 22.2	0.9531	0.0	246.0	176.4	3:13.6

Table 35

ANNULAR SOLAR ECLIPSE OF 10 JUN 2002

SAROS 137   Delta T = 66.7 Sec

UNIVERSAL TIME	NORTHERN LIMIT LATITUDE	LONGITUDE	SOUTHERN LIMIT LATITUDE	LONGITUDE	CENTER LINE LATITUDE	LONGITUDE	DIAMETER RATIO	SUN ALT	SUN AZ	PATH WIDTH	DURATION ANNULARITY
LIMITS	1 37.8	-120-29.6	1 1.4	-120-54.0	1 20.4	-120-43.0	0.9796	0.0	67.0	78.2	1:12.6
22: 0	8 50.7	-135-17.0	8 40.5	-136 -4.7	8 45.6	-135-41.1	0.9846	17.9	68.5	59.5	1: 2.2
22: 6	12 21.2	-141-18.3	12 8.5	-141-50.9	12 14.8	-141-34.7	0.9867	25.9	70.1	51.1	0:56.9
22:12	15 2.8	-145-36.9	14 50.1	-146 -2.4	14 56.4	-145-49.7	0.9883	32.1	71.9	44.9	0:52.6
22:18	17 19.2	-149 -8.9	17 6.9	-149-27.8	17 13.0	-149-17.4	0.9895	37.3	73.8	39.9	0:48.8
22:24	19 19.2	-152 -8.1	19 7.4	-152-25.8	19 13.3	-152-16.9	0.9906	41.9	75.9	35.6	0:45.5
22:30	21 7.1	-154-50.4	20 56.0	-155 -5.5	21 1.5	-154-58.0	0.9915	46.2	78.2	32.0	0:42.4
22:36	22 45.5	-157-19.7	22 35.0	-157-32.8	22 40.2	-157-28.2	0.9923	50.1	80.7	28.8	0:39.6
22:42	24 16.0	-159-39.7	24 6.3	-159-51.1	24 11.0	-159-45.4	0.9930	53.8	83.6	26.1	0:37.0
22:48	25 39.7	-161-53.1	25 30.3	-162 -3.0	25 35.0	-161-58.1	0.9936	57.2	86.8	23.7	0:34.6
22:54	26 57.3	-164 -1.8	26 48.4	-164-10.5	26 52.9	-164 -8.1	0.9941	60.5	90.4	21.6	0:32.4
23: 0	28 9.5	-166 -7.3	28 1.1	-166-14.9	28 5.3	-166-11.1	0.9946	63.6	94.7	19.8	0:30.5
23: 6	29 18.6	-168-10.7	29 8.5	-168-17.4	29 12.5	-168-14.1	0.9950	66.6	99.6	18.3	0:28.7
23:12	30 18.8	-170-13.1	30 11.1	-170-19.0	30 15.0	-170-16.0	0.9953	69.4	105.6	17.0	0:27.2
23:18	31 16.4	-172-15.2	31 9.0	-172-20.3	31 12.7	-172-17.8	0.9956	71.9	113.0	15.9	0:25.9
23:24	32 9.6	-174-17.8	32 2.4	-174-22.2	32 6.0	-174-20.0	0.9958	74.2	122.1	15.0	0:24.8
23:30	32 58.2	-176-21.3	32 51.2	-176-25.2	32 54.7	-176-23.3	0.9960	76.1	133.6	14.3	0:23.9
23:36	33 42.5	-178-26.4	33 35.6	-178-29.8	33 39.1	-178-28.1	0.9961	77.5	147.7	13.8	0:23.3
23:42	34 22.4	179 26.5	34 15.5	179 23.5	34 18.9	179 25.0	0.9962	78.1	163.9	13.5	0:22.8
23:48	34 57.7	177 18.8	34 50.8	177 14.4	34 54.3	177 15.6	0.9963	78.0	180.7	13.3	0:22.6
23:54	35 28.5	175 4.3	35 21.4	175 2.2	35 25.0	175 3.2	0.9962	77.1	196.2	13.4	0:22.7
0: 0	35 54.5	172 48.4	35 47.3	172 46.8	35 50.9	172 47.8	0.9962	75.6	209.4	13.6	0:23.0
0: 6	36 15.7	170 28.7	36 8.2	170 27.6	36 11.9	170 28.2	0.9961	73.6	220.1	14.0	0:23.5
0:12	36 31.7	168 4.9	36 23.9	168 4.3	36 27.8	168 4.6	0.9959	71.2	228.9	14.6	0:24.3
0:18	36 42.4	165 36.3	36 34.1	165 36.3	36 38.3	165 36.3	0.9957	68.6	236.2	15.3	0:25.3
0:24	36 47.4	163 2.6	36 38.6	163 3.3	36 43.0	163 2.9	0.9954	65.8	242.4	16.3	0:26.6
0:30	36 46.3	160 23.0	36 36.9	160 24.5	36 41.6	160 23.8	0.9951	62.8	247.8	17.5	0:28.0
0:36	36 38.7	157 36.8	36 28.6	157 39.3	36 33.7	157 38.0	0.9947	59.6	252.7	19.0	0:29.7
0:42	36 24.0	154 42.9	36 13.1	154 46.6	36 18.5	154 44.8	0.9942	56.3	257.1	20.7	0:31.7
0:48	36 1.3	151 40.1	35 49.6	151 45.3	35 55.5	151 42.7	0.9937	52.8	261.3	22.7	0:33.8
0:54	35 29.8	148 26.5	35 17.1	148 33.5	35 23.4	148 30.0	0.9931	49.1	265.2	25.1	0:36.2
1: 0	34 47.9	144 59.5	34 34.3	145 8.7	34 41.1	145 4.1	0.9923	45.1	269.0	27.9	0:38.9
1: 6	33 53.7	141 15.1	33 39.2	141 27.2	33 46.4	141 21.2	0.9915	40.7	272.7	31.2	0:41.8
1:12	32 44.0	137 6.8	32 28.6	137 22.6	32 36.3	137 14.8	0.9905	36.0	276.4	35.2	0:45.0
1:18	31 13.2	132 22.8	30 57.1	132 44.0	31 5.2	132 33.5	0.9892	30.5	280.2	40.1	0:48.7
1:24	29 9.8	126 37.8	28 53.8	127 7.8	29 1.7	126 53.0	0.9876	24.0	284.1	46.5	0:53.0
1:30	25 57.5	118 30.9	25 45.6	119 22.3	25 51.6	118 57.0	0.9853	15.1	288.7	55.8	0:58.6
LIMITS	20 4.8	104 37.5	19 31.6	105 0.5	19 51.2	104 55.9	0.9811	0.0	294.7	72.3	1: 7.0

Table 36

TOTAL SOLAR ECLIPSE OF 4 DEC 2002

SAROS 142   Delta T = 67.1 Sec

UNIVERSAL TIME	NORTHERN LIMIT LATITUDE	LONGITUDE	SOUTHERN LIMIT LATITUDE	LONGITUDE	CENTER LINE LATITUDE	LONGITUDE	DIAMETER RATIO	SUN ALT	SUN AZ	PATH WIDTH	DURATION TOTALITY
LIMITS	-3-49.3	1 39.2	-4 -3.8	1 48.7	-3-57.6	1 41.3	1.0081	0.0	112.2	31.3	0:26.1
5:54	-9-47.3	-11 -5.2	-9-50.2	-10-14.9	-9-48.9	-10-40.1	1.0125	14.4	110.6	48.6	0:44.0
6: 0	-13-53.8	-18-14.5	-14 -5.1	-17-31.2	-13-59.5	-17-52.9	1.0152	23.8	108.4	59.2	0:57.0
6: 6	-16-56.4	-23 -1.3	-17-12.2	-22-19.6	-17 -4.4	-22-40.5	1.0170	30.5	106.1	66.0	1: 6.8
6:12	-19-30.1	-28-49.1	-19-49.3	-26 -7.9	-19-39.8	-26-28.6	1.0184	36.0	103.6	71.0	1:15.1
6:18	-21-45.9	-30 -3.8	-22 -7.8	-29-22.9	-21-56.9	-29-43.4	1.0195	40.8	100.9	74.8	1:22.3
6:24	-23-48.8	-32-57.6	-24-13.2	-32-17.0	-24 -1.1	-32-37.4	1.0204	45.2	97.9	77.8	1:28.8
6:30	-25-41.8	-35-37.5	-26 -8.5	-34-57.3	-25-55.3	-35-17.5	1.0212	49.1	94.5	80.2	1:34.7
6:36	-27-26.7	-38 -8.0	-27-55.6	-37-28.3	-27-41.2	-37-48.2	1.0219	52.8	90.7	82.1	1:40.0
6:42	-29 -4.7	-40-32.2	-29-35.5	-39-53.1	-29-20.2	-40-12.7	1.0225	56.2	86.4	83.7	1:44.7
6:48	-30-36.5	-42-52.3	-31 -9.2	-42-14.0	-30-52.9	-42-33.2	1.0230	59.3	81.4	84.9	1:48.9
6:54	-32 -2.8	-45-10.2	-32-37.2	-44-33.0	-32-20.0	-44-51.6	1.0234	62.2	75.7	85.8	1:52.7
7: 0	-33-23.8	-47-27.2	-33-59.9	-46-51.3	-33-41.9	-47 -9.3	1.0237	64.9	69.0	86.5	1:55.8
7: 6	-34-39.9	-49-44.6	-35-17.6	-49-10.3	-34-58.8	-49-27.5	1.0240	67.2	61.2	86.9	1:58.5
7:12	-35-51.2	-52 -3.4	-36-30.4	-51-31.0	-36-10.8	-51-47.3	1.0242	69.2	52.1	87.2	2: 0.6
7:18	-36-57.7	-54-24.8	-37-38.3	-53-54.4	-37-18.0	-54 -9.6	1.0243	70.7	41.6	87.3	2: 2.2
7:24	-37-59.6	-56-49.0	-38-41.3	-56-21.3	-38-20.5	-56-35.2	1.0244	71.7	29.9	87.2	2: 3.3
7:30	-38-56.6	-59-17.4	-39-39.4	-58-52.5	-39-18.0	-59 -5.0	1.0244	72.1	17.4	87.0	2: 3.7
7:36	-39-48.7	-61-50.5	-40-32.4	-61-28.7	-40-10.5	-61-39.7	1.0243	71.9	4.7	86.6	2: 3.6
7:42	-40-35.6	-64-29.1	-41-20.0	-64-10.7	-40-57.8	-64-19.9	1.0242	71.1	352.5	86.1	2: 2.9
7:48	-41-17.2	-67-13.8	-42 -2.1	-66-59.2	-41-39.7	-67 -8.5	1.0240	69.7	341.4	85.4	2: 1.6
7:54	-41-53.1	-70 -5.2	-42-38.2	-69-54.8	-42-15.7	-69-60.0	1.0238	67.9	331.3	84.5	1:59.8
8: 0	-42-23.0	-73 -4.2	-43 -8.0	-72-58.1	-42-45.5	-73 -1.1	1.0235	65.7	322.4	83.4	1:57.3
8: 6	-42-46.3	-76-11.5	-43-30.9	-76 -9.9	-43 -8.6	-76-10.8	1.0231	63.2	314.4	82.2	1:54.2
8:12	-43 -2.5	-79-27.9	-43-46.4	-79-31.1	-43-24.5	-79-29.4	1.0226	60.4	307.3	80.8	1:50.6
8:18	-43-10.9	-82-54.3	-43-53.8	-83 -2.4	-43-32.4	-82-58.3	1.0221	57.3	300.7	79.1	1:46.3
8:24	-43-10.7	-86-32.2	-43-52.1	-86-45.1	-43-31.4	-86-38.5	1.0215	54.0	294.6	77.2	1:41.5
8:30	-43 -0.8	-90-23.0	-43-40.2	-90-40.7	-43-20.6	-90-31.7	1.0208	50.4	288.8	74.9	1:36.0
8:36	-42-39.7	-94-29.0	-43-16.7	-94-51.5	-42-58.3	-94-40.1	1.0200	46.6	283.3	72.4	1:30.0
8:42	-42 -5.5	-98-53.8	-42-39.5	-99-20.8	-42-22.6	-99 -7.1	1.0190	42.4	277.9	69.3	1:23.2
8:48	-41-15.1	-103-42.9	-41-45.5	-104-14.1	-41-30.4	-103-58.3	1.0179	37.8	272.5	65.7	1:15.7
8:54	-40 -3.7	-109 -6.0	-40-29.7	-109-41.4	-40-16.8	-109-23.5	1.0165	32.5	267.2	61.2	1: 7.3
9: 0	-38-21.9	-115-23.4	-38-42.1	-116 -2.9	-38-32.1	-115-43.0	1.0148	26.4	261.5	55.5	0:57.7
9: 6	-35-45.2	-123-29.4	-35-57.0	-124-15.0	-35-51.3	-123-52.1	1.0125	18.3	255.3	47.4	0:45.8
LIMITS	-28-25.2	-142-21.6	-28-37.3	-142-31.0	-28-32.5	-142-23.3	1.0069	0.0	244.4	26.8	0:22.3

Table 37

ANNULAR SOLAR ECLIPSE OF 31 MAY 2003

SAROS 147   Delta T = 67.5 Sec

UNIVERSAL TIME	NORTHERN LIMIT LATITUDE	LONGITUDE	SOUTHERN LIMIT LATITUDE	LONGITUDE	CENTER LINE LATITUDE	LONGITUDE	DIAMETER RATIO	SUN ALT	SUN AZ	PATH WIDTH	DURATION ANNULARITY
LIMITS	58 46.0	4 32.5	58 46.0	4 32.5	67 27.1	-33-13.0	0.9404	0.0	81.5	924.6	3:41.2
3:46	57 11.3	5 33.5	59 22.5	0 49.2	67 11.4	-32-53.2	0.9403	10.2	81.4	985.1	3:40.9
3:48	57 53.6	7 19.6	61 27.6	0-52.6	66 47.7	-32-23.2	0.9400	9.2	81.3	1093.1	3:40.3
3:50	58 35.2	9 8.5	63 4.3	-1-33.6	66 23.7	-23-14.2	0.9397	8.2	73.5	1227.5	3:39.8
3:52	59 15.8	11 0.4	64 28.3	-1-43.6	65 59.2	-10-12.6	0.9395	7.1	62.4	1399.8	3:39.2
3:54	59 55.5	12 55.6	65 44.3	-1-31.3	65 33.9	-2-14.7	0.9392	6.0	55.8	1629.6	3:38.6
3:56	60 34.2	14 54.0	66 54.6	-1-0.1	65 6.8	4 9.1	0.9389	4.9	50.6	1953.7	3:38.0
3:58	61 11.9	16 55.9	68 0.4	0-11.5	64 36.6	9 44.8	0.9386	3.7	46.1	2451.4	3:37.4
4: 0	61 48.4	19 1.4	69 2.1	0 54.3	63 58.2	14 57.7	0.9383	2.4	41.9	3338.3	3:36.7
4: 2	62 23.8	21 10.7	70 0.2	2 17.7	62 53.0	20 23.0	0.9378	0.5	37.5	5185.8	3:35.7
4: 4	62 57.9	23 23.8	70 54.7	3 59.8	64 47.8	20 19.6	0.9382	2.2	38.1	4989.8	3:36.5
4: 6	63 30.7	25 40.8	71 45.4	6 1.3	65 50.4	21 50.8	0.9384	2.7	37.2	4685.4	3:36.8
4: 8	64 2.1	28 2.0	72 32.2	8 24.0	66 33.3	24 1.9	0.9384	2.9	35.7	4509.9	3:36.9
4:10	64 32.1	30 27.2	73 14.5	11 9.5	66 59.1	26 46.3	0.9384	2.8	33.7	4482.1	3:36.8
4:12	65 0.5	32 56.7	73 51.8	14 19.0	67 3.5	30 5.5	0.9383	2.2	31.1	4639.7	3:36.5
4:14	65 27.3	35 30.4	74 23.1	17 53.5	66 18.5	34 26.9	0.9379	0.8	27.6	5158.9	3:35.7
4:16	65 52.3	38 8.2	74 47.6	21 53.1	68 14.5	35 7.9	0.9383	2.2	27.5	3241.6	3:36.4
4:18	66 15.5	40 50.2	75 3.9	26 16.7	70 19.1	35 34.9	0.9386	3.6	27.6	2314.9	3:37.1
4:20	66 36.8	43 36.3	75 10.4	31 1.6	72 23.8	35 58.3	0.9390	4.8	27.8	1823.1	3:37.7
4:22	66 56.1	46 26.2	75 5.3	36 3.0	74 38.2	35 56.6	0.9392	5.9	28.4	1510.0	3:38.3
4:24	67 13.3	49 19.8	74 46.1	41 14.8	77 11.1	34 51.5	0.9395	7.0	30.1	1290.7	3:38.9
4:26	67 28.3	52 16.8	74 9.1	46 29.2	80 19.1	30 29.7	0.9398	8.1	35.1	1127.5	3:39.4
4:28	67 41.1	55 17.0	73 7.2	51 39.0	85 4.9	-10-27.0	0.9401	9.1	78.5	1000.9	3:40.0
4:30	67 51.6	58 19.8	71 20.2	58 40.0	90 0.0	158 6.8	0.9404	10.1	270.3	899.7	3:40.6
LIMITS	67 57.3	60 24.3	67 57.3	60 24.3	90 0.0	158 29.3	0.9406	0.0	270.3	836.2	3:41.0

Table 38

TOTAL SOLAR ECLIPSE OF 23 NOV 2003

SAROS 152   Delta T = 67.9 Sec

UNIVERSAL TIME	NORTHERN LIMIT LATITUDE LONGITUDE		SOUTHERN LIMIT LATITUDE LONGITUDE		CENTER LINE LATITUDE LONGITUDE		DIAMETER RATIO	SUN ALT	SUN AZ	PATH WIDTH	DURATION TOTALITY
LIMITS	-50-57.2	-84-26.6	-54-11.5	-78-58.2	-53-39.0	-84-25.5	1.0337	0.0	122.7	515.3	1:36.6
22:28	-60-34.8	-97-23.6	-57-23.7	-84 -9.9	-59-27.8	-91-42.5	1.0360	8.8	115.3	546.1	1:47.3
22:30	-61-57.7	-98-10.3	-59-28.9	-86 -9.6	-61 -0.6	-92-42.9	1.0364	10.1	113.8	543.7	1:49.4
22:32	-63-16.8	-98-43.4	-61 -7.8	-87-13.4	-62-28.4	-93-21.8	1.0367	11.2	112.5	539.4	1:51.1
22:34	-64-32.9	-99 -4.5	-62-37.9	-87-46.5	-63-47.3	-93-43.8	1.0370	12.0	111.5	534.1	1:52.6
22:36	-65-46.8	-99-14.6	-64 -1.2	-87-57.3	-65 -4.5	-93-51.0	1.0372	12.8	110.7	528.4	1:53.8
22:38	-66-58.8	-99-13.7	-65-19.6	-87-49.2	-66-19.0	-93-44.3	1.0374	13.3	110.1	522.7	1:54.8
22:40	-68 -9.4	-99 -1.8	-66-34.1	-87-23.6	-67-31.1	-93-23.8	1.0376	13.8	109.7	517.1	1:55.6
22:42	-69-18.6	-98-38.1	-67-45.4	-86-40.4	-68-41.1	-92-49.0	1.0377	14.2	109.5	511.8	1:56.2
22:44	-70-26.6	-98 -1.4	-68-53.8	-85-39.0	-69-49.3	-91-58.6	1.0378	14.5	109.5	506.9	1:56.7
22:46	-71-33.5	-97-10.1	-69-59.3	-84-17.8	-70-55.7	-90-51.1	1.0378	14.6	109.8	502.4	1:57.0
22:48	-72-39.2	-96 -2.0	-71 -1.8	-82-34.9	-72 -0.1	-89-24.1	1.0379	14.7	110.4	498.4	1:57.1
22:50	-73-43.6	-94-34.0	-72 -1.0	-80-27.3	-73 -2.3	-87-34.4	1.0379	14.7	111.4	494.9	1:57.1
22:52	-74-46.6	-92-42.3	-72-56.4	-77-52.0	-74 -2.1	-85-18.5	1.0378	14.6	112.8	491.9	1:57.0
22:54	-75-47.6	-90-21.9	-73-47.2	-74-45.1	-74-58.7	-82-31.5	1.0378	14.5	114.7	489.4	1:56.7
22:56	-76-46.2	-87-28.5	-74-32.3	-71 -2.9	-75-51.4	-79 -8.1	1.0377	14.2	117.2	487.5	1:56.3
22:58	-77-41.3	-83-48.4	-75-10.2	-66-41.8	-76-39.0	-75 -2.5	1.0376	13.8	120.4	486.3	1:55.7
23: 0	-78-31.9	-79-18.5	-75-38.9	-61-39.8	-77-19.8	-70 -9.1	1.0374	13.4	124.4	485.6	1:54.9
23: 2	-79-15.9	-73-47.8	-75-55.6	-55-57.4	-77-51.6	-64-24.3	1.0372	12.8	129.3	485.6	1:54.0
23: 4	-79-50.9	-67 -9.1	-75-57.1	-49-39.1	-78-11.7	-57-48.4	1.0370	12.1	135.0	486.3	1:52.8
23: 6	-80-13.6	-59-21.9	-75-38.4	-42-54.2	-78-16.7	-50-28.8	1.0368	11.2	141.5	487.8	1:51.4
23: 8	-80-20.2	-50-37.9	-74-52.0	-35-55.6	-78 -2.5	-42-41.1	1.0364	10.2	148.4	490.2	1:49.6
23:10	-80 -6.6	-41-23.6	-73-19.9	-28-54.1	-77-23.9	-34-47.8	1.0360	8.9	155.4	493.5	1:47.5
LIMITS	-69-34.9	-8-51.4	-69 -5.9	-21-42.7	-71-19.4	-16 -3.1	1.0337	0.0	172.0	506.4	1:36.6

Table 39

ANN/TOT SOLAR ECLIPSE OF   8 APR 2005

SAROS 129   Delta T =   69.0 Sec

UNIVERSAL TIME	NORTHERN LIMIT LATITUDE LONGITUDE		SOUTHERN LIMIT LATITUDE LONGITUDE		CENTER LINE LATITUDE LONGITUDE		DIAMETER RATIO	SUN ALT	SUN AZ	PATH WIDTH	DURATION ANNULARITY
LIMITS	-47 -49.9	-175 -21.8	-48 -5.2	-175 -24.1	-47 -56.6	-175 -29.9	0.9923	0.0	78.5	28.3	0:28.3
18:54	-46 -42.0	177 20.8	-47 -4.0	178 10.8	-46 -52.7	177 44.3	0.9937	4.9	73.5	23.0	0:24.0
19: 0	-42 -14.5	160 13.6	-42 -19.6	160 14.3	-42 -17.0	160 14.0	0.9977	19.1	59.8	8.1	0: 9.6
19: 6	-39 -9.2	152 34.5	-39 -9.7	152 34.3	-39 -9.4	152 34.4	0.9997	26.6	53.3	1.0	0: 1.2
19:12	-36 -28.6	147 17.4	-36 -26.4	147 18.4	-36 -27.5	147 17.9	1.0012	32.5	48.4	4.2	0: 5.5T
19:18	-34 -2.5	143 13.4	-33 -58.4	143 15.7	-34 -0.5	143 14.5	1.0023	37.5	44.2	8.3	0:11.2T
19:24	-31 -46.8	139 54.5	-31 -41.1	139 58.3	-31 -43.8	139 56.4	1.0033	41.9	40.4	11.7	0:16.3T
19:30	-29 -38.4	137 6.7	-29 -31.8	137 11.7	-29 -35.1	137 9.2	1.0041	45.9	36.7	14.5	0:20.7T
19:36	-27 -36.4	134 41.2	-27 -29.0	134 47.3	-27 -32.7	134 44.2	1.0048	49.5	33.0	16.9	0:24.7T
19:42	-25 -39.5	132 32.4	-25 -31.4	132 39.5	-25 -35.5	132 36.0	1.0054	52.9	29.2	19.0	0:28.2T
19:48	-23 -47.0	130 36.5	-23 -38.4	130 44.5	-23 -42.7	130 40.5	1.0059	56.0	25.1	20.7	0:31.3T
19:54	-21 -58.3	128 50.6	-21 -49.3	128 59.3	-21 -53.8	128 55.0	1.0063	58.8	20.6	22.2	0:34.0T
20: 0	-20 -12.9	127 12.5	-20 -3.6	127 21.8	-20 -8.2	127 17.2	1.0066	61.4	15.6	23.5	0:36.3T
20: 6	-18 -30.5	125 40.5	-18 -20.9	125 50.3	-18 -25.7	125 45.4	1.0069	63.8	9.9	24.6	0:38.2T
20:12	-16 -50.8	124 13.1	-16 -41.0	124 23.3	-16 -45.9	124 18.2	1.0071	65.8	3.5	25.1	0:39.7T
20:18	-15 -13.6	122 49.2	-15 -3.6	122 59.6	-15 -8.6	122 54.4	1.0073	67.4	356.3	26.1	0:40.9T
20:24	-13 -38.6	121 27.6	-13 -28.6	121 38.3	-13 -33.6	121 33.0	1.0074	68.6	348.2	26.6	0:41.6T
20:30	-12 -5.8	120 7.6	-11 -55.7	120 18.3	-12 -0.7	120 12.9	1.0074	69.4	339.6	26.9	0:42.0T
20:36	-10 -34.9	118 48.1	-10 -24.8	118 58.9	-10 -29.9	118 53.5	1.0074	69.6	330.7	27.0	0:42.0T
20:42	-9 -5.9	117 28.5	-8 -55.9	117 39.2	-9 -0.9	117 33.8	1.0073	69.4	321.9	26.9	0:41.7T
20:48	-7 -38.7	116 7.8	-7 -28.8	116 18.3	-7 -33.8	116 13.1	1.0072	68.6	313.6	26.7	0:40.9T
20:54	-6 -13.3	114 45.3	-6 -3.5	114 55.6	-6 -8.5	114 50.4	1.0070	67.3	306.1	26.2	0:39.8T
21: 0	-4 -49.7	113 20.1	-4 -40.1	113 30.1	-4 -44.9	113 25.1	1.0067	65.6	299.6	25.5	0:38.2T
21: 6	-3 -27.7	111 51.3	-3 -18.4	112 0.9	-3 -23.1	111 56.1	1.0064	63.6	294.1	24.5	0:36.3T
21:12	-2 -7.5	110 17.9	-1 -58.6	110 26.9	-2 -3.1	110 22.4	1.0061	61.2	289.5	23.3	0:34.0T
21:18	0 -49.1	108 38.7	0 -40.7	108 47.0	0 -44.9	108 42.9	1.0057	58.6	285.6	21.7	0:31.2T
21:24	0 27.3	106 52.2	0 35.2	106 59.8	0 31.3	106 56.0	1.0051	55.7	282.5	19.9	0:28.1T
21:30	1 41.8	104 56.8	1 48.9	105 2.8	1 45.3	104 59.8	1.0048	52.6	280.0	17.7	0:24.5T
21:36	2 53.9	102 50.2	3 0.1	102 56.0	2 57.0	102 53.1	1.0039	49.2	278.0	15.1	0:20.5T
21:42	4 3.5	100 29.5	4 8.5	100 34.1	4 6.0	100 31.8	1.0031	45.6	276.5	12.1	0:16.1T
21:48	5 9.9	97 50.5	5 13.6	97 53.9	5 11.7	97 52.2	1.0022	41.6	275.4	8.6	0:11.1T
21:54	6 12.4	94 47.0	6 14.4	94 48.8	6 13.4	94 47.9	1.0012	37.1	274.7	4.5	0: 5.7T
22: 0	7 9.6	91 8.8	7 9.4	91 8.6	7 9.5	91 8.7	0.9999	32.1	274.4	0.4	0: 0.5
22: 6	7 58.9	86 35.7	7 55.8	86 32.8	7 57.4	86 34.3	0.9983	26.1	274.5	6.3	0: 7.5
22:12	8 33.6	80 17.2	8 26.2	80 9.1	8 29.9	80 13.1	0.9962	18.4	275.2	14.3	0:16.1
LIMITS	7 43.9	63 4.6	7 25.5	63 4.6	7 34.9	63 5.8	0.9909	0.0	277.6	33.6	0:33.6

Table 40

ANNULAR SOLAR ECLIPSE OF 3 OCT 2005

SAROS 134 Delta T = 69.4 Sec

UNIVERSAL TIME	NORTHERN LIMIT LATITUDE	LONGITUDE	SOUTHERN LIMIT LATITUDE	LONGITUDE	CENTER LINE LATITUDE	LONGITUDE	DIAMETER RATIO	SUN ALT	SUN AZ	PATH WIDTH	DURATION ANNULARITY
LIMITS	49 12.1	38 54.9	47 11.5	38 55.8	48 9.5	38 26.9	0.9428	0.0	96.7	222.0	3:52.9
8:48	45 50.1	17 5.7	43 41.4	16 9.8	44 44.6	16 33.7	0.9470	16.6	113.7	200.3	4: 3.7
8:54	42 58.8	7 37.7	41 7.3	7 46.2	42 2.4	7 40.8	0.9490	24.5	121.3	190.9	4: 9.0
9: 0	40 29.1	1 30.8	38 48.7	2 4.3	39 38.5	1 47.1	0.9505	30.5	126.7	184.5	4:12.9
9: 6	38 11.6	-3 -3.2	36 39.5	-2 -16.7	37 25.2	-2 -40.1	0.9516	35.6	131.2	179.6	4:16.0
9:12	36 2.5	-6 -41.8	34 37.0	-5 -47.6	35 19.4	-6 -14.8	0.9528	40.0	135.2	175.6	4:18.6
9:18	33 59.8	-9 -43.2	32 39.7	-8 -44.2	33 19.5	-9 -13.7	0.9534	44.0	139.0	172.3	4:20.8
9:24	32 2.3	-12 -18.0	30 46.7	-11 -15.8	31 24.3	-11 -46.9	0.9541	47.6	142.6	169.5	4:22.6
9:30	30 9.1	-14 -32.8	28 57.3	-13 -28.6	29 33.1	-14 -0.6	0.9547	51.0	146.3	167.2	4:24.2
9:36	28 19.7	-16 -32.2	27 11.1	-15 -26.7	27 45.2	-15 -59.4	0.9553	54.2	150.1	165.3	4:25.6
9:42	26 33.4	-18 -19.6	25 27.6	-17 -13.2	26 0.3	-17 -46.3	0.9558	57.1	154.2	163.7	4:26.8
9:48	24 50.0	-19 -57.5	23 46.4	-18 -50.4	24 18.0	-19 -23.9	0.9562	59.8	158.7	162.5	4:27.8
9:54	23 9.0	-21 -27.7	22 7.4	-20 -20.2	22 38.1	-20 -53.9	0.9565	62.3	163.6	161.5	4:28.7
10: 0	21 30.4	-22 -51.9	20 30.4	-21 -44.1	21 0.2	-22 -17.9	0.9568	64.5	169.1	160.9	4:29.4
10: 6	19 53.8	-24 -11.3	18 55.0	-23 -3.2	19 24.2	-23 -37.2	0.9571	66.5	175.4	160.5	4:30.0
10:12	18 19.0	-25 -27.0	17 21.3	-24 -18.7	17 50.0	-24 -52.8	0.9573	68.1	182.5	160.4	4:30.5
10:18	16 46.0	-26 -39.9	15 49.0	-25 -31.5	16 17.3	-26 -5.6	0.9574	69.4	190.3	160.6	4:30.9
10:24	15 14.6	-27 -51.0	14 18.0	-26 -42.3	14 46.1	-27 -16.8	0.9575	70.3	198.8	161.1	4:31.2
10:30	13 44.6	-29 -0.9	12 48.3	-27 -52.0	13 16.3	-28 -26.4	0.9576	70.7	207.8	161.8	4:31.4
10:36	12 16.0	-30 -10.4	11 19.7	-29 -1.3	11 47.7	-29 -35.8	0.9576	70.6	216.8	162.8	4:31.4
10:42	10 48.7	-31 -20.3	9 52.2	-30 -10.8	10 20.3	-30 -45.5	0.9576	70.0	225.3	164.1	4:31.4
10:48	9 22.7	-32 -31.2	8 25.7	-31 -21.3	8 54.0	-31 -56.2	0.9575	68.9	233.2	165.6	4:31.2
10:54	7 57.8	-33 -43.9	7 0.2	-32 -33.5	7 28.8	-33 -8.6	0.9574	67.5	240.1	167.3	4:30.9
11: 0	6 34.1	-34 -59.0	5 35.6	-33 -48.1	6 4.6	-34 -23.4	0.9573	65.7	246.1	169.3	4:30.5
11: 6	5 11.5	-36 -17.5	4 11.9	-35 -5.9	4 41.4	-35 -41.6	0.9570	63.6	251.2	171.4	4:29.9
11:12	3 50.0	-37 -40.3	2 49.1	-36 -27.9	3 19.3	-37 -4.0	0.9568	61.2	255.4	173.7	4:29.1
11:18	2 29.8	-39 -8.4	1 27.2	-37 -55.2	1 58.1	-38 -31.7	0.9565	58.6	258.9	176.1	4:28.1
11:24	1 10.4	-40 -43.1	0 6.1	-39 -28.8	0 38.0	-40 -5.9	0.9561	55.8	261.8	178.7	4:26.9
11:30	0 -7.6	-42 -26.1	-1 -13.9	-41 -10.5	0 -41.1	-41 -48.1	0.9557	52.8	264.1	181.3	4:25.5
11:36	-1 -24.3	-44 -19.1	-2 -33.0	-43 -2.1	-1 -58.9	-43 -40.4	0.9551	49.6	266.0	183.9	4:23.7
11:42	-2 -39.6	-46 -24.9	-3 -50.9	-45 -6.0	-3 -15.5	-45 -45.2	0.9545	46.1	267.5	186.6	4:21.7
11:48	-3 -53.1	-48 -47.1	-5 -7.4	-47 -25.8	-4 -30.6	-48 -6.2	0.9538	42.3	268.5	189.2	4:19.2
11:54	-5 -4.5	-51 -30.9	-6 -22.3	-50 -6.4	-5 -43.7	-50 -48.3	0.9530	38.1	269.2	192.0	4:16.4
12: 0	-6 -13.2	-54 -45.1	-7 -35.0	-53 -15.7	-6 -54.5	-53 -60.0	0.9520	33.5	269.6	194.9	4:13.0
12: 6	-7 -17.8	-58 -45.2	-8 -44.7	-57 -7.6	-8 -1.6	-57 -55.8	0.9508	28.1	269.6	198.1	4: 8.9
12:12	-8 -15.6	-64 -7.1	-9 -49.2	-62 -12.4	-9 -2.9	-63 -8.5	0.9493	21.5	269.1	202.1	4: 3.8
12:18	-8 -54.9	-73 -16.3	-10 -41.6	-70 -11.8	-9 -49.3	-71 -37.9	0.9468	11.6	267.7	208.7	3:56.2
LIMITS	-8 -35.7	-82 -53.2	-10 -34.6	-82 -44.3	-9 -36.9	-82 -26.3	0.9438	0.0	265.8	217.2	3:48.2

Table 41

TOTAL SOLAR ECLIPSE OF 29 MAR 2006

SAROS 139   Delta T = 69.8 Sec

UNIVERSAL TIME	NORTHERN LIMIT LATITUDE	LONGITUDE	SOUTHERN LIMIT LATITUDE	LONGITUDE	CENTER LINE LATITUDE	LONGITUDE	DIAMETER RATIO	SUN ALT	SUN AZ	PATH WIDTH	DURATION TOTALITY
LIMITS	-5-44.1	37 21.8	-6-52.9	37 8.1	-6-17.8	37 4.1	1.0349	0.0	86.5	128.8	1:53.8
8:36	-5-29.7	34 23.8	-6 -9.2	29 22.7	-5-47.4	31 25.8	1.0368	6.0	85.9	136.7	2: 4.4
8:42	-2-55.9	20 17.9	-3-50.5	18 21.7	-3-23.0	19 18.7	1.0410	19.7	85.5	156.1	2:32.0
8:48	0-53.3	14 14.0	-1-50.5	12 34.3	-1-21.8	13 23.3	1.0432	27.2	85.5	166.3	2:48.3
8:54	1 2.1	9 54.1	0 3.4	8 20.8	0 32.9	9 7.1	1.0448	33.1	86.3	173.5	3: 1.2
9: 0	2 53.5	6 27.4	1 54.0	4 57.5	2 23.9	5 42.2	1.0461	38.1	87.6	178.8	3:12.2
9: 6	4 42.5	3 34.1	3 42.4	2 6.2	4 12.5	2 49.9	1.0472	42.6	89.3	182.6	3:21.6
9:12	6 29.8	1 3.6	5 29.1	0-23.0	5 59.5	0 20.1	1.0481	46.5	91.4	185.4	3:30.0
9:18	8 15.8	-1-10.2	7 14.7	-2-35.9	7 45.3	-1-53.2	1.0488	50.1	94.1	187.3	3:37.3
9:24	10 0.9	-3-11.7	8 59.4	-4-36.7	9 30.2	-3-54.4	1.0495	53.4	97.3	188.4	3:43.7
9:30	11 45.5	-5 -3.9	10 43.5	-6-28.2	11 14.5	-5-46.2	1.0500	56.4	101.1	188.9	3:49.2
9:36	13 29.6	-6-49.0	12 27.2	-8-12.7	12 58.4	-7-31.0	1.0505	59.1	105.7	188.9	3:53.9
9:42	15 13.6	-8-28.8	14 10.6	-9-52.0	14 42.1	-9-10.6	1.0508	61.5	111.1	188.5	3:57.9
9:48	16 57.5	-10 -5.0	15 53.8	-11-27.6	16 25.6	-10-46.4	1.0511	63.6	117.3	187.9	4: 1.1
9:54	18 41.4	-11-38.8	17 37.1	-13 -0.8	18 9.2	-12-19.9	1.0513	65.2	124.5	187.0	4: 3.6
10: 0	20 25.7	-13-11.4	19 20.5	-14-32.7	19 53.1	-13-52.2	1.0514	66.4	132.6	185.9	4: 5.3
10: 6	22 10.3	-14-44.0	21 4.2	-16 -4.5	21 37.2	-15-24.4	1.0515	67.1	141.3	184.7	4: 6.4
10:12	23 55.3	-16-17.8	22 48.2	-17-37.2	23 21.7	-16-57.6	1.0515	67.3	150.4	183.3	4: 6.7
10:18	25 41.0	-17-53.8	24 32.7	-19-12.1	25 6.8	-18-33.1	1.0513	67.0	159.5	181.9	4: 6.3
10:24	27 27.4	-19-33.4	26 17.8	-20-50.2	26 52.5	-20-11.9	1.0511	66.1	168.3	180.3	4: 5.1
10:30	29 14.7	-21-17.8	28 3.8	-22-32.9	28 39.1	-21-55.4	1.0511	64.8	176.5	178.7	4: 3.3
10:36	31 2.9	-23 -8.5	29 50.2	-24-21.8	30 26.5	-23-45.2	1.0508	63.0	184.0	177.0	4: 0.6
10:42	32 52.2	-25 -7.5	31 37.8	-26-18.0	32 14.8	-25-42.8	1.0505	60.9	190.9	175.1	3:57.2
10:48	34 42.6	-27-16.7	33 26.4	-28-24.2	34 4.3	-27-50.5	1.0500	58.4	197.1	173.2	3:53.0
10:54	36 34.3	-29-39.0	35 16.1	-30-42.5	35 55.0	-30-10.7	1.0495	55.6	202.9	171.1	3:48.0
11: 0	38 27.3	-32-17.5	37 7.0	-33-18.1	37 47.0	-32-46.7	1.0489	52.5	208.3	168.9	3:42.1
11: 6	40 21.7	-35-16.8	38 59.2	-36 -9.1	39 40.3	-35-42.8	1.0482	49.1	213.5	166.4	3:35.3
11:12	42 17.5	-38-43.0	40 52.7	-39-27.2	41 34.9	-39 -4.9	1.0474	45.4	218.7	163.7	3:27.5
11:18	44 14.4	-42-44.9	42 47.4	-43-18.3	43 30.7	-43 -1.3	1.0464	41.3	224.0	160.7	3:18.6
11:24	46 12.1	-47-38.1	44 43.1	-47-54.7	45 27.3	-47-44.8	1.0452	36.8	229.7	157.2	3: 8.4
11:30	48 9.6	-53-39.4	46 38.9	-53-36.6	47 24.0	-53-37.0	1.0438	31.5	236.0	153.2	2:56.6
11:36	50 4.4	-61-40.3	48 33.1	-61 -3.5	49 18.5	-61-20.3	1.0420	25.3	243.7	148.2	2:42.5
11:42	51 48.7	-73-45.9	50 20.1	-71-58.2	51 4.4	-72-48.4	1.0395	16.9	254.2	141.3	2:24.2
LIMITS	52 6.9	-99 -0.3	51 0.0	-98-36.8	51 34.7	-98-26.9	1.0342	0.0	275.5	126.4	1:51.5

Table 42

ANNULAR SOLAR ECLIPSE OF 22 SEP 2006

SAROS 144   Delta T = 70.2 Sec

UNIVERSAL TIME	NORTHERN LIMIT LATITUDE	LONGITUDE	SOUTHERN LIMIT LATITUDE	LONGITUDE	CENTER LINE LATITUDE	LONGITUDE	DIAMETER RATIO	SUN ALT	SUN AZ	PATH WIDTH	DURATION ANNULARITY
LIMITS	6 41.6	59 27.0	3 46.3	59 58.2	5 14.5	59 5.0	0.9220	0.0	89.8	322.5	5:31.5
9:54	5 46.2	46 11.0	3 33.3	52 22.9	4 41.3	48 53.2	0.9247	11.4	90.7	318.0	5:48.3
10: 0	4 19.2	39 10.1	2 16.3	42 52.0	3 18.7	40 56.6	0.9270	20.8	91.0	315.0	6: 4.2
10: 6	2 52.6	34 35.6	0 56.0	37 40.5	1 55.2	36 5.8	0.9285	27.2	90.7	312.9	6:15.1
10:12	1 26.3	31 5.8	0-25.3	33 52.4	0 31.3	32 27.6	0.9296	32.3	90.0	310.8	6:24.0
10:18	0 0.1	28 14.4	-1-47.3	30 49.8	0-52.9	29 31.0	0.9306	36.8	89.0	308.3	6:31.4
10:24	-1-26.1	25 49.1	-3 -9.8	28 18.7	-2-17.2	27 2.0	0.9313	40.7	87.6	305.3	6:37.8
10:30	-2-52.2	23 42.5	-4-32.8	26 4.4	-3-41.8	24 52.7	0.9320	44.3	85.9	302.0	6:43.4
10:36	-4-18.4	21 50.2	-5-56.2	24 7.7	-5 -6.6	22 58.3	0.9326	47.5	83.9	298.4	6:48.2
10:42	-5-44.7	20 8.9	-7-20.1	22 22.8	-6-31.7	21 15.3	0.9331	50.5	81.5	294.5	6:52.4
10:48	-7-11.1	18 36.4	-8-44.5	20 47.4	-7-57.2	19 41.3	0.9335	53.3	78.6	290.5	6:56.0
10:54	-8-37.7	17 10.8	-10 -9.4	19 19.3	-9-23.0	18 14.6	0.9339	55.8	75.2	286.4	6:59.1
11: 0	-10 -4.7	15 50.8	-11-34.9	17 57.2	-10-49.2	16 53.5	0.9342	58.1	71.4	282.3	7: 1.8
11: 6	-11-31.9	14 35.1	-13 -1.1	16 39.7	-12-15.9	15 37.0	0.9345	60.2	66.9	278.4	7: 4.0
11:12	-12-59.5	13 22.8	-14-27.9	15 25.8	-13-43.1	14 23.9	0.9347	61.9	61.8	274.7	7: 5.8
11:18	-14-27.5	12 13.0	-15-55.4	14 14.5	-15-10.9	13 13.4	0.9349	63.4	56.1	271.2	7: 7.2
11:24	-15-56.0	11 4.9	-17-23.7	13 5.1	-16-39.3	12 4.6	0.9350	64.6	49.7	268.0	7: 8.3
11:30	-17-25.1	9 57.6	-18-53.0	11 56.8	-18 -8.5	10 56.9	0.9351	65.4	42.9	265.1	7: 9.0
11:36	-18-54.9	8 50.6	-20-23.2	10 48.8	-19-38.5	9 49.3	0.9351	65.9	35.6	262.6	7: 9.3
11:42	-20-25.3	7 43.1	-21-54.4	9 40.3	-21 -9.4	8 41.4	0.9352	65.9	28.2	260.4	7: 9.3
11:48	-21-56.6	6 34.4	-23-26.8	8 30.6	-22-41.2	7 32.2	0.9351	65.6	20.8	258.6	7: 9.0
11:54	-23-28.7	5 23.6	-25 -0.4	7 19.0	-24-14.1	6 21.0	0.9350	64.9	13.6	257.1	7: 8.3
12: 0	-25 -1.9	4 10.1	-26-35.5	6 4.4	-25-48.2	5 7.0	0.9349	63.8	6.8	256.1	7: 7.3
12: 6	-26-38.1	2 52.9	-28-12.0	4 46.1	-27-23.5	3 49.2	0.9348	62.4	0.5	255.4	7: 6.0
12:12	-28-11.5	1 31.0	-29-50.1	3 22.9	-29 -0.3	2 26.8	0.9346	60.7	354.6	255.1	7: 4.2
12:18	-29-48.3	0 3.2	-31-30.1	1 53.4	-30-38.7	0 58.0	0.9343	58.8	349.1	255.1	7: 2.1
12:24	-31-26.5	-1-31.8	-33-12.0	0 18.3	-32-18.7	0-38.1	0.9340	56.6	344.1	255.6	6:59.5
12:30	-33 -6.3	-3-15.8	-34-56.2	-1-30.5	-34 -0.7	-2-23.4	0.9337	54.1	339.3	256.4	6:56.5
12:36	-34-48.0	-5-10.9	-36-42.9	-3-29.2	-35-44.8	-4-20.2	0.9332	51.4	334.8	257.7	6:53.0
12:42	-36-31.7	-7-19.7	-38-32.3	-5-43.0	-37-31.3	-6-31.5	0.9328	48.5	330.5	259.4	6:49.0
12:48	-38-17.6	-9-45.7	-40-24.8	-8-16.1	-39-20.4	-9 -0.9	0.9322	45.3	326.2	261.7	6:44.4
12:54	-40 -8.0	-12-33.9	-42-20.9	-11-14.2	-41-12.5	-11-53.8	0.9316	41.8	321.8	264.5	6:39.0
13: 0	-41-57.2	-15-51.3	-44-21.0	-14-46.1	-43 -8.1	-15-18.0	0.9308	38.0	317.3	268.1	6:32.9
13: 6	-43-51.5	-19-48.4	-46-25.9	-19 -5.5	-45 -7.5	-19-25.6	0.9300	33.8	312.3	272.6	6:25.9
13:12	-45-49.4	-24-43.4	-48-36.5	-24-37.0	-47-11.4	-24-37.4	0.9289	28.9	306.7	278.4	6:17.6
13:18	-47-51.3	-31-11.2	-50-54.2	-32-13.0	-49-20.7	-31-35.8	0.9278	23.0	299.7	286.2	6: 7.5
13:24	-49-57.7	-40-42.4	-53-22.2	-44-45.7	-51-36.8	-42-21.8	0.9256	14.9	289.8	298.2	5:53.9
LIMITS	-51-58.5	-65-27.7	-54-53.0	-66-11.0	-53-24.9	-64-44.5	0.9221	0.0	271.0	322.5	5:31.3

Table 43

ANNULAR SOLAR ECLIPSE OF 7 FEB 2008

SAROS 121   Delta T = 71.3 Sec

UNIVERSAL TIME	NORTHERN LIMIT LATITUDE LONGITUDE	SOUTHERN LIMIT LATITUDE LONGITUDE	CENTER LINE LATITUDE LONGITUDE	DIAMETER RATIO	SUN ALT	SUN AZ	PATH WIDTH	DURATION ANNULARITY
LIMITS	-73-43.1 64 27.5	-71-48.3 80 57.8	-74-52.9 76 34.9	0.9609	0.0	207.7	558.3	2:13.9
3:30	-79-48.1 123 25.0	-75-11.2 93 23.8	-78-25.2 109 29.9	0.9630	9.5	238.0	478.4	2:13.2
3:32	-79 -5.3 130 31.8	-76 -0.8 103 25.1	-78 -7.9 117 50.2	0.9634	10.8	245.5	466.3	2:13.0
3:34	-78-14.3 136 8.3	-76 -4.7 111 40.1	-77-34.5 124 43.8	0.9637	11.9	251.8	457.1	2:12.8
3:36	-77-18.1 140 28.4	-75-45.1 118 31.5	-76-50.7 130 19.7	0.9639	12.8	256.3	450.0	2:12.7
3:38	-76-19.0 144 0.1	-75-11.4 124 10.7	-76 -0.3 134 50.2	0.9641	13.5	260.0	444.6	2:12.5
3:40	-75-18.2 146 44.5	-74-28.4 128 48.5	-75 -5.6 138 27.8	0.9643	14.2	262.7	440.7	2:12.4
3:42	-74-16.4 148 55.0	-73-39.4 132 35.5	-74 -8.2 141 23.1	0.9645	14.7	264.8	438.0	2:12.3
3:44	-73-14.2 150 39.0	-72-46.4 135 40.8	-73 -9.1 143 44.4	0.9646	15.1	266.3	436.4	2:12.2
3:46	-72-12.0 152 2.0	-71-50.6 138 11.8	-72 -9.0 145 38.6	0.9647	15.5	267.4	435.9	2:12.1
3:48	-71 -9.7 153 7.9	-70-53.0 140 14.8	-71 -8.2 147 10.6	0.9648	15.8	268.1	436.3	2:12.0
3:50	-70 -7.7 153 59.9	-69-54.0 141 54.6	-70 -7.0 148 24.4	0.9648	16.0	268.5	437.6	2:11.9
3:52	-69 -5.8 154 40.3	-68-54.1 143 14.9	-69 -5.6 149 23.1	0.9649	16.2	268.6	439.8	2:11.8
3:54	-68 -4.1 155 10.8	-67-53.4 144 18.8	-68 -4.1 150 8.9	0.9650	16.2	268.6	442.8	2:11.7
3:56	-67 -2.6 155 32.8	-66-52.1 145 8.5	-67 -2.5 150 43.8	0.9650	16.3	268.3	446.7	2:11.6
3:58	-66 -1.3 155 47.4	-65-50.3 145 45.7	-66 -0.8 151 8.9	0.9650	16.2	268.0	451.4	2:11.6
4: 0	-65 0.0 155 55.2	-64-48.0 146 11.5	-64-58.9 151 25.4	0.9650	16.1	267.5	456.8	2:11.5
4: 2	-63-58.8 155 57.0	-63-45.1 146 27.0	-63-56.9 151 33.9	0.9649	15.9	266.9	463.1	2:11.5
4: 4	-62-57.5 155 53.0	-62-41.5 146 32.8	-62-54.6 151 35.1	0.9649	15.6	266.1	470.2	2:11.4
4: 6	-61-56.1 155 43.6	-61-37.0 146 29.1	-61-51.8 151 29.2	0.9648	15.2	265.3	478.2	2:11.4
4: 8	-60-54.4 155 28.8	-60-31.5 146 15.9	-60-48.5 151 16.3	0.9647	14.8	264.4	486.9	2:11.4
4:10	-59-52.2 155 8.7	-59-24.5 145 52.5	-59-44.5 150 56.2	0.9646	14.3	263.5	498.3	2:11.4
4:12	-58-49.5 154 42.9	-58-15.7 145 17.7	-58-39.5 150 28.6	0.9644	13.7	262.4	506.5	2:11.4
4:14	-57-46.0 154 11.3	-57 -4.1 144 29.6	-57-33.1 149 52.8	0.9642	12.9	261.2	517.2	2:11.4
4:16	-56-41.4 153 33.2	-55-48.7 143 24.0	-56-25.0 149 7.4	0.9640	12.1	259.9	528.4	2:11.5
4:18	-55-35.4 152 47.6	-54-26.8 141 52.7	-55-14.3 148 10.4	0.9637	11.1	258.5	539.7	2:11.5
4:20	-54-27.5 151 53.2	-52-52.0 139 33.6	-54 -0.1 146 58.3	0.9634	9.8	257.0	550.8	2:11.6
LIMITS	-45-47.8 137 32.3	-50 -0.7 132 37.9	-48-44.9 138 14.4	0.9613	0.0	248.6	569.8	2:12.3

Table 44

TOTAL SOLAR ECLIPSE OF 1 AUG 2008

SAROS 128   Delta T = 71.7 Sec

UNIVERSAL TIME	NORTHERN LIMIT LATITUDE	LONGITUDE	SOUTHERN LIMIT LATITUDE	LONGITUDE	CENTER LINE LATITUDE	LONGITUDE	DIAMETER RATIO	SUN ALT	SUN AZ	PATH WIDTH	DURATION TOTALITY
LIMITS	68 43.1	105 15.2	67 48.1	100 55.2	68 49.2	102 0.4	1.0301	0.0	35.3	206.4	1:30.9
9:24	70 43.7	101 44.0	75 53.7	81 20.0	74 16.2	90 34.1	1.0321	7.1	46.5	213.6	1:41.2
9:28	78 36.7	83 7.0	79 50.5	63 12.8	79 26.3	73 12.2	1.0341	13.5	64.8	218.2	1:51.9
9:32	82 16.7	64 39.9	82 0.0	41 31.6	82 18.4	52 30.3	1.0352	17.5	86.8	220.3	1:58.9
9:36	84 19.9	37 31.0	82 45.8	16 36.3	83 38.5	25 29.7	1.0361	20.6	115.2	221.7	2: 4.5
9:40	84 40.7	3 51.0	82 20.7	-6-16.9	83 31.1	-2-16.7	1.0368	23.1	144.6	222.9	2: 9.1
9:44	83 37.3	-22-38.8	81 10.3	-23-27.4	82 22.7	-23-10.5	1.0374	25.3	167.2	224.0	2:12.9
9:48	81 54.9	-38-49.3	79 37.3	-35-27.2	80 45.4	-36-58.8	1.0379	27.1	182.9	225.0	2:16.2
9:52	79 59.7	-48-49.1	77 54.4	-43-57.4	78 56.8	-46 -9.6	1.0383	28.6	194.0	226.1	2:19.0
9:56	78 1.5	-55-32.0	76 7.6	-50-14.6	77 4.7	-52-41.4	1.0386	29.9	202.5	227.3	2:21.3
10: 0	76 3.7	-60-24.6	74 19.9	-55 -5.8	75 12.0	-57-35.5	1.0389	31.0	209.5	228.6	2:23.3
10: 4	74 7.3	-64-10.6	72 32.3	-58-59.8	73 20.2	-61-27.3	1.0391	31.9	215.4	229.9	2:24.8
10: 8	72 12.7	-67-14.0	70 45.4	-62-14.2	71 29.5	-64-37.7	1.0393	32.5	220.8	231.4	2:25.9
10:12	70 19.8	-69-48.6	68 59.3	-65 -0.5	69 40.1	-67-19.3	1.0394	33.0	225.6	233.0	2:26.7
10:16	68 28.5	-72 -3.4	67 14.2	-67-26.4	67 51.9	-69-40.4	1.0394	33.4	230.2	234.6	2:27.1
10:20	66 38.4	-74 -4.0	65 29.7	-69-37.2	66 4.8	-71-46.8	1.0394	33.5	234.4	236.3	2:27.2
10:24	64 49.2	-75-54.7	63 45.9	-71-37.0	64 18.2	-73-42.6	1.0394	33.5	238.5	238.2	2:27.0
10:28	63 0.8	-77-38.4	62 2.4	-73-28.8	62 32.2	-75-30.7	1.0393	33.3	242.4	240.0	2:26.4
10:32	61 12.7	-79-17.6	60 19.0	-75-15.0	60 46.5	-77-13.7	1.0392	32.9	246.2	242.0	2:25.5
10:36	59 24.6	-80-54.3	58 35.5	-76-57.5	59 0.7	-78-53.6	1.0390	32.3	249.9	243.9	2:24.2
10:40	57 36.1	-82-30.2	56 51.5	-78-38.3	57 14.5	-80-32.2	1.0388	31.5	253.4	245.8	2:22.6
10:44	55 46.8	-84 -7.2	55 6.7	-80-18.9	55 27.4	-82-11.1	1.0385	30.6	256.9	247.6	2:20.6
10:48	53 56.1	-85-47.1	53 20.7	-82 -1.1	53 39.2	-83-52.1	1.0382	29.4	260.3	249.2	2:18.3
10:52	52 3.5	-87-31.8	51 33.0	-83-46.8	51 49.0	-85-37.5	1.0377	28.0	263.6	250.5	2:15.6
10:56	50 8.1	-89-24.1	49 42.8	-85-38.3	49 56.3	-87-29.3	1.0373	26.4	266.9	251.3	2:12.4
11: 0	48 8.7	-91-27.3	47 49.3	-87-38.3	47 60.0	-89-30.8	1.0367	24.5	270.1	251.4	2: 8.7
11: 4	46 3.4	-93-46.9	45 51.2	-89-50.9	45 58.4	-91-46.6	1.0360	22.2	273.3	250.4	2: 4.5
11: 8	43 49.2	-96-31.7	43 46.3	-92-22.6	43 49.2	-94-24.2	1.0352	19.4	276.6	247.7	1:59.5
11:12	41 19.6	-100 -0.6	41 30.7	-95-24.8	41 27.3	-97-38.0	1.0341	16.0	280.0	242.3	1:53.5
11:16	38 14.8	-105-11.6	38 55.7	-99-24.1	38 40.1	-102 -5.6	1.0327	11.3	284.0	232.1	1:45.5
LIMITS	34 13.0	-114-35.1	32 43.4	-113-13.9	33 43.0	-113 -8.8	1.0294	0.0	291.3	201.2	1:28.8

Table 45

ANNULAR SOLAR ECLIPSE OF 26 JAN 2009

SAROS 131   Delta T = 72.1 Sec

UNIVERSAL TIME	NORTHERN LIMIT LATITUDE	LONGITUDE	SOUTHERN LIMIT LATITUDE	LONGITUDE	CENTER LINE LATITUDE	LONGITUDE	DIAMETER RATIO	SUN ALT	SUN AZ	PATH WIDTH	DURATION ANNULARITY
LIMITS	-33 -9.2	10 41.9	-35-57.9	12 48.7	-34-43.2	11 15.4	0.9140	0.0	112.4	362.1	5:44.2
6:12	-39-36.8	-11 -7.2	-41-33.7	-5-58.7	-40-38.0	-8-46.4	0.9185	18.4	98.9	334.3	6:17.4
6:18	-41 -4.1	-18-32.5	-43-27.9	-15 -5.9	-42-16.7	-16-56.2	0.9202	25.8	92.1	322.8	6:33.2
6:24	-41-53.3	-24-26.3	-44-28.0	-21-56.8	-43-10.7	-23-18.4	0.9215	31.5	86.3	314.3	6:46.1
6:30	-42-19.5	-29-27.9	-44-59.3	-27-39.8	-43-39.2	-28-37.4	0.9226	36.4	80.9	307.5	6:57.4
6:36	-42-29.7	-33-53.8	-45-11.6	-32-38.9	-43-50.3	-33-19.0	0.9235	40.8	75.7	301.9	7: 7.3
6:42	-42-27.5	-37-52.8	-45 -9.6	-37 -5.8	-43-48.1	-37-31.2	0.9242	44.7	70.7	297.3	7:16.1
6:48	-42-15.4	-41-30.2	-44-56.5	-41 -7.2	-43-35.5	-41-20.1	0.9249	48.4	65.7	293.3	7:24.0
6:54	-41-55.2	-44-49.7	-44-34.3	-44-47.5	-43-14.2	-44-49.6	0.9255	51.8	60.7	290.0	7:31.0
7: 0	-41-27.9	-47-53.9	-44 -4.4	-48-10.0	-42-45.7	-48 -2.5	0.9260	55.0	55.5	287.3	7:37.0
7: 6	-40-54.6	-50-44.8	-43-28.0	-51-16.9	-42-10.8	-51 -1.1	0.9264	58.0	50.2	285.0	7:42.2
7:12	-40-16.0	-53-24.0	-42-45.9	-54-10.1	-41-30.5	-53-47.1	0.9268	60.8	44.5	283.1	7:46.5
7:18	-39-32.7	-55-52.8	-41-58.9	-56-51.2	-40-45.4	-56-21.9	0.9272	63.4	38.4	281.6	7:50.0
7:24	-38-45.1	-58-12.3	-41 -7.5	-59-21.7	-39-56.0	-58-46.7	0.9274	65.8	31.7	280.5	7:52.6
7:30	-37-53.7	-60-23.7	-40-12.1	-61-42.6	-39 -2.6	-61 -2.8	0.9277	68.0	24.2	279.6	7:54.5
7:36	-36-58.8	-62-27.8	-39-13.2	-63-55.0	-38 -5.7	-63-10.9	0.9279	69.9	15.8	279.2	7:55.6
7:42	-36 -0.5	-64-25.3	-38-11.1	-66 0.0	-37 -5.5	-65-12.1	0.9280	71.4	6.4	279.2	7:58.0
7:48	-34-59.2	-66-17.2	-37 -5.9	-67-58.4	-36 -2.3	-67 -7.2	0.9282	72.6	355.9	279.2	7:55.7
7:54	-33-54.9	-68 -4.1	-35-57.9	-69-51.2	-34-56.2	-68-57.0	0.9282	73.2	344.7	279.6	7:54.7
8: 0	-32-47.8	-69-46.7	-34-47.1	-71-39.0	-33-47.3	-70-42.2	0.9283	73.4	333.2	280.4	7:53.1
8: 6	-31-38.0	-71-25.7	-33-33.8	-73-22.7	-32-35.7	-72-23.5	0.9283	73.0	321.9	281.6	7:50.9
8:12	-30-25.6	-73 -1.8	-32-18.0	-75 -3.0	-31-21.6	-74 -1.7	0.9282	72.1	311.4	283.1	7:48.1
8:18	-29-10.4	-74-35.5	-30-59.7	-76-40.7	-30 -4.8	-75-37.4	0.9281	70.8	302.1	284.9	7:44.9
8:24	-27-52.5	-76 -7.6	-29-38.8	-78-16.8	-28-45.5	-77-11.4	0.9280	69.0	294.1	287.0	7:41.1
8:30	-26-31.9	-77-38.9	-28-15.3	-79-51.5	-27-23.4	-78-44.4	0.9278	67.0	287.2	289.6	7:37.0
8:36	-25 -8.5	-79 -9.9	-26-49.2	-81-26.1	-25-58.6	-80-17.3	0.9276	64.7	281.3	292.5	7:32.4
8:42	-23-42.0	-80-41.7	-25-20.2	-83 -1.6	-24-30.9	-81-50.8	0.9273	62.2	276.3	295.7	7:27.4
8:48	-22-12.2	-82-15.1	-23-48.1	-84-38.7	-22-60.0	-83-26.1	0.9270	59.5	272.0	299.3	7:22.1
8:54	-20-39.0	-83-51.1	-22-12.6	-86-18.9	-21-25.6	-85 -4.1	0.9266	56.6	268.3	303.3	7:16.3
9: 0	-19 -1.9	-85-31.1	-20-33.3	-88 -3.5	-19-47.4	-86-46.4	0.9262	53.5	265.1	307.6	7:10.3
9: 6	-17-20.3	-87-16.7	-18-49.7	-89-54.3	-18 -4.8	-88-34.5	0.9257	50.3	262.3	312.2	7: 3.8
9:12	-15-33.7	-89 -9.9	-17 -0.9	-91-53.7	-16-17.2	-90-30.7	0.9251	46.7	259.9	317.0	6:57.0
9:18	-13-41.0	-91-13.6	-15 -6.0	-94 -5.0	-14-23.4	-92-38.0	0.9245	42.9	257.8	322.1	6:49.7
9:24	-11-40.8	-93-31.6	-13 -3.1	-96-32.9	-12-21.9	-95 -0.7	0.9237	38.8	255.9	327.4	6:42.0
9:30	-9-30.7	-96-10.2	-10-49.8	-99-25.1	-10-10.3	-97-45.7	0.9228	34.2	253.0	332.8	6:33.5
9:36	-7 -7.1	-99-20.0	-8-20.3	-102-55.7	-7-44.2	-101 -5.1	0.9217	29.0	251.0	338.4	6:24.2
9:42	-4-22.0	-103-22.4	-5-25.7	-107-36.7	-4-54.6	-105-24.8	0.9202	22.6	251.1	344.2	6:13.4
9:48	0-53.4	-109-18.8	-1-21.0	-115-34.4	-1-12.0	-112 -8.8	0.9181	13.6	251.1	350.7	5:59.2
LIMITS	5 7.8	-123-13.3	2 15.7	-124-51.3	3 33.0	-123-35.0	0.9148	0.0	251.3	358.2	5:40.7

Table 46

TOTAL SOLAR ECLIPSE OF 22 JUL 2009

SAROS 138 Delta T = 72.5 Sec

UNIVERSAL TIME	NORTHERN LIMIT LATITUDE LONGITUDE	SOUTHERN LIMIT LATITUDE LONGITUDE	CENTER LINE LATITUDE LONGITUDE	DIAMETER RATIO	SUN ALT	SUN AZ	PATH WIDTH	DURATION TOTALITY
LIMITS	21 10.5 -70 -8.5	19 30.8 -70-59.7	20 22.4 -70-38.4	1.0610	0.0	68.4	205.5	3: 9.4
0:54	23 52.0 -77-26.4	22 56.4 -80-45.6	23 28.1 -79-15.8	1.0639	8.9	71.8	214.0	3:31.1
1: 0	27 57.2 -90-22.7	26 15.5 -91-58.4	27 7.1 -91-13.0	1.0680	21.6	77.6	225.2	4: 8.8
1: 6	29 44.5 -97-41.3	27 50.0 -98-50.0	28 47.6 -98-17.5	1.0703	29.3	81.8	231.4	4:30.4
1:12	30 48.1 -103-12.9	28 48.1 -104-13.2	29 47.3 -103-49.6	1.0720	35.5	85.5	236.0	4:50.1
1:18	31 26.4 -108-12.0	29 19.4 -108-47.2	30 23.0 -108-30.8	1.0734	40.9	88.8	239.7	5: 7.4
1:24	31 47.1 -112-28.5	29 36.6 -112-48.4	30 41.9 -112-38.5	1.0746	45.8	92.0	242.9	5:22.8
1:30	31 54.2 -116-15.9	29 41.4 -116-25.6	30 47.8 -116-21.6	1.0756	50.2	95.1	245.6	5:38.6
1:36	31 50.4 -119-45.8	29 32.6 -119-44.1	30 43.3 -119-45.7	1.0764	54.4	98.2	248.0	5:49.0
1:42	31 37.5 -122-59.8	29 22.6 -122-47.6	30 30.0 -122-54.2	1.0772	58.4	101.2	250.0	5:59.9
1:48	31 16.6 -126 -0.6	29 1.7 -125-38.6	30 9.1 -125-50.0	1.0778	62.2	104.4	251.8	6: 9.5
1:54	30 48.7 -128-50.0	28 34.4 -128-18.9	29 41.5 -128-34.7	1.0783	65.9	107.7	253.4	6:17.7
2: 0	30 14.6 -131-29.7	28 1.3 -130-50.2	29 7.9 -131-10.1	1.0788	69.4	111.3	254.7	6:24.6
2: 6	29 34.8 -134 -0.8	27 22.9 -133-13.6	28 28.8 -133-37.3	1.0791	72.8	115.5	255.8	6:30.2
2:12	28 49.8 -136-24.5	26 39.6 -135-30.2	27 44.7 -135-57.3	1.0794	76.2	120.9	256.7	6:34.4
2:18	28 0.0 -138-41.7	25 51.8 -137-41.0	26 55.9 -138-11.2	1.0797	79.3	128.3	257.4	6:37.4
2:24	27 5.6 -140-53.3	24 59.7 -139-46.7	26 2.7 -140-19.8	1.0798	82.3	140.6	257.9	6:39.1
2:30	26 6.9 -143 -0.1	24 3.4 -141-48.1	25 5.2 -142-23.8	1.0799	84.7	164.5	258.2	6:39.5
2:36	25 4.1 -145 -2.7	23 3.2 -143-46.0	24 3.7 -144-24.1	1.0799	85.6	207.4	258.5	6:38.7
2:42	23 57.2 -147 -2.1	21 59.0 -145-41.1	22 58.2 -146-21.3	1.0799	84.2	243.4	258.6	6:36.8
2:48	22 46.4 -148-58.9	20 50.9 -147-34.1	21 48.7 -148-16.1	1.0797	81.6	261.2	258.5	6:33.7
2:54	21 31.5 -150-54.0	19 38.8 -149-25.7	20 35.3 -150 -9.5	1.0796	78.6	270.6	258.4	6:29.6
3: 0	20 12.7 -152-48.2	18 22.8 -151-16.8	19 17.9 -152 -2.1	1.0793	75.4	276.4	258.1	6:24.4
3: 6	18 49.7 -154-42.3	17 2.7 -153 -8.2	17 56.4 -153-54.8	1.0790	72.0	280.4	257.7	6:18.1
3:12	17 22.4 -156-37.4	15 38.2 -155 -0.9	16 30.5 -155-48.7	1.0786	68.6	283.4	257.2	6:10.9
3:18	15 50.5 -158-34.6	14 9.2 -156-56.8	15 0.0 -157-44.4	1.0781	65.0	285.8	256.5	6: 2.7
3:24	14 13.6 -160-35.4	12 35.1 -158-54.7	13 24.6 -159-44.6	1.0775	61.3	287.7	255.6	5:53.6
3:30	12 31.1 -162-41.4	10 55.4 -160-58.9	11 43.5 -161-49.6	1.0769	57.4	289.3	254.5	5:43.6
3:36	10 42.2 -164-54.7	9 9.5 -163-10.4	9 56.2 -164 -2.0	1.0761	53.4	290.6	253.0	5:32.5
3:42	8 46.0 -167-18.4	7 16.2 -165-32.2	8 1.4 -166-24.8	1.0752	49.2	291.7	251.1	5:20.5
3:48	6 40.6 -169-56.6	5 13.9 -168 -8.3	5 57.8 -169 -1.8	1.0742	44.6	292.6	248.7	5: 7.3
3:54	4 23.6 -172-55.6	3 0.4 -171 -4.7	3 42.4 -171-59.5	1.0729	39.6	293.3	245.5	4:52.9
4: 0	1 50.7 -176-26.3	0 31.5 -174-31.6	1 11.6 -175-28.2	1.0714	34.1	293.8	241.3	4:36.7
4: 6	-1 -6.7 179 9.7	-2-20.5 -178-48.9	-1-43.0 -179-48.6	1.0696	27.6	293.9	235.5	4:18.1
4:12	-4-51.9 172 53.4	-5-55.5 175 11.8	-5-22.8 174 4.3	1.0670	19.1	293.6	226.9	3:55.1
LIMITS	-12 -5.7 157 15.0	-13-45.9 158 4.6	-12-55.1 157 42.0	1.0608	0.0	290.7	204.7	3: 8.8

Table 47

ANNULAR SOLAR ECLIPSE OF 15 JAN 2010

SAROS 141 Delta T = 72.9 Sec

UNIVERSAL TIME	NORTHERN LIMIT LATITUDE	LONGITUDE	SOUTHERN LIMIT LATITUDE	LONGITUDE	CENTER LINE LATITUDE	LONGITUDE	DIAMETER RATIO	SUN ALT	SUN AZ	PATH WIDTH	DURATION ANNULARITY
LIMITS	8 37.0	-16 -8.8	5 21.3	-15 -14.9	6 42.6	-16 -22.4	0.9063	0.0	111.4	370.0	7:11.3
5:24	3 12.7	-30 -44.6	0-36.2	-32 -8.3	1 15.6	-31 -31.5	0.9106	18.3	112.8	355.0	8: 5.8
5:30	1 27.0	-36 -26.8	-2 -5.9	-37 -24.9	0-21.0	-36 -58.5	0.9122	25.3	113.4	350.8	8:32.5
5:36	0 22.7	-40 -34.3	-3 -2.2	-41 -24.5	-1-21.0	-41 -1.2	0.9134	30.8	114.0	348.4	8:55.0
5:42	0-19.7	-43 -54.2	-3-39.2	-44 -42.3	-2 -0.5	-44 -19.5	0.9143	35.4	114.8	347.1	9:15.2
5:48	0-47.4	-48 -44.4	-4 -2.9	-47 -32.8	-2-28.1	-47 -9.8	0.9151	39.5	115.8	346.3	9:33.5
5:54	-1 -4.3	-49 -14.0	-4-16.3	-50 -4.0	-2-41.2	-49 -39.8	0.9158	43.2	117.0	345.9	9:50.2
6: 0	-1-12.6	-51 -28.2	-4-21.8	-52 -20.6	-2-48.0	-51 -55.0	0.9164	46.6	118.4	345.7	10: 5.4
6: 6	-1-13.8	-53 -30.5	-4-19.9	-54 -25.8	-2-47.7	-53 -58.6	0.9169	49.7	120.2	345.6	10:19.1
6:12	-1 -8.9	-55 -23.3	-4-12.2	-56 -21.6	-2-41.4	-55 -52.8	0.9174	52.6	122.4	345.4	10:31.2
6:18	0-58.6	-57 -8.4	-3-59.2	-58 -9.9	-2-29.8	-57 -39.4	0.9177	55.2	125.0	345.0	10:41.7
6:24	0-43.5	-58 -47.2	-3-41.4	-59 -51.9	-2-13.3	-59 -19.7	0.9181	57.6	128.0	344.4	10:50.6
6:30	0-23.9	-60 -20.7	-3-19.1	-61 -28.5	-1-52.3	-60 -54.7	0.9183	59.8	131.7	343.6	10:57.9
6:36	0 -0.1	-61 -49.8	-2-52.7	-63 -0.7	-1-27.2	-62 -25.3	0.9185	61.7	135.9	342.4	11: 3.6
6:42	0 27.6	-63 -15.4	-2-22.3	-64 -29.1	0-58.2	-63 -52.3	0.9187	63.3	140.7	341.0	11: 7.6
6:48	0 59.1	-64 -37.9	-1-48.2	-65 -54.5	0-25.4	-65 -16.2	0.9189	64.6	146.1	339.3	11:10.0
6:54	1 34.3	-65 -58.1	-1-10.6	-67 -17.3	0 11.0	-66 -37.6	0.9190	65.6	152.1	337.4	11:10.9
7: 0	2 13.1	-67 -16.4	0-29.4	-68 -38.1	0 51.0	-67 -57.2	0.9190	66.2	158.5	335.4	11:10.2
7: 6	2 55.4	-68 -33.5	0 15.2	-69 -57.4	1 34.4	-69 -15.3	0.9190	66.4	165.2	333.3	11: 8.1
7:12	3 41.3	-69 -49.7	1 3.2	-71 -15.7	2 21.4	-70 -32.5	0.9190	66.2	171.9	331.3	11: 4.7
7:18	4 30.8	-71 -5.6	1 54.7	-72 -33.5	3 11.9	-71 -49.4	0.9189	65.7	178.4	329.3	10:60.0
7:24	5 24.0	-72 -21.7	2 49.6	-73 -51.3	4 5.9	-73 -8.3	0.9188	64.7	184.5	327.6	10:54.1
7:30	6 21.0	-73 -38.6	3 48.2	-75 -9.7	5 3.7	-74 -23.9	0.9187	63.5	190.2	326.0	10:47.2
7:36	7 22.0	-74 -56.8	4 50.4	-76 -29.3	6 5.3	-75 -42.9	0.9185	61.9	195.3	324.7	10:39.4
7:42	8 27.2	-76 -17.2	5 56.5	-77 -50.8	7 10.9	-77 -3.7	0.9182	60.0	199.9	323.8	10:30.6
7:48	9 36.9	-77 -40.4	7 6.7	-79 -14.9	8 20.9	-78 -27.4	0.9179	57.9	203.9	323.2	10:21.0
7:54	10 51.5	-79 -7.4	8 21.4	-80 -42.6	9 35.6	-79 -54.7	0.9176	55.5	207.5	323.0	10:10.7
8: 0	12 11.5	-80 -39.3	9 41.1	-82 -14.9	10 55.4	-81 -26.8	0.9172	52.9	210.8	323.2	9:59.7
8: 6	13 37.6	-82 -17.8	11 6.2	-83 -53.4	12 20.9	-83	0.9168	50.1	213.7	323.9	9:48.0
8:12	15 10.6	-84 -4.5	12 37.5	-85 -39.7	13 53.0	-84 -51.6	0.9162	47.0	216.4	325.0	9:35.6
8:18	16 51.8	-86 -2.3	14 16.2	-87 -36.3	15 32.9	-86 -48.7	0.9156	43.7	218.9	328.7	9:22.6
8:24	18 42.9	-88 -14.9	16 3.7	-89 -46.7	17 22.1	-89 -0.1	0.9149	40.0	221.3	328.9	9: 8.8
8:30	20 46.8	-90 -48.1	18 2.3	-92 -16.0	19 23.2	-91 -31.1	0.9141	36.0	223.7	331.9	8:54.1
8:36	23 7.9	-93 -51.7	20 15.7	-95 -12.6	21 40.2	-94 -30.9	0.9132	31.4	226.2	335.7	8:38.3
8:42	25 55.0	-97 -45.0	22 50.7	-98 -52.3	24 20.7	-98 -16.5	0.9120	26.0	228.9	340.8	8:20.8
8:48	29 30.0	-103 -17.1	26 2.6	-103 -51.4	27 42.8	-103 -29.8	0.9104	19.2	232.4	348.1	8: 0.2
8:54	36 20.5	-116 -12.6	30 52.1	-112 -37.9	33 11.9	-113 -33.8	0.9077	7.9	238.6	361.9	7:29.8
LIMITS	38 23.9	-120 -56.5	35 15.9	-122 -29.5	36 30.0	-120 -55.9	0.9059	0.0	243.0	371.7	7:13.1

Table 48

TOTAL SOLAR ECLIPSE OF 11 JUL 2010

SAROS 146   Delta T = 73.4 Sec

UNIVERSAL TIME	NORTHERN LIMIT LATITUDE	LONGITUDE	SOUTHERN LIMIT LATITUDE	LONGITUDE	CENTER LINE LATITUDE	LONGITUDE	DIAMETER RATIO	SUN ALT	SUN AZ	PATH WIDTH	DURATION TOTALITY
LIMITS	-26 -4.4	171 8.6	-27-42.1	170 46.0	-26-37.2	170 19.4	1.0443	0.0	64.7	180.1	2:44.0
18:20	-20-50.1	157 51.0	-23-55.6	161 6.9	-22-18.8	159 18.4	1.0480	12.2	59.6	197.3	3:15.5
18:24	-19-12.8	153 1.4	-21-51.0	155 5.7	-20-30.4	153 59.7	1.0499	18.2	56.9	207.5	3:34.2
18:28	-18 -9.7	149 26.4	-20-38.5	151 4.1	-19-23.2	150 12.9	1.0512	22.5	54.7	215.7	3:48.8
18:32	-17-25.1	146 29.8	-19-49.2	147 53.2	-18-36.4	147 9.7	1.0523	26.1	52.6	222.8	4: 1.3
18:36	-16-52.7	143 57.3	-19-14.1	145 11.7	-18 -2.9	144 33.1	1.0532	29.1	50.6	229.2	4:12.4
18:40	-16-29.5	141 41.7	-18-49.1	142 49.9	-17-38.9	142 14.7	1.0539	31.8	48.5	235.0	4:22.3
18:44	-16-13.5	139 38.6	-18-32.0	140 42.4	-17-22.3	140 9.5	1.0546	34.1	46.4	240.3	4:31.3
18:48	-16 -3.5	137 45.1	-18-21.1	138 45.7	-17-11.9	138 14.6	1.0552	36.2	44.3	245.0	4:39.4
18:52	-15-58.6	135 59.4	-18-15.5	136 57.5	-17 -6.7	136 27.7	1.0557	38.1	42.0	249.2	4:46.8
18:56	-15-58.2	134 19.9	-18-14.6	135 16.1	-17 -6.1	134 47.4	1.0561	39.8	39.7	252.8	4:53.3
19: 0	-16 -1.9	132 45.5	-18-17.9	133 40.2	-17 -9.5	133 12.3	1.0565	41.3	37.2	255.8	4:59.2
19: 4	-16 -9.3	131 15.3	-18-24.9	132 8.8	-17-16.7	131 41.6	1.0568	42.6	34.7	258.3	5: 4.3
19: 8	-16-20.1	129 48.7	-18-35.4	130 41.2	-17-27.4	130 14.5	1.0571	43.8	32.2	260.1	5: 8.6
19:12	-16-34.2	128 25.0	-18-49.1	129 16.6	-17-41.3	128 50.5	1.0574	44.8	29.2	261.4	5:12.3
19:16	-16-51.4	127 3.6	-19 -6.0	127 54.6	-17-58.3	127 28.8	1.0576	45.6	26.4	262.0	5:15.3
19:20	-17-11.6	125 44.3	-19-25.8	126 34.6	-18-18.3	126 9.2	1.0578	46.2	23.4	262.1	5:17.5
19:24	-17-34.6	124 26.4	-19-48.6	125 16.1	-18-41.3	124 51.1	1.0579	46.7	20.4	261.7	5:19.1
19:28	-18 -0.6	123 9.8	-20-14.3	123 58.8	-19 -7.0	123 34.1	1.0580	47.0	17.3	260.7	5:20.0
19:32	-18-29.4	121 53.9	-20-42.8	122 42.2	-19-35.7	122 18.0	1.0580	47.2	14.2	259.3	5:20.2
19:36	-19 -1.0	120 38.4	-21-14.1	121 26.1	-20 -7.1	121 2.2	1.0580	47.1	11.1	257.5	5:19.8
19:40	-19-35.5	119 23.1	-21-48.4	120 9.9	-20-41.5	119 46.5	1.0580	46.9	8.0	255.1	5:18.7
19:44	-20-12.9	118 7.4	-22-25.6	118 53.4	-21-18.8	118 30.5	1.0580	46.6	4.8	252.9	5:17.0
19:48	-20-53.3	116 51.2	-23 -6.0	117 36.1	-21-59.2	117 13.7	1.0579	46.1	1.8	250.2	5:14.6
19:52	-21-36.8	115 33.8	-23-49.5	116 17.6	-22-42.7	115 55.8	1.0577	45.4	358.7	247.3	5:11.6
19:56	-22-23.7	114 15.0	-24-36.4	114 57.3	-23-29.5	114 36.3	1.0576	44.5	355.7	244.3	5: 8.0
20: 0	-23-14.0	112 54.1	-25-27.0	113 34.8	-24-19.9	113 14.7	1.0573	43.5	352.8	241.1	5: 3.7
20: 4	-24 -8.0	111 30.8	-26-21.4	112 9.4	-25-14.1	111 50.3	1.0571	42.2	349.9	237.8	4:58.9
20: 8	-25 -6.1	110 3.8	-27-20.1	110 40.3	-26-12.5	110 22.4	1.0568	40.9	347.1	234.4	4:53.5
20:12	-26 -8.7	108 32.9	-28-23.5	109 6.6	-27-15.4	108 50.1	1.0564	39.3	344.3	230.9	4:47.4
20:16	-27-16.3	106 56.7	-29-32.4	107 27.0	-28-23.6	107 12.4	1.0560	37.5	341.6	227.4	4:40.8
20:20	-28-29.6	105 14.0	-30-47.4	105 40.1	-29-37.7	105 27.6	1.0555	35.6	338.9	223.7	4:33.5
20:24	-29-49.6	103 22.9	-32 -9.7	103 43.5	-30-58.8	103 33.9	1.0549	33.4	336.2	220.0	4:25.5
20:28	-31-17.7	101 20.8	-33-41.0	101 34.4	-32-28.3	101 28.5	1.0543	30.9	333.5	216.2	4:16.7
20:32	-32-55.8	99 4.0	-35-23.7	99 7.9	-34 -8.6	99 7.2	1.0535	28.2	330.7	212.3	4: 7.0
20:36	-34-47.2	96 26.5	-37-21.8	96 16.5	-36 -3.0	96 23.2	1.0526	25.0	327.8	208.1	3:56.3
20:40	-36-57.2	93 17.9	-39-43.0	92 45.2	-38-18.1	93 4.3	1.0515	21.2	324.5	203.5	3:44.0
20:44	-39-37.1	89 15.8	-42-45.1	87 58.5	-41 -7.8	88 42.7	1.0501	16.4	320.5	198.3	3:29.4
20:48	-43-20.8	83 12.6	-47-50.4	79 6.4	-45-21.1	81 38.6	1.0479	9.4	314.6	191.2	3: 9.1
LIMITS	-50 -9.0	70 18.8	-51-36.9	71 33.0	-50-30.9	71 41.4	1.0452	0.0	306.8	183.4	2:46.8

Table 49

ANNULAR SOLAR ECLIPSE OF 20 MAY 2012

SAROS 128

Delta T = 74.9 Sec

UNIVERSAL TIME	NORTHERN LIMIT LATITUDE	LONGITUDE	SOUTHERN LIMIT LATITUDE	LONGITUDE	CENTER LINE LATITUDE	LONGITUDE	DIAMETER RATIO	SUN ALT	SUN AZ	PATH WIDTH	DURATION ANNULARITY
LIMITS	22 22.3	-107-47.9	19 55.0	-109-40.7	21 22.7	-109-23.7	0.9311	0.0	68.6	323.8	4:23.3
22:12	25 35.1	-115-55.8	25 55.9	-123-45.0	25 56.6	-120-23.3	0.9340	12.3	73.4	312.6	4:35.2
22:18	29 53.8	-125-27.7	29 9.0	-130-17.2	29 33.4	-128 -0.2	0.9361	21.0	78.2	301.7	4:45.5
22:24	32 39.4	-131 -3.6	31 35.2	-134-59.6	32 8.1	-133 -6.0	0.9378	27.0	82.3	293.0	4:53.4
22:30	34 52.8	-135-25.8	33 37.8	-138-52.2	34 15.7	-137-12.2	0.9387	31.9	86.3	285.3	5: 0.2
22:36	36 47.6	-139 -9.9	35 25.1	-142-16.2	36 6.5	-140-45.7	0.9396	36.1	90.4	278.4	5: 6.3
22:42	38 29.5	-142-31.2	37 1.3	-145-22.1	37 45.4	-143-58.9	0.9403	39.8	94.6	272.2	5:11.9
22:48	40 1.5	-145-37.7	38 28.5	-148-15.8	39 15.0	-146-58.8	0.9410	43.1	99.1	266.6	5:17.0
22:54	41 25.6	-148-34.4	39 48.3	-151 -1.4	40 36.8	-149-49.7	0.9415	46.1	103.9	261.6	5:21.8
23: 0	42 42.7	-151-24.8	41 1.6	-153-41.4	41 52.0	-152-34.8	0.9420	48.9	109.0	257.0	5:26.1
23: 6	43 53.8	-154-11.3	42 9.1	-156-18.0	43 1.1	-155-16.2	0.9424	51.3	114.5	253.0	5:30.1
23:12	44 58.9	-156-55.7	43 11.1	-158-52.6	44 4.8	-157-55.6	0.9428	53.5	120.5	249.5	5:33.7
23:18	45 58.9	-159-39.3	44 8.0	-161-26.5	45 3.2	-160-34.2	0.9431	55.5	126.8	246.4	5:36.8
23:24	46 53.8	-162-23.4	45 0.1	-164 -0.5	45 56.6	-163-13.2	0.9433	57.1	133.3	243.8	5:39.6
23:30	47 43.8	-165 -8.9	45 47.3	-166-35.6	46 45.2	-165-53.3	0.9435	58.5	140.9	241.6	5:41.9
23:36	48 28.8	-167-56.5	46 29.8	-169-12.3	47 28.9	-168-35.4	0.9437	59.6	148.6	239.8	5:43.7
23:42	49 8.8	-170-46.9	47 7.6	-171-51.3	48 7.8	-171-19.9	0.9438	60.4	156.7	238.4	5:45.1
23:48	49 43.9	-173-40.6	47 40.6	-174-32.9	48 41.8	-174 -7.4	0.9439	60.8	165.0	237.4	5:46.0
23:54	50 13.8	-176-38.0	48 8.7	-177-17.7	49 10.8	-176-58.2	0.9439	60.9	173.5	236.8	5:46.4
0: 0	50 38.5	-179-39.5	48 31.9	179 54.2	49 34.7	-179-52.9	0.9439	60.6	182.0	236.6	5:46.3
0: 6	50 57.7	177 14.6	48 49.9	177 2.3	49 53.3	177 8.5	0.9438	60.0	190.4	236.7	5:45.7
0:12	51 11.3	174	49 2.7	174 6.4	50 6.5	174 5.8	0.9437	59.1	198.6	237.3	5:44.5
0:18	51 18.9	170 48.4	49 9.9	171 6.2	50 13.9	170 57.8	0.9435	57.9	206.6	238.3	5:42.9
0:24	51 20.4	167 27.7	49 11.4	168 1.4	50 15.4	167 45.3	0.9433	56.4	214.2	239.7	5:40.7
0:30	51 15.3	164 1.3	49 6.7	164 51.6	50 10.5	164 27.5	0.9431	54.6	221.4	241.6	5:38.0
0:36	51 3.2	160 28.9	48 55.7	161 36.4	49 59.0	161 4.1	0.9428	52.5	228.4	243.9	5:34.9
0:42	50 43.6	156 49.8	48 37.7	158 15.2	49 40.3	157 34.2	0.9424	50.2	235.0	246.7	5:31.2
0:48	50 15.9	153 3.1	48 12.3	154 47.0	49 13.8	153 57.1	0.9420	47.6	241.3	250.1	5:27.0
0:54	49 39.2	149 7.2	47 38.8	151 10.7	48 38.7	150 11.3	0.9415	44.7	247.4	254.1	5:22.4
1: 0	48 52.4	144 59.8	46 56.1	147 24.3	47 54.1	146 14.9	0.9409	41.6	253.2	258.7	5:17.2
1: 6	47 53.8	140 37.5	46 2.8	143 25.0	46 58.3	142 4.6	0.9402	38.1	258.9	264.1	5:11.6
1:12	46 40.7	135 54.2	44 57.0	139 8.3	45 49.1	137 35.2	0.9394	34.2	264.5	270.4	5: 5.3
1:18	45 8.7	130 39.0	43 35.3	134 28.6	44 22.7	132 37.8	0.9384	29.7	270.1	277.8	4:58.4
1:24	43 8.5	124 29.1	41 51.6	129 5.1	42 31.4	126 54.3	0.9372	24.4	275.9	286.7	4:50.6
1:30	40 13.4	116 17.1	39 31.7	122 29.6	39 56.7	119 37.9	0.9358	17.4	282.2	298.0	4:41.2
LIMITS	34 6.2	100 3.7	31 42.6	102 10.7	33 10.3	101 51.1	0.9314	0.0	294.1	322.3	4:22.0

Table 50

TOTAL SOLAR ECLIPSE OF 13 NOV 2012

SAROS 133   Delta T = 75.4 Sec

UNIVERSAL TIME	NORTHERN LIMIT LATITUDE LONGITUDE	SOUTHERN LIMIT LATITUDE LONGITUDE	CENTER LINE LATITUDE LONGITUDE	DIAMETER RATIO	SUN ALT	SUN AZ	PATH WIDTH	DURATION TOTALITY
LIMITS	-11-26.8 -133-24.6	-12-27.4 -132-49.5	-12 -0.9 -133-18.5	1.0327	0.0	108.5	126.2	1:41.1
20:42	-18-17.4 -151 -9.0	-18-48.1 -148-56.3	-18-33.5 -150 -4.1	1.0387	18.9	103.4	150.2	2:15.7
20:48	-21 -9.3 -157-20.3	-21-53.2 -155-30.7	-21-31.7 -156-26.3	1.0410	26.7	100.1	158.8	2:32.1
20:54	-23-28.6 -162 -0.3	-24-18.1 -160-20.8	-23-52.7 -161-11.1	1.0427	32.8	96.9	164.5	2:45.5
21: 0	-25-24.7 -165-54.3	-26-21.9 -164-21.2	-25-53.6 -165 -8.3	1.0441	37.9	93.5	168.6	2:57.1
21: 6	-27 -9.7 -169-20.7	-28-11.7 -167-52.4	-27-40.9 -168-37.0	1.0452	42.4	89.8	171.7	3: 7.3
21:12	-28-44.9 -172-28.9	-29-51.1 -171 -4.7	-29-18.2 -171-47.3	1.0462	46.4	85.8	174.0	3:16.6
21:18	-30-12.2 -175-24.7	-31-22.1 -174 -4.3	-30-47.3 -174-45.0	1.0470	50.1	81.4	175.7	3:24.9
21:24	-31-32.7 -178-11.9	-32-46.1 -176-55.3	-32 -9.5 -177-34.1	1.0477	53.5	76.5	177.0	3:32.4
21:30	-32-47.4 179 6.8	-34 -3.9 -179-40.6	-33-25.8 179 42.7	1.0482	56.6	71.1	178.0	3:39.0
21:36	-33-58.7 176 29.3	-35-16.2 177 37.8	-34-36.5 177 3.1	1.0487	59.3	64.9	178.7	3:44.8
21:42	-35 -1.0 173 54.0	-36-23.2 174 58.0	-35-42.2 174 25.6	1.0491	61.8	57.9	179.1	3:49.9
21:48	-36 -0.5 171 19.6	-37-25.4 172 18.8	-36-43.0 171 48.9	1.0495	63.9	50.0	179.4	3:54.0
21:54	-36-55.5 168 45.0	-38-22.7 169 39.0	-37-39.1 169 11.7	1.0497	65.6	41.2	179.5	3:57.4
22: 0	-37-48.0 166 9.3	-39-15.3 166 57.6	-38-30.7 166 33.2	1.0499	66.9	31.4	179.5	3:59.9
22: 6	-38-32.0 163 31.5	-40 -3.2 164 13.8	-39-17.6 163 52.4	1.0500	67.7	20.9	179.3	4: 1.5
22:12	-39-13.5 160 50.9	-40-46.2 161 26.6	-39-59.8 161 8.6	1.0500	67.9	10.0	179.0	4: 2.2
22:18	-39-50.3 158 6.8	-41-24.3 158 35.3	-40-37.3 158 20.9	1.0500	67.6	359.1	178.6	4: 2.0
22:24	-40-22.4 155 18.3	-41-57.4 155 39.3	-41 -9.8 155 28.8	1.0499	66.8	348.6	178.0	4: 1.0
22:30	-40-49.4 152 24.8	-42-25.0 152 37.7	-41-37.2 152 31.3	1.0498	65.4	338.7	177.4	3:59.0
22:36	-41-11.3 149 25.5	-42-47.1 149 29.8	-41-59.1 149 27.8	1.0495	63.6	329.6	176.8	3:56.1
22:42	-41-27.5 146 19.5	-43 -3.1 146 14.9	-42-15.3 146 17.4	1.0492	61.5	321.3	175.7	3:52.2
22:48	-41-37.8 143 5.9	-43-12.6 142 51.8	-42-25.2 142 59.2	1.0488	59.0	313.7	174.6	3:47.5
22:54	-41-41.6 139 43.5	-43-15.1 139 19.6	-42-28.3 139 32.0	1.0483	58.2	306.7	173.3	3:41.8
23: 0	-41-38.2 136 10.8	-43 -9.9 135 36.8	-42-24.1 135 54.4	1.0478	53.1	300.2	171.9	3:35.2
23: 6	-41-26.9 132 26.0	-42-56.0 131 41.5	-42-11.5 132 4.4	1.0471	49.7	294.2	170.2	3:27.7
23:12	-41 -6.3 128 26.4	-42-32.1 127 31.0	-41-49.3 127 59.5	1.0463	45.9	288.4	168.2	3:19.2
23:18	-40-35.1 124 8.1	-41-56.6 123 1.1	-41-16.0 123 35.5	1.0453	41.8	282.8	165.8	3: 9.6
23:24	-39-50.6 119 24.9	-41 -6.8 118 5.4	-40-28.8 118 46.2	1.0442	37.3	277.4	162.8	2:58.9
23:30	-38-48.9 114 6.2	-39-57.7 112 32.3	-39-23.6 113 20.4	1.0428	32.1	271.9	159.1	2:46.8
23:36	-37-22.0 107 50.6	-38-20.5 105 57.5	-37-51.8 106 55.4	1.0411	25.9	266.2	154.2	2:32.7
23:42	-35 -9.9 99 40.8	-35-49.7 97 10.7	-35-30.7 98 28.0	1.0386	17.6	259.7	146.7	2:15.0
LIMITS	-29 -2.4 80 16.5	-30 -2.6 79 35.4	-29-36.4 80 7.8	1.0331	0.0	248.8	127.6	1:42.1

Table 51

ANNULAR SOLAR ECLIPSE OF 10 MAY 2013

SAROS 138   Delta T = 75.8 Sec

UNIVERSAL TIME	NORTHERN LIMIT LATITUDE	LONGITUDE	SOUTHERN LIMIT LATITUDE	LONGITUDE	CENTER LINE LATITUDE	LONGITUDE	DIAMETER RATIO	SUN ALT	SUN AZ	PATH WIDTH	DURATION ANNULARITY
LIMITS	-23-31.6	-118-57.8	-25-27.4	-119-34.8	-24-23.0	-119-36.2	0.9405	0.0	70.3	222.2	4:10.9
22:36	-18-45.3	-132-11.7	-20-44.9	-132-19.4	-19-44.2	-132-17.4	0.9439	13.7	65.2	207.9	4:24.7
22:42	-15-38.8	-139-25.8	-17-22.2	-139-56.2	-16-28.9	-139-41.7	0.9461	22.8	62.1	199.0	4:35.8
22:48	-13-20.6	-144 -9.0	-14-59.4	-144-45.7	-14 -9.4	-144-27.8	0.9476	29.2	60.0	193.1	4:44.6
22:54	-11-27.5	-147-47.6	-13 -2.9	-148-27.0	-12-14.9	-148 -7.6	0.9488	34.5	58.2	188.7	4:52.5
23: 0	-9-49.8	-150-48.7	-11-22.5	-151-29.2	-10-35.9	-151 -9.2	0.9497	39.1	56.6	185.1	4:59.7
23: 6	-8-22.8	-153-25.0	-9-53.4	-154 -5.9	-9 -7.9	-153-45.6	0.9505	43.3	54.9	182.1	5: 6.6
23:12	-7 -4.0	-155-43.6	-8-33.0	-156-24.5	-7-48.3	-156 -4.2	0.9512	47.1	53.2	179.6	5:13.1
23:18	-5-51.9	-157-49.1	-7-19.7	-158-29.7	-6-35.6	-158 -9.5	0.9518	50.7	51.4	177.8	5:19.3
23:24	-4-45.3	-159-44.5	-6-12.3	-160-24.7	-5-28.6	-160 -4.7	0.9523	54.0	49.3	175.9	5:25.2
23:30	-3-43.6	-161-32.1	-5-10.0	-162-11.7	-4-26.8	-161-52.0	0.9528	57.1	46.9	174.5	5:30.8
23:36	-2-46.2	-163-13.6	-4-12.2	-163-52.4	-3-29.2	-163-33.1	0.9531	60.1	44.1	173.4	5:36.1
23:42	-1-52.7	-164-50.2	-3-18.6	-165-28.1	-2-35.4	-165 -9.2	0.9535	62.8	40.8	172.6	5:41.1
23:48	-1 -2.8	-166-23.0	-2-28.7	-166-60.0	-1-45.5	-166-41.5	0.9537	65.4	36.9	172.0	5:45.7
23:54	0-16.2	-167-52.8	-1-42.3	-168-28.8	0-59.1	-168-10.8	0.9540	67.8	32.1	171.7	5:49.9
0: 0	0 27.2	-169-20.4	0-59.2	-169-55.3	0-15.8	-169-37.9	0.9541	69.8	26.3	171.5	5:53.6
0: 6	1 7.5	-170-46.4	0-19.3	-171-20.1	0 24.3	-171 -3.3	0.9543	71.6	19.3	171.5	5:56.8
0:12	1 44.9	-172-11.4	0 17.6	-172-43.9	1 1.4	-172-27.7	0.9544	73.0	11.0	171.7	5:59.6
0:18	2 19.4	-173-35.9	0 51.5	-174 -7.1	1 35.6	-173-51.5	0.9544	74.0	1.5	172.1	6: 1.7
0:24	2 51.1	-175 -0.5	1 22.4	-175-30.2	2 6.9	-175-15.4	0.9544	74.3	351.2	172.5	6: 3.3
0:30	3 20.0	-176-25.5	1 50.5	-176-53.8	2 35.4	-176-39.7	0.9544	74.2	341.0	173.1	6: 4.2
0:36	3 46.0	-177-51.4	2 15.6	-178-18.2	3 1.0	-178 -4.9	0.9543	73.4	331.4	173.8	6: 4.4
0:42	4 9.1	-179-18.8	2 37.9	-179-44.0	3 23.7	-179-31.5	0.9542	72.2	323.0	174.5	6: 4.0
0:48	4 29.3	179 11.8	2 57.1	178 48.3	3 43.4	179 0.0	0.9540	70.5	316.0	175.3	6: 2.9
0:54	4 46.5	177 40.1	3 13.2	177 18.2	4 0.0	177 29.1	0.9538	68.5	310.2	176.2	6: 1.0
1: 0	5 0.5	176 5.2	3 26.1	175 45.2	4 13.5	175 55.2	0.9536	66.3	305.5	177.1	5:58.5
1: 6	5 11.1	174 26.7	3 35.7	174 8.5	4 23.6	174 17.6	0.9533	63.8	301.8	178.1	5:55.2
1:12	5 18.3	172 43.8	3 41.7	172 27.3	4 30.1	172 35.6	0.9529	61.1	298.8	179.2	5:51.3
1:18	5 21.6	170 55.4	3 43.8	170 40.8	4 32.8	170 48.2	0.9525	58.2	296.4	180.5	5:46.8
1:24	5 20.7	169 0.6	3 41.6	168 47.8	4 31.4	168 54.3	0.9520	55.1	294.5	181.8	5:41.6
1:30	5 15.3	166 57.9	3 34.6	166 46.9	4 25.2	166 52.5	0.9515	51.9	293.0	183.4	5:35.8
1:36	5 4.6	164 45.5	3 22.6	164 36.2	4 13.7	164 41.0	0.9509	48.4	291.9	185.2	5:29.5
1:42	4 47.8	162 20.7	3 4.1	162 13.1	3 56.1	162 17.1	0.9502	44.7	290.9	187.3	5:22.5
1:48	4 23.7	159 40.1	2 38.1	159 34.0	3 31.1	159 37.3	0.9494	40.7	290.2	189.9	5:15.0
1:54	3 50.4	156 38.0	2 2.5	156 32.9	2 56.7	156 35.8	0.9485	36.2	289.7	192.9	5: 6.9
2: 0	3 4.8	153 5.1	1 14.0	153 0.1	2 9.7	153 3.0	0.9473	31.2	289.3	196.7	4:58.0
2: 6	2 0.7	148 42.7	0 5.9	148 36.1	1 3.6	148 40.0	0.9460	25.3	289.0	201.5	4:48.2
2:12	0 22.4	142 42.7	-1-39.8	142 27.6	0-38.2	142 36.3	0.9441	17.6	288.7	208.4	4:36.4
LIMITS	-4-28.3	126 48.5	-6-27.2	127 19.0	-5-22.7	127 19.4	0.9396	0.0	287.7	225.5	4:14.7

Table 52

ANN/TOT SOLAR ECLIPSE OF 3 NOV 2013

SAROS 143   Delta T = 76.2 Sec

UNIVERSAL TIME	NORTHERN LIMIT LATITUDE	NORTHERN LIMIT LONGITUDE	SOUTHERN LIMIT LATITUDE	SOUTHERN LIMIT LONGITUDE	CENTER LINE LATITUDE	CENTER LINE LONGITUDE	DIAMETER RATIO	SUN ALT	SUN AZ	PATH WIDTH	DURATION ANNULARITY
LIMITS	30 27.8	71 12.9	30 25.7	71 13.5	30 26.5	71 12.2	0.9989	0.0	107.9	4.0	0: 4.1
11: 6	28 14.6	63 54.1	28 17.3	63 56.5	28 15.9	63 55.3	1.0009	6.9	111.5	3.4	0: 3.6T
11:12	23 43.8	51 43.2	23 53.9	51 43.4	23 48.8	51 43.3	1.0048	20.0	117.9	16.8	0:20.4T
11:18	21 1.8	45 35.5	21 15.3	45 33.3	21 8.6	45 34.4	1.0069	27.4	121.3	24.2	0:31.1T
11:24	18 52.4	41 8.0	19 8.5	41 4.0	19 0.5	41 5.9	1.0085	33.3	124.1	29.6	0:40.2T
11:30	17 1.5	37 32.6	17 19.6	37 27.0	17 10.5	37 29.8	1.0097	38.3	126.6	34.0	0:48.1T
11:36	15 23.0	34 29.7	15 42.7	34 22.9	15 32.8	34 26.3	1.0108	42.8	129.1	37.8	0:55.4T
11:42	13 53.8	31 49.2	14 15.0	31 41.3	14 4.4	31 45.2	1.0117	46.8	131.7	41.0	1: 2.0T
11:48	12 32.0	29 24.9	12 54.4	29 16.1	12 43.2	29 20.5	1.0125	50.5	134.5	43.8	1: 8.1T
11:54	11 16.3	27 12.7	11 39.9	27 3.2	11 28.1	27 7.9	1.0132	54.0	137.5	46.3	1:13.6T
12: 0	10 6.0	25 9.8	10 30.5	24 59.7	10 18.2	25 4.7	1.0137	57.1	140.9	48.4	1:18.7T
12: 6	9 0.3	23 14.1	9 25.7	23 3.5	9 13.0	23 8.7	1.0142	60.1	144.9	50.3	1:23.2T
12:12	7 58.8	21 23.9	8 25.0	21 12.9	8 11.9	21 18.3	1.0147	62.7	149.5	52.0	1:27.2T
12:18	7 1.1	19 37.9	7 28.1	19 26.7	7 14.6	19 32.2	1.0150	65.1	154.9	53.5	1:30.7T
12:24	6 7.1	17 55.0	6 34.6	17 43.7	6 20.9	17 49.3	1.0153	67.2	161.3	54.7	1:33.7T
12:30	5 16.4	16 14.3	5 44.5	16 2.9	5 30.5	16 8.6	1.0156	68.8	168.6	55.8	1:36.1T
12:36	4 29.1	14 35.1	4 57.0	14 23.7	4 43.3	14 29.4	1.0157	70.1	177.0	56.6	1:37.9T
12:42	3 44.9	12 56.4	4 13.8	12 45.2	3 59.3	12 50.8	1.0158	70.8	186.1	57.2	1:39.1T
12:48	3 3.8	11 17.8	3 33.0	11 6.7	3 18.4	11 12.2	1.0159	70.9	195.5	57.6	1:39.7T
12:54	2 25.8	9 38.3	2 55.2	9 27.5	2 40.5	9 32.9	1.0159	70.4	204.6	57.8	1:39.6T
13: 0	1 50.9	7 57.5	2 20.4	7 47.0	2 5.7	7 52.2	1.0158	69.4	213.1	57.7	1:39.0T
13: 6	1 19.2	6 14.6	1 48.7	6 4.4	1 34.0	6 9.5	1.0157	68.0	220.5	57.4	1:37.7T
13:12	0 50.8	4 28.8	1 20.1	4 19.1	1 5.5	4 23.9	1.0155	66.1	226.9	56.8	1:35.7T
13:18	0 25.8	2 39.3	0 54.8	2 30.1	0 40.3	2 34.7	1.0153	63.8	232.2	55.9	1:33.2T
13:24	0 4.4	0 45.3	0 33.0	0 36.5	0 18.7	0 40.9	1.0149	61.3	236.5	54.8	1:29.9T
13:30	0-13.2	-1-14.5	0 14.9	-1-22.7	0 0.9	-1-18.6	1.0146	58.5	240.1	53.3	1:26.1T
13:36	0-26.5	-3-21.4	0 0.9	-3-29.0	0-12.8	-3-25.2	1.0141	55.4	243.4	51.6	1:21.7T
13:42	0-35.2	-5-37.0	0 -8.7	-5-44.1	0-21.9	-5-40.5	1.0135	52.1	245.4	49.4	1:16.6T
13:48	0-38.5	-8 -3.6	0-13.1	-8-10.1	0-25.8	-8 -6.8	1.0129	48.5	247.3	46.9	1:10.9T
13:54	0-35.7	-10-44.2	0-11.6	-10-50.2	0-23.6	-10-47.2	1.0121	44.7	248.9	44.0	1: 4.6T
14: 0	0-25.3	-13-43.4	0 -2.8	-13-48.9	0-14.0	-13-46.1	1.0112	40.4	250.1	40.5	0:57.7T
14: 6	0 -5.2	-17 -8.2	0 15.2	-17-13.4	0 5.0	-17-10.7	1.0101	35.7	251.2	36.4	0:49.9T
14:12	0 28.2	-21-11.2	0 46.2	-21-16.2	0 37.2	-21-13.6	1.0088	30.2	252.0	31.5	0:41.3T
14:18	1 22.8	-26-19.4	1 37.7	-26-24.3	1 30.2	-26-21.8	1.0070	23.7	252.7	25.2	0:31.2T
14:24	3 3.6	-33-59.2	3 13.7	-34 -4.5	3 8.6	-34 -1.8	1.0044	14.4	253.4	15.8	0:18.2T
LIMITS	6 31.5	-47-12.7	6 31.9	-47-12.7	6 31.6	-47-12.2	1.0002	0.0	254.7	0.7	0: 0.7

Table 53

TOTAL SOLAR ECLIPSE OF 20 MAR 2015

SAROS 120 Delta T = 77.4 Sec

UNIVERSAL TIME	NORTHERN LIMIT LATITUDE	LONGITUDE	SOUTHERN LIMIT LATITUDE	LONGITUDE	CENTER LINE LATITUDE	LONGITUDE	DIAMETER RATIO	SUN ALT	SUN AZ	PATH WIDTH	DURATION TOTALITY
LIMITS	55 32.9	46 44.2	51 51.2	45 12.1	53 38.7	43 1.0	1.0393	0.0	93.0	416.7	2: 9.7
9:16	55 35.7	44 16.3	53 31.4	26 46.7	54 34.2	32 34.6	1.0414	8.2	102.3	457.8	2:23.2
9:18	56 12.3	36 40.2	54 5.0	24 21.1	55 9.0	29 22.3	1.0420	10.1	105.5	469.0	2:27.4
9:20	56 49.4	32 47.9	54 39.2	22 12.8	55 44.5	26 45.5	1.0425	11.7	108.3	476.6	2:30.9
9:22	57 27.1	29 51.8	55 14.1	20 16.9	56 20.7	24 30.0	1.0429	12.9	110.8	481.7	2:33.7
9:24	58 5.5	27 24.4	55 49.8	18 30.1	56 57.5	22 28.9	1.0432	14.0	113.2	484.8	2:36.2
9:26	58 44.5	25 15.6	56 25.7	16 50.3	57 35.0	20 38.2	1.0435	14.9	115.4	486.4	2:38.3
9:28	59 24.3	23 19.8	57 2.5	15 16.0	58 13.2	18 55.5	1.0437	15.7	117.6	486.8	2:40.1
9:30	60 5.0	21 33.0	57 40.0	13 46.1	58 52.1	17 18.8	1.0439	16.3	119.7	486.0	2:41.7
9:32	60 46.4	19 53.7	58 18.2	12 19.6	59 31.9	15 47.0	1.0441	16.9	121.8	484.5	2:43.0
9:34	61 28.8	18 20.0	58 57.2	10 55.9	60 12.5	14 19.0	1.0442	17.4	123.9	482.3	2:44.2
9:36	62 12.1	16 50.8	59 37.0	9 34.3	60 53.9	12 54.1	1.0443	17.8	125.9	479.5	2:45.1
9:38	62 56.5	15 25.2	60 17.7	8 14.4	61 36.3	11 31.5	1.0444	18.1	127.9	476.3	2:45.8
9:40	63 42.1	14 2.4	60 59.3	6 55.7	62 19.8	10 10.7	1.0445	18.3	129.9	472.7	2:46.4
9:42	64 28.9	12 41.8	61 41.8	5 37.7	63 4.3	8 51.2	1.0445	18.4	131.9	469.0	2:46.7
9:44	65 17.0	11 23.0	62 25.5	4 20.0	63 50.0	7 32.5	1.0446	18.5	133.9	465.0	2:46.9
9:46	66 6.7	10 5.4	63 10.2	3 2.3	64 37.1	6 14.2	1.0446	18.5	135.9	461.0	2:46.8
9:48	66 58.1	8 48.7	63 56.2	1 44.0	65 25.5	4 56.0	1.0445	18.5	137.9	458.9	2:46.4
9:50	67 51.3	7 32.4	64 43.6	0 24.9	66 15.6	3 37.2	1.0445	18.3	139.9	452.8	2:46.4
9:52	68 46.6	6 16.2	65 32.4	0-55.6	67 7.4	2 17.5	1.0444	18.1	141.9	448.7	2:45.9
9:54	69 44.3	4 59.7	66 22.9	-2-17.9	68 1.1	0 58.3	1.0443	17.9	144.0	444.2	2:45.2
9:56	70 44.8	3 42.5	67 15.2	-3-42.8	68 57.0	0-26.8	1.0442	17.5	146.1	440.7	2:44.3
9:58	71 48.4	2 24.3	68 9.5	-5-10.8	69 55.5	-1-52.7	1.0441	17.0	148.3	436.9	2:43.2
10: 0	72 55.9	1 4.6	69 6.2	-6-42.8	70 56.9	-3-22.2	1.0439	16.5	150.5	433.3	2:41.9
10: 2	74 7.9	0-16.8	70 5.5	-8-19.9	72 1.6	-4-56.2	1.0437	15.9	152.7	429.8	2:40.4
10: 4	75 25.7	-1-40.2	71 7.9	-10 -3.6	73 10.5	-6-36.2	1.0435	15.1	155.1	426.4	2:38.6
10: 6	76 50.9	-3 -5.4	72 14.0	-11-55.6	74 24.3	-8-24.2	1.0432	14.3	157.6	423.3	2:36.5
10: 8	78 26.2	-4-31.1	73 24.4	-13-58.7	75 44.4	-10-23.0	1.0429	13.3	160.3	420.4	2:34.2
10:10	80 16.7	-5-51.9	74 40.3	-16-17.0	77 12.9	-12-37.3	1.0425	12.1	163.2	417.7	2:31.4
10:12	82 33.2	-6-43.3	76 3.1	-18-56.9	78 52.9	-15-15.4	1.0420	10.6	166.5	415.3	2:28.1
10:14	85 56.2	-3 -2.2	77 35.3	-22-10.0	80 51.2	-18-35.6	1.0414	8.8	170.4	413.2	2:24.1
LIMITS	88 39.5	53 32.6	87 38.6	-111-34.2	87 58.8	-42-17.0	1.0392	0.0	195.2	409.6	2: 9.2

109

Table 54

TOTAL SOLAR ECLIPSE OF   9 MAR 2016

SAROS 130   Delta T =   78.3 Sec

UNIVERSAL TIME	NORTHERN LIMIT LATITUDE	LONGITUDE	SOUTHERN LIMIT LATITUDE	LONGITUDE	CENTER LINE LATITUDE	LONGITUDE	DIAMETER RATIO	SUN ALT	SUN AZ	PATH WIDTH	DURATION TOTALITY
LIMITS	-1-50.3	-88-22.2	-2-41.3	-88-18.4	-2-16.4	-88-26.4	1.0269	0.0	94.4	93.8	1:30.3
0:18	-2-15.5	-96-37.1	-3-14.3	-97-57.8	-2-44.9	-97-19.4	1.0298	9.4	94.0	104.2	1:47.4
0:24	-2 -8.4	-107-29.3	-3-10.0	-108-15.7	-2-39.1	-107-53.0	1.0335	21.4	93.7	117.8	2:12.8
0:30	-1-39.0	-113-32.5	-2-43.2	-114-14.2	-2-11.0	-113-53.7	1.0357	28.8	93.8	125.9	2:30.2
0:36	-1 -2.1	-118 -4.7	-2 -8.2	-118-45.0	-1-35.1	-118-26.2	1.0374	34.8	94.3	132.1	2:44.7
0:42	0-20.8	-121-47.7	-1-28.5	-122-27.7	0-54.6	-122 -8.0	1.0384	39.9	95.0	137.1	2:57.4
0:48	0 23.5	-124-58.9	0-45.4	-125-39.2	0-10.9	-125-19.3	1.0398	44.5	96.0	141.2	3: 8.8
0:54	1 10.0	-127-47.9	0 0.2	-128-28.6	0 35.2	-128 -8.5	1.0408	48.7	97.4	144.6	3:19.1
1: 0	1 58.5	-130-20.4	0 47.9	-131 -1.6	1 23.2	-130-41.2	1.0416	52.6	99.2	147.4	3:28.4
1: 6	2 48.5	-132-40.3	1 37.4	-133-21.9	2 13.0	-133 -1.2	1.0423	56.3	101.4	149.8	3:36.6
1:12	3 39.9	-134-50.2	2 28.3	-135-32.3	3 4.2	-135-11.4	1.0429	59.7	104.2	151.6	3:44.0
1:18	4 32.6	-136-52.4	3 20.7	-137-34.8	3 56.7	-137-13.7	1.0434	62.8	107.8	153.1	3:50.4
1:24	5 26.5	-138-48.4	4 14.3	-139-31.2	4 50.4	-139 -9.9	1.0438	65.7	112.3	154.1	3:55.8
1:30	6 21.6	-140-39.6	5 9.1	-141-22.6	5 45.4	-141 -1.2	1.0442	68.4	118.0	154.9	4: 0.4
1:36	7 17.7	-142-27.1	6 5.1	-143-10.2	6 41.4	-142-48.7	1.0445	70.7	125.2	155.4	4: 4.0
1:42	8 14.9	-144-12.0	7 2.2	-144-55.1	7 38.6	-144-33.6	1.0447	72.6	134.2	155.5	4: 6.7
1:48	9 13.2	-145-55.1	8 0.4	-146-38.1	8 36.8	-146-16.6	1.0449	74.0	145.0	155.5	4: 8.5
1:54	10 12.5	-147-37.4	8 59.1	-148-20.0	9 36.1	-147-58.7	1.0450	74.8	157.2	155.3	4: 9.4
2: 0	11 13.0	-149-19.5	10 0.1	-150 -1.7	10 36.5	-149-40.6	1.0450	74.8	170.0	154.8	4: 9.3
2: 6	12 14.5	-151 -2.3	11 1.8	-151-44.0	11 38.0	-151-23.1	1.0450	74.0	182.1	154.2	4: 8.4
2:12	13 17.2	-152-46.7	12 4.3	-153-27.7	12 40.7	-153 -7.2	1.0449	72.6	192.8	153.4	4: 6.5
2:18	14 21.2	-154-33.6	13 8.2	-155-13.6	13 44.7	-154-53.6	1.0447	70.7	201.7	152.4	4: 3.8
2:24	15 26.4	-156-23.9	14 13.5	-157 -2.9	14 49.9	-156-43.3	1.0445	68.4	209.1	151.2	4: 0.2
2:30	16 33.0	-158-18.8	15 20.1	-158-56.4	15 56.5	-158-37.5	1.0441	65.7	215.1	149.9	3:55.6
2:36	17 41.1	-160-19.6	16 28.2	-160-55.6	17 4.6	-160-37.5	1.0438	62.8	220.1	148.4	3:50.3
2:42	18 50.8	-162-27.8	17 38.0	-163 -1.8	18 14.3	-162-44.7	1.0433	59.6	224.4	146.6	3:44.0
2:48	20 2.3	-164-45.3	18 49.5	-165-17.1	19 25.8	-165 -1.0	1.0428	56.3	228.2	144.7	3:36.9
2:54	21 15.8	-167-14.8	20 3.2	-167-43.7	20 39.4	-167-29.0	1.0421	52.6	231.6	142.4	3:28.8
3: 0	22 31.6	-169-59.0	21 19.2	-170-24.9	21 55.3	-170-11.7	1.0413	48.7	234.7	139.9	3:19.9
3: 6	23 50.1	-173 -3.1	22 38.0	-173-25.2	23 14.0	-173-13.9	1.0404	44.5	237.8	136.9	3: 9.9
3:12	25 12.0	-176-34.0	24 0.3	-176-51.2	24 36.0	-176-42.3	1.0394	39.9	240.8	133.5	2:58.9
3:18	26 38.0	179 17.1	25 27.0	179 6.2	26 2.4	179 12.0	1.0381	34.8	244.0	129.4	2:46.5
3:24	28 10.1	174 8.7	27 0.0	174 6.9	27 34.9	174 8.3	1.0365	28.6	247.6	124.3	2:32.3
3:30	29 52.1	167 10.2	28 43.4	167 23.8	29 17.6	167 17.8	1.0344	21.3	252.6	117.6	2:15.3
3:36	32 2.2	154 19.3	30 54.3	155 31.6	31 28.1	154 58.3	1.0307	9.2	259.3	106.1	1:50.0
LIMITS	33 0.2	144 30.7	32 7.6	144 30.9	32 33.6	144 35.1	1.0278	0.0	265.0	96.9	1:33.1

Table 55

ANNULAR SOLAR ECLIPSE OF 1 SEP 2016

SAROS 135   Delta T =   78.7 Sec

UNIVERSAL TIME	NORTHERN LIMIT		SOUTHERN LIMIT		CENTER LINE		DIAMETER RATIO	SUN ALT	SUN AZ	PATH WIDTH	DURATION ANNULARITY
	LATITUDE	LONGITUDE	LATITUDE	LONGITUDE	LATITUDE	LONGITUDE					
LIMITS	-2 -23.0	19 21.4	-3 -46.5	19 20.7	-3 -2.3	19 4.5	0.9596	0.0	81.9	152.2	2:41.3
7:24	0 -49.9	3 39.7	-2 -8.6	4 48.4	-1 -28.9	4 13.0	0.9639	16.4	81.1	137.1	2:46.2
7:30	0 -40.3	-2 -48.0	-1 -51.2	-2 -0.7	-1 -15.5	-2 -24.7	0.9659	24.5	80.5	130.2	2:49.1
7:36	0 -47.7	-7 -23.3	-1 -54.1	-6 -43.7	-1 -20.8	-7 -3.7	0.9674	30.5	79.8	125.4	2:51.5
7:42	-1 -4.1	-11 -3.4	-2 -7.2	-10 -27.8	-1 -35.5	-10 -45.8	0.9685	35.6	78.8	121.7	2:53.5
7:48	-1 -26.4	-14 -9.4	-2 -28.8	-13 -36.2	-1 -58.5	-13 -52.9	0.9694	40.0	77.7	118.6	2:55.4
7:54	-1 -53.0	-16 -51.9	-2 -51.2	-16 -20.3	-2 -22.0	-16 -36.1	0.9702	44.1	76.3	115.9	2:57.0
8: 0	-2 -23.0	-19 -17.0	-3 -19.4	-18 -46.5	-2 -51.1	-19 -1.8	0.9709	47.7	74.6	113.5	2:58.5
8: 6	-2 -55.9	-21 -28.8	-3 -50.6	-20 -59.1	-3 -23.1	-21 -14.0	0.9714	51.1	72.5	111.4	2:59.8
8:12	-3 -31.3	-23 -30.1	-4 -24.5	-23 -1.1	-3 -57.8	-23 -15.8	0.9719	54.3	70.1	109.5	3: 1.0
8:18	-4 -8.9	-25 -23.2	-5 -0.8	-24 -54.5	-4 -34.7	-25 -8.9	0.9723	57.2	67.2	107.8	3: 2.0
8:24	-4 -48.4	-27 -9.4	-5 -39.3	-26 -41.1	-5 -13.7	-26 -55.3	0.9727	59.9	63.7	106.3	3: 2.9
8:30	-5 -29.8	-28 -50.2	-6 -19.7	-28 -22.2	-5 -54.6	-28 -36.3	0.9730	62.4	59.5	104.9	3: 3.7
8:36	-6 -12.8	-30 -28.7	-7 -2.0	-29 -58.9	-6 -37.3	-30 -12.8	0.9732	64.6	54.5	103.6	3: 4.3
8:42	-6 -57.5	-31 -59.6	-7 -46.0	-31 -32.0	-7 -21.7	-31 -45.8	0.9734	66.6	48.6	102.5	3: 4.8
8:48	-7 -43.8	-33 -29.8	-8 -31.7	-33 -2.3	-8 -7.7	-33 -16.1	0.9735	68.2	41.8	101.6	3: 5.1
8:54	-8 -31.6	-34 -58.1	-9 -19.1	-34 -30.7	-8 -55.3	-34 -44.4	0.9736	69.4	34.1	100.8	3: 5.4
9: 0	-9 -21.0	-36 -25.0	-10 -8.1	-35 -57.8	-9 -44.5	-36 -11.4	0.9738	70.2	25.6	100.2	3: 5.5
9: 6	-10 -11.8	-37 -51.2	-10 -58.8	-37 -24.1	-10 -35.2	-37 -37.7	0.9738	70.5	16.6	99.8	3: 5.5
9:12	-11 -4.2	-39 -17.4	-11 -51.2	-38 -50.4	-11 -27.6	-39 -4.0	0.9736	70.3	7.6	99.5	3: 5.5
9:18	-11 -58.1	-40 -44.2	-12 -45.2	-40 -17.3	-12 -21.6	-40 -30.8	0.9735	69.6	358.9	99.4	3: 5.3
9:24	-12 -53.6	-42 -12.2	-13 -41.0	-41 -45.5	-13 -17.2	-41 -58.9	0.9733	68.5	351.0	99.6	3: 5.1
9:30	-13 -50.8	-43 -42.2	-14 -38.6	-43 -15.6	-14 -14.6	-43 -28.9	0.9731	67.0	343.9	99.9	3: 4.7
9:36	-14 -49.7	-45 -14.8	-15 -38.1	-44 -48.5	-15 -13.8	-45 -1.6	0.9729	65.1	337.6	100.5	3: 4.3
9:42	-15 -50.4	-46 -51.0	-16 -39.6	-46 -24.9	-16 -14.9	-46 -38.0	0.9726	62.9	332.1	101.3	3: 3.8
9:48	-16 -53.2	-48 -31.7	-17 -43.3	-48 -6.0	-17 -18.2	-48 -18.9	0.9722	60.5	327.3	102.3	3: 3.3
9:54	-17 -58.1	-50 -18.2	-18 -49.3	-49 -52.9	-18 -23.6	-50 -5.6	0.9718	57.9	323.1	103.6	3: 2.6
10: 0	-19 -5.4	-52 -11.7	-19 -58.0	-51 -47.2	-19 -31.6	-51 -59.5	0.9713	55.0	319.3	105.2	3: 1.9
10: 6	-20 -15.3	-54 -14.3	-21 -9.6	-53 -50.8	-20 -42.4	-54 -2.4	0.9708	51.9	315.9	107.0	3: 1.1
10:12	-21 -28.3	-56 -28.1	-22 -24.5	-56 -5.6	-21 -56.3	-56 -16.9	0.9702	48.5	312.7	109.2	3: 0.2
10:18	-22 -44.8	-58 -58.3	-23 -43.3	-58 -35.6	-23 -13.9	-58 -45.9	0.9695	44.9	309.6	111.8	2:59.2
10:24	-24 -5.5	-61 -43.5	-25 -6.9	-61 -25.2	-24 -36.1	-61 -34.3	0.9686	41.0	306.6	114.9	2:58.1
10:30	-25 -31.5	-64 -58.4	-26 -36.4	-64 -41.8	-26 -3.8	-64 -49.0	0.9677	36.6	303.6	118.6	2:56.8
10:36	-27 -4.6	-68 -46.3	-28 -13.9	-68 -37.6	-27 -39.1	-68 -41.8	0.9665	31.7	300.3	123.0	2:55.4
10:42	-28 -48.0	-73 -34.8	-30 -3.2	-73 -36.6	-29 -25.4	-73 -35.4	0.9651	25.9	296.7	128.7	2:53.7
10:48	-30 -49.4	-80 -15.3	-32 -14.3	-80 -41.6	-31 -31.4	-80 -27.5	0.9632	18.5	292.0	136.5	2:51.6
LIMITS	-34 -55.7	-100 -38.3	-36 -21.4	-100 -31.7	-35 -36.2	-100 -19.1	0.9583	0.0	279.9	157.6	2:46.7

Table 56

ANNULAR SOLAR ECLIPSE OF 26 FEB 2017

SAROS 140 Delta T = 79.1 Sec

UNIVERSAL TIME	NORTHERN LIMIT LATITUDE	NORTHERN LIMIT LONGITUDE	SOUTHERN LIMIT LATITUDE	SOUTHERN LIMIT LONGITUDE	CENTER LINE LATITUDE	CENTER LINE LONGITUDE	DIAMETER RATIO	SUN ALT	SUN AZ	PATH WIDTH	DURATION ANNULARITY
LIMITS	-42-43.9	113 37.6	-43-33.1	114 4.0	-43-10.5	113 36.8	0.9769	0.0	101.3	96.2	1:22.4
13:18	-44-37.1	98 45.9	-45 -7.4	101 19.3	-44-52.6	99 57.8	0.9798	10.5	91.4	82.6	1:17.9
13:24	-45-25.3	86 18.7	-46 -1.1	87 18.8	-45-43.1	86 48.0	0.9827	20.7	80.5	69.6	1:12.6
13:30	-45-28.9	78 14.2	-46 -2.9	78 48.8	-45-45.8	78 31.2	0.9845	27.3	72.8	61.7	1: 8.7
13:36	-45-14.0	71 54.4	-45-45.5	72 15.6	-45-29.7	72 4.9	0.9859	32.6	66.1	55.7	1: 5.5
13:42	-44-48.0	66 36.5	-45-17.1	66 49.2	-45 -2.5	66 42.8	0.9870	37.2	59.9	51.0	1: 2.5
13:48	-44-14.4	62 0.8	-44-41.3	62 7.8	-44-27.8	62 4.3	0.9879	41.2	54.1	47.1	0:59.8
13:54	-43-35.1	57 56.6	-43-60.0	57 59.4	-43-47.5	57 58.0	0.9887	44.7	48.3	43.8	0:57.4
14: 0	-42-51.3	54 16.8	-43-14.4	54 16.6	-43 -2.8	54 16.7	0.9894	48.0	42.5	41.0	0:55.1
14: 6	-42 -4.1	50 56.6	-42-25.5	50 54.1	-42-14.8	50 55.4	0.9900	50.9	36.6	38.6	0:53.1
14:12	-41-13.8	47 52.5	-41-33.8	47 48.3	-41-23.8	47 50.4	0.9905	53.5	30.4	36.6	0:51.2
14:18	-40-21.1	45 1.8	-40-39.8	44 56.3	-40-30.4	44 59.0	0.9909	55.8	24.0	35.0	0:49.6
14:24	-39-26.2	42 22.2	-39-43.9	42 15.7	-39-35.0	42 19.0	0.9913	57.8	17.3	33.6	0:48.1
14:30	-38-29.3	39 52.0	-38-46.1	39 44.7	-38-37.7	39 48.4	0.9916	59.5	10.1	32.5	0:46.9
14:36	-37-30.8	37 29.7	-37-46.8	37 21.6	-37-38.3	37 25.7	0.9919	60.9	2.6	31.7	0:45.8
14:42	-36-30.7	35 13.4	-36-46.0	35 5.2	-36-38.3	35 9.6	0.9920	61.8	354.8	31.1	0:45.0
14:48	-35-29.1	33 3.4	-35-44.0	32 54.1	-35-36.5	32 58.8	0.9922	62.4	346.8	30.7	0:44.4
14:54	-34-26.2	30 57.2	-34-40.6	30 47.3	-34-33.4	30 52.3	0.9922	62.6	338.8	30.6	0:44.0
15: 0	-33-21.8	28 54.1	-33-36.0	28 43.7	-33-28.9	28 48.9	0.9922	62.3	330.8	30.8	0:43.8
15: 6	-32-16.2	26 53.2	-32-30.2	26 42.2	-32-23.2	26 47.7	0.9922	61.7	323.2	31.2	0:43.9
15:12	-31 -9.2	24 53.6	-31-23.2	24 41.9	-31-16.2	24 47.7	0.9921	60.6	315.9	31.8	0:44.2
15:18	-30 -0.8	22 54.1	-30-14.9	22 41.6	-30 -7.8	22 47.9	0.9919	59.2	309.2	32.7	0:44.7
15:24	-28-51.0	20 53.7	-29 -5.2	20 40.3	-28-58.0	20 47.1	0.9917	57.4	303.1	33.8	0:45.4
15:30	-27-39.6	18 51.2	-27-54.0	18 36.8	-27-46.8	18 44.0	0.9914	55.3	297.6	35.3	0:46.4
15:36	-26-28.4	16 45.2	-26-41.3	16 29.6	-26-33.8	16 37.4	0.9910	52.9	292.5	37.1	0:47.6
15:42	-25-11.4	14 34.0	-25-26.7	14 18.9	-25-19.0	14 25.5	0.9906	50.2	288.0	39.2	0:49.0
15:48	-23-54.1	12 15.7	-24-10.1	11 56.8	-24 -2.0	12 6.3	0.9901	47.3	284.0	41.7	0:50.7
15:54	-22-34.2	9 47.4	-22-50.9	9 26.3	-22-42.5	9 36.9	0.9894	44.0	280.3	44.6	0:52.6
16: 0	-21-11.1	7 5.6	-21-28.7	6 41.6	-21-19.8	6 53.6	0.9887	40.3	277.0	47.9	0:54.8
16: 6	-19-43.9	4 4.5	-20 -2.5	3 36.8	-19-53.2	3 50.7	0.9878	36.2	273.9	51.9	0:57.3
16:12	-18-11.2	0 35.3	-18-31.0	0 2.2	-18-21.0	0 18.9	0.9867	31.5	271.1	56.6	1: 0.2
16:18	-16-30.2	-3-39.0	-16-51.2	-4-20.9	-16-40.6	-3-59.7	0.9854	26.0	268.5	62.4	1: 3.5
16:24	-14-34.5	-9-19.0	-14-56.1	-10-19.4	-14-45.2	-9-48.7	0.9835	18.9	265.9	70.0	1: 7.6
16:30	-11-51.8	-19-48.5	-11-57.9	-23-19.3	-11-56.9	-21-17.2	0.9799	6.0	262.5	83.8	1:14.5
LIMITS	-10-32.9	-27 -1.0	-11-19.7	-27-20.1	-10-58.2	-26-57.5	0.9783	0.0	261.4	90.1	1:17.3

Table 57

TOTAL SOLAR ECLIPSE OF 21 AUG 2017

SAROS 145   Delta T =   79.6 Sec

UNIVERSAL TIME	NORTHERN LIMIT LATITUDE	LONGITUDE	SOUTHERN LIMIT LATITUDE	LONGITUDE	CENTER LINE LATITUDE	LONGITUDE	DIAMETER RATIO	SUN ALT	SUN AZ	PATH WIDTH	DURATION TOTALITY
LIMITS	39 59.6	171 42.3	39 28.8	171 23.3	39 45.4	171 27.1	1.0156	0.0	74.7	62.2	0:51.6
16:54	43 38.8	151 40.2	43 7.0	150 8.2	43 23.2	150 53.2	1.0205	16.7	89.5	80.1	1:17.3
17: 0	44 46.4	141 44.7	44 3.9	140 40.2	44 25.3	141 11.9	1.0228	24.7	97.9	87.6	1:31.3
17: 6	45 14.4	134 26.8	44 26.0	133 37.7	44 50.3	134 1.8	1.0243	30.6	104.9	92.8	1:42.4
17:12	45 21.9	128 26.2	44 29.5	127 48.9	44 55.8	128 7.2	1.0256	35.6	111.2	96.7	1:51.9
17:18	45 15.8	123 14.8	44 20.6	122 47.4	44 48.2	123 0.7	1.0266	40.0	117.3	99.8	2: 0.3
17:24	44 59.7	118 38.4	44 2.5	118 19.9	44 31.1	118 28.8	1.0274	43.8	123.3	102.5	2: 7.7
17:30	44 35.8	114 29.0	43 37.3	114 18.3	44 6.6	114 23.4	1.0281	47.3	129.3	104.7	2:14.2
17:36	44 5.6	110 41.1	43 6.2	110 37.6	43 35.9	110 39.2	1.0287	50.5	135.4	106.6	2:20.0
17:42	43 30.1	107 11.0	42 30.2	107 13.9	43 0.1	107 12.3	1.0292	53.3	141.8	108.3	2:25.0
17:48	42 50.0	103 55.7	41 50.1	104 4.5	42 20.0	103 60.0	1.0296	55.9	148.4	109.7	2:29.3
17:54	42 6.0	100 53.0	41 6.2	101 7.1	41 36.1	101 0.0	1.0300	58.1	155.5	110.9	2:32.9
18: 0	41 18.4	98 1.1	40 19.1	98 20.0	40 48.8	98 10.5	1.0302	60.0	162.9	112.0	2:35.8
18: 6	40 27.8	95 18.3	39 29.1	95 41.5	39 58.4	95 29.9	1.0304	61.6	170.9	112.9	2:37.9
18:12	39 34.2	92 43.3	38 36.3	93 10.4	39 5.2	92 56.9	1.0305	62.8	179.2	113.6	2:39.4
18:18	38 38.0	90 14.9	37 41.0	90 45.5	38 9.5	90 30.2	1.0306	63.6	187.9	114.2	2:40.2
18:24	37 39.3	87 51.8	36 43.4	88 25.5	37 11.3	88 8.8	1.0306	63.9	196.7	114.7	2:40.2
18:30	36 38.1	85 33.1	35 43.4	86 9.6	36 10.8	85 51.5	1.0305	63.8	205.4	115.0	2:39.6
18:36	35 34.7	83 17.8	34 41.3	83 56.7	35 8.0	83 37.4	1.0304	63.2	213.9	115.1	2:38.3
18:42	34 29.0	81 4.7	33 36.9	81 45.9	34 3.0	81 25.4	1.0302	62.2	221.9	115.1	2:36.4
18:48	33 20.9	78 52.8	32 30.3	79 35.9	32 55.7	79 14.5	1.0300	60.9	229.3	114.9	2:33.8
18:54	32 10.5	76 41.0	31 21.4	77 25.9	31 46.0	77 3.6	1.0297	59.1	236.1	114.5	2:30.6
19: 0	30 57.8	74 28.1	30 10.1	75 14.5	30 33.9	74 51.4	1.0293	57.0	242.2	113.8	2:26.7
19: 6	29 42.0	72 12.6	28 56.3	73 0.5	29 19.2	72 36.7	1.0288	54.6	247.8	112.9	2:22.2
19:12	28 23.6	69 53.0	27 39.6	70 42.2	28 1.7	70 17.8	1.0283	51.9	252.7	111.6	2:17.1
19:18	27 1.8	67 27.2	26 19.8	68 17.6	26 40.9	67 52.6	1.0277	48.9	257.1	109.9	2:11.4
19:24	25 36.3	64 52.6	24 56.3	65 44.2	25 16.4	65 18.5	1.0269	45.6	261.1	107.8	2: 5.0
19:30	24 6.2	62 5.5	23 28.5	62 58.4	23 47.5	62 32.1	1.0261	41.9	264.8	105.0	1:57.9
19:36	22 30.5	59 0.6	21 55.2	59 55.1	22 13.0	59 28.0	1.0251	37.8	268.1	101.5	1:50.0
19:42	20 47.1	55 29.4	20 14.8	56 26.1	20 31.1	55 57.9	1.0239	33.1	271.1	97.1	1:41.2
19:48	18 52.6	51 16.2	18 24.0	52 16.4	18 38.5	51 46.5	1.0224	27.7	274.0	91.3	1:31.1
19:54	16 38.9	45 44.5	16 15.6	46 52.1	16 27.4	46 18.6	1.0205	20.9	276.8	83.3	1:19.0
20: 0	13 31.5	36 18.0	13 21.5	38 0.6	13 27.0	37 10.7	1.0173	10.3	279.9	69.9	1: 1.8
LIMITS	11 15.5	27 17.1	10 46.9	27 30.5	11 2.8	27 31.2	1.0142	0.0	282.1	56.8	0:47.3

Table 58

TOTAL SOLAR ECLIPSE OF 2 JUL 2019

SAROS 127

Delta T = 81.3 Sec

UNIVERSAL TIME	NORTHERN LIMIT LATITUDE	LONGITUDE	SOUTHERN LIMIT LATITUDE	LONGITUDE	CENTER LINE LATITUDE	LONGITUDE	DIAMETER RATIO	SUN ALT	SUN AZ	PATH WIDTH	DURATION TOTALITY
LIMITS	-37 -7.1	160 36.7	-38-12.4	160 8.3	-37-27.1	159 55.0	1.0318	0.0	59.9	126.2	2: 0.4
18: 4	-31-51.6	149 32.4	-34-35.0	152 15.6	-33 -8.7	150 43.4	1.0346	9.5	54.3	136.2	2:21.0
18: 8	-28-51.3	143 28.4	-30-56.5	144 39.8	-29-52.8	144 1.6	1.0369	16.7	49.9	144.9	2:39.8
18:12	-26-50.1	139 25.0	-28-45.7	140 10.1	-27-47.2	139 46.1	1.0383	21.5	46.8	151.2	2:53.6
18:16	-25-15.7	136 13.3	-27 -7.0	138 44.9	-26-10.9	136 28.1	1.0394	25.4	44.2	156.4	3: 5.4
18:20	-23-57.9	133 31.9	-25-48.9	133 27.7	-24-52.0	133 42.6	1.0404	28.6	41.8	161.1	3:15.8
18:24	-22-51.7	131 10.6	-24-39.5	131 16.5	-23-45.3	131 18.5	1.0412	31.5	39.4	165.4	3:25.3
18:28	-21-54.4	129 3.7	-23-41.6	129 18.5	-22-47.7	129 9.5	1.0419	34.1	37.1	169.4	3:34.0
18:32	-21 -4.3	127 7.7	-22-51.2	127 17.2	-21-57.5	127 12.0	1.0425	36.4	34.7	173.2	3:42.1
18:36	-20-20.2	125 20.2	-22 -7.0	125 27.1	-21-13.3	125 23.2	1.0431	38.4	32.4	176.7	3:49.5
18:40	-19-41.3	123 39.4	-21-28.1	123 44.3	-20-34.4	123 41.4	1.0435	40.3	29.9	180.1	3:56.3
18:44	-19 -6.9	122 4.0	-20-53.8	122 7.2	-20 -0.2	122 5.3	1.0440	42.0	27.4	183.2	4: 2.5
18:48	-18-36.6	120 33.0	-20-23.7	120 34.9	-19-30.0	120 33.7	1.0443	43.5	24.7	186.2	4: 8.2
18:52	-18-10.1	119 5.8	-19-57.4	119 6.5	-19 -3.5	119 5.8	1.0447	44.8	22.0	188.9	4:13.3
18:56	-17-47.0	117 41.3	-19-34.5	117 41.3	-18-40.6	117 41.0	1.0450	46.0	19.2	191.4	4:17.8
19: 0	-17-27.1	116 19.3	-19-14.9	116 18.8	-18-20.8	116 18.8	1.0452	47.0	16.3	193.7	4:21.7
19: 4	-17-10.3	114 59.3	-18-58.4	114 58.0	-18 -4.2	114 58.5	1.0454	47.9	13.3	195.6	4:25.1
19: 8	-16-58.4	113 40.8	-18-44.7	113 39.0	-17-50.4	113 39.8	1.0456	48.8	10.2	197.3	4:27.8
19:12	-16-45.4	112 23.4	-18-33.9	112 21.3	-17-39.5	112 22.3	1.0457	49.1	7.1	198.6	4:30.0
19:16	-16-37.1	111 6.9	-18-25.7	111 4.4	-17-31.3	111 5.6	1.0458	49.4	3.9	199.7	4:31.5
19:20	-16-31.5	109 50.8	-18-20.3	109 48.1	-17-25.8	109 49.4	1.0459	49.6	0.7	200.3	4:32.4
19:24	-16-28.6	108 35.0	-18-17.5	108 32.0	-17-22.9	108 33.5	1.0459	49.7	357.5	200.6	4:32.7
19:28	-16-28.4	107 19.0	-18-17.4	107 15.7	-17-22.7	107 17.4	1.0459	49.5	354.3	200.5	4:32.4
19:32	-16-30.9	106 2.6	-18-19.9	105 59.0	-17-25.2	106 0.9	1.0459	49.2	351.1	200.1	4:31.5
19:36	-16-36.1	104 45.5	-18-25.1	104 41.5	-17-30.4	104 43.6	1.0458	48.8	348.0	199.3	4:29.9
19:40	-16-44.1	103 27.3	-18-33.0	103 22.9	-17-38.4	103 25.2	1.0457	48.1	344.9	198.1	4:27.7
19:44	-16-55.0	102 7.7	-18-43.8	102 2.8	-17-49.3	102 5.4	1.0456	47.4	341.9	196.6	4:25.0
19:48	-17 -8.9	100 46.2	-18-57.6	100 40.8	-18 -3.1	100 43.7	1.0454	46.4	339.1	194.8	4:21.6
19:52	-17-25.9	99 22.5	-19-14.6	99 16.3	-18-20.1	99 19.7	1.0452	45.3	336.3	192.7	4:17.6
19:56	-17-46.3	97 56.0	-19-34.8	97 48.9	-18-40.4	97 52.8	1.0449	44.0	333.6	190.3	4:13.0
20: 0	-18-10.2	96 28.2	-19-58.7	96 17.9	-19 -4.2	96 22.4	1.0446	42.6	331.0	187.7	4: 7.8
20: 4	-18-38.1	94 52.2	-20-26.5	94 42.5	-19-32.0	94 47.7	1.0443	41.0	328.5	184.8	4: 2.0
20: 8	-19-10.2	93 13.2	-20-58.6	93 1.7	-20 -4.2	93 7.9	1.0438	39.2	326.0	181.7	3:55.6
20:12	-19-47.2	91 28.0	-21-35.7	91 14.3	-20-41.2	91 21.6	1.0434	37.2	323.7	178.3	3:48.7
20:16	-20-29.7	89 35.1	-22-18.6	89 18.5	-21-23.9	89 27.4	1.0428	35.0	321.4	174.8	3:41.0
20:20	-21-18.8	87 32.4	-23 -8.3	87 12.0	-22-13.2	87 22.9	1.0422	32.5	319.2	171.0	3:32.7
20:24	-22-15.9	85 17.0	-24 -6.5	84 51.4	-23-10.8	85 4.9	1.0415	29.8	316.9	166.8	3:23.7
20:28	-23-23.2	82 44.0	-25-15.6	82 11.1	-24-19.0	82 28.5	1.0407	26.7	314.6	162.4	3:13.7
20:32	-24-44.5	79 45.8	-26-40.3	79 1.7	-25-41.8	79 25.0	1.0397	23.1	312.2	157.4	3: 2.5
20:36	-26-27.1	76 6.9	-28-29.7	75 2.9	-27-27.5	75 36.9	1.0384	18.7	309.5	151.6	2:49.6
20:40	-28-49.7	71 7.2	-31-11.9	69 13.6	-29-58.7	70 15.4	1.0366	12.7	306.1	144.2	2:33.4
LIMITS	-35-13.9	57 26.3	-36-21.8	57 53.3	-35-36.0	58 5.7	1.0329	0.0	298.9	130.1	2: 4.1

Table 59

ANNULAR SOLAR ECLIPSE OF 26 DEC 2019

Delta T = 81.7 Sec

SAROS 132

UNIVERSAL TIME	NORTHERN LIMIT LATITUDE LONGITUDE		SOUTHERN LIMIT LATITUDE LONGITUDE		CENTER LINE LATITUDE LONGITUDE		DIAMETER RATIO	SUN ALT	SUN AZ	PATH WIDTH	DURATION ANNULARITY
LIMITS	26 40.2	-48-33.0	25 16.7	-47-58.3	25 48.8	-48-37.4	0.9564	0.0	116.5	163.4	2:58.8
3:42	18 49.2	-64-26.9	17 16.6	-64-33.4	18 2.3	-64-30.9	0.9612	18.3	123.9	143.9	3: 4.7
3:48	15 40.3	-70-13.4	14 19.2	-70 -5.9	14 59.4	-70-10.0	0.9631	25.7	125.9	136.9	3: 8.1
3:54	13 21.9	-74-21.4	12 6.5	-74 -9.1	12 44.0	-74-15.5	0.9645	31.4	128.1	132.0	3:11.1
4: 0	11 30.3	-77-40.9	10 18.6	-77-28.4	10 54.2	-77-33.8	0.9655	36.2	130.1	128.3	3:14.0
4: 6	9 56.4	-80-30.8	8 47.2	-80-15.4	9 21.6	-80-23.2	0.9664	40.4	132.1	125.4	3:16.7
4:12	8 35.6	-83 -0.6	7 28.3	-82-45.0	8 1.8	-82-52.9	0.9672	44.2	134.3	123.1	3:19.4
4:18	7 25.2	-85-16.0	6 19.2	-85 -0.6	6 52.0	-85 -8.3	0.9678	47.6	136.7	121.3	3:22.0
4:24	6 23.3	-87-20.6	5 18.4	-87 -5.6	5 50.7	-87-13.2	0.9683	50.8	139.4	119.9	3:24.5
4:30	5 28.8	-89-17.0	4 24.7	-89 -2.6	4 56.6	-89 -9.8	0.9688	53.6	142.5	118.8	3:26.9
4:36	4 40.9	-91 -7.1	3 37.3	-90-53.3	4 8.9	-91 -0.2	0.9692	56.3	146.0	118.0	3:29.1
4:42	3 58.8	-92-52.1	2 55.6	-92-39.2	3 27.0	-92-45.7	0.9695	58.6	149.9	117.5	3:31.3
4:48	3 22.1	-94-33.4	2 19.1	-94-21.3	2 50.5	-94-27.4	0.9697	60.6	154.5	117.2	3:33.2
4:54	2 50.5	-96-11.7	1 47.6	-96 -0.6	2 18.9	-96 -6.2	0.9699	62.4	159.8	117.0	3:34.9
5: 0	2 23.7	-97-47.9	1 20.8	-97-37.8	1 52.1	-97-42.9	0.9700	63.8	165.3	117.0	3:36.4
5: 6	2 1.5	-99-22.7	58.5	-99-13.6	1 29.9	-99-18.2	0.9701	64.8	171.5	117.2	3:37.7
5:12	1 43.9	-100-56.8	40.8	-100-48.6	1 12.1	-100-52.6	0.9702	65.4	178.1	117.4	3:38.6
5:18	1 30.6	-102-30.3	27.1	-102-23.3	58.7	-102-26.8	0.9701	65.6	184.9	117.8	3:39.3
5:24	1 21.7	-104 -4.2	17.8	-103-58.4	49.6	-104 -1.3	0.9701	65.3	191.7	118.2	3:39.7
5:30	1 17.3	-105-39.1	13.0	-105-34.3	45.8	-105-36.7	0.9699	64.6	198.2	118.7	3:39.7
5:36	1 17.0	-107-15.3	12.6	-107-11.7	44.8	-107-13.5	0.9698	63.6	204.1	119.2	3:39.5
5:42	1 22.0	-108-53.6	16.7	-108-51.1	49.2	-108-52.3	0.9695	62.1	209.6	119.8	3:38.9
5:48	1 31.5	-110-34.7	25.5	-110-33.2	58.4	-110-33.9	0.9692	60.3	214.4	120.5	3:38.1
5:54	1 46.1	-112-19.2	39.3	-112-18.8	1 12.6	-112-19.0	0.9689	58.2	218.5	121.4	3:36.9
6: 0	2 6.1	-114 -8.2	58.5	-114 -8.7	1 32.1	-114 -8.4	0.9685	55.8	222.2	122.4	3:35.5
6: 6	2 32.0	-116 -2.7	1 23.3	-116 -4.2	1 57.5	-116 -3.4	0.9680	53.2	225.3	123.6	3:33.8
6:12	3 4.4	-118 -4.3	1 54.6	-118 -6.5	2 29.3	-118 -5.4	0.9675	50.2	228.0	125.0	3:31.8
6:18	3 44.2	-120-14.8	2 33.0	-120-17.8	3 8.4	-120-16.1	0.9669	47.0	230.3	126.7	3:29.7
6:24	4 32.7	-122-36.7	3 19.7	-122-39.9	3 56.0	-122-38.3	0.9661	43.5	232.4	128.8	3:27.3
6:30	5 31.5	-125-14.0	4 16.4	-125-17.1	4 53.8	-125-15.5	0.9653	39.7	234.2	131.4	3:24.7
6:36	6 43.7	-128-12.5	5 25.8	-128-14.8	6 4.5	-128-13.5	0.9643	35.4	235.8	134.5	3:21.9
6:42	8 13.9	-131-42.4	6 52.1	-131-42.6	7 32.7	-131-42.3	0.9631	30.5	237.3	138.5	3:18.8
6:48	10 12.1	-136 -4.9	8 44.2	-135-59.7	9 27.8	-136 -1.9	0.9616	24.6	238.9	143.7	3:15.4
6:54	13 6.2	-142-20.1	11 25.2	-141-57.5	12 15.0	-142 -7.8	0.9595	16.7	240.9	151.2	3:11.4
LIMITS	19 36.9	-156-30.5	18 9.7	-157 -1.8	18 45.3	-156-28.1	0.9550	0.0	245.2	168.8	3: 4.7

Table 60

ANNULAR SOLAR ECLIPSE OF 21 JUN 2020

SAROS 137   Delta T = 82.2 Sec

UNIVERSAL TIME	NORTHERN LIMIT LATITUDE	LONGITUDE	SOUTHERN LIMIT LATITUDE	LONGITUDE	CENTER LINE LATITUDE	LONGITUDE	DIAMETER RATIO	SUN ALT	SUN AZ	PATH WIDTH	DURATION ANNULARITY
LIMITS	1 36.0	-17-39.2	0 56.0	-18 -5.6	1 17.5	-17-55.2	0.9772	0.0	66.6	85.2	1:22.4
4:54	8 55.2	-32-51.6	8 39.3	-33-38.1	8 47.2	-33-15.1	0.9823	18.3	68.1	66.3	1:13.4
5: 0	12 12.8	-38-53.1	11 55.1	-39-26.0	12 3.9	-39 -9.7	0.9844	26.3	69.6	58.2	1: 8.8
5: 6	14 42.1	-43-15.5	14 24.4	-43-41.7	14 33.2	-43-28.6	0.9860	32.5	71.2	52.2	1: 5.1
5:12	16 46.0	-46-50.0	16 28.8	-47-11.9	16 37.4	-47 -1.0	0.9872	37.7	72.9	47.3	1: 1.9
5:18	18 33.1	-49-55.9	18 16.6	-50-14.7	18 24.8	-50 -5.3	0.9883	42.4	74.7	43.1	0:58.9
5:24	20 7.8	-52-42.8	19 52.0	-52-59.1	19 59.8	-52-50.9	0.9892	46.7	76.7	39.6	0:56.2
5:30	21 32.6	-55-16.3	21 17.4	-55-30.5	21 25.0	-55-23.4	0.9900	50.7	78.8	36.5	0:53.8
5:36	22 49.2	-57-40.1	22 34.7	-57-52.5	22 41.9	-57-46.3	0.9907	54.4	81.1	33.9	0:51.5
5:42	23 58.7	-59-58.6	23 44.7	-60 -7.6	23 51.7	-60 -2.1	0.9913	58.0	83.6	31.5	0:49.4
5:48	25 1.9	-62 -7.8	24 48.4	-62-17.5	24 55.1	-62-12.7	0.9918	61.4	86.5	29.5	0:47.4
5:54	25 59.3	-64-15.1	25 46.3	-64-23.5	25 52.8	-64-19.3	0.9923	64.7	89.7	27.7	0:45.7
6: 0	26 51.4	-66-19.4	26 38.8	-66-26.8	26 45.1	-66-23.1	0.9927	67.9	93.5	26.2	0:44.1
6: 6	27 38.4	-68-21.6	27 26.2	-68-28.1	27 32.3	-68-24.9	0.9930	70.9	98.2	24.9	0:42.7
6:12	28 20.6	-70-22.0	28 8.7	-70-28.2	28 14.6	-70-25.4	0.9933	73.8	104.0	23.8	0:41.4
6:18	28 58.1	-72-22.7	28 46.4	-72-27.6	28 52.3	-72-25.2	0.9935	76.6	111.6	22.9	0:40.4
6:24	29 31.1	-74-22.7	29 19.6	-74-26.8	29 25.3	-74-24.8	0.9937	79.0	122.3	22.2	0:39.6
6:30	29 59.8	-76-22.9	29 48.2	-76-26.3	29 53.9	-76-24.6	0.9939	81.1	137.9	21.7	0:38.9
6:36	30 23.6	-78-23.7	30 12.3	-78-26.5	30 17.9	-78-25.1	0.9940	82.4	160.0	21.3	0:38.5
6:42	30 43.1	-80-25.5	30 31.8	-80-27.6	30 37.5	-80-26.6	0.9940	82.6	186.0	21.2	0:38.2
6:48	30 58.2	-82-28.7	30 46.8	-82-30.1	30 52.5	-82-29.4	0.9940	81.5	208.6	21.2	0:38.2
6:54	31 8.6	-84-33.5	30 57.1	-84-34.3	31 2.9	-84-33.9	0.9940	79.7	224.8	21.3	0:38.4
7: 0	31 14.4	-86-40.5	31 2.7	-86-40.6	31 8.6	-86-40.5	0.9939	77.3	235.9	21.7	0:38.7
7: 6	31 15.4	-88-49.8	31 3.4	-88-49.2	31 9.4	-88-49.5	0.9937	74.6	243.9	22.2	0:39.3
7:12	31 11.4	-91 -2.0	30 59.1	-91 -0.6	31 5.3	-91 -1.3	0.9935	71.8	250.1	22.9	0:40.1
7:18	31 2.2	-93-17.5	30 49.5	-93-15.2	30 55.8	-93-16.4	0.9933	68.7	255.1	23.8	0:41.0
7:24	30 47.5	-95-36.9	30 34.3	-95-33.7	30 40.9	-95-35.3	0.9930	65.6	259.3	24.9	0:42.2
7:30	30 27.0	-98 -1.0	30 13.2	-97-56.7	30 20.1	-97-58.9	0.9926	62.4	263.0	26.2	0:43.6
7:36	30 0.1	-100-30.6	29 45.8	-100-25.1	29 53.0	-100-27.9	0.9922	59.0	266.4	27.8	0:45.1
7:42	29 26.4	-103 -6.9	29 11.4	-103 0.0	29 18.9	-103 -3.5	0.9917	55.5	269.5	29.7	0:46.9
7:48	28 45.1	-105-51.5	28 29.3	-105-43.0	28 37.2	-105-47.2	0.9911	51.7	272.4	31.9	0:48.8
7:54	27 55.0	-108-46.5	27 38.5	-108-36.0	27 46.7	-108-41.3	0.9904	47.8	275.2	34.5	0:51.0
8: 0	26 54.7	-111-55.1	26 37.3	-111-42.3	26 46.0	-111-48.7	0.9896	43.6	277.8	37.5	0:53.3
8: 6	25 41.8	-115-22.2	25 23.6	-115 -6.4	25 32.7	-115-14.2	0.9887	39.1	280.4	41.0	0:56.0
8:12	24 12.5	-119-15.8	23 53.6	-118-56.0	24 3.0	-119 -5.9	0.9876	34.0	283.5	45.3	0:58.9
8:18	22 19.7	-123-51.4	22 0.4	-123-25.9	22 10.0	-123-38.6	0.9862	28.2	285.7	50.6	1: 2.3
8:24	19 46.4	-129-46.4	19 27.8	-129-10.5	19 37.1	-129-28.3	0.9844	20.9	288.5	57.8	1: 6.4
8:30	15 13.4	-139-59.8	15 7.7	-138-39.9	15 11.0	-139-18.4	0.9813	9.2	292.0	69.8	1:12.5
LIMITS	11 46.7	-147-49.9	11 9.6	-147-26.8	11 30.9	-147-32.0	0.9787	0.0	294.0	79.4	1:16.9

116

Table 61

TOTAL SOLAR ECLIPSE OF 14 DEC 2020

SAROS 142   Delta T =   82.6 Sec

UNIVERSAL TIME	NORTHERN LIMIT LATITUDE	LONGITUDE	SOUTHERN LIMIT LATITUDE	LONGITUDE	CENTER LINE LATITUDE	LONGITUDE	DIAMETER RATIO	SUN ALT	SUN AZ	PATH WIDTH	DURATION TOTALITY
LIMITS	-7-38.9	132 42.3	-7-54.9	132 53.0	-7-48.0	132 45.1	1.0090	0.0	113.4	34.7	0:28.8
14:36	-13-42.9	120 0.2	-13-48.0	120 54.4	-13-44.8	120 27.2	1.0133	14.2	110.9	51.6	0:46.8
14:42	-17-49.6	112 36.2	-18 -2.3	113 21.9	-17-58.0	112 59.0	1.0160	23.7	108.1	62.1	1: 0.3
14:48	-20-47.7	107 36.3	-21 -5.5	108 19.9	-20-58.7	107 58.0	1.0179	30.5	105.3	68.8	1:10.4
14:54	-23-14.8	103 35.5	-23-36.5	104 18.1	-23-25.8	103 56.7	1.0193	36.0	102.4	73.7	1:19.0
15: 0	-25-22.6	100 7.6	-25-47.5	100 49.4	-25-35.2	100 28.5	1.0204	40.9	99.3	77.5	1:26.5
15: 6	-27-16.5	97 0.4	-27-44.3	97 41.4	-27-30.5	97 20.8	1.0214	45.2	96.0	80.5	1:33.3
15:12	-28-59.5	94 6.8	-29-29.9	94 47.0	-29-14.8	94 26.8	1.0222	49.2	92.3	82.9	1:39.4
15:18	-30-33.5	91 22.5	-31 -8.3	92 1.6	-30-50.0	91 42.0	1.0229	52.9	88.2	84.8	1:45.0
15:24	-31-59.8	88 44.4	-32-34.8	89 22.3	-32-17.3	89 3.3	1.0235	56.3	83.6	86.4	1:49.9
15:30	-33-19.1	86 10.4	-33-56.1	86 46.8	-33-37.7	86 58.0	1.0240	59.5	78.4	87.6	1:54.4
15:36	-34-32.1	83 38.8	-35-11.0	84 13.4	-34-51.6	83 56.0	1.0244	62.5	72.4	88.6	1:58.3
15:42	-35-39.0	81 8.2	-36-19.7	81 40.8	-35-59.4	81 24.4	1.0247	65.1	65.5	89.4	2: 1.6
15:48	-36-40.3	78 37.7	-37-22.6	79 7.9	-37 -1.5	78 52.7	1.0250	67.5	57.4	89.9	2: 4.4
15:54	-37-36.0	76 6.2	-38-19.7	76 33.7	-37-57.9	76 19.9	1.0252	69.5	48.0	90.3	2: 6.6
16: 0	-38-26.2	73 33.0	-39-11.2	73 57.6	-38-48.7	73 45.3	1.0253	71.1	37.1	90.4	2: 8.2
16: 6	-39-10.9	70 57.6	-39-56.9	71 18.9	-39-34.0	71 8.2	1.0254	72.1	25.0	90.5	2: 9.2
16:12	-39-50.1	68 19.3	-40-37.0	68 37.1	-40-13.6	68 28.2	1.0254	72.6	12.0	90.3	2: 9.7
16:18	-40-23.7	65 37.6	-41-11.1	65 51.6	-40-47.4	65 44.6	1.0253	72.4	358.9	90.0	2: 9.5
16:24	-40-51.5	62 52.1	-41-39.3	63 2.0	-41-15.4	62 57.1	1.0252	71.6	346.3	89.5	2: 8.7
16:30	-41-13.3	60 2.3	-42 -1.2	60 8.0	-41-37.2	60 5.2	1.0250	70.2	334.9	88.9	2: 7.3
16:36	-41-28.9	57 7.8	-42-16.5	57 9.1	-41-52.7	57 8.5	1.0248	68.3	324.7	88.1	2: 5.3
16:42	-41-37.9	54 8.0	-42-25.0	54 4.8	-42 -1.5	54 6.5	1.0245	66.1	315.8	87.1	2: 2.7
16:48	-41-40.0	51 2.6	-42-26.3	50 54.9	-42 -3.2	50 58.8	1.0241	63.5	307.9	86.0	1:59.5
16:54	-41-34.7	47 50.8	-42-19.8	47 38.6	-41-57.3	47 44.8	1.0237	60.7	300.8	84.6	1:55.6
17: 0	-41-21.4	44 31.8	-42 -5.1	44 15.2	-41-43.3	44 23.7	1.0231	57.6	294.4	83.0	1:51.2
17: 6	-40-59.5	41 4.7	-41-41.2	40 43.8	-41-20.4	40 54.4	1.0225	54.3	288.6	81.1	1:46.2
17:12	-40-27.9	37 27.8	-41 -7.4	37 2.8	-40-47.7	37 15.5	1.0218	50.7	283.2	79.0	1:40.6
17:18	-39-45.4	33 38.9	-40-22.2	33 10.2	-40 -3.9	33 24.7	1.0210	46.9	278.0	76.4	1:34.3
17:24	-38-50.2	29 34.8	-39-23.8	29 2.6	-39 -7.1	29 18.8	1.0200	42.7	273.1	73.4	1:27.4
17:30	-37-39.5	25 15.0	-38 -9.4	24 34.5	-37-54.5	24 52.4	1.0189	38.0	268.3	69.8	1:19.8
17:36	-36 -8.6	20 16.0	-36-34.0	19 36.4	-36-21.4	19 55.9	1.0175	32.8	263.6	65.4	1:11.3
17:42	-34 -8.5	14 30.7	-34-28.2	13 48.9	-34-18.5	14 10.0	1.0158	26.7	258.8	59.7	1: 1.5
17:48	-31-15.9	7 6.5	-31-27.3	6 19.2	-31-21.7	6 43.0	1.0135	18.6	253.5	51.5	0:49.5
LIMITS	-23-30.9	-11 -1.8	-23-44.6	-11-12.1	-23-39.1	-11 -3.9	1.0078	0.0	244.3	30.3	0:25.2

Table 62

ANNULAR SOLAR ECLIPSE OF 10 JUN 2021

SAROS 147   Delta T = 83.1 Sec

UNIVERSAL TIME	NORTHERN LIMIT LATITUDE	LONGITUDE	SOUTHERN LIMIT LATITUDE	LONGITUDE	CENTER LINE LATITUDE	LONGITUDE	DIAMETER RATIO	SUN ALT	SUN AZ	PATH WIDTH	DURATION ANNULARITY
LIMITS	52 19.5	93 35.4	48 9.3	85 57.5	51 41.7	86 32.8	0.9381	0.0	54.9	691.6	3:38.1
10: 4	58 49.2	83 43.8	61 48.7	67 37.7	60 50.4	74 39.9	0.9410	13.6	67.3	639.7	3:44.7
10: 8	62 4.1	80 31.2	64 5.8	65 35.2	63 25.0	72 24.3	0.9416	16.0	70.7	618.3	3:46.2
10:12	64 46.1	78 29.6	66 14.8	63 48.4	65 45.9	70 38.4	0.9421	17.8	73.8	599.4	3:47.4
10:16	67 11.9	77 7.9	68 18.0	62 12.0	67 58.2	69 13.0	0.9425	19.3	76.7	582.8	3:48.4
10:20	69 27.9	76 16.0	70 17.3	60 42.6	70 4.8	68 3.6	0.9428	20.5	79.4	568.6	3:49.3
10:24	71 37.4	75 50.9	72 13.7	59 17.7	72 7.4	67 8.2	0.9430	21.5	81.9	556.5	3:49.9
10:28	73 42.3	75 54.0	74 8.2	57 54.9	74 7.2	66 26.7	0.9432	22.2	84.3	546.5	3:50.5
10:32	75 43.7	76 30.7	76 1.4	58 32.0	76 5.2	66 0.4	0.9433	22.8	86.5	538.5	3:50.9
10:36	77 42.4	77 52.6	77 53.9	55 6.4	78 2.2	65 53.5	0.9434	23.1	88.5	532.3	3:51.1
10:40	79 38.2	80 21.2	79 46.3	53 34.1	79 58.6	66 14.7	0.9435	23.3	89.9	527.9	3:51.3
10:44	81 30.4	84 38.4	81 39.2	51 48.6	81 54.9	67 22.5	0.9435	23.2	90.6	525.3	3:51.2
10:48	83 15.9	92 8.2	83 32.9	49 36.6	83 51.1	70 0.4	0.9434	23.0	89.8	524.4	3:51.1
10:52	84 46.7	105 36.4	85 27.9	46 22.5	85 46.0	76 14.7	0.9434	22.7	85.4	525.4	3:50.8
10:56	85 42.7	128 42.2	87 24.0	39 44.7	87 31.9	94 18.7	0.9432	22.1	69.2	528.3	3:50.4
11: 0	85 33.9	157 45.1	89 13.3	3 17.6	88 18.1	151 15.4	0.9431	21.3	14.1	533.2	3:49.8
11: 4	84 17.5	179 47.5	88 20.0	-109 -12.3	86 57.5	-184 -16.8	0.9428	20.3	331.3	540.3	3:49.1
11: 8	82 16.7	-167 -48.4	86 13.7	-122 -3.4	84 50.5	-151 -26.1	0.9425	19.0	320.2	550.0	3:48.2
11:12	79 44.6	-161 -22.8	83 58.0	-126 -48.2	82 26.0	-147 -3.5	0.9422	17.5	317.5	562.7	3:47.0
11:16	76 38.5	-158 -28.3	81 32.5	-129 -58.9	79 45.0	-145 -44.1	0.9417	15.5	317.7	579.1	3:45.7
11:20	72 31.6	-158 -17.3	78 53.8	-132 -46.5	76 40.8	-146 -4.7	0.9411	13.1	319.5	600.4	3:44.0
LIMITS	64 46.4	-163 -38.0	62 18.1	-150 -39.8	65 43.9	-154 -4.9	0.9384	0.0	329.6	681.0	3:37.3

Table 63

TOTAL SOLAR ECLIPSE OF 4 DEC 2021

SAROS 152 Delta T = 83.5 Sec

UNIVERSAL TIME	NORTHERN LIMIT LATITUDE LONGITUDE		SOUTHERN LIMIT LATITUDE LONGITUDE		CENTER LINE LATITUDE LONGITUDE		DIAMETER RATIO	SUN ALT	SUN AZ	PATH WIDTH	DURATION TOTALITY
LIMITS	-51-54.0	48 57.2	-54-24.1	53 46.0	-54 -7.0	49 15.8	1.0317	0.0	127.3	424.0	1:29.9
7: 6	-80-25.0	37 0.0	-55-53.1	51 12.3	-58-58.5	42 38.3	1.0337	7.5	120.9	450.6	1:39.0
7: 8	-82 -0.9	35 43.2	-59 -1.1	47 13.6	-60-51.8	40 56.4	1.0343	9.5	118.9	452.8	1:41.9
7:10	-83-29.6	34 44.8	-61 -4.5	45 26.6	-62-31.0	39 46.8	1.0348	10.9	117.2	452.2	1:44.2
7:12	-84-53.2	34 0.5	-62-48.6	44 22.5	-64 -1.8	38 58.6	1.0352	12.1	115.8	450.1	1:46.1
7:14	-66-13.1	33 28.1	-84-22.1	43 44.8	-65-26.8	38 26.9	1.0355	13.1	114.6	447.3	1:47.8
7:16	-67-30.1	33 6.5	-65-48.8	43 27.0	-66-47.7	38 9.5	1.0357	14.0	113.6	444.0	1:49.2
7:18	-88-44.9	32 55.3	-67-10.4	43 26.4	-68 -5.3	38 5.2	1.0360	14.7	112.8	440.6	1:50.3
7:20	-89-57.7	32 54.5	-68-28.1	43 41.9	-69-20.2	38 13.9	1.0362	15.3	112.2	437.2	1:51.3
7:22	-71 -9.0	33 4.5	-69-42.5	44 13.6	-70-32.8	38 36.0	1.0363	15.8	111.8	433.8	1:52.2
7:24	-72-18.8	33 26.5	-70-54.1	45 2.2	-71-43.5	39 12.6	1.0364	16.2	111.5	430.7	1:52.9
7:26	-73-27.5	34 2.0	-72 -3.0	46 9.4	-72-52.4	40 5.3	1.0365	16.6	111.6	427.7	1:53.4
7:28	-74-35.0	34 53.1	-73 -9.2	47 37.3	-73-59.4	41 16.6	1.0366	16.8	111.9	425.0	1:53.8
7:30	-75-41.3	36 2.7	-74-12.7	49 28.9	-75 -4.6	42 49.5	1.0367	17.0	112.6	422.5	1:54.1
7:32	-76-46.3	37 34.9	-75-13.2	51 48.1	-76 -7.6	44 48.2	1.0367	17.0	113.7	420.3	1:54.3
7:34	-77-49.8	39 35.2	-76-10.2	54 39.1	-77 -8.3	47 18.1	1.0367	17.0	115.3	418.4	1:54.3
7:36	-78-51.5	42 10.9	-77 -2.8	58 7.4	-78 -5.8	50 25.9	1.0367	17.0	117.5	416.8	1:54.2
7:38	-79-50.8	45 31.8	-77-50.2	62 18.1	-78-59.5	54 19.5	1.0366	16.8	120.5	415.5	1:53.9
7:40	-80-46.2	49 50.4	-78-31.0	67 16.1	-79-48.1	59 8.0	1.0366	16.6	124.4	414.5	1:53.5
7:42	-81-36.8	55 22.2	-79 -3.3	73 3.7	-80-29.8	64 59.6	1.0365	16.3	129.3	413.8	1:53.0
7:44	-82-20.2	62 23.4	-79-25.2	79 38.6	-81 -2.4	71 59.4	1.0363	15.9	135.4	413.4	1:52.4
7:46	-82-53.4	71 4.8	-79-34.4	86 51.9	-81-23.2	80 3.6	1.0362	15.4	142.6	413.4	1:51.6
7:48	-83-12.8	81 19.9	-79-28.6	94 26.5	-81-29.5	88 55.4	1.0360	14.8	150.6	413.8	1:50.6
7:50	-83-14.6	92 33.7	-79 -6.1	102 0.0	-81-19.0	98 4.1	1.0358	14.1	158.8	414.5	1:49.5
7:52	-82-58.6	103 45.2	-78-24.9	109 9.4	-80-50.2	106 52.4	1.0355	13.3	166.8	415.5	1:48.1
7:54	-82-18.5	113 53.0	-77-23.1	115 35.7	-80 -2.4	114 48.6	1.0352	12.3	173.8	416.9	1:46.5
7:56	-81-21.2	122 21.0	-75-56.8	121 6.8	-78-55.0	121 33.9	1.0348	11.1	179.8	418.7	1:44.6
7:58	-80 -5.7	129 1.4	-73-55.3	125 34.9	-77-26.0	127 1.9	1.0344	9.6	184.4	420.7	1:42.3
8: 0	-78-31.2	134 2.6	-70-30.0	128 44.2	-75-29.3	131 12.6	1.0338	7.8	187.8	422.9	1:39.5
LIMITS	-67 -3.5	138 36.2	-67-36.1	128 56.3	-68-54.0	134 33.6	1.0318	0.0	190.1	421.2	1:29.9

Table 64

ANN/TOT SOLAR ECLIPSE OF 20 APR 2023

SAROS 129

Delta T = 84.8 Sec

UNIVERSAL TIME	NORTHERN LIMIT LATITUDE	LONGITUDE	SOUTHERN LIMIT LATITUDE	LONGITUDE	CENTER LINE LATITUDE	LONGITUDE	DIAMETER RATIO	SUN ALT	SUN AZ	PATH WIDTH	DURATION ANNULARITY
LIMITS	-48-25.7	-63-40.6	-48-29.1	-63-41.6	-48-28.8	-63-43.9	0.9983	0.0	72.4	6.2	0: 6.2
2:42	-42-10.5	-84-45.2	-42 -2.7	-84-47.0	-42 -6.8	-84-46.1	1.0032	16.9	56.2	11.5	0:13.3T
2:48	-38-27.3	-92-45.8	-38-16.0	-92-43.8	-38-21.6	-92-44.7	1.0054	25.0	49.5	19.1	0:23.7T
2:54	-35-25.8	-98 -2.7	-35-12.5	-97-57.3	-35-19.2	-98 0.0	1.0069	31.0	44.7	24.5	0:31.8T
3: 0	-32-46.4	-102 -2.6	-32-31.8	-101-54.6	-32-39.1	-101-58.7	1.0081	36.1	40.6	28.7	0:38.7T
3: 6	-30-21.5	-105-16.7	-30 -5.9	-105 -6.7	-30-13.7	-105-11.7	1.0091	40.5	36.9	32.2	0:44.8T
3:12	-28 -7.4	-108 -0.3	-27-51.0	-107-48.6	-27-59.2	-107-54.5	1.0100	44.4	33.2	35.1	0:50.1T
3:18	-26 -1.6	-110-22.3	-25-44.8	-110 -9.2	-25-53.1	-110-15.8	1.0107	48.0	29.6	37.6	0:54.9T
3:24	-24 -2.6	-112-28.4	-23-45.2	-112-14.1	-23-53.9	-112-21.3	1.0112	51.3	25.7	39.8	0:59.1T
3:30	-22 -9.4	-114-22.4	-21-51.5	-114 -7.1	-22 -0.5	-114-14.8	1.0117	54.3	21.6	41.7	1: 2.9T
3:36	-20-21.1	-116 -7.2	-20 -2.9	-115-51.1	-20-12.0	-115-59.2	1.0122	57.0	17.2	43.3	1: 6.1T
3:42	-18-37.2	-117-44.9	-18-18.7	-117-28.2	-18-27.9	-117-36.6	1.0125	59.5	12.2	44.7	1: 8.9T
3:48	-16-57.1	-119-17.3	-16-38.4	-119	-16-47.7	-119 -8.7	1.0128	61.6	6.7	45.9	1:11.3T
3:54	-15-20.6	-120-45.6	-15 -1.6	-120-28.0	-15-11.1	-120-36.8	1.0130	63.5	0.8	46.9	1:13.2T
4: 0	-13-47.3	-122-11.1	-13-28.1	-121-53.2	-13-37.7	-122 -2.2	1.0131	64.9	353.9	47.7	1:14.6T
4: 6	-12-17.1	-123-34.8	-11-57.7	-123-16.7	-12 -7.4	-123-25.7	1.0132	66.0	346.7	48.4	1:15.6T
4:12	-10-49.6	-124-57.6	-10-30.1	-124-39.3	-10-39.9	-124-48.5	1.0132	66.6	339.0	48.9	1:16.2T
4:18	-9-24.9	-126-20.3	-9 -5.3	-126 -2.0	-9-15.1	-126-11.2	1.0132	66.7	331.3	49.1	1:16.2T
4:24	-8 -2.8	-127-43.7	-7-43.2	-127-25.6	-7-53.0	-127-34.7	1.0131	66.3	323.7	49.2	1:15.9T
4:30	-6-43.3	-129 -8.7	-6-23.6	-128-50.8	-6-33.5	-128-59.7	1.0130	65.5	316.5	49.1	1:15.0T
4:36	-5-26.3	-130-36.1	-5 -6.7	-130-18.4	-5-16.5	-130-27.3	1.0128	64.3	310.0	48.7	1:13.6T
4:42	-4-11.9	-132 -6.7	-3-52.3	-131-49.4	-4 -2.1	-131-58.1	1.0125	62.6	304.2	48.1	1:11.8T
4:48	-3 0.0	-133-41.6	-2-40.7	-133-24.8	-2-50.4	-133-33.2	1.0122	60.7	299.2	47.1	1: 9.5T
4:54	-1-50.9	-135-21.7	-1-31.8	-135 -5.5	-1-41.4	-135-13.6	1.0118	58.4	294.9	45.9	1: 6.6T
5: 0	0-44.6	-137 -8.4	0-25.9	-136-52.8	0-35.3	-137 -0.6	1.0114	55.8	291.4	44.3	1: 3.3T
5: 6	0 18.6	-139 -3.2	0 36.8	-138-48.4	0 27.7	-138-55.8	1.0108	52.9	288.4	42.3	0:59.4T
5:12	1 18.4	-141 -8.0	1 35.9	-140-54.0	1 27.1	-141 -1.0	1.0102	49.8	286.0	39.9	0:55.0T
5:18	2 14.3	-143-25.3	2 30.9	-143-12.3	2 22.6	-143-18.8	1.0095	46.4	284.0	37.0	0:50.0T
5:24	3 5.7	-145-58.6	3 21.2	-145-46.7	3 13.4	-145-52.7	1.0086	42.6	282.5	33.6	0:44.4T
5:30	3 51.5	-148-53.0	4 5.6	-148-42.3	3 58.5	-148-47.6	1.0076	38.4	281.4	29.6	0:38.2T
5:36	4 30.1	-152-16.6	4 42.4	-152 -7.2	4 36.3	-152-11.9	1.0065	33.8	280.7	24.9	0:31.3T
5:42	4 58.4	-156-23.9	5 8.3	-156-16.0	5 3.3	-156-19.9	1.0050	28.3	280.3	19.3	0:23.4T
5:48	5 9.7	-161-47.6	5 16.2	-161-41.8	5 12.9	-161-44.7	1.0032	21.6	280.3	12.0	0:14.0T
5:54	4 39.0	-170-30.0	4 39.6	-170-29.2	4 39.3	-170-29.6	1.0003	11.4	280.8	1.0	0: 1.1T
LIMITS	2 58.5	178 44.7	2 52.3	178 45.1	2 55.6	178 46.1	0.9969	0.0	281.5	11.3	0:11.4

Table 65

ANNULAR SOLAR ECLIPSE OF 14 OCT 2023

SAROS 134

Delta T = 85.3 Sec

UNIVERSAL TIME	NORTHERN LIMIT LATITUDE	LONGITUDE	SOUTHERN LIMIT LATITUDE	LONGITUDE	CENTER LINE LATITUDE	LONGITUDE	DIAMETER RATIO	SUN ALT	SUN AZ	PATH WIDTH	DURATION ANNULARITY
LIMITS	50 25.9	146 38.9	48 14.1	147 2.9	49 14.3	146 14.4	0.9376	0.0	103.4	244.2	4:19.8
16:18	45 0.5	124 14.2	42 35.5	123 40.1	43 46.3	123 53.1	0.9421	17.6	121.0	220.6	4:34.7
16:24	41 37.1	115 57.5	39 33.7	116 21.0	40 34.4	116 8.0	0.9439	25.1	127.7	211.7	4:41.4
16:30	38 47.7	110 32.0	36 57.2	111 17.9	37 51.8	110 54.4	0.9453	30.9	132.7	205.5	4:46.5
16:36	36 16.8	106 27.5	34 35.4	107 25.0	35 25.6	106 56.0	0.9464	35.7	136.9	200.8	4:50.8
16:42	33 58.2	103 11.5	32 23.8	104 15.8	33 10.6	103 43.5	0.9473	40.0	140.6	197.0	4:54.5
16:48	31 48.8	100 28.1	30 20.0	101 36.5	31 4.1	101 2.2	0.9481	43.8	144.2	193.9	4:57.7
16:54	29 46.6	98 7.8	28 22.3	99 18.8	29 4.2	98 43.4	0.9488	47.3	147.8	191.3	5: 0.6
17: 0	27 50.3	96 4.9	26 29.7	97 17.6	27 9.7	96 41.3	0.9494	50.6	151.5	189.2	5: 3.1
17: 6	25 59.0	94 15.1	24 41.4	95 28.9	25 20.0	94 52.1	0.9499	53.5	155.3	187.6	5: 5.4
17:12	24 12.0	92 35.6	22 56.9	93 50.1	23 34.2	93 13.0	0.9504	56.3	159.4	186.3	5: 7.5
17:18	22 28.8	91 4.2	21 15.7	92 19.1	21 52.0	91 41.7	0.9507	58.8	163.9	185.3	5: 9.3
17:24	20 48.9	89 39.0	19 37.4	90 54.3	20 12.9	90 16.7	0.9511	61.1	168.9	184.7	5:11.0
17:30	19 12.0	88 18.8	18 1.8	89 34.3	18 36.6	88 56.6	0.9514	63.1	174.3	184.4	5:12.5
17:36	17 37.9	87 2.3	16 28.5	88 17.9	17 2.9	87 40.2	0.9516	64.8	180.4	184.4	5:13.8
17:42	16 6.2	85 48.5	14 57.5	87 4.3	15 31.6	86 28.5	0.9518	66.2	187.1	184.8	5:15.0
17:48	14 37.0	84 36.7	13 28.5	85 52.6	14 2.5	85 14.7	0.9519	67.2	194.4	185.4	5:15.9
17:54	13 9.9	83 25.9	12 1.5	84 42.0	12 35.4	84 4.0	0.9520	67.8	202.0	186.4	5:16.7
18: 0	11 44.9	82 15.5	10 36.3	83 31.8	11 10.3	82 53.7	0.9520	67.9	209.8	187.6	5:17.3
18: 6	10 22.0	81 4.8	9 12.8	82 21.3	9 47.1	81 43.1	0.9520	67.7	217.4	189.1	5:17.7
18:12	9 0.9	79 53.0	7 51.1	81 9.8	8 25.7	80 31.5	0.9520	67.0	224.7	190.8	5:17.8
18:18	7 41.8	78 39.6	6 30.9	79 56.6	7 6.0	79 18.2	0.9519	65.9	231.4	192.7	5:17.7
18:24	6 24.5	77 23.7	5 12.4	78 41.1	5 48.1	78 2.5	0.9518	64.4	237.4	194.9	5:17.4
18:30	5 9.1	76 4.5	3 55.6	77 22.4	4 32.0	76 43.6	0.9516	62.6	242.6	197.2	5:16.7
18:36	3 55.5	74 41.3	2 40.3	75 59.7	3 17.6	75 20.6	0.9514	60.6	247.0	199.6	5:15.8
18:42	2 44.0	73 13.0	1 26.8	74 32.0	2 5.0	73 52.6	0.9511	58.2	250.8	202.1	5:14.5
18:48	1 34.5	71 38.5	0 15.2	72 58.1	0 54.4	72 18.5	0.9507	55.7	254.0	204.6	5:12.8
18:54	0 27.2	69 56.3	0-54.6	71 16.8	0-14.1	70 36.7	0.9503	52.9	256.7	207.1	5:10.8
19: 0	0-37.7	68 4.7	-2 -2.2	69 26.3	-1-20.4	68 45.7	0.9499	49.8	258.8	209.5	5: 8.2
19: 6	-1-39.9	66 1.5	-3 -7.4	67 24.5	-2-24.1	66 43.2	0.9493	46.5	260.6	211.9	5: 5.3
19:12	-2-38.9	63 43.5	-4 -9.8	65 8.3	-3-24.8	64 26.2	0.9487	42.9	261.9	214.2	5: 1.7
19:18	-3-34.0	61 6.1	-5 -8.8	62 33.7	-4-21.8	61 50.3	0.9479	39.0	262.9	216.4	4:57.6
19:24	-4-24.2	58 2.5	-6 -3.5	59 34.0	-5-14.3	58 48.8	0.9470	34.6	263.6	218.7	4:52.7
19:30	-5 -7.5	54 20.2	-6-52.3	55 58.7	-6 -0.4	55 10.2	0.9459	29.6	263.9	221.2	4:47.0
19:36	-5-40.0	49 33.8	-7-32.1	51 26.1	-6-36.6	50 31.2	0.9445	23.6	263.8	224.3	4:40.0
19:42	-5-49.1	42 23.9	-7-54.3	44 57.7	-6-52.8	43 44.5	0.9425	15.5	263.3	228.9	4:30.8
LIMITS	-4-35.9	29 17.9	-6-47.8	29 19.3	-5-46.0	29 45.7	0.9386	0.0	261.7	239.7	4:15.4

Table 66

TOTAL SOLAR ECLIPSE OF 8 APR 2024

SAROS 139   Delta T =   85.7 Sec

UNIVERSAL TIME	NORTHERN LIMIT LATITUDE	LONGITUDE	SOUTHERN LIMIT LATITUDE	LONGITUDE	CENTER LINE LATITUDE	LONGITUDE	DIAMETER RATIO	SUN ALT	SUN AZ	PATH WIDTH	DURATION TOTALITY
LIMITS	-7 -11.9	158 40.7	-8 -27.5	158 17.7	-7 -48.3	158 17.5	1.0396	0.0	82.3	144.3	2: 6.6
16:42	-5 -22.6	149 10.7	-6 -4.8	146 8.1	-5 -42.9	147 35.0	1.0432	11.6	81.0	160.1	2:28.6
16:48	-2 -21.9	139 44.8	-3 -15.2	137 43.3	-2 -48.3	138 43.1	1.0464	22.2	80.6	174.8	2:51.2
16:54	0 3.5	134 21.5	0 -53.2	132 33.5	0 -24.6	133 26.9	1.0485	29.3	81.0	183.7	3: 6.9
17: 0	2 18.0	130 22.0	1 17.4	128 40.0	1 46.8	129 30.6	1.0500	35.0	82.0	190.0	3:19.7
17: 6	4 21.0	127 8.3	3 21.0	125 29.7	3 51.1	126 18.7	1.0513	39.9	83.3	194.6	3:30.6
17:12	6 20.9	124 24.0	5 19.8	122 47.5	5 50.5	123 35.4	1.0523	44.3	85.1	197.8	3:40.3
17:18	8 17.2	121 59.9	7 15.1	120 24.9	7 46.2	121 12.2	1.0532	48.3	87.3	200.1	3:48.8
17:24	10 10.6	119 50.6	9 7.6	118 16.7	9 39.2	119 3.4	1.0539	51.9	90.1	201.5	3:56.3
17:30	12 1.8	117 52.2	10 58.0	116 19.2	11 29.9	117 5.5	1.0545	55.3	93.4	202.4	4: 2.9
17:36	13 51.3	116 1.8	12 46.5	114 29.6	13 18.9	115 15.5	1.0551	58.4	97.4	202.7	4: 8.7
17:42	15 39.3	114 17.5	14 33.5	112 46.0	15 6.4	113 31.6	1.0555	61.2	102.2	202.6	4:13.7
17:48	17 26.1	112 37.3	16 19.3	111 8.6	16 52.7	111 51.8	1.0559	63.6	107.9	202.1	4:18.0
17:54	19 12.0	110 59.9	18 4.0	109 30.0	18 38.0	110 14.8	1.0561	65.8	114.7	201.5	4:21.5
18: 0	20 57.1	109 24.0	19 47.9	107 55.0	20 22.5	108 39.3	1.0563	67.6	122.6	200.6	4:24.3
18: 6	22 41.6	107 48.3	21 31.1	106 20.3	22 6.3	107 4.1	1.0565	68.9	131.6	199.6	4:26.3
18:12	24 25.6	106 11.7	23 13.6	104 44.8	23 49.6	105 28.1	1.0565	69.6	141.5	198.5	4:27.6
18:18	26 9.2	104 33.1	24 55.6	103 7.6	25 32.3	103 50.2	1.0566	69.8	151.9	197.3	4:28.2
18:24	27 52.5	102 51.2	26 37.2	101 27.3	27 14.8	102 9.1	1.0565	69.4	162.2	196.0	4:27.9
18:30	29 35.5	101 4.9	28 18.4	99 42.9	28 56.8	100 23.8	1.0563	68.5	172.0	194.6	4:27.0
18:36	31 18.3	99 12.7	29 59.2	97 53.0	30 38.7	98 32.7	1.0561	67.0	181.0	193.1	4:25.2
18:42	33 0.9	97 13.1	31 39.8	95 56.2	32 20.2	96 34.6	1.0559	65.1	189.1	191.6	4:22.6
18:48	34 43.3	95 4.3	33 20.0	93 50.9	34 1.5	94 27.5	1.0555	62.8	196.4	189.9	4:19.2
18:54	36 25.3	92 44.1	34 59.8	91 34.9	35 42.4	92 9.4	1.0551	60.2	202.9	188.1	4:15.0
19: 0	38 7.0	90 9.9	36 39.1	89 5.9	37 22.9	89 37.9	1.0546	57.3	208.9	186.2	4: 9.8
19: 6	39 47.9	87 18.5	38 17.8	86 20.9	39 2.7	86 49.8	1.0540	54.2	214.4	184.1	4: 3.7
19:12	41 27.9	84 5.7	39 55.4	83 16.1	40 41.5	83 41.1	1.0533	50.7	219.8	181.8	3:56.7
19:18	43 6.3	80 25.8	41 31.7	79 46.3	42 18.8	80 6.8	1.0525	46.9	225.1	179.3	3:48.5
19:24	44 42.2	76 11.0	43 5.8	75 44.4	43 53.8	75 58.2	1.0515	42.8	230.5	176.5	3:39.3
19:30	46 14.3	71 9.8	44 36.5	71 0.1	45 25.2	71 5.7	1.0503	38.3	236.3	173.3	3:28.7
19:36	47 40.0	65 3.5	46 1.8	65 16.7	46 50.7	65 11.2	1.0490	33.2	242.6	169.5	3:16.5
19:42	48 54.2	57 17.3	47 17.5	58 4.1	48 5.7	57 42.5	1.0473	27.1	250.0	164.9	3: 2.1
19:48	49 44.2	46 25.0	48 13.8	48 11.2	48 59.1	47 21.4	1.0449	19.3	259.4	158.7	2:44.1
LIMITS	48 13.4	19 26.4	47 0.1	20 1.9	47 39.2	20 3.5	1.0389	0.0	281.4	142.1	2: 4.5

Table 67

ANNULAR SOLAR ECLIPSE OF 2 OCT 2024

SAROS 144   Delta T = 86.2 Sec

UNIVERSAL TIME	NORTHERN LIMIT LATITUDE	LONGITUDE	SOUTHERN LIMIT LATITUDE	LONGITUDE	CENTER LINE LATITUDE	LONGITUDE	DIAMETER RATIO	SUN ALT	SUN AZ	PATH WIDTH	DURATION ANNULARITY
LIMITS	9 49.9	165 5.3	6 52.7	165 52.8	8 19.4	164 56.8	0.9192	0.0	94.1	331.8	5:39.7
17: 0	6 25.8	146 35.3	4 35.1	150 47.0	5 31.0	148 35.3	0.9237	18.6	96.1	324.4	6: 8.3
17: 6	4 32.3	141 29.6	2 44.7	144 51.0	3 39.1	143 7.7	0.9253	25.7	96.2	321.6	6:20.3
17:12	2 45.8	137 46.0	1 1.8	140 44.2	1 54.5	139 13.0	0.9265	31.3	95.8	318.8	6:29.7
17:18	1 3.6	134 45.4	0-37.2	137 31.5	0 13.8	136 7.2	0.9275	36.0	95.1	315.7	6:37.7
17:24	0-35.8	132 15.0	-2-13.7	134 52.2	-1-24.1	133 32.7	0.9284	40.2	94.0	312.4	6:44.5
17:30	-2-13.1	130 5.3	-3-48.4	132 36.0	-3 -0.2	131 19.8	0.9291	44.0	92.6	308.8	6:50.6
17:36	-3-48.8	128 11.0	-5-21.8	130 36.5	-4-34.7	129 23.0	0.9297	47.5	90.9	304.9	6:55.9
17:42	-5-23.1	126 28.3	-6-54.3	128 49.8	-6 -8.1	127 38.5	0.9302	50.7	88.7	300.8	7: 0.7
17:48	-6-56.4	124 54.9	-8-26.0	127 13.0	-7-40.7	126 3.4	0.9307	53.7	86.2	296.7	7: 4.9
17:54	-8-28.8	123 28.7	-9-57.2	125 44.0	-9-12.5	124 35.8	0.9311	56.5	83.1	292.6	7: 8.7
18: 0	-10 -0.6	122 8.2	-11-28.0	124 21.1	-10-43.8	123 14.1	0.9315	59.0	79.5	288.6	7:12.0
18: 6	-11-31.6	120 52.1	-12-58.5	123 2.9	-12-14.6	121 57.1	0.9317	61.4	75.3	284.8	7:14.9
18:12	-13 -2.5	119 39.4	-14-28.9	121 48.4	-13-45.2	120 43.5	0.9320	63.5	70.4	281.1	7:17.5
18:18	-14-32.9	118 29.2	-15-59.2	120 36.6	-15-15.5	119 32.5	0.9322	65.3	64.7	277.8	7:19.7
18:24	-16 -3.0	117 20.7	-17-29.5	119 26.8	-16-45.8	118 23.3	0.9324	66.9	58.1	274.9	7:21.5
18:30	-17-33.0	116 13.1	-18-60.0	118 17.7	-18-16.0	117 15.0	0.9325	68.1	50.8	271.9	7:23.0
18:36	-19 -2.8	115 5.6	-20-30.7	117 9.0	-19-46.3	116 6.9	0.9325	68.9	42.7	269.5	7:24.1
18:42	-20-32.7	113 57.8	-22 -1.7	115 59.9	-21-16.8	114 58.4	0.9326	69.3	34.2	267.4	7:24.9
18:48	-22 -2.6	112 48.5	-23-33.0	114 49.5	-22-47.4	113 48.6	0.9326	69.3	25.4	265.7	7:25.4
18:54	-23-32.8	111 37.3	-25 -4.8	113 37.3	-24-18.3	112 36.9	0.9325	68.8	16.9	264.3	7:25.5
19: 0	-25 -2.8	110 23.5	-26-37.2	112 22.2	-25-49.6	111 22.5	0.9325	67.9	8.8	263.2	7:25.2
19: 6	-26-33.2	109 6.2	-28-10.1	111 3.5	-27-21.3	110 4.5	0.9323	66.7	1.2	262.5	7:24.6
19:12	-28 -3.9	107 44.5	-29-43.7	109 40.1	-28-53.4	108 42.0	0.9322	65.1	354.3	262.2	7:23.5
19:18	-29-34.9	106 17.3	-31-18.0	108 11.0	-30-26.1	107 13.8	0.9320	63.2	348.1	262.2	7:22.0
19:24	-31 -6.2	104 43.5	-32-53.2	106 34.8	-31-59.3	105 38.9	0.9317	61.1	342.5	262.6	7:20.0
19:30	-32-38.0	103 1.7	-34-29.1	104 50.4	-33-33.1	103 55.6	0.9314	58.7	337.3	263.3	7:17.6
19:36	-34-10.1	101 10.3	-36 -6.0	102 54.7	-35 -7.6	102 2.2	0.9310	56.1	332.5	264.4	7:14.6
19:42	-35-42.7	99 7.1	-37-43.9	100 46.5	-36-42.8	99 56.6	0.9306	53.3	327.9	265.7	7:11.0
19:48	-37-15.6	96 49.6	-39-22.8	98 22.4	-38-18.6	97 35.9	0.9301	50.3	323.5	267.7	7: 6.8
19:54	-38-48.8	94 14.5	-41 -2.3	95 38.4	-39-54.9	94 56.5	0.9296	47.0	319.2	270.1	7: 1.9
20: 0	-40-22.0	91 17.2	-42-42.9	92 29.1	-41-31.7	91 53.5	0.9290	43.5	314.8	273.0	6:56.3
20: 6	-41-55.1	87 51.6	-44-23.9	88 46.8	-43 -8.7	88 19.9	0.9282	39.8	310.2	276.6	6:49.7
20:12	-43-27.5	83 48.2	-46 -5.1	84 19.4	-44-45.3	84 5.3	0.9274	35.4	305.3	281.0	6:42.1
20:18	-44-58.1	78 52.2	-47-45.3	78 46.9	-46-20.5	78 52.3	0.9264	30.6	299.7	286.6	6:33.3
20:24	-46-25.1	72 36.0	-49-22.3	71 29.5	-47-52.3	72 8.3	0.9251	24.9	293.1	294.0	6:22.6
20:30	-47-43.4	63 55.7	-50-49.1	60 38.2	-49-14.8	62 31.4	0.9233	17.5	284.3	304.6	6: 8.9
LIMITS	-48 -2.1	37 31.4	-50-58.2	36 26.5	-49-33.1	37 49.0	0.9191	0.0	264.2	332.2	5:39.9

123

Table 68

ANNULAR SOLAR ECLIPSE OF 17 FEB 2028

Delta T = 87.5 Sec

SAROS 121

UNIVERSAL TIME	NORTHERN LIMIT LATITUDE	LONGITUDE	SOUTHERN LIMIT LATITUDE	LONGITUDE	CENTER LINE LATITUDE	LONGITUDE	DIAMETER RATIO	SUN ALT	SUN AZ	PATH WIDTH	DURATION ANNULARITY
LIMITS	-73-55.0	-144-41.0	-69 -5.7	-128 -9.3	-73-30.8	-127-15.8	0.9603	0.0	236.8	711.4	2:20.8
11:56	-72-45.6	-91-10.8	-70-19.7	-119-14.5	-72-21.1	-101-43.5	0.9620	9.1	259.2	616.5	2:20.4
11:58	-71-44.6	-88-39.4	-70-12.8	-112 -5.9	-71-28.8	-98 -3.7	0.9622	9.9	262.0	607.9	2:20.3
12: 0	-70-42.8	-86-38.8	-69-38.9	-107-14.5	-70-32.9	-95 -9.3	0.9624	10.6	264.1	602.7	2:20.2
12: 2	-89-40.5	-85 -2.5	-68-53.9	-103-35.0	-69-34.7	-92-50.1	0.9626	11.1	265.6	600.2	2:20.1
12: 4	-68-38.3	-83-45.8	-68 -2.9	-100-44.7	-68-35.1	-90-59.0	0.9627	11.6	266.7	600.2	2:20.0
12: 6	-67-36.2	-82-45.2	-67 -8.2	-98-31.2	-67-34.8	-89-30.5	0.9628	11.9	267.4	602.2	2:19.9
12: 8	-66-34.3	-81-58.1	-66-11.0	-96-47.0	-66-33.9	-88-20.7	0.9629	12.1	267.8	606.2	2:19.8
12:10	-65-32.6	-81-22.3	-65-12.0	-95-26.6	-65-32.8	-87-26.7	0.9630	12.2	267.9	612.1	2:19.7
12:12	-64-31.2	-80-56.6	-64-11.8	-94-26.6	-64-31.4	-86-48.4	0.9630	12.3	267.8	619.8	2:19.7
12:14	-63-30.1	-80-39.5	-63-10.4	-93-44.5	-63-29.9	-86-18.1	0.9630	12.2	267.6	629.1	2:19.6
12:16	-62-29.0	-80-30.5	-62 -7.8	-93-18.8	-62-28.2	-86 -0.7	0.9630	12.1	267.2	640.2	2:19.6
12:18	-61-28.1	-80-28.9	-61 -4.0	-93 -9.0	-61-26.2	-85-53.5	0.9629	11.8	266.6	652.8	2:19.5
12:20	-60-27.1	-80-34.3	-59-58.7	-93-15.5	-60-23.7	-85-56.1	0.9628	11.5	266.0	867.0	2:19.5
12:22	-59-26.0	-80-46.8	-58-51.2	-93-40.0	-59-20.7	-86 -8.6	0.9627	11.1	265.1	682.6	2:19.4
12:24	-58-24.7	-81 -5.9	-57-40.4	-94-27.2	-58-16.9	-86-31.5	0.9626	10.5	264.2	699.2	2:19.4
12:26	-57-22.8	-81-32.5	-56-23.8	-95-48.2	-57-11.8	-87 -5.8	0.9624	9.9	263.1	716.5	2:19.4
12:28	-56-20.2	-82 -7.1	-54-51.9	-98-28.6	-56 -4.9	-87-53.9	0.9622	9.0	261.9	733.7	2:19.4
LIMITS	-47-30.5	-96-29.8	-53-32.0	-102-48.3	-51-26.1	-94-46.9	0.9806	0.0	254.5	762.4	2:19.6

Table 69

TOTAL SOLAR ECLIPSE OF 12 AUG 2026

SAROS 126   Delta T = 88.0 Sec

UNIVERSAL TIME	NORTHERN LIMIT LATITUDE	LONGITUDE	SOUTHERN LIMIT LATITUDE	LONGITUDE	CENTER LINE LATITUDE	LONGITUDE	DIAMETER RATIO	SUN ALT	SUN AZ	PATH WIDTH	DURATION TOTALITY
LIMITS	75 10.3	-108-57.8	74 54.3	-118-13.2	76 1.8	-114 -8.2	1.0314	0.0	7.9	269.4	1:35.9
17: 4	82 46.1	-101-58.2	88 7.7	-105 -1.2	85 44.3	-102-18.5	1.0344	10.5	357.5	273.9	1:51.6
17: 8	86 31.2	-60-28.4	87 31.1	23 9.0	87 46.2	-24-22.3	1.0355	14.4	281.1	274.5	1:58.3
17:12	85 33.4	-5 -7.7	84 18.6	32 53.1	85 10.1	16 57.2	1.0364	17.2	241.4	275.2	2: 3.3
17:16	82 55.1	13 8.0	81 25.1	34 4.6	82 16.0	24 52.5	1.0370	19.3	235.2	276.2	2: 7.2
17:20	80 13.9	19 21.4	78 46.4	33 51.9	79 33.6	27 15.7	1.0375	21.1	234.6	277.5	2:10.4
17:24	77 40.5	21 55.9	76 18.5	33 13.4	77 1.9	27 58.2	1.0379	22.5	235.7	279.1	2:12.9
17:28	75 15.0	23 1.2	73 58.9	32 24.6	74 38.9	27 58.8	1.0382	23.6	237.6	281.1	2:14.9
17:32	72 56.1	23 22.3	71 45.7	31 30.9	72 22.5	27 38.1	1.0384	24.5	239.8	283.4	2:16.4
17:36	70 42.8	23 18.1	69 37.5	30 34.2	70 11.5	27 5.0	1.0386	25.1	242.3	286.1	2:17.5
17:40	68 33.5	22 57.7	67 33.4	29 35.4	68 4.8	26 23.7	1.0386	25.6	244.9	289.1	2:18.2
17:44	66 27.7	22 25.7	65 32.5	28 34.4	66 1.4	25 36.0	1.0387	25.8	247.5	292.4	2:18.4
17:48	64 24.7	21 44.6	63 34.1	27 31.0	64 0.6	24 42.8	1.0386	25.8	250.3	295.9	2:18.2
17:52	62 23.5	20 55.2	61 37.8	26 24.8	62 1.7	23 44.4	1.0386	25.5	253.1	299.6	2:17.7
17:56	60 23.7	19 58.1	59 42.3	25 14.4	60 4.1	22 40.3	1.0384	25.1	255.9	303.5	2:16.7
18: 0	58 24.4	18 52.8	57 47.7	23 59.8	58 7.2	21 29.9	1.0382	24.5	258.8	307.5	2:15.4
18: 4	56 25.0	17 37.9	55 53.2	22 38.8	56 10.3	20 12.0	1.0379	23.6	261.6	311.3	2:13.6
18: 8	54 24.5	16 12.0	53 58.2	21 10.3	54 12.6	18 44.9	1.0376	22.4	264.5	314.8	2:11.3
18:12	52 22.0	14 32.1	52 1.7	19 31.9	52 11.0	17 5.9	1.0371	21.0	267.5	317.6	2: 8.5
18:16	50 15.6	12 33.1	50 2.7	17 40.0	50 10.8	15 11.1	1.0366	19.2	270.5	319.2	2: 5.2
18:20	48 2.7	10 6.0	47 59.5	15 29.1	48 3.1	12 53.4	1.0369	17.0	273.6	318.6	2: 1.0
18:24	45 37.4	6 51.1	45 49.2	12 49.2	45 46.4	9 59.3	1.0350	14.1	277.1	314.3	1:55.9
18:28	42 41.0	1 41.0	43 25.7	9 18.0	43 10.8	5 53.4	1.0337	10.1	281.0	303.2	1:49.0
LIMITS	39 39.9	-6-22.5	37 38.9	-4-34.6	38 57.9	-4-17.4	1.0309	0.0	288.6	265.2	1:34.2

Table 70

ANNULAR SOLAR ECLIPSE OF   6 FEB 2027

Delta T =   88.5 Sec

SAROS 131

UNIVERSAL TIME	NORTHERN LIMIT LATITUDE	LONGITUDE	SOUTHERN LIMIT LATITUDE	LONGITUDE	CENTER LINE LATITUDE	LONGITUDE	DIAMETER RATIO	SUN ALT	SUN AZ	PATH WIDTH	DURATION ANNULARITY
LIMITS	-37-38.9	130 16.3	-40-32.0	132 18.3	-39-14.5	130 43.1	0.9139	0.0	109.7	360.6	5:48.0
14:12	-42-33.6	108 35.7	-44-44.4	114 9.0	-43-41.8	111 4.4	0.9180	16.7	95.5	334.6	6:18.0
14:18	-43-34.1	100 9.3	-46-11.6	103 26.9	-44-53.2	101 39.9	0.9199	24.5	87.5	322.3	6:34.6
14:24	-43-55.7	93 40.8	-46-41.3	95 48.9	-45-18.2	94 39.6	0.9212	30.4	81.0	313.5	6:47.6
14:30	-43-55.4	88 17.2	-46-43.1	89 37.7	-45-18.8	88 54.0	0.9223	35.4	75.1	306.7	6:58.7
14:36	-43-40.3	83 38.0	-46-27.3	84 21.9	-45 -3.2	83 57.6	0.9232	39.9	69.7	301.2	7: 8.4
14:42	-43-14.3	79 31.8	-45-58.9	79 46.5	-44-36.0	79 37.5	0.9240	43.9	64.5	296.6	7:17.0
14:48	-42-39.7	75 51.8	-45-21.0	75 42.4	-43-59.8	75 46.0	0.9247	47.5	59.4	292.9	7:24.4
14:54	-41-58.3	72 33.2	-44-35.6	72 3.8	-43-16.4	72 17.9	0.9253	51.0	54.4	289.7	7:31.0
15: 0	-41-11.2	69 32.7	-43-44.2	68 46.5	-42-27.2	69 9.3	0.9258	54.2	49.3	287.1	7:36.6
15: 6	-40-19.4	66 47.6	-42-47.8	65 47.3	-41-33.1	66 17.4	0.9263	57.1	44.1	284.9	7:41.3
15:12	-39-23.5	64 15.8	-41-47.2	63 33.3	-40-34.9	63 39.8	0.9266	60.0	38.6	283.2	7:45.2
15:18	-38-24.0	61 55.5	-40-43.1	60 33.3	-39-33.2	61 14.7	0.9270	62.6	32.7	281.8	7:48.4
15:24	-37-21.4	59 45.3	-39-35.9	58 14.5	-38-28.3	59 0.3	0.9273	65.0	26.2	280.8	7:50.7
15:30	-36-16.0	57 43.9	-38-26.1	56 5.8	-37-20.7	56 55.3	0.9275	67.1	19.1	280.1	7:52.3
15:36	-35 -8.0	55 50.1	-37-13.8	54 5.8	-36-10.6	54 58.4	0.9277	69.0	11.2	279.8	7:53.2
15:42	-33-57.7	54 3.0	-35-59.3	52 13.3	-34-58.2	53 8.7	0.9279	70.6	2.3	279.7	7:53.5
15:48	-32-45.2	52 21.6	-34-42.9	50 27.2	-33-43.8	51 25.0	0.9280	71.7	352.5	280.1	7:53.2
15:54	-31-30.5	50 45.1	-33-24.6	48 46.7	-32-27.3	49 46.5	0.9281	72.4	342.0	280.7	7:52.2
16: 0	-30-13.9	49 12.8	-32 -4.5	47 10.7	-31 -8.9	48 12.3	0.9281	72.6	331.2	281.7	7:50.7
16: 6	-28-55.3	47 43.8	-30-42.6	45 38.4	-29-48.7	46 41.7	0.9281	72.3	320.5	283.0	7:48.7
16:12	-27-34.8	46 17.5	-29-19.1	44 9.1	-28-26.7	45 13.9	0.9281	71.5	310.6	284.7	7:46.3
16:18	-26-12.2	44 53.1	-27-53.8	42 41.9	-27 -2.8	43 48.1	0.9280	70.3	301.6	286.7	7:43.3
16:24	-24-47.7	43 30.0	-26-26.7	41 16.0	-25-37.0	42 23.7	0.9279	68.7	293.8	289.1	7:40.0
16:30	-23-21.1	42 7.4	-24-57.3	39 50.6	-24 -9.2	40 59.7	0.9277	66.7	287.1	291.8	7:36.3
16:36	-21-52.3	40 44.5	-23-26.9	38 24.9	-22-39.3	39 35.4	0.9275	64.5	281.3	294.9	7:32.2
16:42	-20-21.1	39 20.4	-21-53.9	36 57.9	-21 -7.2	38 9.9	0.9272	62.1	276.5	298.4	7:27.7
16:48	-18-47.4	37 54.3	-20-18.5	35 28.6	-19-32.7	36 42.2	0.9269	59.4	272.3	302.1	7:22.9
16:54	-17-10.9	36 25.0	-18-40.5	33 55.7	-17-55.4	35 11.2	0.9265	56.6	268.7	306.2	7:17.7
17: 0	-15-31.2	34 51.3	-18-59.5	32 17.9	-16-15.1	33 35.5	0.9261	53.6	265.6	310.6	7:12.2
17: 6	-13-48.0	33 11.4	-15-15.1	30 33.4	-14-31.3	31 53.4	0.9256	50.3	263.0	315.2	7: 6.3
17:12	-12 -0.5	31 23.5	-13-26.8	28 39.7	-12-43.3	30 2.7	0.9251	46.9	260.7	320.0	6:60.0
17:18	-10 -8.1	29 24.6	-11-33.0	26 33.7	-10-50.3	28 0.4	0.9244	43.1	258.8	324.9	6:53.2
17:24	-8 -9.3	27 10.8	-9-33.1	24 10.6	-8-51.0	25 42.2	0.9237	39.0	257.1	329.9	6:45.9
17:30	-6 -2.5	24 36.0	-7-24.5	21 23.0	-6-43.4	23 1.3	0.9228	34.5	255.8	334.8	6:37.9
17:36	-3-44.4	21 29.9	-5 -3.6	17 57.3	-4-24.0	19 46.1	0.9217	29.4	254.7	339.8	6:29.0
17:42	-1 -8.8	17 31.7	-2-21.7	13 23.5	-1-18.6	15 31.9	0.9203	23.1	253.9	344.7	6:18.6
17:48	2 1.1	11 45.1	1 11.5	5 52.8	1 33.5	9 3.4	0.9182	14.6	253.6	350.1	6: 4.9
LIMITS	7 39.3	-3 -3.8	4 42.9	-4-32.4	6 3.5	-3-19.2	0.9147	0.0	254.4	357.0	5:44.7

126

Table 71

TOTAL SOLAR ECLIPSE OF 2 AUG 2027

SAROS 136   Delta T = 88.9 Sec

UNIVERSAL TIME	NORTHERN LIMIT LATITUDE	LONGITUDE	SOUTHERN LIMIT LATITUDE	LONGITUDE	CENTER LINE LATITUDE	LONGITUDE	DIAMETER RATIO	SUN ALT	SUN AZ	PATH WIDTH	DURATION TOTALITY
LIMITS	28 48.7	44 51.7	27 6.0	43 56.5	27 58.1	44 18.3	1.0603	0.0	70.0	205.8	3: 6.0
8:30	33 57.5	26 26.9	32 20.9	24 18.8	33 10.2	25 19.1	1.0662	18.2	80.5	222.1	3:52.5
8:36	35 39.5	17 18.2	33 45.4	15 56.8	34 42.8	16 35.2	1.0688	26.8	86.5	228.8	4:17.5
8:42	36 26.8	10 33.9	34 24.2	9 39.7	35 25.6	10 5.0	1.0707	33.4	91.4	233.6	4:37.5
8:48	36 45.8	5 0.5	34 38.2	4 27.2	35 42.0	4 42.5	1.0722	39.0	95.9	237.4	4:54.6
8:54	36 46.0	0 12.4	34 35.4	0 -3.3	35 40.7	0 3.4	1.0734	43.9	100.1	240.6	5: 9.6
9: 0	36 32.5	-4 -3.1	34 20.3	-4 -3.4	35 26.3	-4 -4.1	1.0745	48.5	104.1	243.3	5:23.0
9: 6	36 8.1	-7 -53.3	33 55.4	-7 -39.9	35 1.6	-7 -47.3	1.0753	52.7	108.1	245.7	5:34.9
9:12	35 35.0	-11 -23.0	33 22.4	-10 -57.4	34 28.6	-11 -10.7	1.0761	56.6	112.1	247.8	5:45.3
9:18	34 54.4	-14 -35.7	32 42.7	-13 -59.1	33 48.4	-14 -17.7	1.0768	60.4	116.3	249.6	5:54.4
9:24	34 7.4	-17 -33.9	31 57.2	-16 -47.5	33 2.2	-17 -10.8	1.0773	64.0	120.8	251.2	6: 2.2
9:30	33 14.9	-20 -19.6	31 6.5	-19 -24.5	32 10.5	-19 -52.1	1.0778	67.4	125.8	252.6	6: 8.8
9:36	32 17.4	-22 -54.6	30 11.1	-21 -51.7	31 14.2	-22 -23.1	1.0782	70.7	131.6	253.8	6:14.1
9:42	31 15.5	-25 -20.2	29 11.6	-24 -10.4	30 13.5	-24 -45.2	1.0785	73.8	138.8	254.9	6:18.1
9:48	30 9.8	-27 -37.8	28 8.2	-26 -22.0	29 8.8	-26 -59.7	1.0787	76.6	148.3	255.8	6:21.0
9:54	28 59.9	-29 -48.5	27 1.2	-28 -27.3	28 0.5	-29 -7.7	1.0789	79.1	161.4	256.5	6:22.7
10: 0	27 46.6	-31 -53.3	25 50.8	-30 -27.3	26 48.7	-31 -10.1	1.0790	80.9	179.7	257.1	6:23.3
10: 6	26 30.1	-33 -53.2	24 37.0	-32 -23.1	25 33.6	-33 -7.8	1.0790	81.6	202.8	257.7	6:22.7
10:12	25 10.3	-35 -49.1	23 20.0	-34 -15.4	24 15.2	-35 -1.9	1.0790	81.0	225.7	258.1	6:21.1
10:18	23 47.3	-37 -41.9	21 59.8	-36 -5.0	22 53.6	-36 -53.1	1.0789	79.4	243.6	258.4	6:18.5
10:24	22 21.1	-39 -32.6	20 36.4	-37 -53.0	21 28.8	-38 -42.4	1.0787	77.0	256.0	258.6	6:14.9
10:30	20 51.6	-41 -22.2	19 9.8	-39 -40.0	20 0.7	-40 -30.7	1.0785	74.2	264.7	258.6	6:10.3
10:36	19 18.9	-43 -11.5	17 39.7	-41 -27.2	18 29.4	-42 -18.9	1.0782	71.2	271.0	258.8	6: 4.8
10:42	17 42.5	-45 -1.7	16 6.0	-43 -15.6	16 54.4	-44 -8.2	1.0778	67.9	275.7	258.6	5:58.3
10:48	16 2.4	-46 -54.1	14 28.4	-45 -6.1	15 15.6	-45 -59.7	1.0773	64.5	279.3	258.3	5:51.0
10:54	14 18.1	-48 -50.1	12 46.7	-47 -0.5	13 32.6	-47 -54.8	1.0768	61.0	282.2	257.8	5:42.7
11: 0	12 29.2	-50 -51.4	11 0.2	-49 -0.2	11 44.9	-49 -55.3	1.0761	57.2	284.6	257.0	5:33.6
11: 6	10 34.8	-53 -0.3	9 8.4	-51 -7.5	9 51.9	-52 -3.4	1.0754	53.3	286.5	255.8	5:23.5
11:12	8 34.1	-55 -19.7	7 10.3	-53 -25.1	7 52.5	-54 -21.9	1.0745	49.1	288.0	254.1	5:12.3
11:18	6 25.7	-57 -54.0	5 4.5	-55 -57.2	5 45.4	-56 -55.0	1.0735	44.6	289.2	251.8	5: 0.1
11:24	4 7.2	-60 -49.4	2 49.1	-58 -49.7	3 28.5	-59 -48.9	1.0723	39.7	290.2	248.7	4:46.6
11:30	1 35.0	-64 -16.7	0 20.4	-62 -12.7	0 58.1	-63 -13.9	1.0708	34.2	290.9	244.4	4:31.4
11:36	-1 -18.4	-68 -37.3	-2 -27.9	-66 -25.3	-1 -52.6	-67 -30.3	1.0690	27.8	291.2	238.4	4:13.9
11:42	-4 -52.1	-74 -47.9	-5 -52.2	-72 -16.0	-5 -21.3	-73 -30.2	1.0665	19.6	290.9	229.4	3:52.1
LIMITS	-11 -39.4	-90 -54.1	-13 -21.2	-90 -6.8	-12 -28.4	-90 -24.4	1.0602	0.0	288.1	205.4	3: 5.8

Table 72

ANNULAR SOLAR ECLIPSE OF 26 JAN 2028

Delta T = 89.4 Sec

SAROS 141

UNIVERSAL TIME	NORTHERN LIMIT LATITUDE	LONGITUDE	SOUTHERN LIMIT LATITUDE	LONGITUDE	CENTER LINE LATITUDE	LONGITUDE	DIAMETER RATIO	SUN ALT	SUN AZ	PATH WIDTH	DURATION ANNULARITY
LIMITS	4 21.0	105 9.5	1 6.2	105 50.6	2 29.5	104 49.8	0.9080	0.0	108.8	363.5	6:59.1
13:24	0 1.7	91 13.3	-3 -42.6	89 18.0	-1 -53.0	90 10.2	0.9120	17.4	109.1	350.5	7:47.5
13:30	-1 -23.9	85 5.6	-4 -50.8	83 40.8	-3 -8.8	84 20.6	0.9137	24.7	109.2	346.6	8:13.0
13:36	-2 -8.6	80 45.6	-5 -26.7	79 29.9	-3 -48.8	80 6.0	0.9150	30.3	109.6	344.5	8:34.1
13:42	-2 -32.4	77 17.8	-5 -44.4	76 4.7	-4 -9.4	76 40.0	0.9160	35.1	110.1	343.3	8:52.5
13:48	-2 -42.7	74 22.1	-5 -49.7	73 9.1	-4 -17.1	73 44.7	0.9168	39.2	110.9	342.7	9: 9.1
13:54	-2 -42.9	71 48.7	-5 -45.7	70 34.5	-4 -15.2	71 10.9	0.9175	43.0	111.9	342.2	9:23.9
14: 0	-2 -35.3	69 31.9	-5 -34.1	68 15.8	-4 -5.5	68 53.3	0.9181	46.4	113.2	341.8	9:37.2
14: 6	-2 -21.2	67 27.9	-5 -16.3	66 9.5	-3 -49.5	66 48.3	0.9186	49.6	114.9	341.3	9:49.0
14:12	-2 -1.5	65 34.1	-4 -53.0	64 13.4	-3 -28.1	64 53.5	0.9191	52.5	116.9	340.6	9:59.2
14:18	-1 -36.9	63 48.7	-4 -25.0	62 25.5	-3 -1.8	63 6.9	0.9195	55.2	119.4	339.6	10: 8.0
14:24	-1 -7.9	62 10.2	-3 -52.7	60 44.5	-2 -31.1	61 27.2	0.9198	57.7	122.4	338.4	10:15.3
14:30	0 -34.8	60 37.4	-3 -16.4	59 9.4	-1 -56.4	59 53.3	0.9201	59.9	126.0	336.8	10:21.2
14:36	0 2.0	59 9.4	-2 -36.4	57 39.2	-1 -38.1	58 24.3	0.9203	61.9	130.2	335.0	10:25.6
14:42	0 42.4	57 45.4	-1 -53.0	56 13.1	0 -36.2	56 59.2	0.9205	63.6	135.0	332.9	10:28.6
14:48	1 26.2	58 24.6	-1 -6.3	54 50.4	0 9.1	55 37.5	0.9206	65.0	140.5	330.7	10:30.3
14:54	2 13.3	55 6.5	0 -16.4	53 30.5	0 57.6	54 18.6	0.9207	66.0	146.6	328.3	10:30.6
15: 0	3 3.7	53 50.5	0 36.5	52 12.8	1 49.2	53 1.7	0.9208	66.7	153.2	325.9	10:29.8
15: 6	3 57.3	52 36.0	1 32.5	50 56.9	2 44.1	51 46.5	0.9208	67.0	160.1	323.6	10:27.8
15:12	4 54.2	51 22.5	2 31.5	49 42.0	3 42.0	50 32.4	0.9208	66.9	167.1	321.4	10:24.8
15:18	5 54.4	50 9.5	3 33.6	48 27.8	4 43.1	49 18.7	0.9207	66.4	173.9	319.4	10:20.7
15:24	6 58.0	48 56.4	4 38.7	47 13.6	5 47.5	48 5.1	0.9206	65.5	180.4	317.6	10:15.8
15:30	8 5.0	47 42.6	5 47.1	45 58.9	6 55.2	46 50.9	0.9205	64.3	186.3	316.1	10:10.0
15:36	9 15.8	46 27.5	6 58.7	44 43.0	8 6.4	45 35.4	0.9203	62.7	191.7	314.9	10: 3.4
15:42	10 30.4	45 10.4	8 13.9	43 25.3	9 21.3	44 18.0	0.9200	60.9	196.5	314.1	9:56.2
15:48	11 49.1	43 50.4	9 32.8	42 4.9	10 40.2	42 57.8	0.9198	58.8	200.8	313.7	9:48.2
15:54	13 12.4	42 26.6	10 55.8	40 40.8	12 3.3	41 33.9	0.9194	56.4	204.7	313.6	9:39.7
16: 0	14 40.7	40 57.9	12 23.2	39 12.0	13 31.1	40 5.2	0.9190	53.8	208.1	314.0	9:30.5
16: 6	16 14.6	39 22.6	13 55.6	37 37.1	15 4.2	38 30.1	0.9186	51.0	211.2	314.8	9:20.8
16:12	17 54.9	37 39.0	15 33.7	35 54.2	16 43.4	36 46.9	0.9181	48.0	214.1	316.1	9:10.4
16:18	19 42.8	35 44.3	17 18.5	34 1.0	18 29.7	34 53.1	0.9175	44.7	216.8	317.9	8:59.5
16:24	21 39.7	33 35.4	19 11.3	31 54.0	20 24.4	32 45.0	0.9168	41.1	219.4	320.2	8:47.8
16:30	23 48.2	31 5.3	21 14.0	29 28.3	22 29.9	30 17.5	0.9160	37.1	222.0	323.2	8:35.4
16:36	26 12.1	28 6.0	23 29.8	26 36.2	24 49.5	27 22.0	0.9151	32.6	224.7	327.0	8:21.8
16:42	28 58.5	24 19.9	26 4.2	23 2.7	27 22.0	23 43.0	0.9140	27.5	227.7	332.0	8: 6.8
16:48	32 23.8	19 5.9	29 8.9	18 17.6	30 43.5	18 45.2	0.9125	21.1	231.4	338.8	7:49.2
16:54	37 36.4	9 26.3	33 20.8	10 38.6	35 19.9	10 19.3	0.9102	11.6	237.2	350.4	7:25.4
LIMITS	42 6.0	-1 -22.4	38 59.1	-2 -51.1	40 14.8	-1 -19.1	0.9076	0.0	244.8	365.5	7: 1.1

Table 73

TOTAL SOLAR ECLIPSE OF 22 JUL 2028

SAROS 146   Delta T = 89.9 Sec

UNIVERSAL TIME	NORTHERN LIMIT LATITUDE	LONGITUDE	SOUTHERN LIMIT LATITUDE	LONGITUDE	CENTER LINE LATITUDE	LONGITUDE	DIAMETER RATIO	SUN ALT	SUN AZ	PATH WIDTH	DURATION TOTALITY
LIMITS	-17-41.7	-75-35.0	-19 -8.0	-75-49.5	-18-14.2	-76-10.5	1.0411	0.0	68.4	158.1	2:29.4
1:36	-12-48.4	-90 -2.2	-15 -8.1	-87-47.0	-13-56.9	-88-59.0	1.0455	14.4	64.3	177.8	3: 5.4
1:42	-11-11.0	-96-28.5	-13-16.0	-95 -0.8	-12-13.1	-95-46.4	1.0480	22.4	61.7	190.7	3:29.5
1:48	-10-21.0	-101 -2.3	-12-22.0	-99-49.8	-11-21.2	-100-27.2	1.0497	28.1	59.4	200.5	3:48.1
1:54	-9-55.2	-104-42.6	-11-54.5	-103-37.6	-10-54.6	-104-10.9	1.0510	32.7	57.0	208.7	4: 3.8
2: 0	-9-45.3	-107-50.6	-11-43.8	-106-49.5	-10-44.3	-107-20.7	1.0520	36.6	54.5	215.7	4:17.3
2: 6	-9-47.3	-110-38.5	-11-45.2	-109-37.7	-10-46.1	-110 -7.6	1.0529	40.0	51.6	221.6	4:29.1
2:12	-9-58.9	-113 -8.4	-11-56.4	-112 -8.9	-10-57.5	-112-38.0	1.0537	42.9	48.5	226.4	4:39.2
2:18	-10-18.6	-115-24.1	-12-15.6	-114-27.3	-11-16.9	-114-56.1	1.0543	45.4	45.0	230.0	4:47.9
2:24	-10-45.4	-117-32.5	-12-41.8	-116-36.0	-11-43.4	-117 -4.5	1.0548	47.5	41.2	232.5	4:55.1
2:30	-11-18.6	-119-33.5	-13-14.4	-118-37.3	-12-16.3	-119 -5.6	1.0552	49.3	37.1	234.0	5: 0.8
2:36	-11-57.7	-121-28.8	-13-52.9	-120-32.7	-12-55.1	-121 -0.9	1.0555	50.7	32.6	234.4	5: 5.1
2:42	-12-42.4	-123-19.8	-14-37.0	-122-23.8	-13-39.5	-122-51.9	1.0558	51.8	27.8	233.9	5: 8.0
2:48	-13-32.6	-125 -7.7	-15-26.6	-124-11.8	-14-29.3	-124-39.8	1.0559	52.4	22.8	232.7	5: 9.6
2:54	-14-28.1	-126-53.4	-16-21.5	-125-57.7	-15-24.5	-126-25.6	1.0560	52.7	17.7	230.7	5: 9.8
3: 0	-15-29.0	-128-38.0	-17-21.8	-127-42.8	-16-25.1	-128-10.4	1.0560	52.6	12.5	228.2	5: 8.7
3: 6	-16-35.4	-130-22.6	-18-27.6	-129-28.0	-17-31.2	-129-55.3	1.0558	52.0	7.4	225.2	5: 6.4
3:12	-17-47.5	-132 -8.2	-19-39.3	-131-14.5	-18-43.1	-131-41.3	1.0556	51.1	2.3	222.0	5: 2.9
3:18	-19 -5.6	-133-56.1	-20-57.1	-133 -3.5	-20 -1.0	-133-29.7	1.0553	49.9	357.5	218.4	4:58.1
3:24	-20-30.2	-135-47.6	-22-21.6	-134-56.8	-21-25.5	-135-21.9	1.0549	48.2	352.8	214.7	4:52.2
3:30	-22 -2.0	-137-44.4	-23-53.8	-136-55.4	-22-57.4	-137-19.7	1.0544	46.2	348.4	210.9	4:45.2
3:36	-23-41.9	-139-48.7	-25-34.0	-139 -2.4	-24-37.5	-139-25.3	1.0537	43.9	344.1	206.9	4:37.0
3:42	-25-31.2	-142 -3.2	-27-24.4	-141-20.6	-26-27.3	-141-41.6	1.0530	41.1	340.1	202.8	4:27.6
3:48	-27-32.2	-144-32.1	-29-26.9	-143-54.6	-28-28.8	-144-12.9	1.0521	37.9	336.1	198.6	4:17.0
3:54	-29-47.1	-147-21.5	-31-44.8	-146-51.1	-30-45.3	-147 -5.7	1.0509	34.2	332.2	194.1	4: 5.1
4: 0	-32-21.5	-150-41.5	-34-23.9	-150-22.0	-33-30.9	-150-30.9	1.0509	29.9	328.1	189.2	3:51.4
4: 6	-35-24.6	-154-51.9	-37-35.4	-154-51.4	-36-28.9	-154-50.2	1.0495	24.7	323.7	183.7	3:35.5
4:12	-39-18.7	-160-41.1	-41-48.8	-161-24.4	-40-31.8	-160-59.4	1.0474	17.8	318.1	177.0	3:15.7
LIMITS	-49-51.8	178 57.4	-51-11.0	179 56.4	-50-16.2	-179-57.1	1.0421	0.0	303.0	161.7	2:32.6

Table 74

ANNULAR SOLAR ECLIPSE OF 1 JUN 2030

SAROS 128 Delta T = 91.8 Sec

UNIVERSAL TIME	NORTHERN LIMIT LATITUDE	LONGITUDE	SOUTHERN LIMIT LATITUDE	LONGITUDE	CENTER LINE LATITUDE	LONGITUDE	DIAMETER RATIO	SUN ALT	SUN AZ	PATH WIDTH	DURATION ANNULARITY
LIMITS	30 40.1	-2-38.2	28 11.4	-4-58.1	29 45.6	-4-38.1	0.9322	0.0	65.1	344.2	4:13.4
4:54	35 15.7	-13 -3.5	35 45.9	-21-13.9	35 41.6	-17-37.7	0.9355	13.9	72.9	328.0	4:25.9
5: 0	39 31.9	-21-56.0	38 57.2	-27-33.4	39 17.4	-24-54.3	0.9374	21.5	78.8	315.4	4:34.3
5: 6	42 23.8	-27-45.4	41 25.0	-32-27.3	41 55.7	-30-12.4	0.9387	27.0	84.1	305.2	4:41.0
5:12	44 42.0	-32-31.2	43 28.7	-36-40.5	44 6.1	-34-40.5	0.9397	31.5	89.2	296.4	4:46.7
5:18	46 39.9	-36-44.8	45 16.1	-40-30.7	45 58.4	-38-41.7	0.9406	35.4	94.5	288.7	4:51.9
5:24	48 23.0	-40-40.1	46 50.9	-44 -6.9	47 37.2	-42-27.0	0.9413	38.7	99.9	281.9	4:56.5
5:30	49 54.5	-44-24.6	48 15.5	-47-34.7	49 5.1	-46 -2.9	0.9418	41.7	105.6	275.9	5: 0.8
5:36	51 16.2	-48 -3.1	49 31.1	-50-57.5	50 23.6	-49-33.3	0.9424	44.4	111.5	270.6	5: 4.6
5:42	52 29.2	-51-38.8	50 38.8	-54-17.9	51 33.8	-53 -1.1	0.9428	46.8	117.8	266.0	5: 8.0
5:48	53 34.2	-55-14.0	51 38.9	-57-37.5	52 36.3	-56-28.3	0.9432	48.9	124.5	262.0	5:11.1
5:54	54 31.5	-58-50.3	52 31.9	-60-57.6	53 31.4	-59-56.2	0.9435	50.7	131.5	258.7	5:13.8
6: 0	55 21.5	-62-28.6	53 18.1	-64-19.1	54 19.4	-63-25.9	0.9437	52.2	138.9	255.9	5:16.0
6: 6	56 4.2	-66 -9.8	53 57.5	-67-42.6	55 0.3	-66-58.0	0.9439	53.5	146.6	253.6	5:17.9
6:12	56 39.6	-69-54.3	54 30.2	-71 -8.5	55 34.4	-70-32.8	0.9441	54.4	154.7	251.8	5:19.3
6:18	57 7.8	-73-42.3	54 56.2	-74-36.9	56 1.4	-74-10.6	0.9442	55.1	163.0	250.6	5:20.3
6:24	57 28.7	-77-33.6	55 15.4	-78 -7.9	56 21.4	-77-51.3	0.9442	55.4	171.4	249.5	5:20.8
6:30	57 42.1	-81-28.0	55 27.8	-81-41.2	56 34.3	-81-34.8	0.9443	55.5	180.0	249.0	5:20.9
6:36	57 47.9	-85-25.1	55 33.3	-85-16.8	56 39.9	-85-20.7	0.9442	55.2	188.6	249.0	5:20.6
6:42	57 46.0	-89-24.5	55 31.6	-88-54.3	56 38.2	-89 -8.7	0.9441	54.6	197.0	250.4	5:19.8
6:48	57 36.1	-93-25.7	55 22.8	-92-33.4	56 28.9	-92-58.4	0.9440	53.7	205.3	251.6	5:18.6
6:54	57 18.3	-97-28.2	55 6.7	-96-14.1	56 11.9	-96-49.5	0.9438	52.5	213.4	253.3	5:16.9
7: 0	56 52.1	-101-31.8	54 43.0	-99-56.1	55 47.1	-100-41.9	0.9436	51.1	221.2	255.5	5:14.8
7: 6	56 17.5	-105-36.5	54 11.6	-103-39.5	55 14.1	-104-35.5	0.9433	49.3	228.7	258.3	5:12.2
7:12	55 33.9	-109-42.6	53 32.0	-107-24.6	54 32.7	-108-30.7	0.9430	47.3	235.9	261.6	5: 9.2
7:18	54 41.1	-113-50.8	52 43.9	-111-12.1	53 42.3	-112-28.2	0.9426	45.0	242.7	265.6	5: 5.9
7:24	53 38.2	-118 -2.6	51 46.7	-115 -3.1	52 42.4	-116-29.3	0.9421	42.4	249.3	270.4	5: 2.1
7:30	52 24.1	-122-20.3	50 39.6	-118-59.6	51 32.0	-120-36.0	0.9415	39.4	255.6	275.8	4:57.8
7:36	50 57.2	-126-48.1	49 21.1	-123 -4.2	50 9.5	-124-51.7	0.9409	36.2	261.7	282.2	4:53.1
7:42	49 14.7	-131-32.2	47 49.2	-127-21.9	48 32.7	-129-21.9	0.9401	32.4	267.6	289.5	4:48.0
7:48	47 11.5	-136-44.8	46 0.4	-132 -0.3	46 37.2	-134-16.2	0.9391	28.1	273.4	298.1	4:42.2
7:54	44 36.2	-142-51.6	43 47.9	-137-14.6	44 14.6	-139-53.3	0.9379	22.9	279.3	308.2	4:35.6
8: 0	40 53.2	-151-15.5	40 55.9	-143-40.8	41 2.0	-147 -7.1	0.9363	15.9	285.8	320.8	4:27.5
LIMITS	34 54.1	-164-59.3	32 27.6	-162-32.1	34 1.9	-162-53.5	0.9324	0.0	296.6	343.0	4:12.5

Table 75

TOTAL SOLAR ECLIPSE OF 25 NOV 2030

SAROS 133   Delta T =   92.2 Sec

UNIVERSAL TIME	NORTHERN LIMIT LATITUDE	NORTHERN LIMIT LONGITUDE	SOUTHERN LIMIT LATITUDE	SOUTHERN LIMIT LONGITUDE	CENTER LINE LATITUDE	CENTER LINE LONGITUDE	DIAMETER RATIO	SUN ALT	SUN AZ	PATH WIDTH	DURATION TOTALITY
LIMITS	-15-46.0	-1-58.8	-16-40.3	-1-22.4	-16-17.3	-1-51.5	1.0297	0.0	111.5	116.8	1:31.1
5:18	-21-19.7	-15-23.5	-21-26.9	-12-41.0	-21-24.7	-14 -5.0	1.0339	13.4	107.2	133.9	1:53.6
5:24	-25 -1.0	-23-18.7	-25-33.2	-21-21.4	-25-17.7	-22-21.0	1.0369	23.1	102.8	145.0	2:12.2
5:30	-27-39.5	-28-45.8	-28-22.5	-27 -2.4	-28 -1.4	-27-54.7	1.0388	29.8	98.9	151.5	2:26.0
5:36	-29-49.3	-33-11.8	-30-39.6	-31-36.5	-30-14.7	-32-24.7	1.0403	35.3	94.9	156.1	2:37.7
5:42	-31-40.9	-37 -4.2	-32-37.1	-35-34.7	-32 -9.3	-36-20.0	1.0415	40.0	90.8	159.5	2:48.0
5:48	-33-19.3	-40-35.5	-34-20.5	-39-10.9	-33-50.1	-39-53.7	1.0426	44.2	86.4	162.1	2:57.2
5:54	-34-47.3	-43-52.7	-35-52.9	-42-32.9	-35-20.3	-43-13.3	1.0434	48.1	81.7	164.1	3: 5.5
6: 0	-36 -8.5	-47 -0.5	-37-16.2	-45-45.4	-36-23.4	-46-23.4	1.0442	51.5	76.5	165.6	3:12.9
6: 6	-37-18.1	-50 -1.9	-38-31.3	-48-51.7	-37-54.8	-49-27.3	1.0448	54.7	70.8	166.8	3:19.5
6:12	-38-22.7	-52-59.1	-39-39.3	-51-54.1	-39 -1.1	-52-27.1	1.0453	57.5	64.4	167.7	3:25.4
6:18	-39-20.9	-55-53.8	-40-40.5	-54-54.4	-40 -0.7	-55-24.5	1.0458	60.1	57.3	168.4	3:30.4
6:24	-40-13.0	-58-47.3	-41-35.2	-57-54.0	-40-54.1	-58-21.1	1.0461	62.3	49.4	168.9	3:34.6
6:30	-40-59.2	-61-40.7	-42-23.8	-60-53.9	-41-41.5	-61-17.6	1.0464	64.1	40.7	169.2	3:38.1
6:36	-41-39.6	-64-34.7	-43 -6.3	-63-54.9	-42-22.9	-64-15.1	1.0466	65.6	31.0	169.4	3:40.7
6:42	-42-14.3	-67-30.2	-43-42.7	-66-57.8	-42-58.5	-67-14.2	1.0468	66.5	20.7	169.4	3:42.5
6:48	-42-43.3	-70-27.6	-44-13.0	-70 -3.1	-43-28.1	-70-15.5	1.0468	67.0	9.8	169.4	3:43.4
6:54	-43 -6.5	-73-27.6	-44-37.1	-73-11.4	-43-51.7	-73-19.8	1.0468	66.9	358.9	169.2	3:43.5
7: 0	-43-23.7	-76-30.6	-44-54.8	-76-23.1	-44 -9.3	-76-26.8	1.0468	66.3	348.1	168.9	3:42.7
7: 6	-43-34.9	-79-37.1	-45 -6.1	-79-38.5	-44-20.5	-79-37.7	1.0466	65.1	337.9	168.4	3:41.1
7:12	-43-39.8	-82-47.5	-45-10.5	-82-58.1	-44-25.1	-82-52.6	1.0464	63.6	328.3	167.9	3:38.6
7:18	-43-38.1	-86 -2.5	-45 -7.9	-86-22.4	-44-23.0	-86-12.1	1.0462	61.6	319.6	167.2	3:35.3
7:24	-43-29.5	-89-22.6	-44-57.8	-89-51.9	-44-13.6	-89-36.9	1.0458	59.3	311.5	166.4	3:31.0
7:30	-43-13.5	-92-48.7	-44-39.8	-93-27.4	-43-56.7	-93 -7.6	1.0454	56.6	304.2	165.4	3:26.0
7:36	-42-49.5	-96-21.9	-44-13.3	-97 -9.8	-43-31.5	-96-45.3	1.0448	53.7	297.4	164.2	3:20.1
7:42	-42-16.9	-100 -3.5	-43-37.4	-101 -0.8	-42-57.3	-100-31.5	1.0442	50.4	291.1	162.7	3:13.3
7:48	-41-34.6	-103-55.9	-42-51.2	-105 -2.3	-42-13.0	-104-28.4	1.0434	46.9	285.3	160.9	3: 5.6
7:54	-40-41.1	-108 -2.0	-41-52.9	-109-17.7	-41-17.2	-108-39.1	1.0425	42.9	279.7	158.8	2:57.0
8: 0	-39-34.2	-112-26.9	-40-40.4	-113-52.2	-40 -7.5	-113 -8.8	1.0415	38.6	274.3	156.1	2:47.4
8: 6	-38-10.4	-117-19.0	-39 -9.4	-118-54.9	-38-40.3	-118 -6.1	1.0402	33.6	269.0	152.7	2:36.5
8:12	-36-23.0	-122-54.2	-37-12.6	-124-43.6	-36-48.2	-123-47.9	1.0386	27.8	263.7	148.1	2:24.1
8:18	-33-56.5	-129-50.9	-34-31.0	-132 -3.5	-34-14.5	-130-55.8	1.0365	20.5	258.1	141.4	2: 8.9
8:24	-29-44.4	-140-57.7	-29-20.5	-145-11.9	-29-37.5	-142-53.5	1.0327	8.3	250.6	128.3	1:45.9
LIMITS	-25-53.6	-150-56.0	-26-47.9	-151-35.9	-26-24.8	-151 -5.2	1.0301	0.0	246.6	118.4	1:32.3

Table 76

ANNULAR SOLAR ECLIPSE OF 21 MAY 2031

SAROS 138 Delta T = 92.7 Sec

UNIVERSAL TIME	NORTHERN LIMIT LATITUDE	LONGITUDE	SOUTHERN LIMIT LATITUDE	LONGITUDE	CENTER LINE LATITUDE	LONGITUDE	DIAMETER RATIO	SUN ALT	SUN AZ	PATH WIDTH	DURATION ANNULARITY
LIMITS	-15-58.2	-15 -7.2	-17-42.0	-15-47.7	-16-44.8	-15-41.8	0.9446	0.0	68.7	204.3	3:49.0
5:24	-10-46.8	-27-56.5	-12-20.1	-28-41.0	-11-33.0	-28-19.7	0.9483	14.4	65.3	190.0	4: 0.6
5:30	-7-32.0	-34-54.1	-8-59.4	-35-39.7	-8-15.5	-35-17.4	0.9505	23.3	63.7	181.5	4: 9.8
5:36	-5-13.7	-39-30.1	-6-38.0	-40-14.8	-5-55.7	-39-52.8	0.9520	29.8	62.6	175.7	4:18.9
5:42	-3-21.5	-43 -5.2	-4-43.7	-43-48.5	-4 -2.5	-43-27.1	0.9531	35.1	61.7	171.3	4:23.6
5:48	-1-45.6	-46 -5.0	-3 -8.3	-46-46.8	-2-25.8	-46-26.1	0.9541	39.8	60.9	167.6	4:29.6
5:54	0-21.4	-48-41.3	-1-40.9	-49-21.7	-1 -1.0	-49 -1.7	0.9549	44.1	60.0	164.5	4:35.5
6: 0	0 53.9	-51 -1.1	0-24.8	-51-39.9	0 14.6	-51-20.6	0.9556	48.0	59.1	161.9	4:41.1
6: 6	2 1.9	-53 -8.5	0 43.7	-53-45.8	1 22.5	-53-27.2	0.9562	51.7	58.1	159.7	4:46.5
6:12	3 3.6	-55 -6.4	1 45.8	-55-42.2	2 24.8	-55-24.4	0.9567	55.1	56.8	157.9	4:51.7
6:18	4 0.0	-56-57.0	3 42.4	-57-31.2	3 21.3	-57-14.2	0.9572	58.4	55.3	156.3	4:56.8
6:24	4 51.6	-58-41.8	4 33.9	-59-14.5	4 12.8	-58-58.2	0.9576	61.5	53.5	155.0	5: 1.3
6:30	5 38.7	-60-22.1	4 21.0	-60-53.1	4 60.0	-60-37.7	0.9579	64.5	51.1	154.0	5: 5.7
6:36	6 21.9	-61-58.8	5 4.0	-62-28.1	5 43.0	-62-13.5	0.9582	67.3	48.1	153.2	5: 9.7
6:42	7 1.2	-63-32.7	5 43.0	-64 -0.3	6 22.2	-63-46.5	0.9584	70.0	44.2	152.6	5:13.4
6:48	7 36.9	-65 -4.5	6 18.4	-65-30.3	6 57.8	-65-17.4	0.9586	72.5	39.0	152.2	5:16.7
6:54	8 9.2	-66-34.8	6 50.2	-66-58.8	7 29.8	-66-46.8	0.9588	74.6	32.1	152.2	5:19.6
7: 0	8 38.1	-68 -4.1	7 18.7	-68-26.2	7 58.5	-68-15.2	0.9589	76.5	23.1	151.9	5:22.0
7: 6	9 3.7	-69-32.9	7 43.7	-69-53.1	8 23.8	-69-43.0	0.9589	77.8	11.6	152.0	5:23.9
7:12	9 26.1	-71 -1.6	8 5.5	-71-19.8	8 45.9	-71-10.7	0.9589	78.5	358.0	152.1	5:25.2
7:18	9 45.2	-72-30.6	8 24.1	-72-46.8	9 4.7	-72-38.8	0.9589	78.5	344.0	152.4	5:26.0
7:24	10 1.0	-74 -0.5	8 39.3	-74-14.6	9 20.3	-74 -7.6	0.9588	77.7	331.1	152.8	5:26.2
7:30	10 13.6	-75-31.5	8 51.3	-75-43.4	9 32.6	-75-37.5	0.9587	76.2	320.7	153.3	5:25.8
7:36	10 22.8	-77 -4.2	8 59.9	-77-13.9	9 41.5	-77 -9.1	0.9586	74.3	312.6	153.9	5:24.8
7:42	10 28.6	-78-39.0	9 5.9	-78-46.4	9 46.9	-78-42.7	0.9584	72.1	306.6	154.6	5:23.2
7:48	10 30.8	-80-16.4	9 6.5	-80-21.5	9 48.8	-80-18.9	0.9581	69.6	302.2	155.4	5:21.0
7:54	10 29.2	-81-56.9	9 4.3	-81-59.7	9 46.9	-81-58.3	0.9578	66.9	298.9	156.3	5:18.2
8: 0	10 23.7	-83-41.3	8 58.1	-83-41.7	9 41.0	-83-41.5	0.9575	64.1	296.4	157.3	5:14.9
8: 6	10 14.0	-85-30.4	8 47.7	-85-28.3	9 30.9	-85-29.3	0.9571	61.1	294.6	158.5	5:11.0
8:12	9 59.7	-87-25.0	8 32.6	-87-20.5	9 16.2	-87-22.7	0.9567	57.9	293.3	160.0	5: 6.7
8:18	9 40.3	-89-26.4	8 12.5	-89-19.4	8 56.5	-89-22.8	0.9561	54.6	292.3	161.6	5: 1.9
8:24	9 15.3	-91-36.2	7 46.8	-91-26.6	8 31.1	-91-31.3	0.9556	51.1	291.6	163.5	4:56.7
8:30	8 43.8	-93-56.4	7 14.2	-93-44.2	7 59.1	-93-50.2	0.9549	47.4	291.1	165.8	4:51.1
8:36	8 4.8	-96-30.0	6 34.1	-96-15.2	7 19.5	-96-22.5	0.9541	43.5	290.9	168.4	4:45.1
8:42	7 16.3	-99-21.4	5 44.5	-99 -3.9	6 30.5	-99-12.5	0.9532	39.2	290.7	171.5	4:38.8
8:48	6 15.8	-102-37.6	4 42.5	-102-17.3	5 29.3	-102-27.2	0.9522	34.4	290.7	175.3	4:31.7
8:54	4 58.3	-106-31.4	3 23.0	-106 -8.3	4 10.8	-106-19.5	0.9510	28.9	290.7	179.9	4:24.1
9: 0	3 13.0	-111-31.0	1 35.0	-111 -5.2	2 24.2	-111-17.6	0.9494	22.3	290.8	186.0	4:15.6
9: 6	0 22.7	-119-15.3	-1-20.5	-118-48.8	0-28.4	-119 -0.9	0.9469	12.6	290.8	195.3	4: 4.7
LIMITS	-3-47.2	-130-32.0	-5-33.6	-129-53.9	-4-36.9	-130 -3.4	0.9437	0.0	290.2	208.0	3:53.0

132

Table 77

ANN/TOT SOLAR ECLIPSE OF 14 NOV 2031

SAROS 143 Delta T = 93.2 Sec

UNIVERSAL TIME	NORTHERN LIMIT LATITUDE	NORTHERN LIMIT LONGITUDE	SOUTHERN LIMIT LATITUDE	SOUTHERN LIMIT LONGITUDE	CENTER LINE LATITUDE	CENTER LINE LONGITUDE	DIAMETER RATIO	SUN ALT	SUN AZ	PATH WIDTH	DURATION ANNULARITY
LIMITS	25 53.1	-164-25.0	25 41.3	-184-21.1	25 46.8	-164-24.3	0.9938	0.0	110.6	22.7	0:23.5
19:24	24 31.2	-168-16.5	24 13.8	-168-35.4	24 22.3	-168-26.2	0.9948	4.0	112.4	18.5	0:19.7
19:30	18 49.5	177 50.6	18 48.0	177 50.7	18 48.8	177 50.6	0.9992	19.2	118.2	2.7	0: 3.2
19:36	15 55.7	171 45.5	15 58.4	171 44.9	15 57.1	171 45.2	1.0014	26.9	121.0	4.9	0: 6.4T
19:42	13 41.8	167 22.3	13 47.4	167 20.6	13 44.6	167 21.4	1.0030	32.9	123.1	10.5	0:14.4T
19:48	11 49.6	163 50.4	11 57.4	163 47.8	11 53.5	163 49.0	1.0043	38.0	125.1	15.0	0:21.5T
19:54	10 11.9	160 50.1	10 21.6	160 46.7	10 16.8	160 48.4	1.0054	42.5	127.1	18.8	0:27.9T
20: 0	8 45.0	158 11.4	8 56.3	158 7.3	8 50.6	158 9.3	1.0063	46.7	129.2	22.1	0:33.8T
20: 6	7 26.6	155 48.3	7 39.4	155 43.5	7 33.0	155 45.9	1.0071	50.4	131.5	24.9	0:39.3T
20:12	6 15.3	153 36.7	6 29.4	153 31.5	6 22.3	153 34.0	1.0078	53.9	134.1	27.4	0:44.2T
20:18	5 10.1	151 33.9	5 25.3	151 28.3	5 17.7	151 31.1	1.0084	57.2	137.0	29.5	0:48.8T
20:24	4 10.3	149 37.9	4 26.5	149 32.1	4 18.4	149 35.0	1.0089	60.2	140.5	31.4	0:52.9T
20:30	3 15.4	147 47.2	3 32.4	147 41.1	3 23.9	147 44.2	1.0093	63.0	144.7	33.1	0:56.6T
20:36	2 24.9	146 0.5	2 42.6	145 54.3	2 33.8	145 57.4	1.0097	65.5	149.6	34.5	0:59.8T
20:42	1 38.6	144 16.7	1 57.0	144 10.5	1 47.8	144 13.6	1.0100	67.7	155.6	35.7	1: 2.6T
20:48	0 56.2	142 35.1	1 15.2	142 29.0	1 5.7	142 32.0	1.0102	69.5	162.7	36.7	1: 4.8T
20:54	0 17.7	140 54.8	0 37.1	140 48.8	0 27.4	140 51.8	1.0104	70.9	171.0	37.5	1: 6.5T
21: 0	0-17.2	139 15.1	0 2.6	139 9.3	0 -7.3	139 12.2	1.0105	71.8	180.2	38.0	1: 7.7T
21: 6	0-48.4	137 35.5	0-28.3	137 30.0	0-38.3	137 32.7	1.0106	72.1	190.0	38.4	1: 8.4T
21:12	-1-15.9	135 55.3	0-55.7	135 50.1	-1 -5.8	135 52.7	1.0106	71.8	199.6	38.5	1: 8.5T
21:18	-1-39.9	134 13.9	-1-19.8	134 9.0	-1-29.7	134 11.5	1.0106	70.9	208.6	38.4	1: 8.1T
21:24	-2 -0.1	132 30.7	-1-39.9	132 28.1	-1-50.0	132 28.4	1.0105	69.5	216.1	38.1	1: 7.0T
21:30	-2-16.5	130 44.9	-1-56.5	130 40.8	-2 -6.5	130 42.9	1.0103	67.6	223.1	37.5	1: 5.5T
21:36	-2-28.9	128 58.0	-2 -9.3	128 52.3	-2-19.1	128 54.1	1.0101	65.4	228.5	36.7	1: 3.3T
21:42	-2-37.2	127 3.0	-2-18.0	126 59.7	-2-27.6	127 1.3	1.0098	62.9	233.0	35.6	1: 0.6T
21:48	-2-41.0	125 4.9	-2-22.5	125 2.1	-2-31.7	125 3.5	1.0094	60.1	236.6	34.3	0:57.4T
21:54	-2-40.0	123 0.7	-2-22.2	122 58.3	-2-31.1	122 59.5	1.0090	57.1	239.5	32.6	0:53.7T
22: 0	-2-33.7	120 48.9	-2-16.9	120 46.9	-2-25.3	120 47.9	1.0085	53.9	241.8	30.7	0:49.4T
22: 6	-2-21.4	118 27.5	-2 -5.8	118 25.8	-2-13.6	118 26.7	1.0079	50.4	243.6	28.4	0:44.7T
22:12	-2 -2.3	115 54.0	-1-48.0	115 52.7	-1-55.1	115 53.4	1.0072	46.6	245.1	25.7	0:39.4T
22:18	-1-34.9	113 4.8	-1-22.3	113 3.7	-1-28.6	113 4.3	1.0063	42.5	246.3	22.6	0:33.5T
22:24	0-57.3	109 54.1	0-46.6	109 53.3	0-52.0	109 53.7	1.0053	37.9	247.3	19.0	0:27.1T
22:30	0 -6.0	106 12.7	0 2.3	106 12.0	0 -1.9	106 12.3	1.0041	32.8	248.1	14.6	0:20.0T
22:36	1 5.5	101 42.0	1 10.8	101 41.4	1 8.2	101 41.7	1.0026	26.8	248.8	9.2	0:11.9T
22:42	2 54.5	95 34.9	2 55.0	95 34.7	2 54.5	95 34.8	1.0005	19.1	249.5	1.7	0: 2.1T
22:48	7 22.2	81 55.2	7 8.4	82 12.4	7 15.2	82 4.1	0.9960	3.5	251.0	14.2	0:15.1
LIMITS	8 26.5	78 45.5	8 17.1	78 43.2	8 20.7	78 47.6	0.9950	0.0	251.5	17.9	0:18.5

Table 78

ANNULAR SOLAR ECLIPSE OF 9 MAY 2032

SAROS 148   Delta T = 93.7 Sec

UNIVERSAL TIME	NORTHERN LIMIT LATITUDE	LONGITUDE	SOUTHERN LIMIT LATITUDE	LONGITUDE	CENTER LINE LATITUDE	LONGITUDE	DIAMETER RATIO	SUN ALT	SUN AZ	PATH WIDTH	DURATION ANNULARITY
LIMITS	-69-46.4	43 16.6	-70-11.9	41 20.6	-69-35.9	41 43.8	0.9904	0.0	27.0	85.8	0:40.6
12:50	-63-34.0	33 3.2	-65-32.4	34 26.5	-64-28.4	33 37.9	0.9922	6.6	19.0	68.9	0:35.1
12:52	-61-48.4	30 6.5	-63 -8.2	30 44.4	-62-25.7	30 23.5	0.9928	9.1	15.5	63.0	0:32.8
12:54	-60-25.9	27 45.6	-61-28.6	28 6.2	-60-55.8	27 55.1	0.9933	10.9	12.7	58.8	0:31.0
12:56	-59-17.0	25 45.2	-60 -9.6	25 56.7	-59-42.3	25 50.6	0.9937	12.3	10.3	55.6	0:29.6
12:58	-58-17.4	23 58.2	-59 -3.1	24 4.3	-58-39.5	24 1.0	0.9941	13.6	8.1	53.0	0:28.3
13: 0	-57-24.6	22 20.9	-58 -5.4	22 23.3	-57-44.4	22 22.0	0.9944	14.6	6.0	51.0	0:27.2
13: 2	-56-37.4	20 50.8	-57-14.3	20 50.6	-56-55.3	20 50.7	0.9946	15.5	4.0	49.2	0:26.3
13: 4	-55-54.5	19 28.2	-56-28.5	19 24.2	-56-11.1	19 25.2	0.9948	16.3	2.1	47.8	0:25.5
13: 6	-55-15.5	18 6.0	-55-47.0	18 2.6	-55-30.9	18 4.3	0.9950	17.0	0.3	46.6	0:24.8
13: 8	-54-39.8	16 49.4	-55 -9.2	16 44.8	-54-54.1	16 47.1	0.9951	17.6	358.5	45.7	0:24.2
13:10	-54 -6.6	15 35.6	-54-34.7	15 30.0	-54-20.4	15 32.9	0.9953	18.2	356.8	44.9	0:23.6
13:12	-53-36.2	14 24.0	-54 -3.0	14 17.7	-53-49.4	14 20.9	0.9954	18.6	355.0	44.3	0:23.2
13:14	-53 -8.1	13 14.3	-53-33.9	13 7.3	-53-20.8	13 10.9	0.9955	19.0	353.4	43.9	0:22.8
13:16	-52-42.2	12 6.1	-53 -7.1	11 58.3	-52-54.4	12 2.3	0.9956	19.3	351.7	43.6	0:22.6
13:18	-52-18.3	10 58.9	-52-42.5	10 50.5	-52-30.2	10 54.8	0.9956	19.6	350.0	43.5	0:22.3
13:20	-51-56.2	9 52.5	-52-20.0	9 43.6	-52 -7.9	9 48.1	0.9956	19.7	348.4	43.5	0:22.2
13:22	-51-35.9	8 46.7	-51-59.4	8 37.1	-51-47.4	8 42.0	0.9957	19.9	346.8	43.7	0:22.1
13:24	-51-17.4	7 41.1	-51-40.6	7 30.8	-51-28.8	7 36.1	0.9957	19.9	345.1	44.0	0:22.1
13:26	-51 -0.5	6 35.5	-51-23.7	6 24.5	-51-11.9	6 30.1	0.9957	19.9	343.5	44.4	0:22.2
13:28	-50-45.3	5 29.7	-51 -8.5	5 17.9	-50-56.7	5 23.9	0.9956	19.9	341.9	44.9	0:22.3
13:30	-50-31.7	4 23.4	-50-55.1	4 10.7	-50-43.2	4 17.2	0.9956	19.8	340.3	45.6	0:22.5
13:32	-50-19.7	3 16.3	-50-43.5	3 2.6	-50-31.4	3 9.6	0.9955	19.6	338.7	46.4	0:22.8
13:34	-50 -9.4	2 8.2	-50-33.6	1 53.4	-50-21.3	2 1.0	0.9954	19.4	337.0	47.3	0:23.2
13:36	-50 -0.7	0 58.8	-50-25.6	0 42.8	-50-12.9	0 50.9	0.9953	19.1	335.4	48.3	0:23.6
13:38	-49-53.8	0-12.3	-50-19.5	0-30.0	-50 -6.4	0-21.0	0.9952	18.7	333.7	49.4	0:24.1
13:40	-49-48.7	-1-25.5	-50-15.4	-1-45.0	-50 -1.8	-1-35.0	0.9951	18.3	332.1	50.6	0:24.6
13:42	-49-45.5	-2-41.2	-50-13.4	-3 -2.8	-49-59.2	-2-51.7	0.9949	17.8	330.4	52.0	0:25.3
13:44	-49-44.5	-3-60.0	-50-13.8	-4-24.1	-49-58.8	-4-11.7	0.9948	17.2	328.7	53.5	0:26.0
13:46	-49-45.7	-5-22.5	-50-16.8	-5-49.7	-50 -0.9	-5-35.7	0.9948	16.5	326.9	55.2	0:26.8
13:48	-49-49.4	-6-49.6	-50-22.7	-7-20.7	-50 -5.6	-7 -4.7	0.9943	15.7	325.1	57.0	0:27.7
13:50	-49-56.1	-8-22.4	-50-32.1	-8-58.5	-50-13.6	-8-39.8	0.9941	14.9	323.2	59.0	0:28.8
13:52	-50 -6.2	-10 -2.6	-50-45.7	-10-45.2	-50-25.3	-10-23.1	0.9938	13.8	321.3	61.3	0:29.9
13:54	-50-20.6	-11-52.3	-51 -4.7	-12-44.0	-50-41.8	-12-17.0	0.9934	12.7	319.2	63.8	0:31.2
13:56	-50-40.4	-13-55.0	-51-31.0	-15 -0.4	-51 -4.6	-14-25.9	0.9930	11.3	316.9	66.8	0:32.7
13:58	-51 -7.9	-16-17.0	-52 -8.8	-17-45.1	-51-36.6	-16-57.8	0.9925	9.6	314.4	70.4	0:34.5
14: 0	-51-47.4	-19-10.7	-53 -8.6	-21-27.1	-52-24.4	-20-10.9	0.9919	7.3	311.3	75.2	0:36.7
LIMITS	-55-20.0	-30-55.5	-56 -8.3	-30-26.3	-55-29.9	-30 -2.1	0.9899	0.0	302.7	90.4	0:42.8

Table 79

TOTAL SOLAR ECLIPSE OF 30 MAR 2033

SAROS 120   Delta T =   94.6 Sec

UNIVERSAL TIME	NORTHERN LIMIT LATITUDE	NORTHERN LIMIT LONGITUDE	SOUTHERN LIMIT LATITUDE	SOUTHERN LIMIT LONGITUDE	CENTER LINE LATITUDE	CENTER LINE LONGITUDE	DIAMETER RATIO	SUN ALT	SUN AZ	PATH WIDTH	DURATION TOTALITY
LIMITS	84 47.3	-175 -28.1	58 1.7	179 22.7	61 43.0	174 50.1	1.0436	0.0	87.9	787.9	2:20.2
17:48	65 11.9	-178 -35.7	64 4.0	155 34.8	64 58.2	184 5.1	1.0454	8.7	99.4	829.9	2:31.4
17:50	66 28.3	175 51.3	64 56.1	153 58.4	65 54.8	162 11.5	1.0456	9.5	101.8	826.4	2:33.2
17:52	67 34.2	173 12.7	65 48.7	152 29.6	66 51.4	160 34.2	1.0458	10.1	103.9	819.9	2:34.6
17:54	68 37.9	171 20.9	66 42.1	151 7.2	67 48.6	159 9.7	1.0460	10.6	105.9	811.6	2:35.6
17:56	69 41.2	170 1.8	67 36.5	149 50.1	68 48.7	157 58.2	1.0461	11.0	107.7	802.3	2:36.4
17:58	70 44.9	169 9.1	68 32.0	148 37.8	69 45.9	156 52.7	1.0461	11.2	109.4	792.5	2:36.8
18: 0	71 49.6	168 41.3	69 28.8	147 30.0	70 48.5	155 59.1	1.0462	11.3	110.9	782.6	2:37.0
18: 2	72 55.5	168 40.6	70 27.1	146 26.4	71 48.9	155 16.0	1.0462	11.3	112.3	773.1	2:37.0
18: 4	74 3.1	169 12.2	71 27.3	145 27.2	72 53.2	154 44.7	1.0461	11.1	113.5	764.1	2:36.7
18: 6	75 12.3	170 26.6	72 29.5	144 32.9	73 60.0	154 27.8	1.0460	10.9	114.4	755.8	2:36.2
18: 8	76 23.1	172 43.2	73 34.2	143 44.1	75 9.5	154 29.5	1.0459	10.5	115.1	748.5	2:35.4
18:10	77 34.3	176 39.4	74 41.8	143 2.3	76 22.4	154 57.2	1.0457	10.0	115.3	742.3	2:34.2
18:12	78 41.7	-178 -8.7	75 52.7	142 30.1	77 39.1	156 3.6	1.0455	9.4	114.9	737.5	2:32.8
18:14	75 52.7	142 30.1	77 7.9	142 11.8	79 0.2	158 12.8	1.0453	8.5	113.5	734.3	2:30.9
LIMITS	79 1.1	-155 -54.2	84 41.2	-124 -6.7	83 12.1	-169 -56.4	1.0436	0.0	83.8	742.3	2:20.0

Table 80

TOTAL SOLAR ECLIPSE OF 20 MAR 2034

SAROS 130 Delta T = 95.7 Sec

UNIVERSAL TIME	NORTHERN LIMIT LATITUDE	LONGITUDE	SOUTHERN LIMIT LATITUDE	LONGITUDE	CENTER LINE LATITUDE	LONGITUDE	DIAMETER RATIO	SUN ALT	SUN AZ	PATH WIDTH	DURATION TOTALITY
LIMITS	0-30.5	37 34.1	-1-24.1	37 28.0	0-57.3	37 20.5	1.0279	0.0	90.1	98.9	1:33.0
8:42	0 16.3	22 32.2	0-40.8	21 20.0	0-11.9	21 55.2	1.0331	16.7	90.0	118.3	2: 5.0
8:48	1 18.4	15 11.1	0 17.8	14 13.0	0 48.2	14 41.6	1.0367	25.4	90.5	128.3	2:24.5
8:54	2 20.0	10 8.6	1 17.0	9 14.8	1 48.6	9 41.3	1.0376	31.9	91.2	135.4	2:39.8
9: 0	3 21.5	6 9.5	2 16.6	5 17.6	2 49.1	5 43.3	1.0390	37.4	92.3	140.9	2:53.0
9: 6	4 22.9	2 48.5	3 16.6	1 57.5	3 49.8	2 22.8	1.0402	42.1	93.6	145.4	3: 4.7
9:12	5 24.3	0 -6.9	4 16.7	0-57.4	4 50.6	0-32.4	1.0413	46.5	95.3	149.0	3:15.3
9:18	6 25.7	-2-43.9	5 17.2	-3-34.2	5 51.5	-3 -9.2	1.0421	50.4	97.3	151.9	3:24.7
9:24	7 27.2	-5 -7.2	6 17.8	-5-57.2	6 52.6	-5-32.4	1.0429	54.1	99.8	154.3	3:33.2
9:30	8 28.8	-7-19.9	7 18.7	-8 -9.7	7 53.8	-7-44.9	1.0435	57.5	102.9	156.1	3:40.8
9:36	9 30.5	-9-24.4	8 19.7	-10-14.0	8 55.2	-9-49.3	1.0441	60.7	106.6	157.6	3:47.5
9:42	10 32.4	-11-22.6	9 21.0	-12-12.0	9 56.8	-11-47.4	1.0445	63.6	111.2	158.6	3:53.3
9:48	11 34.5	-13-16.0	10 22.6	-14 -5.0	10 58.5	-13-40.6	1.0449	66.2	116.8	159.3	3:58.2
9:54	12 36.8	-15 -5.8	11 24.3	-15-54.4	12 0.6	-15-30.2	1.0452	68.6	123.6	159.7	4: 2.3
10: 0	13 39.3	-16-53.1	12 26.4	-17-41.1	13 2.8	-17-17.2	1.0455	70.5	131.9	159.9	4: 5.4
10: 6	14 42.1	-18-38.9	13 28.7	-19-26.2	14 5.4	-19 -2.6	1.0457	72.0	141.7	159.8	4: 7.7
10:12	15 45.1	-20-24.1	14 31.3	-21-10.6	15 8.2	-20-47.4	1.0458	72.9	152.8	159.6	4: 9.0
10:18	16 48.5	-22 -9.5	15 34.3	-22-55.1	16 11.4	-22-32.3	1.0458	73.1	164.5	159.1	4: 9.5
10:24	17 52.2	-23-56.1	16 37.6	-24-40.5	17 14.9	-24-18.3	1.0458	72.7	176.1	158.5	4: 9.1
10:30	18 56.3	-25-44.6	17 41.2	-26-27.7	18 18.7	-26 -8.2	1.0458	71.6	186.8	157.7	4: 7.7
10:36	20 0.8	-27-36.2	18 45.3	-28-17.7	19 23.0	-27-56.9	1.0456	70.0	196.1	156.7	4: 5.5
10:42	21 5.7	-29-31.7	19 49.8	-30-11.3	20 27.7	-29-51.5	1.0454	67.9	204.1	155.6	4: 2.4
10:48	22 11.0	-31-32.3	20 54.8	-32 -9.9	21 32.8	-31-51.0	1.0451	65.5	210.8	154.3	3:58.4
10:54	23 16.8	-33-39.3	22 0.3	-34-14.6	22 38.4	-33-56.9	1.0448	62.8	216.6	152.9	3:53.5
11: 0	24 23.1	-35-54.4	23 6.3	-36-26.9	23 44.6	-36-10.5	1.0443	59.8	221.6	151.2	3:47.7
11: 6	25 29.9	-38-19.5	24 12.9	-38-48.7	24 51.3	-38-34.0	1.0438	56.5	226.1	149.4	3:41.0
11:12	26 37.3	-40-57.0	25 20.2	-41-22.5	25 58.6	-41 -9.6	1.0432	53.0	230.1	147.2	3:33.3
11:18	27 45.3	-43-50.2	26 28.1	-44-11.3	27 6.6	-44 -0.5	1.0425	49.3	233.9	144.8	3:24.8
11:24	28 53.8	-47 -3.5	27 36.8	-47-19.4	28 15.2	-47-11.2	1.0416	45.2	237.6	142.1	3:15.2
11:30	30 3.0	-50-43.6	28 46.4	-50-53.1	29 24.6	-50-48.0	1.0406	40.8	241.2	138.8	3: 4.5
11:36	31 12.8	-55 -1.0	29 57.0	-55 -2.3	30 34.8	-55 -1.2	1.0394	35.8	245.0	135.1	2:52.5
11:42	32 23.2	-60-14.5	31 8.6	-60 -4.7	31 45.8	-60 -9.0	1.0379	30.2	249.2	130.4	2:38.9
11:48	33 34.1	-67 -5.5	32 21.6	-66-38.1	32 57.8	-66-50.9	1.0360	23.2	254.1	124.4	2:22.7
11:54	34 45.0	-78 -3.7	33 36.6	-76-50.7	34 10.8	-77-25.1	1.0330	13.1	261.1	115.1	2: 0.8
LIMITS	35 13.8	-92-35.4	34 18.5	-92-33.7	34 46.1	-92-31.2	1.0289	0.0	270.1	102.2	1:35.9

Table 81

ANNULAR SOLAR ECLIPSE OF 12 SEP 2034

SAROS 135   Delta T = 96.2 Sec

UNIVERSAL TIME	NORTHERN LIMIT LATITUDE	LONGITUDE	SOUTHERN LIMIT LATITUDE	LONGITUDE	CENTER LINE LATITUDE	LONGITUDE	DIAMETER RATIO	SUN ALT	SUN AZ	PATH WIDTH	DURATION ANNULARITY
LIMITS	-5 -4.3	128 43.7	-6 -30.0	128 55.0	-5 -45.7	128 31.0	0.9600	0.0	85.9	156.1	2:38.1
14:38	-4 -42.8	116 14.8	-6 -1.4	118 26.9	-5 -21.5	117 17.1	0.9632	12.2	84.7	145.0	2:41.0
14:42	-5 -3.7	108 35.5	-6 -12.9	109 50.7	-5 -38.0	109 12.3	0.9656	21.6	83.4	138.8	2:43.7
14:48	-5 -36.0	103 36.3	-6 -40.5	104 35.6	-6 -8.0	104 5.5	0.9672	28.1	82.1	131.5	2:45.7
14:54	-6 -13.5	99 44.2	-7 -14.6	100 35.8	-6 -43.8	100 9.7	0.9684	33.4	80.6	127.4	2:47.4
15: 0	-6 -54.1	96 31.4	-7 -52.6	97 17.8	-7 -23.2	96 54.4	0.9693	37.9	79.0	124.0	2:48.9
15: 6	-7 -37.1	93 44.6	-8 -33.4	94 27.6	-8 -5.1	94 5.9	0.9701	41.9	77.0	121.0	2:50.2
15:12	-8 -21.9	91 16.5	-9 -16.5	91 57.0	-8 -49.0	91 36.6	0.9708	45.6	74.8	118.3	2:51.4
15:18	-9 -8.2	89 2.4	-10 -1.3	89 41.1	-9 -34.6	89 21.7	0.9714	48.9	72.3	115.9	2:52.4
15:24	-9 -55.9	86 59.3	-10 -47.7	87 36.4	-10 -21.7	87 17.8	0.9719	52.0	69.4	113.7	2:53.4
15:30	-10 -44.8	85 4.8	-11 -35.4	85 40.5	-11 -10.0	85 22.5	0.9723	54.8	66.0	111.7	2:54.2
15:36	-11 -34.8	83 18.8	-12 -24.5	83 51.6	-11 -59.5	83 34.1	0.9727	57.4	62.0	109.9	2:55.0
15:42	-12 -25.9	81 34.2	-13 -14.8	82 8.1	-12 -50.2	81 51.1	0.9730	59.7	57.5	108.2	2:55.6
15:48	-13 -17.9	79 55.8	-14 -6.2	80 28.9	-13 -41.9	80 12.3	0.9732	61.7	52.3	106.7	2:56.2
15:54	-14 -10.9	78 20.6	-14 -58.8	78 53.0	-14 -34.7	78 36.8	0.9734	63.5	46.5	105.4	2:56.6
16: 0	-15 -4.9	76 47.7	-15 -52.4	77 19.5	-15 -28.5	77 3.6	0.9735	64.9	39.9	104.3	2:57.0
16: 6	-15 -59.8	75 16.4	-16 -47.1	75 47.5	-16 -23.4	75 31.9	0.9736	65.9	32.7	103.4	2:57.3
16:12	-16 -55.7	73 45.8	-17 -43.0	74 16.3	-17 -19.2	74 1.0	0.9736	66.6	25.0	102.6	2:57.6
16:18	-17 -52.5	72 15.3	-18 -39.9	72 45.3	-18 -16.1	72 30.2	0.9736	66.8	17.1	102.1	2:57.8
16:24	-18 -50.3	70 44.1	-19 -38.0	71 13.5	-19 -14.1	70 58.8	0.9736	66.5	9.1	101.8	2:57.8
16:30	-19 -49.2	69 11.6	-20 -37.3	69 40.4	-20 -13.2	69 26.0	0.9735	65.9	1.4	101.7	2:57.9
16:36	-20 -49.1	67 38.9	-21 -37.9	68 5.1	-21 -13.4	67 51.0	0.9733	64.8	354.1	101.8	2:57.8
16:42	-21 -50.2	65 59.3	-22 -39.7	66 26.9	-22 -14.9	66 13.1	0.9731	63.4	347.4	102.2	2:57.7
16:48	-22 -52.5	64 31.5	-23 -43.0	64 44.6	-23 -17.7	64 31.2	0.9728	61.6	341.3	102.8	2:57.5
16:54	-23 -58.1	62 31.5	-24 -47.7	62 57.3	-24 -21.8	62 44.4	0.9725	59.5	335.8	103.6	2:57.3
17: 0	-25 -1.1	60 39.0	-25 -54.0	61 3.7	-25 -27.5	60 51.3	0.9721	57.2	330.7	104.7	2:56.9
17: 6	-26 -7.6	58 39.0	-27 -2.1	59 2.3	-26 -34.8	58 50.6	0.9717	54.6	326.1	106.2	2:56.5
17:12	-27 -15.8	56 29.5	-28 -12.2	56 51.0	-27 -43.9	56 40.3	0.9712	51.8	321.8	107.9	2:56.0
17:18	-28 -25.9	54 8.3	-29 -24.5	54 27.6	-28 -55.0	54 18.0	0.9706	48.7	317.8	110.0	2:55.4
17:24	-29 -38.1	51 32.2	-30 -39.2	51 48.6	-30 -8.5	51 40.5	0.9699	45.3	313.9	112.5	2:54.7
17:30	-30 -52.8	48 37.0	-31 -57.0	48 49.4	-31 -24.7	48 43.3	0.9692	41.7	310.1	115.4	2:53.9
17:36	-32 -10.5	45 16.4	-33 -18.3	45 23.1	-32 -44.2	45 19.9	0.9683	37.6	306.3	119.0	2:52.9
17:42	-33 -32.1	41 20.1	-34 -44.3	41 18.5	-34 -7.9	41 19.6	0.9672	33.0	302.2	123.3	2:51.8
17:48	-34 -58.7	36 30.3	-36 -16.5	36 14.8	-35 -37.3	36 23.0	0.9659	27.7	297.8	128.7	2:50.5
17:54	-36 -33.3	30 7.3	-37 -59.1	29 24.4	-37 -15.8	29 47.1	0.9642	21.1	292.4	136.0	2:48.7
18: 0	-38 -25.1	19 54.1	-40 -7.2	17 29.1	-39 -15.0	18 48.8	0.9616	11.2	284.3	147.8	2:46.2
LIMITS	-39 -58.6	4 59.6	-41 -26.8	4 53.8	-40 -41.1	5 18.0	0.9586	0.0	275.2	161.6	2:43.5

Table 82

ANNULAR SOLAR ECLIPSE OF 9 MAR 2035

SAROS 140   Delta T = 96.7 Sec

UNIVERSAL TIME	NORTHERN LIMIT LATITUDE	LONGITUDE	SOUTHERN LIMIT LATITUDE	LONGITUDE	CENTER LINE LATITUDE	LONGITUDE	DIAMETER RATIO	SUN ALT	SUN AZ	PATH WIDTH	DURATION ANNULARITY
LIMITS	-42-58.1	-127-18.8	-43-48.5	-128-59.8	-43-24.2	-127-22.9	0.9764	0.0	95.5	95.2	1:25.8
21:30	-43-39.9	-148-11.4	-44-22.3	-148-53.7	-44 -0.9	-147-33.9	0.9808	15.5	80.5	75.7	1:19.0
21:36	-43-12.5	-157-54.4	-43-51.1	-157-18.9	-43-31.7	-157-37.0	0.9831	23.7	72.0	65.9	1:14.6
21:42	-42-30.8	-164-45.3	-43 -5.7	-164-28.4	-42-48.2	-164-36.0	0.9847	29.8	65.3	59.2	1:11.1
21:48	-41-42.3	-170-12.1	-42-14.0	-170 -2.3	-41-58.1	-170 -7.3	0.9859	34.8	59.5	54.0	1: 8.0
21:54	-40-49.7	-174-46.3	-41-18.6	-174-42.1	-41 -4.1	-174-44.2	0.9870	39.1	54.2	49.7	1: 5.2
22: 0	-39-54.4	-178-43.6	-40-20.8	-178-43.3	-40 -7.5	-178-43.4	0.9879	43.0	49.0	46.2	1: 2.6
22: 6	-38-57.1	177 46.5	-39-21.4	177 44.2	-39 -9.2	177 45.3	0.9886	46.5	43.8	43.2	1: 0.2
22:12	-37-58.3	174 37.9	-38-20.8	174 33.7	-38 -9.5	174 35.8	0.9893	49.6	38.6	40.7	0:58.0
22:18	-36-58.5	171 46.2	-37-19.4	171 40.6	-37 -8.9	171 43.4	0.9898	52.5	33.1	38.6	0:56.0
22:24	-35-57.7	169 8.2	-36-17.3	169 1.6	-36 -7.5	169 4.9	0.9903	55.1	27.4	36.7	0:54.2
22:30	-34-56.3	166 41.5	-35-14.7	166 34.1	-35 -5.5	166 37.8	0.9907	57.4	21.4	35.2	0:52.6
22:36	-33-54.3	164 24.0	-34-11.7	164 16.0	-34 -3.0	164 20.0	0.9911	59.4	14.9	34.0	0:51.2
22:42	-32-51.8	162 14.1	-33 -8.3	162 5.6	-33 -0.1	162 9.9	0.9914	61.1	8.0	33.0	0:50.0
22:48	-31-48.9	160 10.4	-32 -4.7	160 1.5	-31-56.8	160 6.0	0.9916	62.4	0.6	32.2	0:49.0
22:54	-30-45.5	158 11.7	-31 -0.8	158 2.5	-30-53.2	158 7.1	0.9918	63.3	352.9	31.7	0:48.3
23: 0	-29-41.8	156 17.0	-29-56.7	156 7.3	-29-49.2	156 12.2	0.9919	63.9	345.0	31.5	0:47.7
23: 6	-28-37.7	154 25.2	-28-52.3	154 15.1	-28-45.0	154 20.2	0.9919	64.0	337.0	31.4	0:47.4
23:12	-27-33.2	152 35.3	-27-47.6	152 24.9	-27-40.4	152 30.1	0.9919	63.6	329.2	31.6	0:47.2
23:18	-26-28.3	150 46.5	-26-42.6	150 35.6	-26-35.5	150 41.1	0.9919	62.9	321.7	32.0	0:47.3
23:24	-25-23.0	148 57.8	-25-37.3	148 46.4	-25-30.1	148 52.1	0.9918	61.7	314.7	32.7	0:47.6
23:30	-24-17.2	147 8.4	-24-31.7	146 56.3	-24-24.4	147 2.3	0.9916	60.2	308.3	33.6	0:48.2
23:36	-23-10.9	145 17.0	-23-25.6	145 4.2	-23-18.2	145 10.6	0.9914	58.4	302.5	34.7	0:49.0
23:42	-22 -3.9	143 22.6	-22-19.1	143 8.9	-22-11.5	143 15.8	0.9911	56.2	297.3	36.2	0:50.0
23:48	-20-56.3	141 24.0	-21-12.0	141 9.2	-21 -4.1	141 16.6	0.9907	53.7	292.6	37.9	0:51.2
23:54	-19-47.8	139 19.3	-20 -4.2	139 3.3	-19-56.0	139 11.3	0.9903	51.0	288.5	39.9	0:52.6
0: 0	-18-38.4	137 6.8	-18-55.5	136 49.2	-18-46.9	136 58.0	0.9897	48.0	284.8	42.3	0:54.3
0: 6	-17-27.7	134 43.6	-17-45.8	134 24.2	-17-36.7	134 34.0	0.9891	44.6	281.6	45.1	0:56.2
0:12	-16-15.5	132 6.4	-16-34.8	131 44.4	-16-25.1	131 55.5	0.9884	40.9	278.6	48.3	0:58.4
0:18	-15 -1.4	129 9.5	-15-22.0	128 44.3	-15-11.7	128 57.0	0.9875	36.8	276.0	52.0	1: 0.8
0:24	-13-44.6	125 44.4	-14 -6.8	125 14.5	-13-55.6	125 29.6	0.9864	32.1	273.7	56.5	1: 3.6
0:30	-12-23.7	121 35.0	-12-47.9	120 57.6	-12-35.7	121 16.4	0.9851	26.6	271.5	61.8	1: 6.9
0:36	-10-55.6	116 3.4	-11-22.1	115 10.6	-11 -8.8	115 37.4	0.9833	19.7	269.4	68.9	1:10.8
0:42	-9 -6.9	106 22.7	-9-33.1	104 2.8	-9-20.1	105 18.5	0.9801	8.0	266.9	81.0	1:16.9
LIMITS	-8 -9.0	97 40.7	-8-56.6	97 26.3	-8-33.7	97 46.2	0.9779	0.0	265.7	89.1	1:20.4

Table 83

TOTAL SOLAR ECLIPSE OF 2 SEP 2035

SAROS 145   Delta T = 97.2 Sec

UNIVERSAL TIME	NORTHERN LIMIT LATITUDE	NORTHERN LIMIT LONGITUDE	SOUTHERN LIMIT LATITUDE	SOUTHERN LIMIT LONGITUDE	CENTER LINE LATITUDE	CENTER LINE LONGITUDE	DIAMETER RATIO	SUN ALT	SUN AZ	PATH WIDTH	DURATION TOTALITY
LIMITS	38 19.7	-79-30.7	37 47.5	-79-44.9	38 4.1	-79-41.4	1.0164	0.0	80.1	62.4	0:55.3
0:18	39 50.4	-91-51.1	39 19.8	-93-37.6	39 35.5	-92-46.2	1.0197	10.7	88.6	73.6	1:11.7
0:24	40 39.6	-104-48.7	39 55.4	-105-43.3	40 17.7	-105-16.8	1.0228	21.4	98.1	83.9	1:30.4
0:30	40 41.1	-112-44.0	39 51.4	-113-20.9	40 16.3	-113 -2.9	1.0246	28.3	104.7	90.1	1:43.6
0:36	40 24.7	-118-50.7	39 31.6	-119-15.8	39 58.2	-119 -3.6	1.0261	33.9	110.4	94.6	1:54.6
0:42	39 57.9	-123-55.4	39 2.4	-124-11.4	39 30.1	-124 -3.7	1.0272	38.7	115.7	98.2	2: 4.2
0:48	39 23.8	-128-18.5	38 26.8	-128-26.9	38 55.3	-128-22.9	1.0282	43.0	120.8	101.2	2:12.8
0:54	38 44.5	-132-11.3	37 46.5	-132-13.0	38 15.5	-132-12.3	1.0290	46.8	125.9	103.8	2:20.4
1: 0	38 1.1	-135-40.6	37 2.4	-135-36.5	37 31.7	-135-38.7	1.0296	50.4	131.1	106.0	2:27.2
1: 6	37 14.4	-138-51.3	36 15.4	-138-42.1	36 44.8	-138-46.8	1.0302	53.6	136.4	107.9	2:33.1
1:12	36 24.9	-141-46.8	35 25.8	-141-33.0	35 55.3	-141-40.0	1.0307	56.8	142.1	109.6	2:38.4
1:18	35 33.0	-144-29.7	34 34.1	-144-11.9	35 3.8	-144-40.9	1.0311	59.2	148.2	111.0	2:42.8
1:24	34 39.1	-147 -2.2	33 40.5	-146-40.8	34 9.8	-146-51.6	1.0314	61.6	154.9	112.3	2:46.5
1:30	33 43.3	-149-26.1	32 45.2	-149 -1.4	33 14.3	-149-13.8	1.0317	63.7	162.2	113.4	2:49.5
1:36	32 46.0	-151-42.6	31 48.5	-151-15.1	32 17.2	-151-28.9	1.0318	65.4	170.2	114.3	2:51.8
1:42	31 47.1	-153-53.2	30 50.3	-153-23.1	31 18.7	-153-38.1	1.0320	66.7	178.9	115.1	2:53.3
1:48	30 46.8	-155-58.9	29 50.8	-155-26.5	30 18.8	-155-42.6	1.0320	67.6	188.2	115.7	2:54.1
1:54	29 45.2	-158 -0.7	28 50.0	-157-26.3	29 17.8	-157-43.5	1.0320	68.0	197.7	116.1	2:54.2
2: 0	28 42.3	-159-59.7	27 48.0	-159-23.4	28 15.2	-159-41.5	1.0320	67.8	207.2	116.4	2:53.5
2: 6	27 38.1	-161-56.7	26 44.8	-161-18.8	27 11.5	-161-37.7	1.0318	67.2	216.3	116.6	2:52.2
2:12	26 32.6	-163-52.7	25 40.3	-163-13.4	26 6.5	-163-33.0	1.0317	66.0	224.8	116.6	2:50.2
2:18	25 25.8	-165-48.6	24 34.6	-165 -8.1	25 0.2	-165-28.2	1.0314	64.5	232.4	116.3	2:47.5
2:24	24 17.6	-167-45.5	23 27.5	-167 -3.8	23 52.6	-167-24.5	1.0311	62.5	239.1	115.9	2:44.1
2:30	23 7.9	-169-44.4	22 19.0	-169 -1.7	22 43.5	-169-22.9	1.0307	60.3	245.1	115.2	2:40.1
2:36	21 56.7	-171-46.6	21 9.0	-171 -3.0	21 32.9	-171-24.7	1.0303	57.7	250.2	114.2	2:35.4
2:42	20 43.6	-173-53.8	19 57.3	-173 -9.2	20 20.5	-173-31.3	1.0297	54.9	254.7	112.9	2:30.0
2:48	19 28.5	-176 -7.2	18 43.6	-175-22.1	19 6.1	-175-44.5	1.0291	51.7	258.6	111.2	2:24.0
2:54	18 11.1	-178-29.7	17 27.7	-177-43.9	17 49.5	-178 -6.6	1.0284	48.3	262.1	109.1	2:17.3
3: 0	16 50.8	178 55.7	16 9.2	179 42.3	16 30.1	179 19.1	1.0278	44.6	265.1	106.4	2: 9.9
3: 6	15 27.1	178 4.5	14 47.3	176 52.1	15 7.3	176 28.4	1.0268	40.5	267.7	103.0	2: 1.7
3:12	13 58.8	172 49.7	13 21.3	173 38.6	13 40.1	173 14.4	1.0255	36.0	270.1	98.8	1:52.6
3:18	12 24.1	168 59.4	11 49.3	169 50.4	12 6.8	169 25.0	1.0241	30.8	272.2	93.5	1:42.4
3:24	10 39.3	164 8.4	10 8.0	165 3.6	10 23.8	164 36.3	1.0223	24.5	274.2	86.6	1:30.5
3:30	8 33.1	157 2.8	8 7.7	158 10.6	8 20.6	157 37.1	1.0199	16.0	276.0	76.5	1:15.3
LIMITS	5 41.4	142 47.9	5 11.6	142 58.5	5 27.4	142 59.9	1.0151	0.0	278.1	57.2	0:50.8

FIFTY YEAR CANON OF SOLAR ECLIPSES : 1986 - 2035

SECTION 4 - <u>GLOBAL MAPS OF SOLAR ECLIPSES : 1986 - 2035</u>

PARTIAL SOLAR ECLIPSE — 9 APR 1986

GREATEST = 6:20:26.7 UT
CONJUNCTION = 5:13:53.7 UT
GAMMA = -1.08235
MAG = 0.82205

CONTACTS
P1 = 4: 9:42.3 UT
P4 = 8:31:41.8 UT

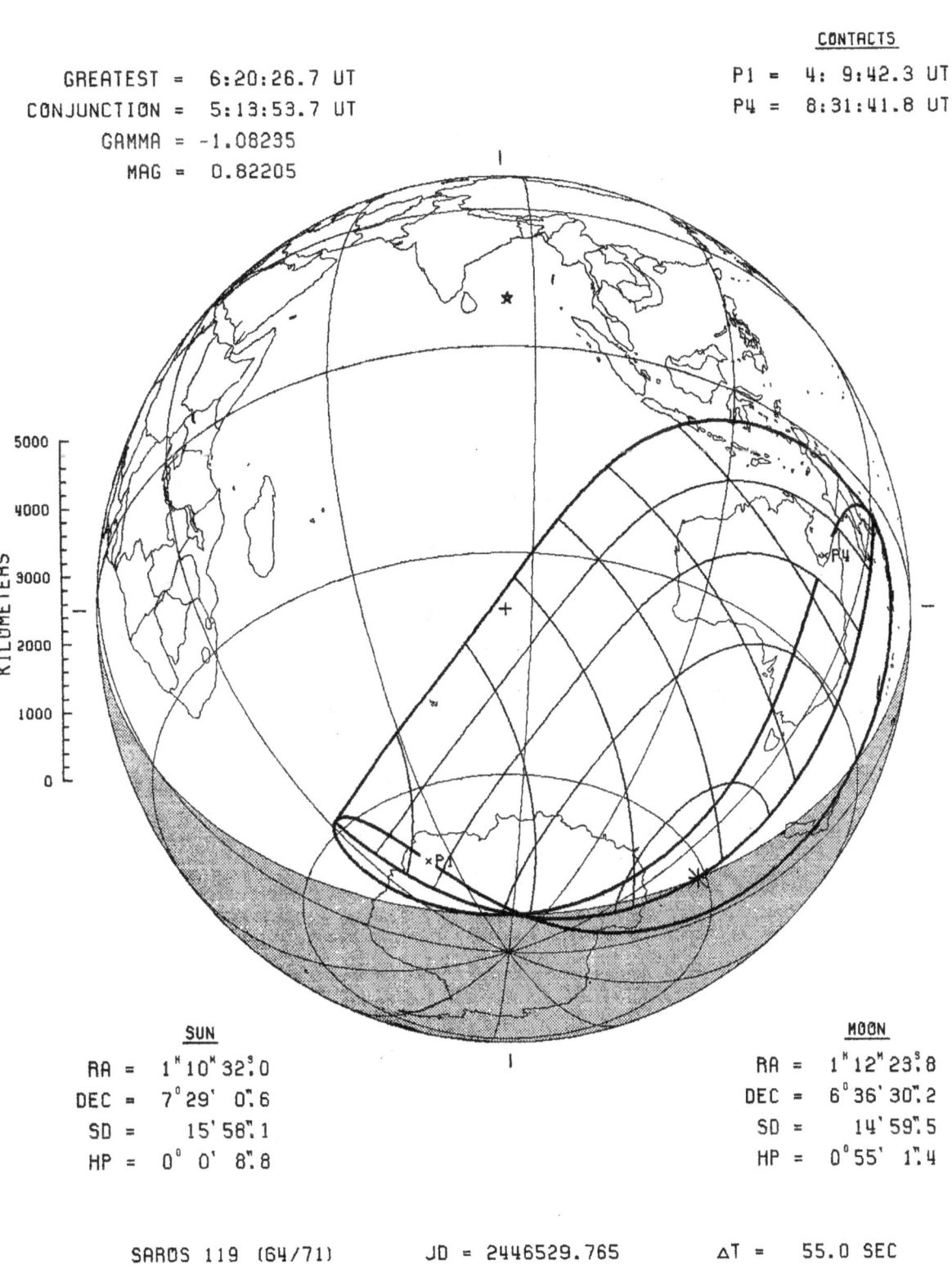

SUN
RA = 1ᴴ10ᴹ32ˢ0
DEC = 7°29' 0".6
SD = 15'58".1
HP = 0° 0' 8".8

MOON
RA = 1ᴴ12ᴹ23ˢ8
DEC = 6°36'30".2
SD = 14'59".5
HP = 0°55' 1".4

SAROS 119 (64/71)   JD = 2446529.765   ΔT = 55.0 SEC

Figure 31

143

ANN/TOT SOLAR ECLIPSE — 3 OCT 1986

GREATEST = 19: 5:18.6 UT
CONJUNCTION = 18: 6:26.7 UT
GAMMA = 0.99280
RATIO = 1.00006

P1 = 16:57:23.3 UT
U1 = 18:54:42.5 UT
U4 = 19:16:28.5 UT
P4 = 21:13:32.4 UT

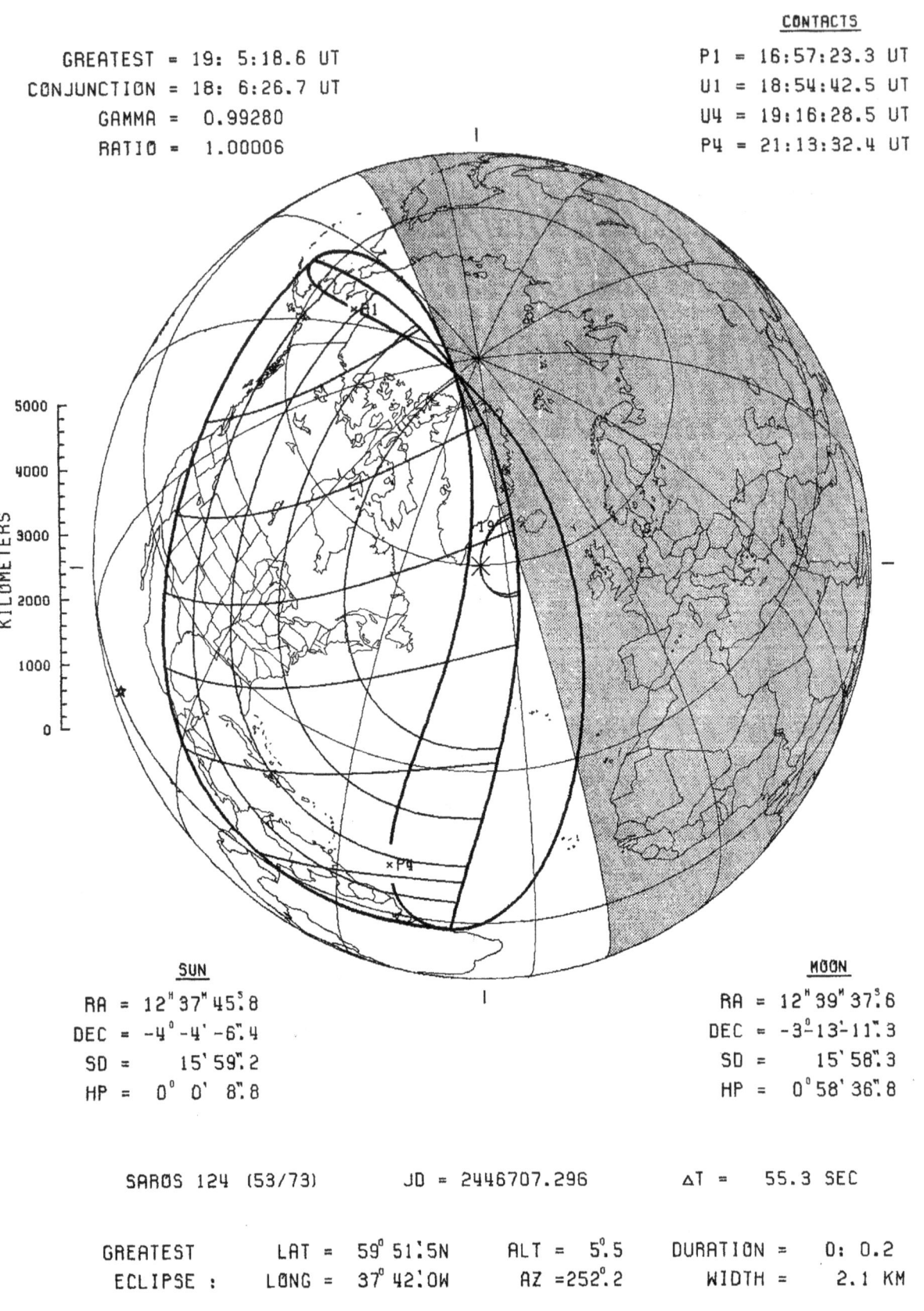

KILOMETERS

5000
4000
3000
2000
1000
0

SUN
RA = $12^h 37^m 45^s.8$
DEC = $-4^0 -4' -6".4$
SD = $15' 59".2$
HP = $0^0 0' 8".8$

MOON
RA = $12^h 39^m 37^s.6$
DEC = $-3^0 13' -11".3$
SD = $15' 58".3$
HP = $0^0 58' 36".8$

SAROS 124 (53/73)   JD = 2446707.296   ΔT =   55.3 SEC

GREATEST   LAT = $59^0 51'.5N$ ALT = $5^0.5$ DURATION =   0: 0.2
ECLIPSE :   LONG = $37^0 42'.0W$ AZ = $252^0.2$ WIDTH =   2.1 KM

Figure 32
144

ANN/TOT SOLAR ECLIPSE — 29 MAR 1987

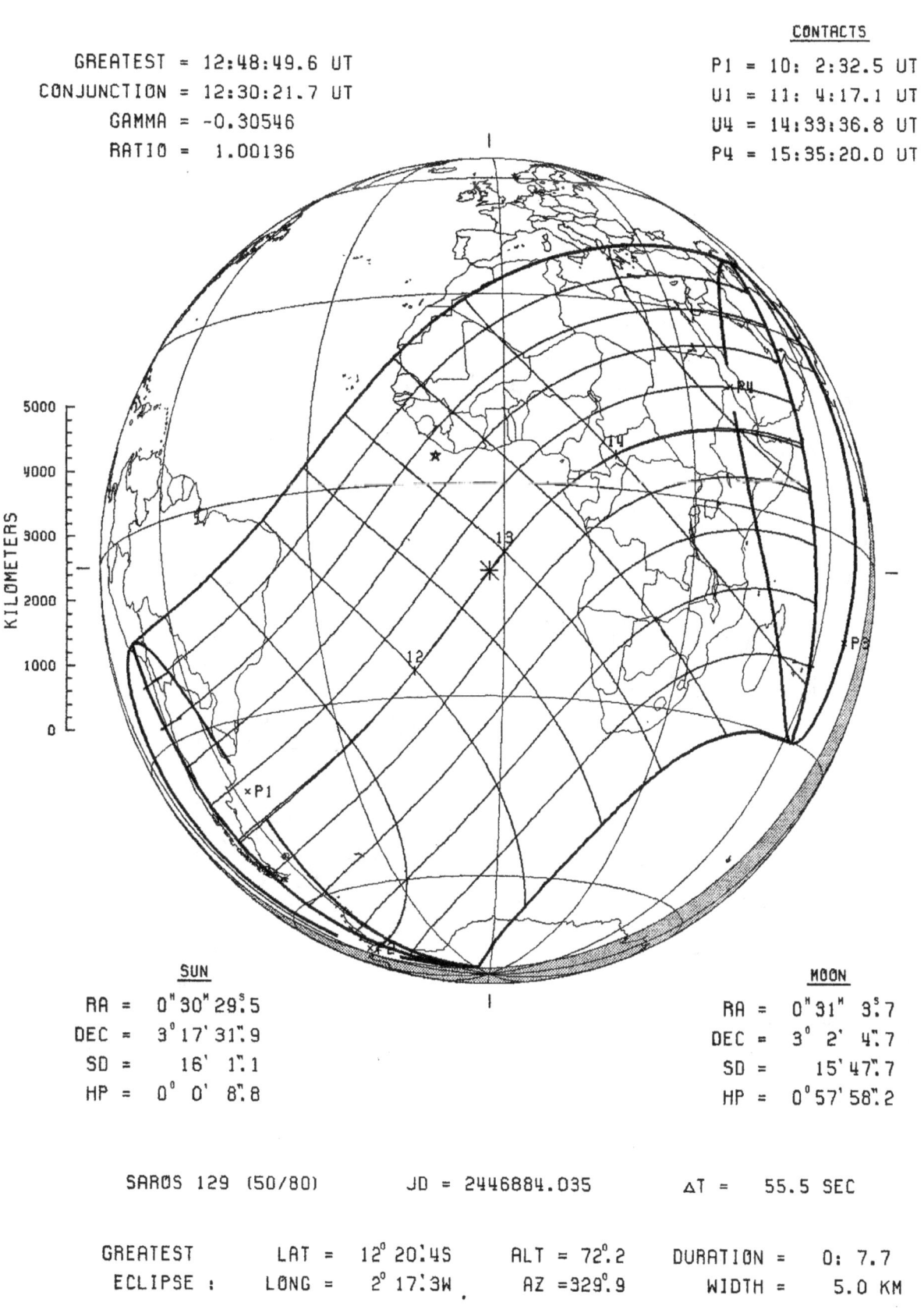

GREATEST = 12:48:49.6 UT
CONJUNCTION = 12:30:21.7 UT
GAMMA = -0.30546
RATIO = 1.00136

CONTACTS

P1 = 10: 2:32.5 UT
U1 = 11: 4:17.1 UT
U4 = 14:33:36.8 UT
P4 = 15:35:20.0 UT

SUN

RA =   0ʰ30ᴹ29ˢ5
DEC =   3°17'31".9
SD =   16' 1".1
HP =   0° 0' 8".8

MOON

RA =   0ʰ31ᴹ 9ˢ7
DEC =   3° 2' 4".7
SD =   15'47".7
HP =   0°57'58".2

SAROS 129 (50/80)   JD = 2446884.035   ΔT =   55.5 SEC

GREATEST   LAT = 12° 20'.45   ALT = 72°.2   DURATION =   0: 7.7
ECLIPSE :   LONG =   2° 17'.3W   AZ =329°.9   WIDTH =   5.0 KM

Figure 33
145

ANNULAR SOLAR ECLIPSE — 23 SEP 1987

GREATEST   = 3:11:24.0 UT
CONJUNCTION = 2:53:31.5 UT
GAMMA   = 0.27841
RATIO   = 0.96336

CONTACTS
P1 = 0:14:52.7 UT
U1 = 1:19:21.4 UT
U4 = 5: 3:34.1 UT
P4 = 6: 7:56.6 UT

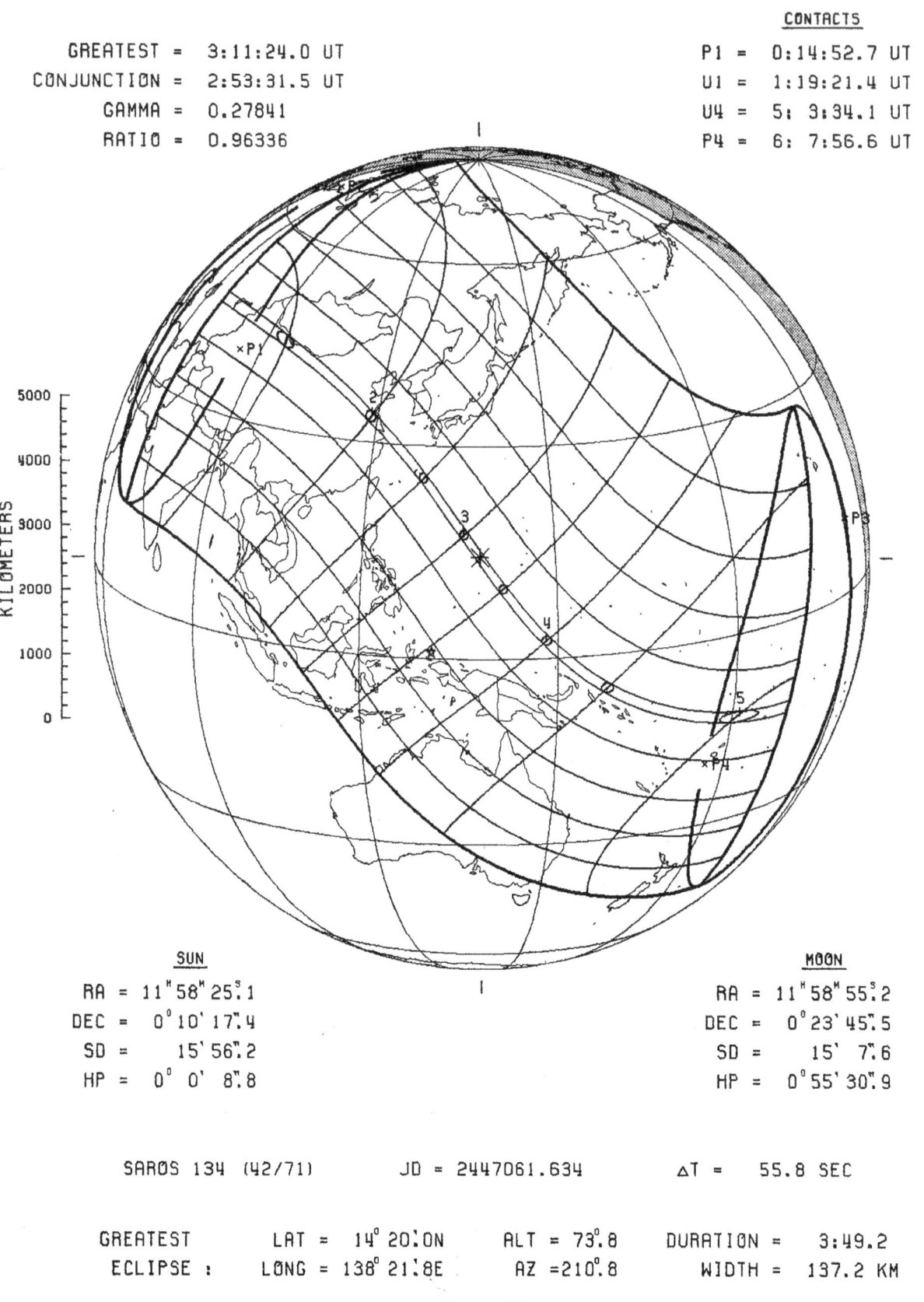

SUN

RA = 11ʰ58ᴹ25ˢ.1
DEC = 0°10' 17".4
SD =   15' 56".2
HP = 0° 0' 8".8

MOON

RA = 11ʰ58ᴹ55ˢ.2
DEC = 0°23' 45".5
SD =   15' 7".6
HP = 0°55' 30".9

SAROS 134 (42/71)   JD = 2447061.634   ΔT =   55.8 SEC

GREATEST   LAT = 14° 20'.0N   ALT = 73°.8   DURATION =   3:49.2
ECLIPSE : LONG = 138° 21'.8E   AZ =210°.8   WIDTH = 137.2 KM

Figure 34
146

TOTAL SOLAR ECLIPSE - 18 MAR 1988

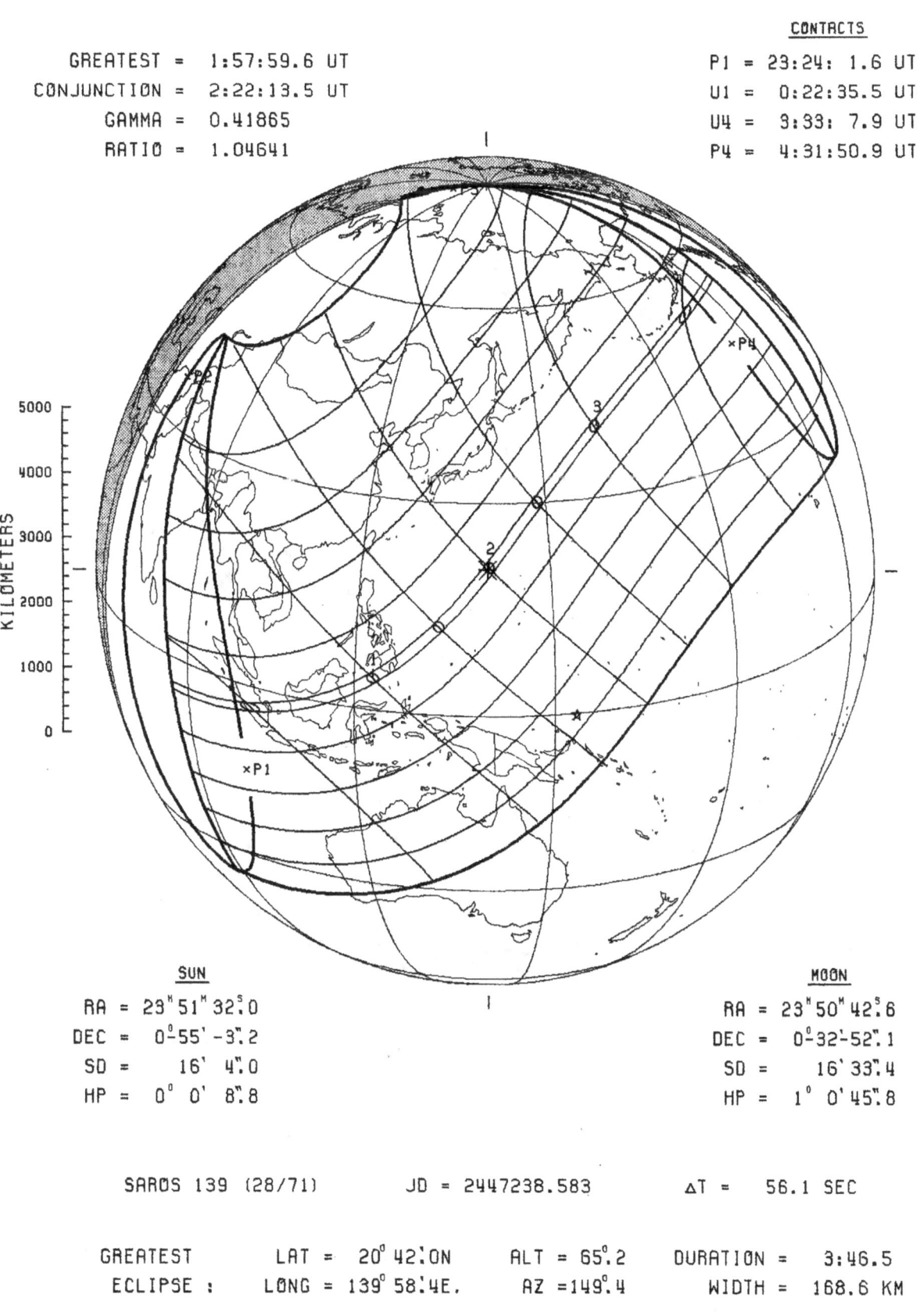

GREATEST = 1:57:59.6 UT
CONJUNCTION = 2:22:13.5 UT
GAMMA = 0.41865
RATIO = 1.04641

CONTACTS

P1 = 23:24: 1.6 UT
U1 = 0:22:35.5 UT
U4 = 3:33: 7.9 UT
P4 = 4:31:50.9 UT

KILOMETERS

5000
4000
3000
2000
1000
0

×P4

×P1

SUN

RA = 23ʰ51ᵐ32ˢ.0
DEC = 0°55'-3".2
SD = 16' 4".0
HP = 0° 0' 8".8

MOON

RA = 23ʰ50ᵐ42ˢ.6
DEC = 0°32'-52".1
SD = 16' 33".4
HP = 1° 0'45".8

SAROS 139 (28/71)   JD = 2447238.583   ΔT = 56.1 SEC

GREATEST   LAT = 20° 42'.0N   ALT = 65°.2   DURATION =   3:46.5
ECLIPSE :   LONG = 139° 58'.4E.   AZ =149°.4   WIDTH = 168.6 KM

Figure 35
147

ANNULAR SOLAR ECLIPSE – 11 SEP 1988

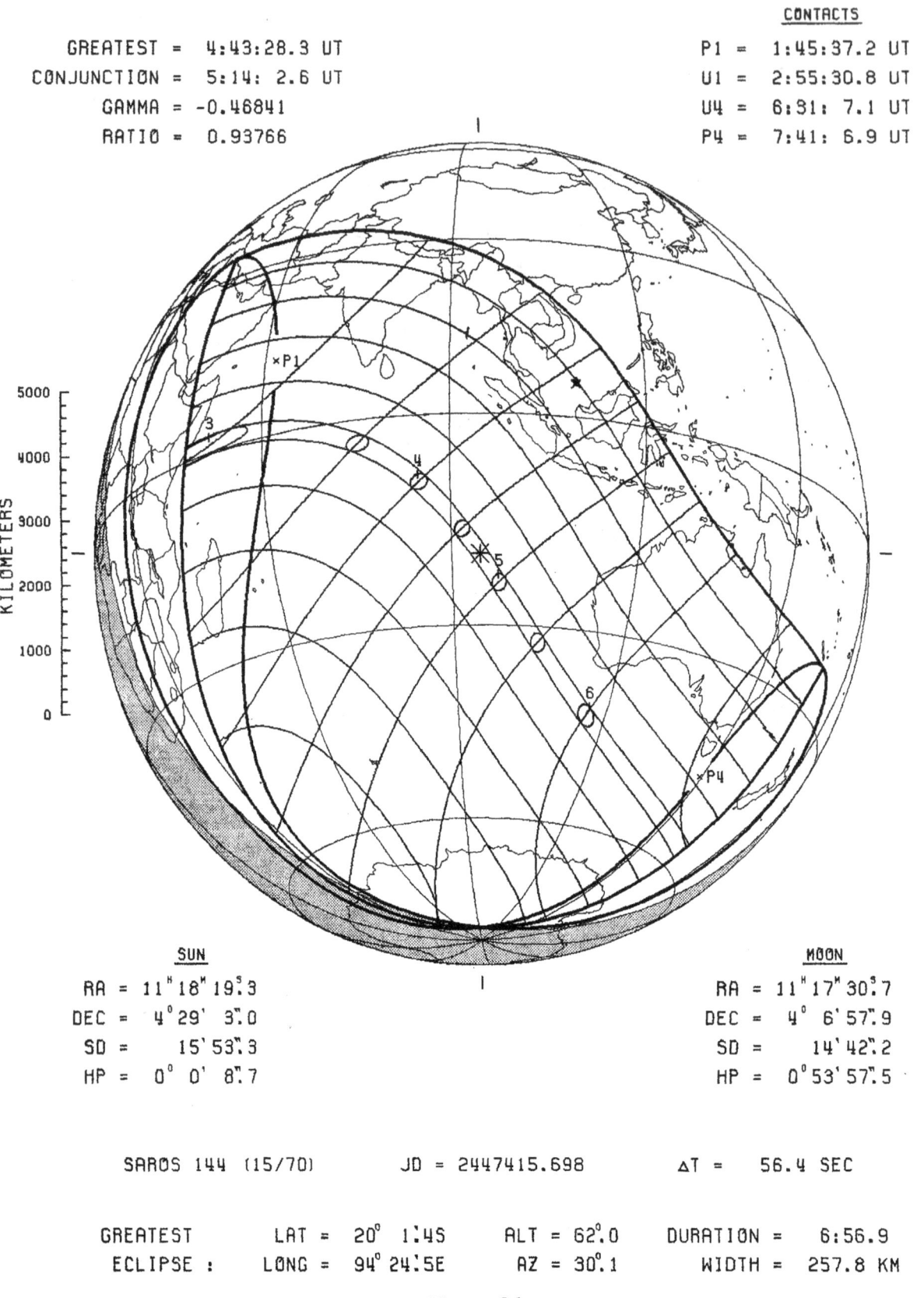

GREATEST = 4:43:28.3 UT
CONJUNCTION = 5:14: 2.6 UT
GAMMA = -0.46841
RATIO = 0.93766

CONTACTS

P1 = 1:45:37.2 UT
U1 = 2:55:30.8 UT
U4 = 6:31: 7.1 UT
P4 = 7:41: 6.9 UT

KILOMETERS

5000
4000
3000
2000
1000
0

SUN

RA = 11ʰ18ᵐ19ˢ.3
DEC = 4°29' 3".0
SD = 15'53".3
HP = 0° 0' 8".7

MOON

RA = 11ʰ17ᵐ30ˢ.7
DEC = 4° 6'57".9
SD = 14'42".2
HP = 0°53'57".5

SAROS 144 (15/70)   JD = 2447415.698   ΔT = 56.4 SEC

GREATEST   LAT = 20° 1.4S   ALT = 62°.0   DURATION = 6:56.9
ECLIPSE :   LONG = 94° 24.5E   AZ = 30°.1   WIDTH = 257.8 KM

Figure 36
148

PARTIAL SOLAR ECLIPSE   -   7 MAR 1989

GREATEST = 18: 7:44.2 UT

CONJUNCTION = 19: 9: 3.0 UT

GAMMA = 1.09805

MAG = 0.82553

CONTACTS

P1 = 16:16:51.5 UT

P4 = 19:58:10.7 UT

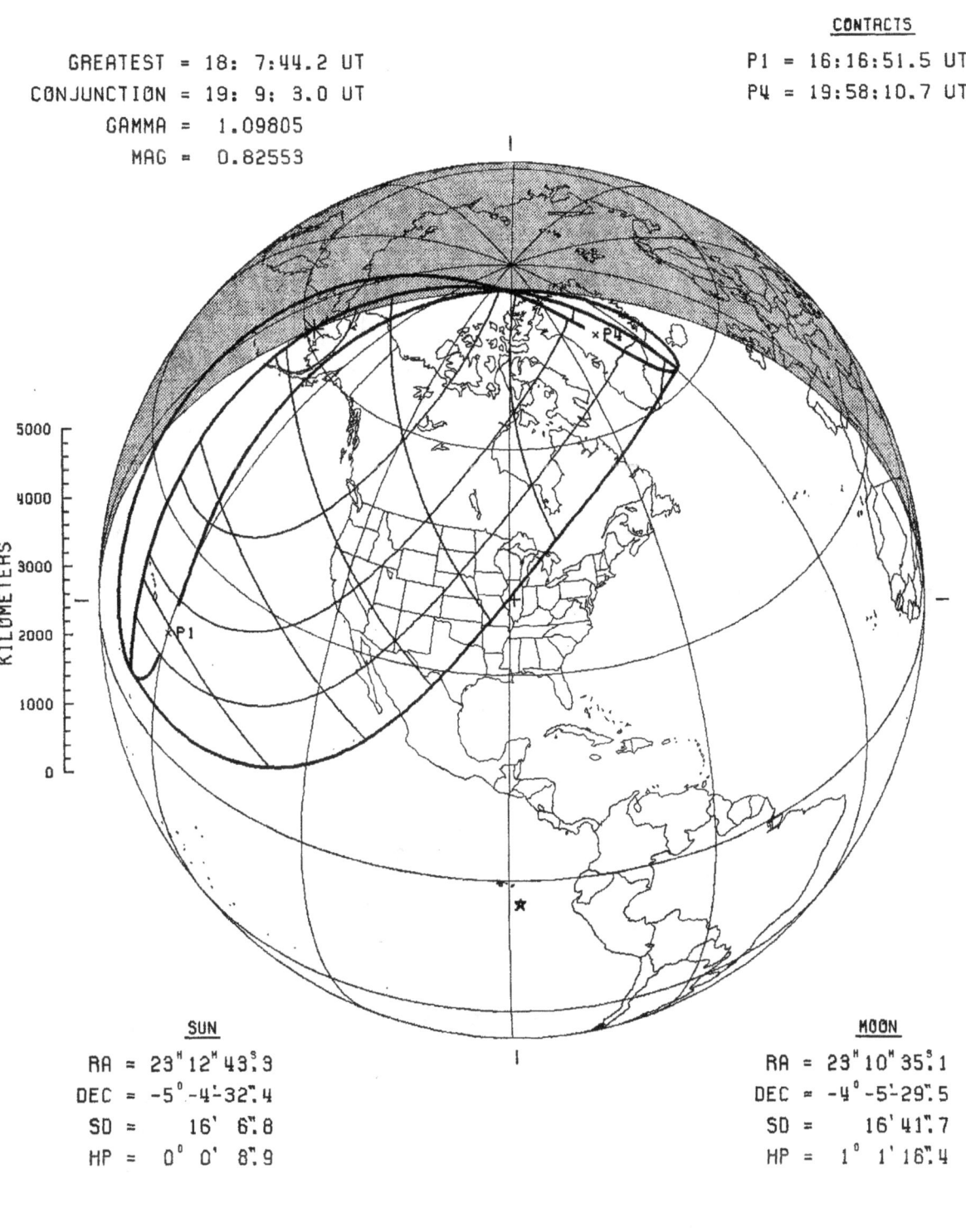

SUN

RA = 23^{h}12^{m}43^{s}3

DEC = -5$^°$-4'-32".4

SD =   16' 6".8

HP = 0$^°$ 0' 8".9

MOON

RA = 23^{h}10^{m}35^{s}1

DEC = -4$^°$-5'-29".5

SD =   16' 41".7

HP = 1$^°$ 1' 16".4

SAROS 149 (19/71)   JD = 2447593.256   ΔT =   56.7 SEC

Figure 37

PARTIAL SOLAR ECLIPSE   -   31 AUG 1989

GREATEST = 5:30:48.5 UT
CONJUNCTION = 6:43: 3.9 UT
GAMMA = -1.19313
MAG = 0.63262

<u>CONTACTS</u>
P1 = 3:33:39.1 UT
P4 = 7:27:33.6 UT

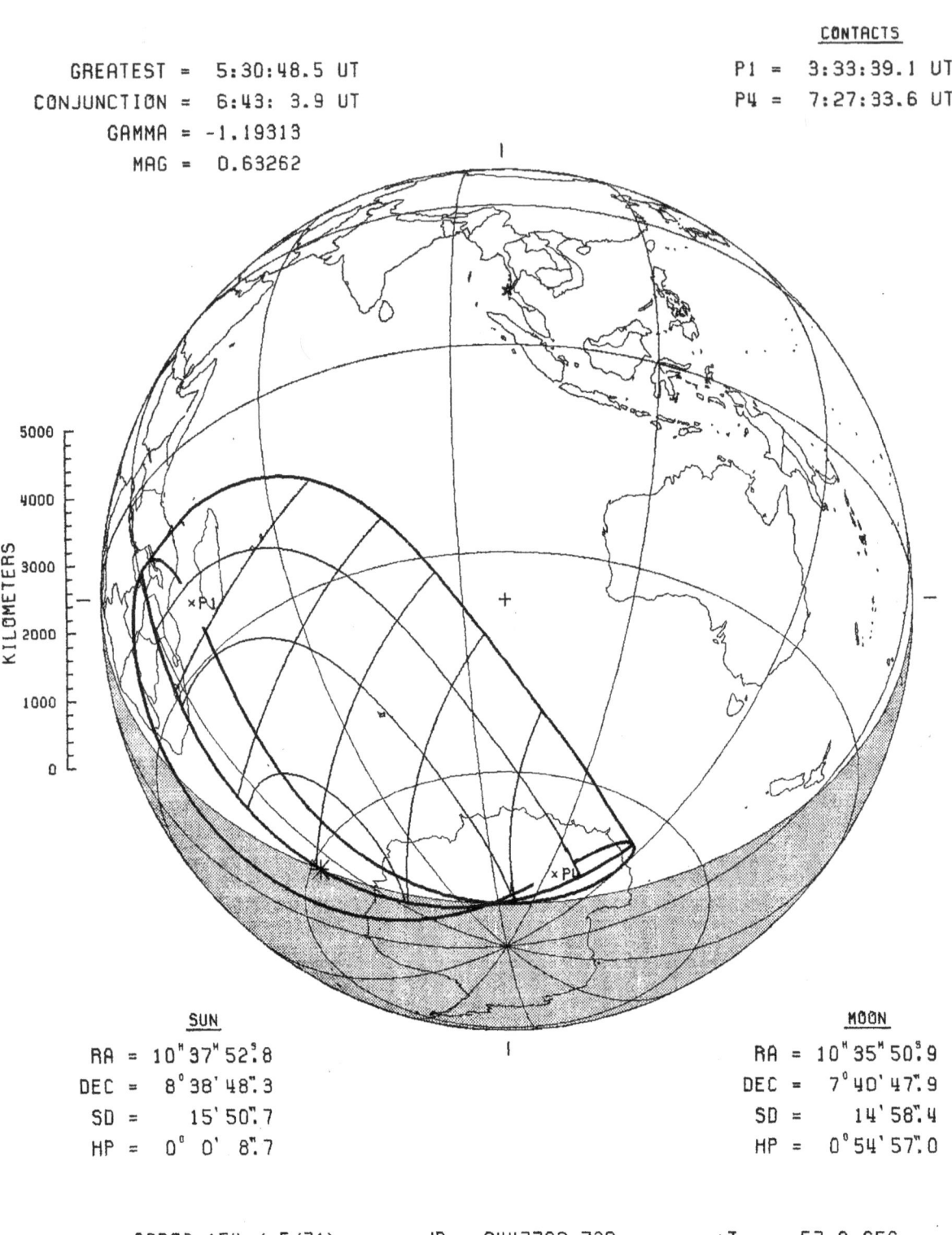

<u>SUN</u>
RA = $10^h 37^m 52\overset{s}{.}8$
DEC = $8° 38' 48\overset{"}{.}3$
SD = $15' 50\overset{"}{.}7$
HP = $0° 0' 8\overset{"}{.}7$

<u>MOON</u>
RA = $10^h 35^m 50\overset{s}{.}9$
DEC = $7° 40' 47\overset{"}{.}9$
SD = $14' 58\overset{"}{.}4$
HP = $0° 54' 57\overset{"}{.}0$

SAROS 154 (5/71)   JD = 2447769.730   ΔT =   57.0 SEC

Figure 38

150

ANNULAR SOLAR ECLIPSE — 26 JAN 1990

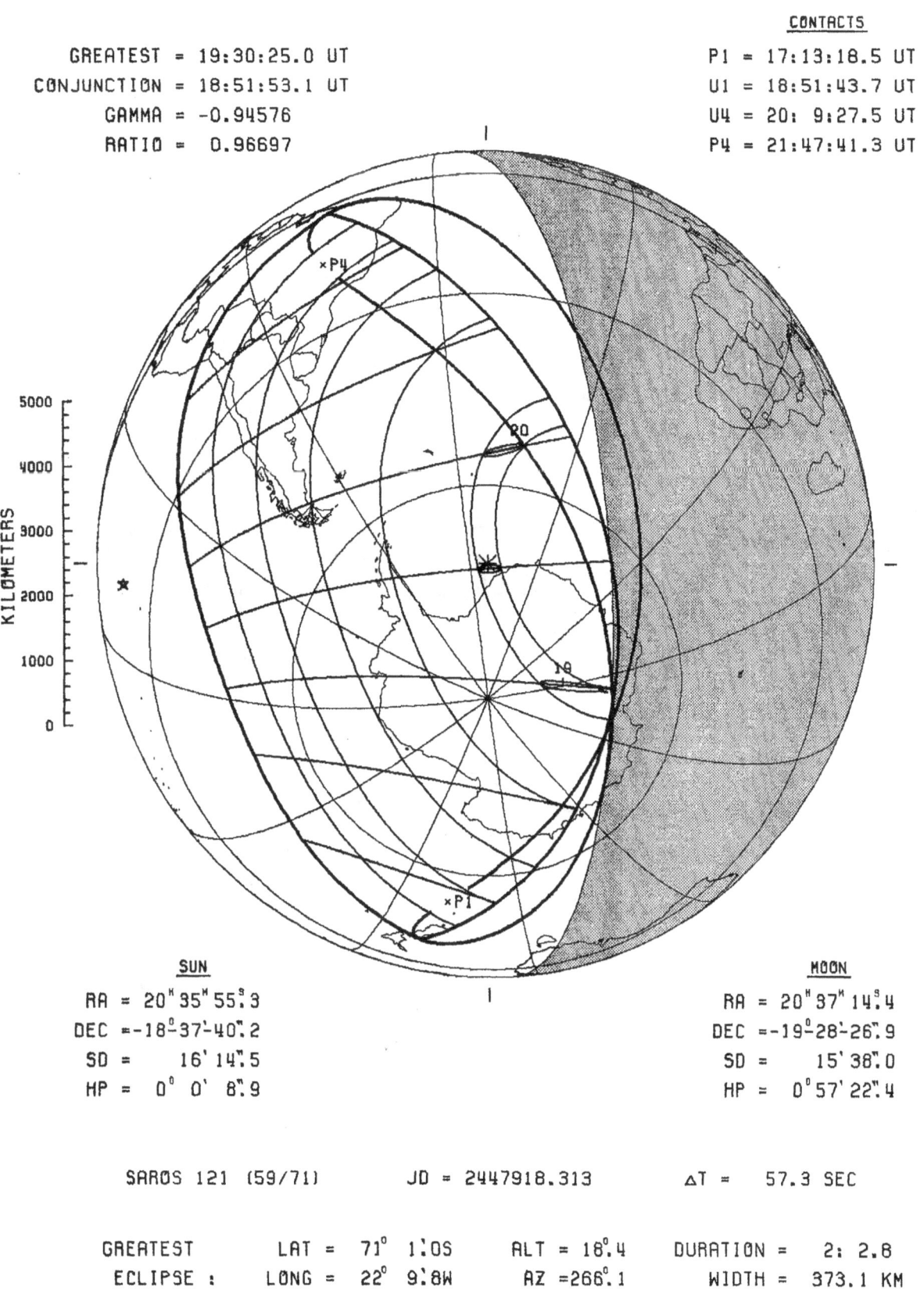

GREATEST = 19:30:25.0 UT
CONJUNCTION = 18:51:53.1 UT
GAMMA = -0.94576
RATIO = 0.96697

CONTACTS

P1 = 17:13:18.5 UT
U1 = 18:51:43.7 UT
U4 = 20: 9:27.5 UT
P4 = 21:47:41.3 UT

SUN

RA = $20^H 35^M 55^S.3$
DEC = $-18^0 37'40".2$
SD = $16' 14".5$
HP = $0^0 0' 8".9$

MOON

RA = $20^H 37^M 14^S.4$
DEC = $-19^0 28'26".9$
SD = $15' 38".0$
HP = $0^0 57' 22".4$

SAROS 121 (59/71)   JD = 2447918.313   ΔT =   57.3 SEC

GREATEST   LAT = $71^0 1'.0S$ ALT = $18^0.4$ DURATION =   2: 2.8
ECLIPSE :   LONG = $22^0 9'.8W$ AZ = $266^0.1$ WIDTH =   373.1 KM

Figure 39
151

TOTAL SOLAR ECLIPSE   -   22 JUL 1990

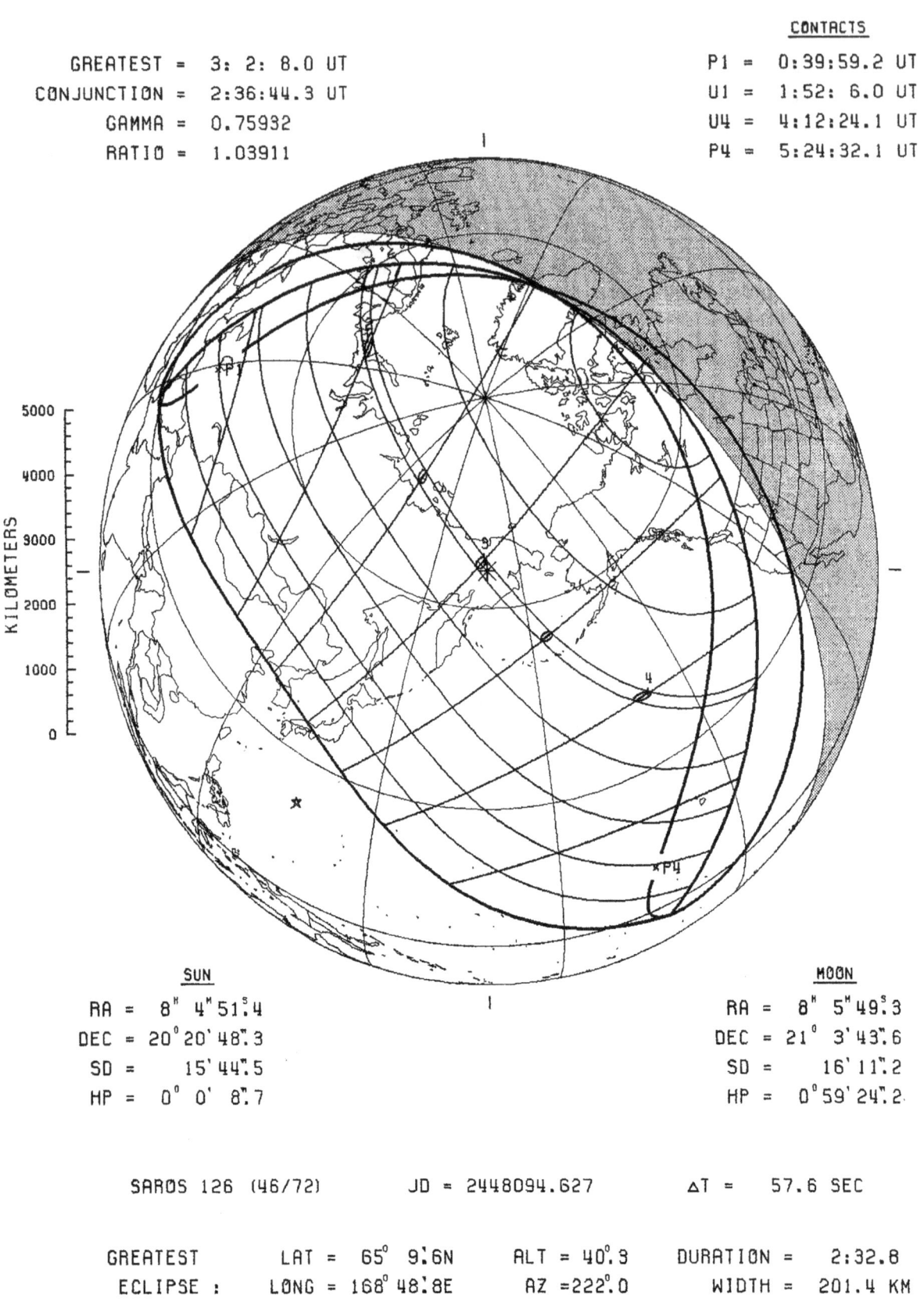

GREATEST = 3: 2: 8.0 UT
CONJUNCTION = 2:36:44.3 UT
GAMMA = 0.75932
RATIO = 1.03911

P1 = 0:39:59.2 UT
U1 = 1:52: 6.0 UT
U4 = 4:12:24.1 UT
P4 = 5:24:32.1 UT

KILOMETERS
5000
4000
3000
2000
1000
0

SUN
RA = 8ʰ 4ᵐ51ˢ4
DEC = 20°20' 48".3
SD =   15' 44".5
HP = 0° 0' 8".7

MOON
RA = 8ʰ 5ᵐ49ˢ3
DEC = 21° 3' 43".6
SD =   16' 11".2
HP = 0°59' 24".2

SAROS 126 (46/72)   JD = 2448094.627   ΔT =   57.6 SEC

GREATEST   LAT = 65° 9:6N   ALT = 40°.9   DURATION = 2:32.8
ECLIPSE :   LONG = 168° 48:8E   AZ =222°.0   WIDTH = 201.4 KM

Figure 40
152

ANNULAR SOLAR ECLIPSE — 15 JAN 1991

GREATEST = 23:52:51.8 UT
CONJUNCTION = 23:43:29.9 UT
GAMMA = -0.27277
RATIO = 0.92899

P1 = 20:50:58.7 UT
U1 = 21:56:46.2 UT
U4 = 1:49: 0.6 UT
P4 = 2:54:44.1 UT

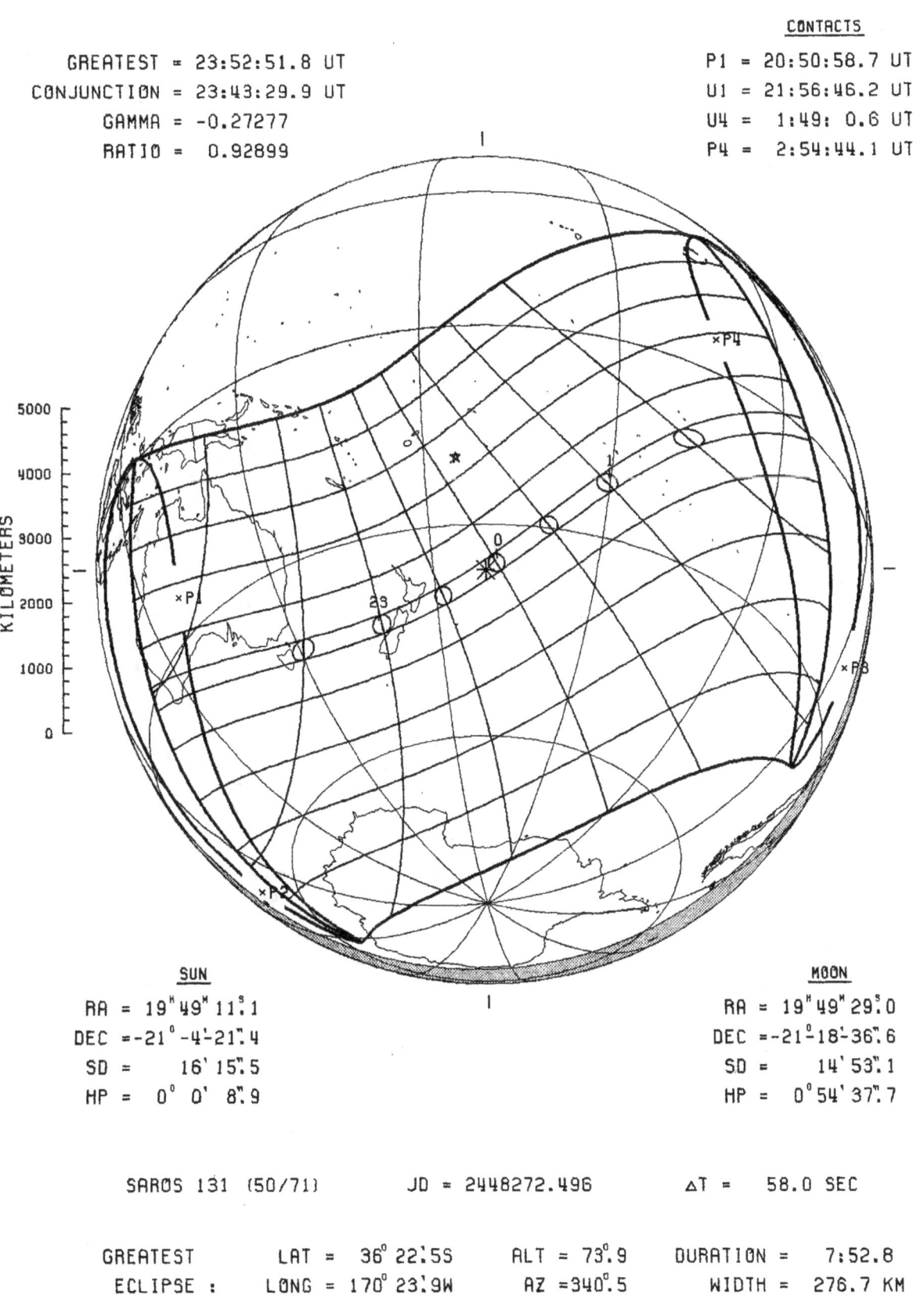

KILOMETERS

5000
4000
3000
2000
1000
0

SUN

RA = $19^h 49^m 11^s.1$
DEC = $-21^\circ 4' 21''.4$
SD = $16' 15''.5$
HP = $0^\circ 0' 8''.9$

MOON

RA = $19^h 49^m 29^s.0$
DEC = $-21^\circ 18' 36''.6$
SD = $14' 53''.1$
HP = $0^\circ 54' 37''.7$

SAROS 131 (50/71)   JD = 2448272.496 ΔT = 58.0 SEC

GREATEST   LAT = $36^\circ 22'.5S$ ALT = $73^\circ.9$ DURATION = 7:52.8
ECLIPSE :   LONG = $170^\circ 23'.9W$ AZ = $340^\circ.5$ WIDTH = 276.7 KM

Figure 41
153

TOTAL SOLAR ECLIPSE — 11 JUL 1991

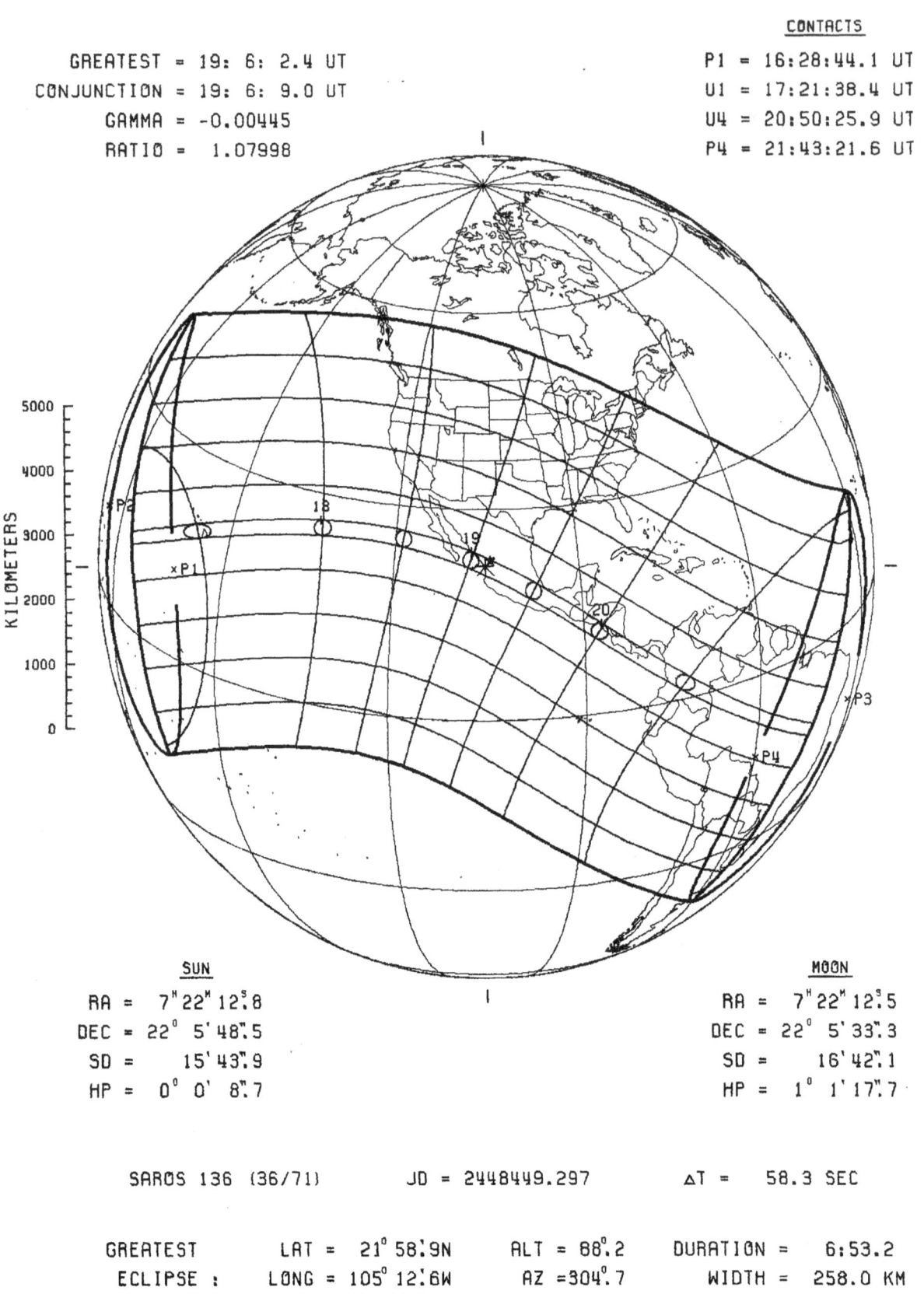

GREATEST = 19: 6: 2.4 UT
CONJUNCTION = 19: 6: 9.0 UT
GAMMA = -0.00445
RATIO = 1.07998

CONTACTS

P1 = 16:28:44.1 UT
U1 = 17:21:38.4 UT
U4 = 20:50:25.9 UT
P4 = 21:43:21.6 UT

SUN

RA = $7^h 22^m 12^s.8$
DEC = $22° 5' 48".5$
SD = $15' 43".9$
HP = $0° 0' 8".7$

MOON

RA = $7^h 22^m 12^s.5$
DEC = $22° 5' 33".3$
SD = $16' 42".1$
HP = $1° 1' 17".7$

SAROS 136 (36/71)   JD = 2448449.297   ΔT = 58.3 SEC

GREATEST   LAT = 21° 58.9N   ALT = 88°.2   DURATION = 6:53.2
ECLIPSE :   LONG = 105° 12.6W   AZ = 304°.7   WIDTH = 258.0 KM

Figure 42
154

ANNULAR SOLAR ECLIPSE   –   4 JAN 1992

GREATEST = 23: 4:35.5 UT
CONJUNCTION = 23:14:41.2 UT
GAMMA = 0.40910
RATIO = 0.91789

P1 = 20: 3:33.7 UT
U1 = 21:12:15.8 UT
U4 = 0:56:49.8 UT
P4 = 2: 5:34.9 UT

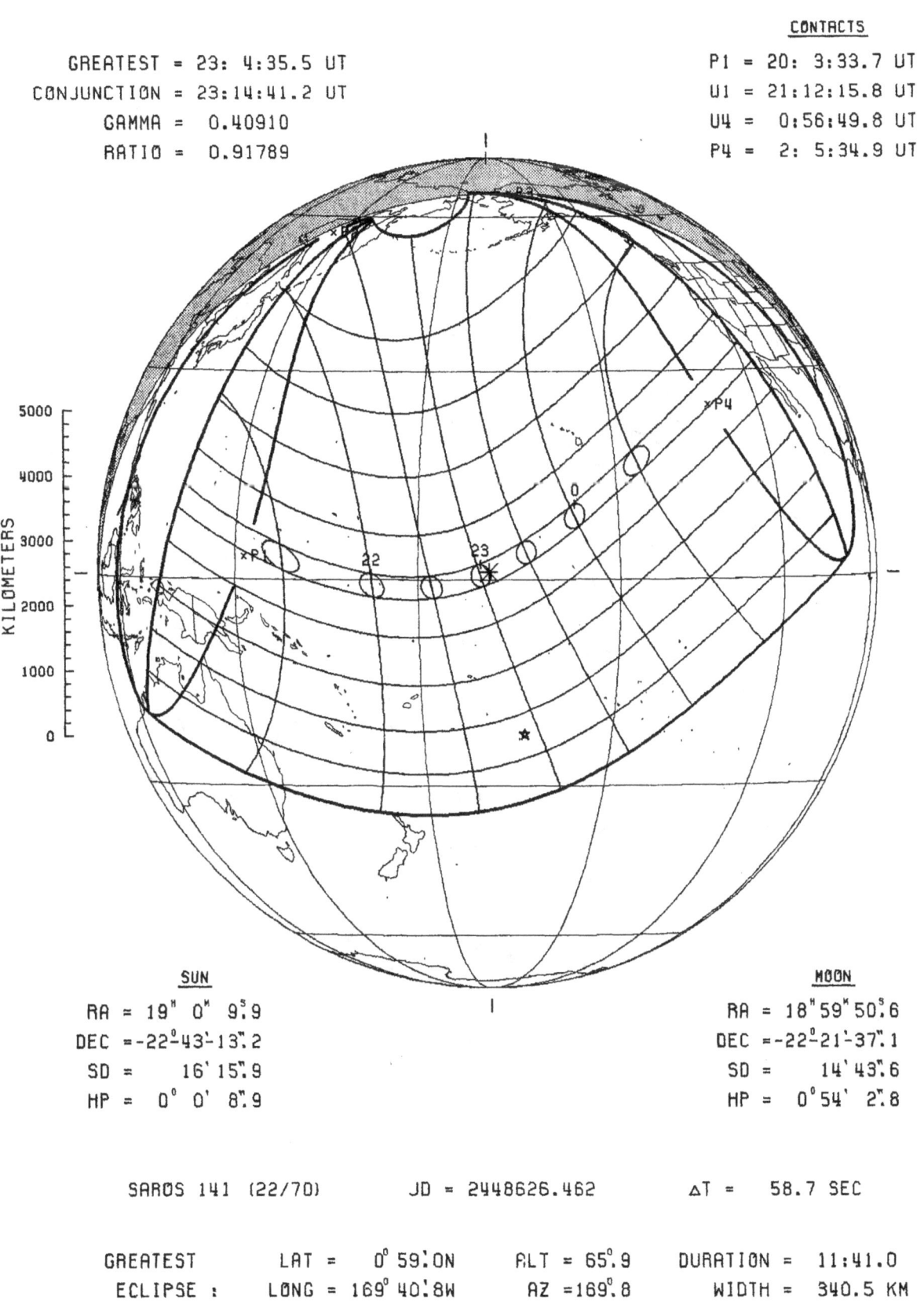

KILOMETERS

5000
4000
3000
2000
1000
0

SUN

RA = 19^h 0^m $9\overset{s}{.}9$
DEC = $-22^{\circ}43'13''.2$
SD = $16'15''.9$
HP = 0° $0'$ $8''.9$

MOON

RA = $18^h59^m50\overset{s}{.}6$
DEC = $-22^{\circ}21'37''.1$
SD = $14'43''.6$
HP = $0^{\circ}54'$ $2''.8$

SAROS 141 (22/70)   JD = 2448626.462 ΔT = 58.7 SEC

GREATEST   LAT = 0° $59'.0$N   ALT = $65^{\circ}.9$ DURATION = 11:41.0
ECLIPSE :   LONG = 169° $40'.8$W   AZ = $169^{\circ}.8$ WIDTH = 340.5 KM

Figure 43
155

TOTAL SOLAR ECLIPSE – 30 JUN 1992

GREATEST = 12:10:23.6 UT
CONJUNCTION = 12:23:23.9 UT
GAMMA = -0.75147
RATIO = 1.05915

CONTACTS

P1 = 9:50:56.6 UT
U1 = 10:59:50.6 UT
U4 = 13:20:49.5 UT
P4 = 14:29:41.7 UT

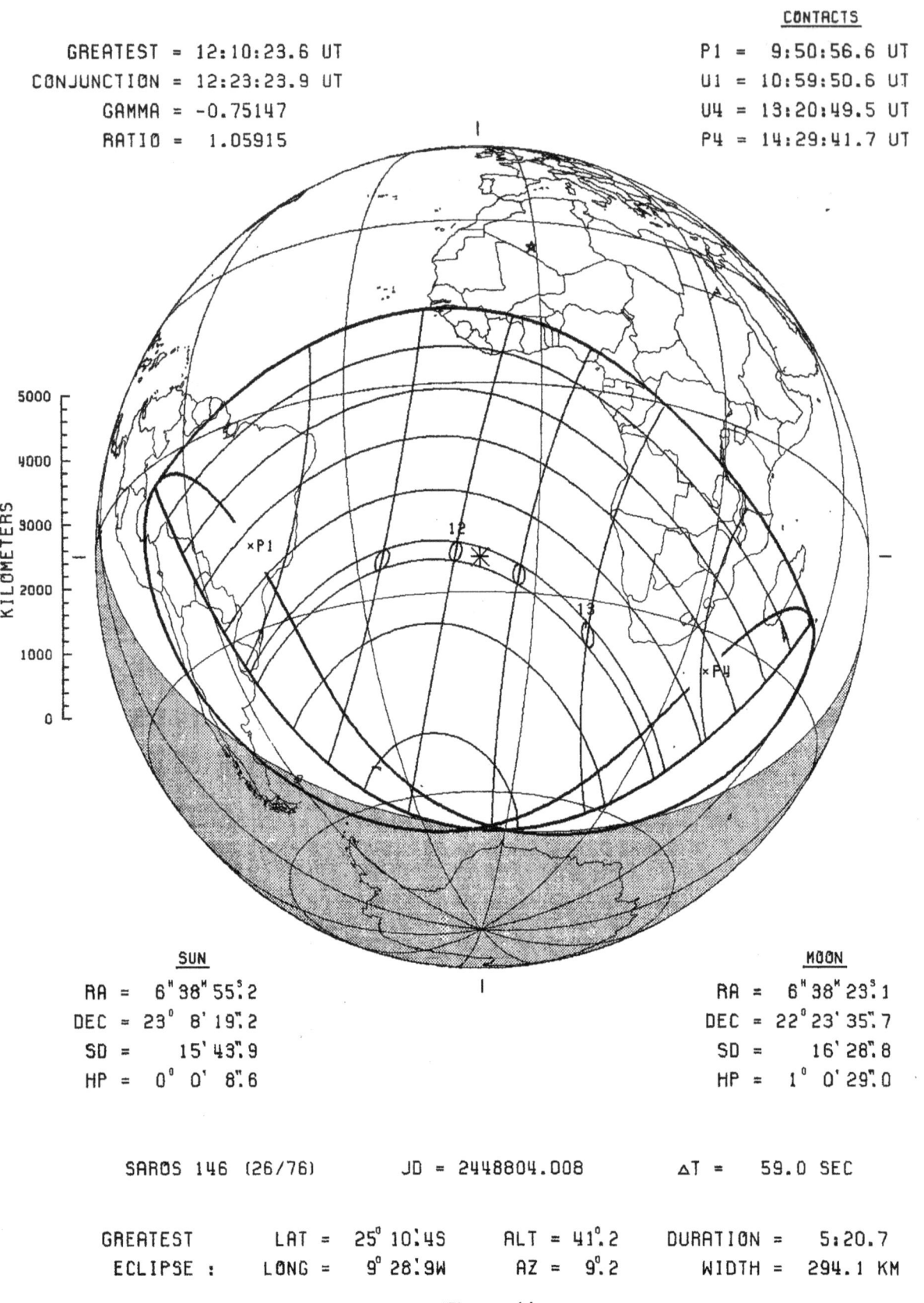

KILOMETERS

SUN
RA = 6^h 38^m 55^s.2
DEC = 23° 8' 19".2
SD = 15' 43".9
HP = 0° 0' 8".6

MOON
RA = 6^h 38^m 23^s.1
DEC = 22° 23' 35".7
SD = 16' 28".8
HP = 1° 0' 29".0

SAROS 146 (26/76)   JD = 2448804.008   ΔT = 59.0 SEC

GREATEST   LAT = 25° 10'.4S   ALT = 41°.2   DURATION = 5:20.7
ECLIPSE :   LONG = 9° 28'.9W   AZ = 9°.2   WIDTH = 294.1 KM

Figure 44
156

PARTIAL SOLAR ECLIPSE — 24 DEC 1992

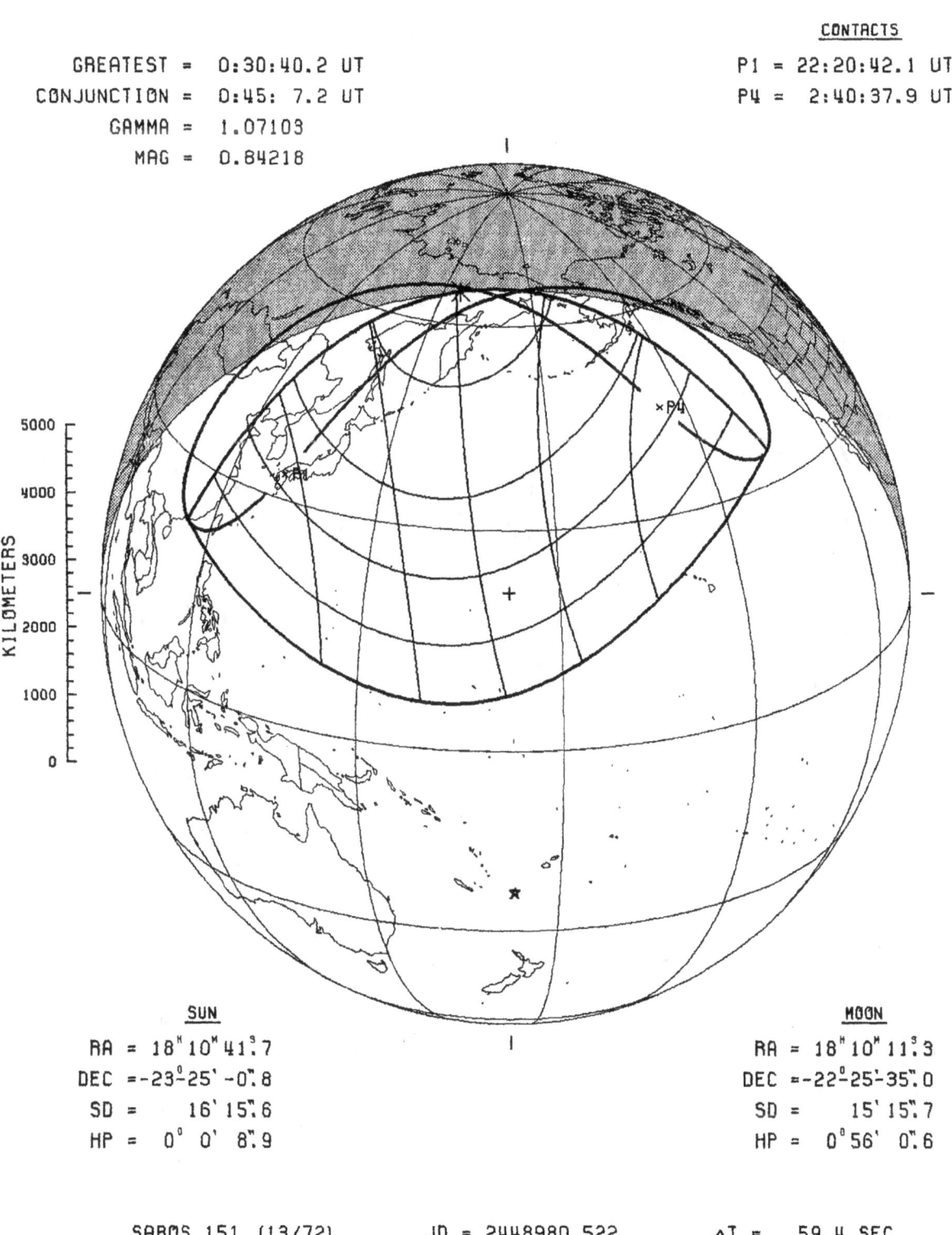

GREATEST = 0:30:40.2 UT

CONJUNCTION = 0:45: 7.2 UT

GAMMA = 1.07103

MAG = 0.84218

CONTACTS

P1 = 22:20:42.1 UT

P4 = 2:40:37.9 UT

KILOMETERS

5000

4000

3000

2000

1000

0

SUN

RA = $18^h10^m41\overset{s}{.}7$

DEC = $-23°25'-0\overset{''}{.}8$

SD = $16'15\overset{''}{.}6$

HP = $0°0'8\overset{''}{.}9$

MOON

RA = $18^h10^m11\overset{s}{.}3$

DEC = $-22°25'-35\overset{''}{.}0$

SD = $15'15\overset{''}{.}7$

HP = $0°56'0\overset{''}{.}6$

SAROS 151 (13/72)   JD = 2448980.522   ΔT = 59.4 SEC

Figure 45

157

PARTIAL SOLAR ECLIPSE — 21 MAY 1993

GREATEST = 14:19:13.3 UT
CONJUNCTION = 14:33:43.5 UT
GAMMA = 1.13694
MAG = 0.73557

P1 = 12:18:42.6 UT
P4 = 16:19:31.9 UT

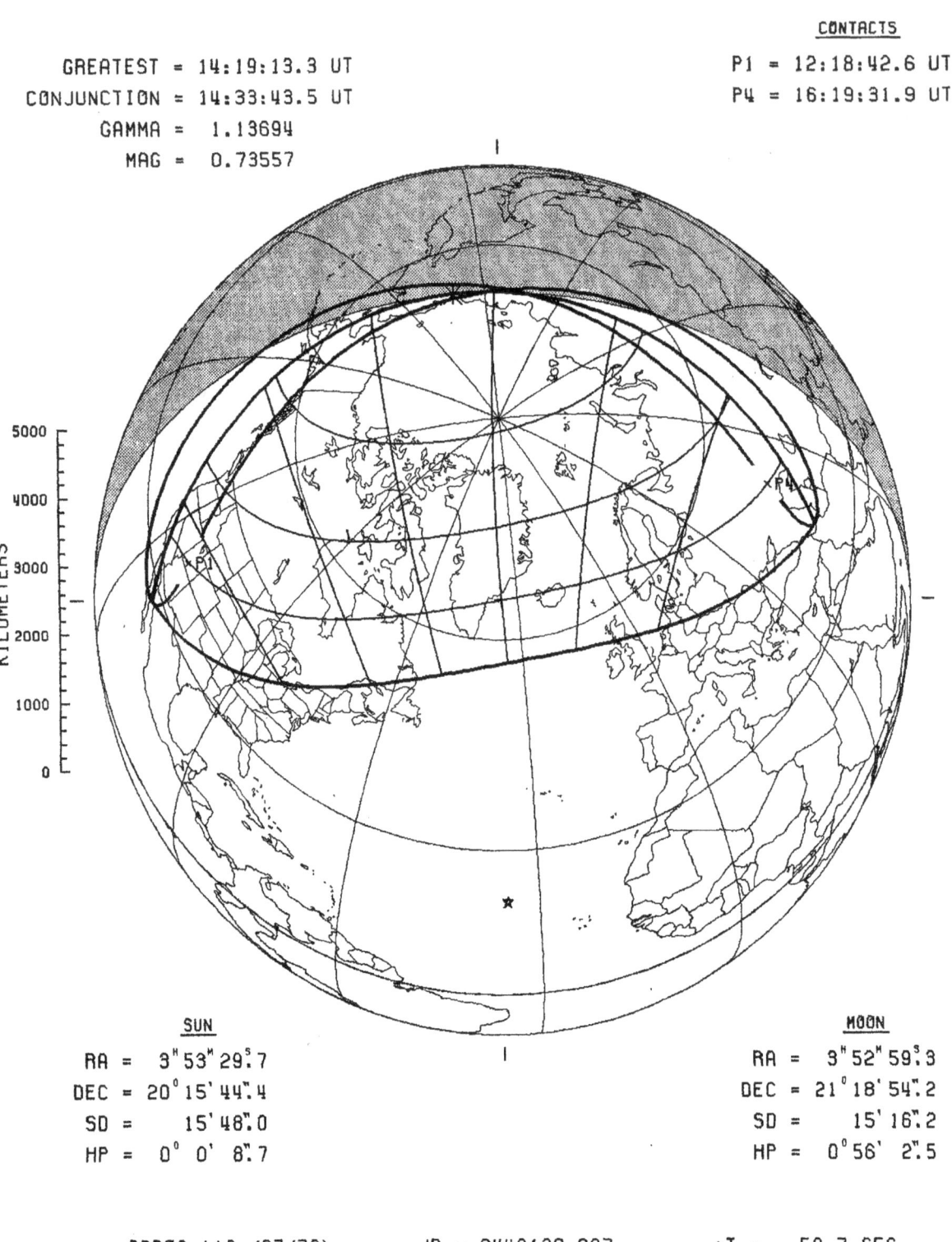

KILOMETERS

5000
4000
3000
2000
1000
0

SUN

RA = 3ʰ53ᵐ29ˢ.7
DEC = 20°15'44".4
SD = 15'48".0
HP = 0°0'8".7

MOON

RA = 3ʰ52ᵐ59ˢ.3
DEC = 21°18'54".2
SD = 15'16".2
HP = 0°56'2".5

SAROS 118 (67/72)   JD = 2449129.097   ΔT = 59.7 SEC

Figure 46

158

PARTIAL SOLAR ECLIPSE — 13 NOV 1993

GREATEST = 21:44:48.5 UT
CONJUNCTION = 22: 3: 8.1 UT
GAMMA = -1.04117
MAG = 0.92778

CONTACTS
P1 = 19:46:21.6 UT
P4 = 23:43:10.2 UT

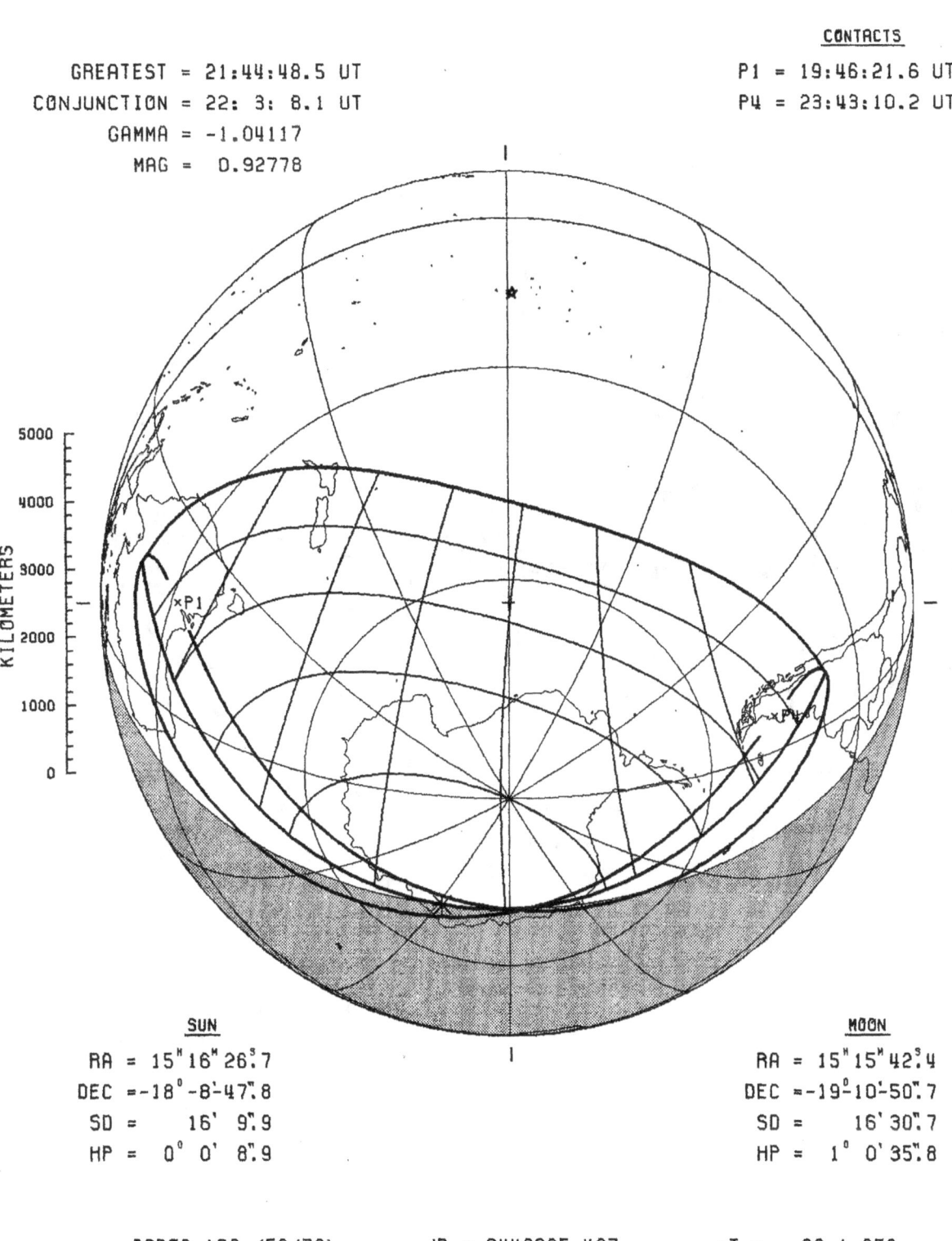

SUN
RA = 15ʰ16ᵐ26ˢ.7
DEC = -18°-8'-47".8
SD = 16' 9".9
HP = 0° 0' 8".9

MOON
RA = 15ʰ15ᵐ42ˢ.4
DEC = -19°-10'-50".7
SD = 16' 30".7
HP = 1° 0' 35".8

SAROS 123 (52/70)   JD = 2449305.407   ΔT = 60.1 SEC

Figure 47

159

ANNULAR SOLAR ECLIPSE – 10 MAY 1994

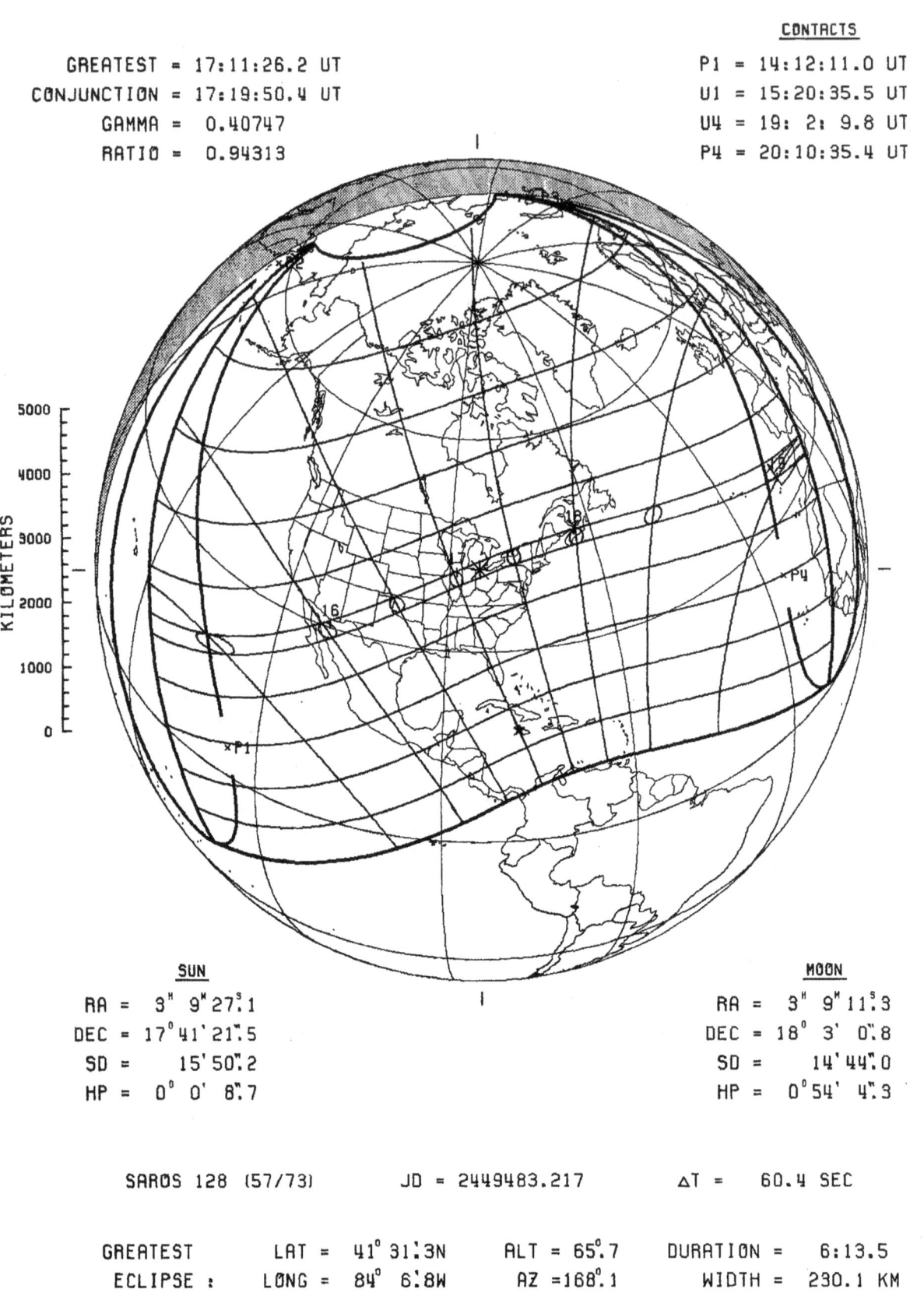

GREATEST = 17:11:26.2 UT
CONJUNCTION = 17:19:50.4 UT
GAMMA = 0.40747
RATIO = 0.94313

CONTACTS

P1 = 14:12:11.0 UT
U1 = 15:20:35.5 UT
U4 = 19: 2: 9.8 UT
P4 = 20:10:35.4 UT

KILOMETERS

5000
4000
3000
2000
1000
0

SUN

RA = 3ʰ 9ᴹ 27ˢ.1
DEC = 17° 41' 21".5
SD = 15' 50".2
HP = 0° 0' 8".7

MOON

RA = 3ʰ 9ᴹ 11ˢ.3
DEC = 18° 3' 0".8
SD = 14' 44".0
HP = 0° 54' 4".3

SAROS 128 (57/73)   JD = 2449483.217   ΔT =   60.4 SEC

GREATEST   LAT = 41° 31'.3N   ALT = 65°.7   DURATION =   6:13.5
ECLIPSE :   LONG = 84° 6'.8W   AZ =168°.1   WIDTH = 230.1 KM

Figure 48
160

TOTAL SOLAR ECLIPSE - 3 NOV 1994

GREATEST = 13:39: 3.5 UT

CONJUNCTION = 13:47: 5.2 UT

GAMMA = -0.35226

RATIO = 1.05352

P1 = 11: 4:58.0 UT

U1 = 12: 1:36.3 UT

U4 = 15:16:25.3 UT

P4 = 16:13: 4.3 UT

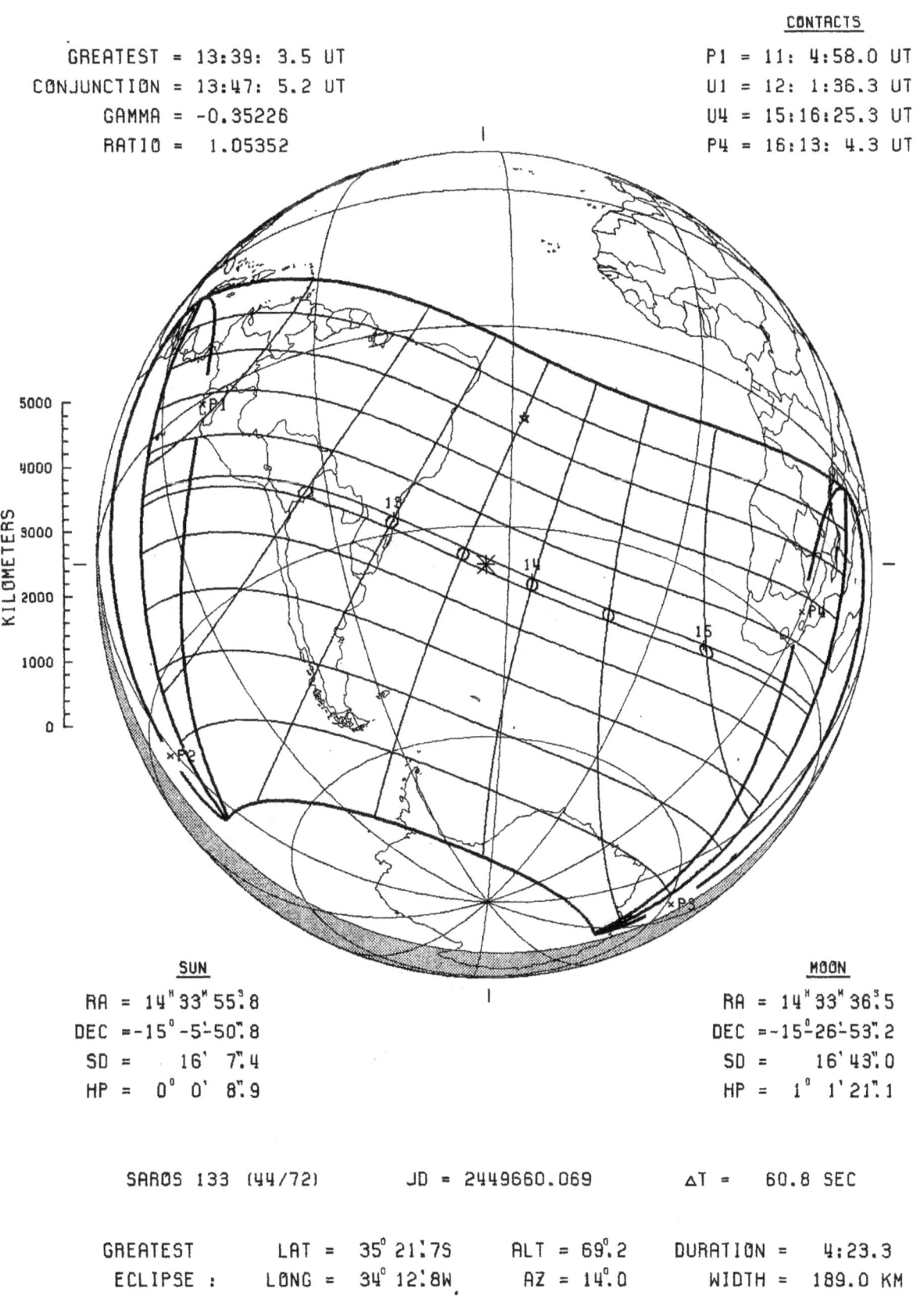

KILOMETERS

5000
4000
3000
2000
1000
0

SUN

RA = 14^{h}33^{m}55^{s}8

DEC = -15^0-5'-50".8

SD = 16' 7".4

HP = 0^0 0' 8".9

MOON

RA = 14^{h}33^{m}36^{s}5

DEC = -15^0-26'-53".2

SD = 16'43".0

HP = 1^0 1'21".1

SAROS 133 (44/72)   JD = 2449660.069   ΔT = 60.8 SEC

GREATEST   LAT = 35^0 21'.7S   ALT = 69^0.2   DURATION = 4:23.3

ECLIPSE :   LONG = 34^0 12'.8W   AZ = 14^0.0   WIDTH = 189.0 KM

Figure 49
161

ANNULAR SOLAR ECLIPSE - 29 APR 1995

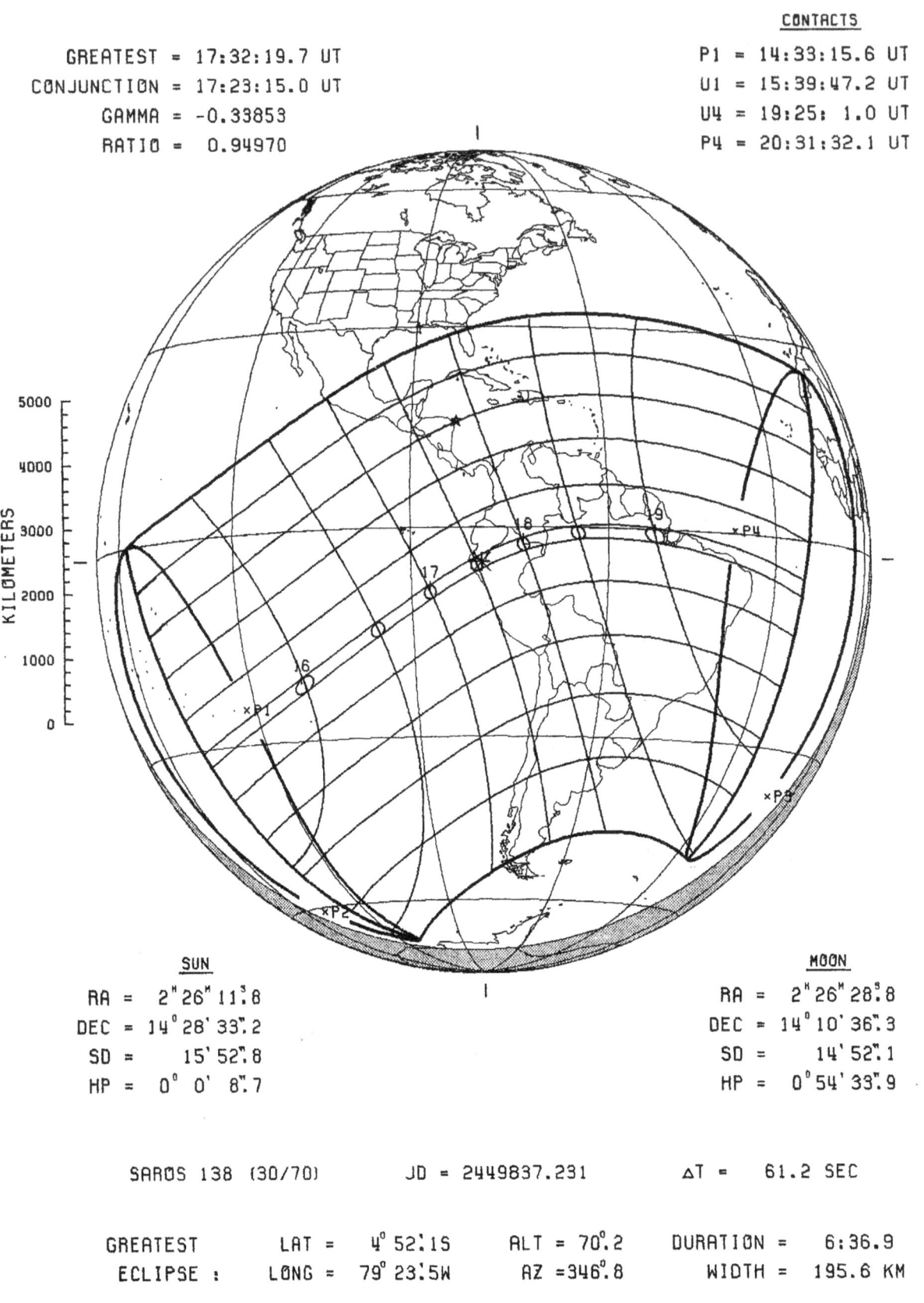

GREATEST = 17:32:19.7 UT
CONJUNCTION = 17:23:15.0 UT
GAMMA = -0.33853
RATIO = 0.94970

CONTACTS

P1 = 14:33:15.6 UT
U1 = 15:39:47.2 UT
U4 = 19:25: 1.0 UT
P4 = 20:31:32.1 UT

KILOMETERS

5000
4000
3000
2000
1000
0

SUN

RA = 2ʰ26ᵐ11ˢ.8
DEC = 14°28'33".2
SD = 15'52".8
HP = 0° 0' 8".7

MOON

RA = 2ʰ26ᵐ28ˢ.8
DEC = 14°10'36".3
SD = 14'52".1
HP = 0°54'33".9

SAROS 138 (30/70)   JD = 2449837.231   ΔT = 61.2 SEC

GREATEST   LAT = 4°52'.1S   ALT = 70°.2   DURATION =   6:36.9
ECLIPSE :   LONG = 79°23'.5W   AZ =346°.8   WIDTH = 195.6 KM

Figure 50
162

TOTAL SOLAR ECLIPSE — 24 OCT 1995

GREATEST = 4:32:27.8 UT
CONJUNCTION = 4:22:29.6 UT
GAMMA = 0.35169
RATIO = 1.02136

P1 = 1:51:51.7 UT
U1 = 2:52:28.7 UT
U4 = 6:12:36.0 UT
P4 = 7:13: 3.6 UT

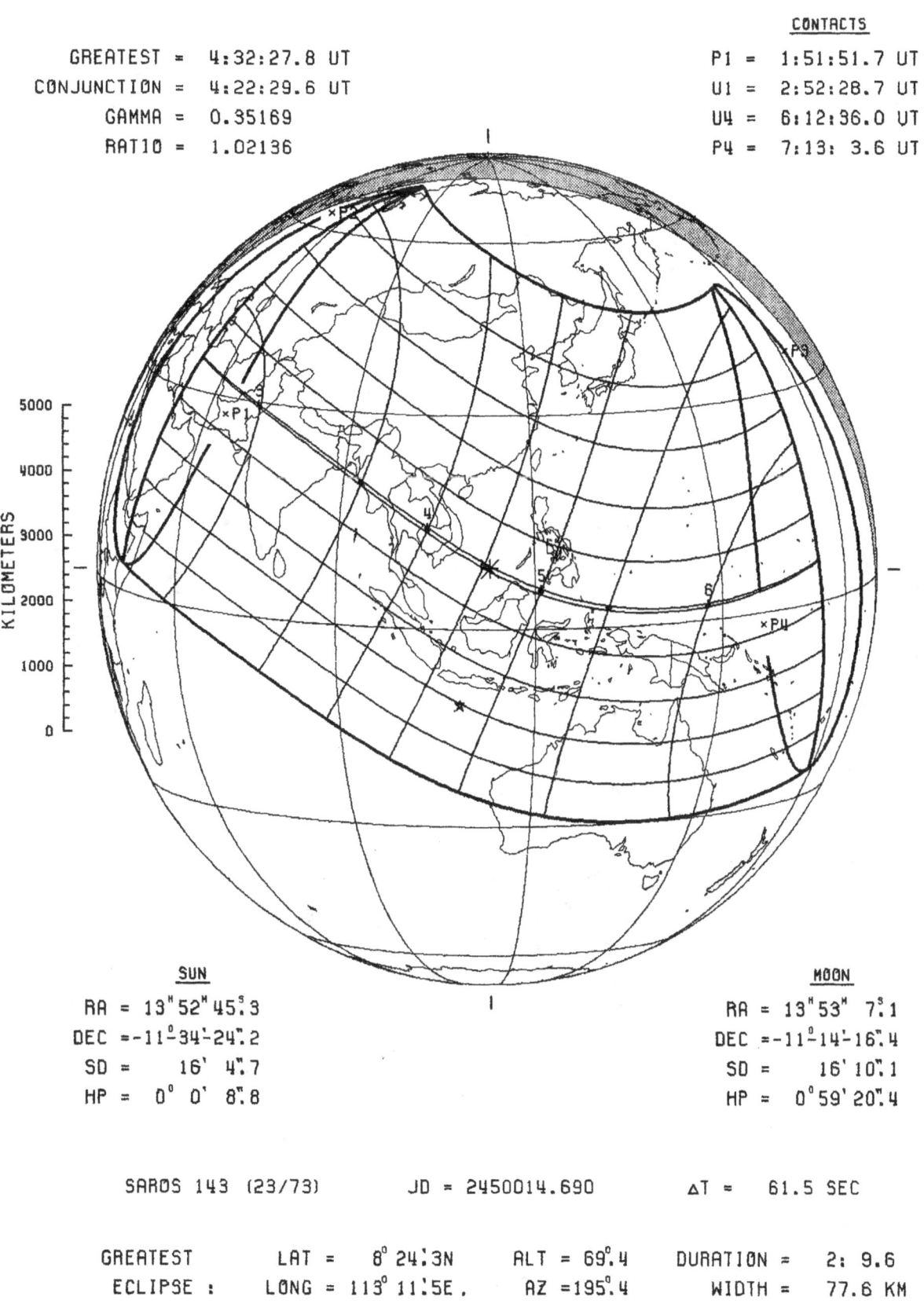

KILOMETERS

5000
4000
3000
2000
1000
0

SUN

RA = 13ʰ52ᵐ45ˢ.3
DEC =-11°34'24".2
SD =   16' 4".7
HP = 0° 0' 8".8

MOON

RA = 13ʰ53ᵐ 7ˢ.1
DEC =-11°14'16".4
SD =   16'10".1
HP = 0°59'20".4

SAROS 143 (23/73)   JD = 2450014.690   ΔT =   61.5 SEC

GREATEST   LAT = 8° 24'.3N   ALT = 69°.4   DURATION = 2: 9.6
ECLIPSE :   LONG = 113° 11'.5E.   AZ =195°.4   WIDTH =   77.6 KM

Figure 51
163

PARTIAL SOLAR ECLIPSE - 17 APR 1996

GREATEST = 22:37: 9.6 UT
CONJUNCTION = 22: 5: 2.6 UT
GAMMA = -1.05825
MAG = 0.87899

P1 = 20:31:24.3 UT
P4 = 0:43:15.4 UT

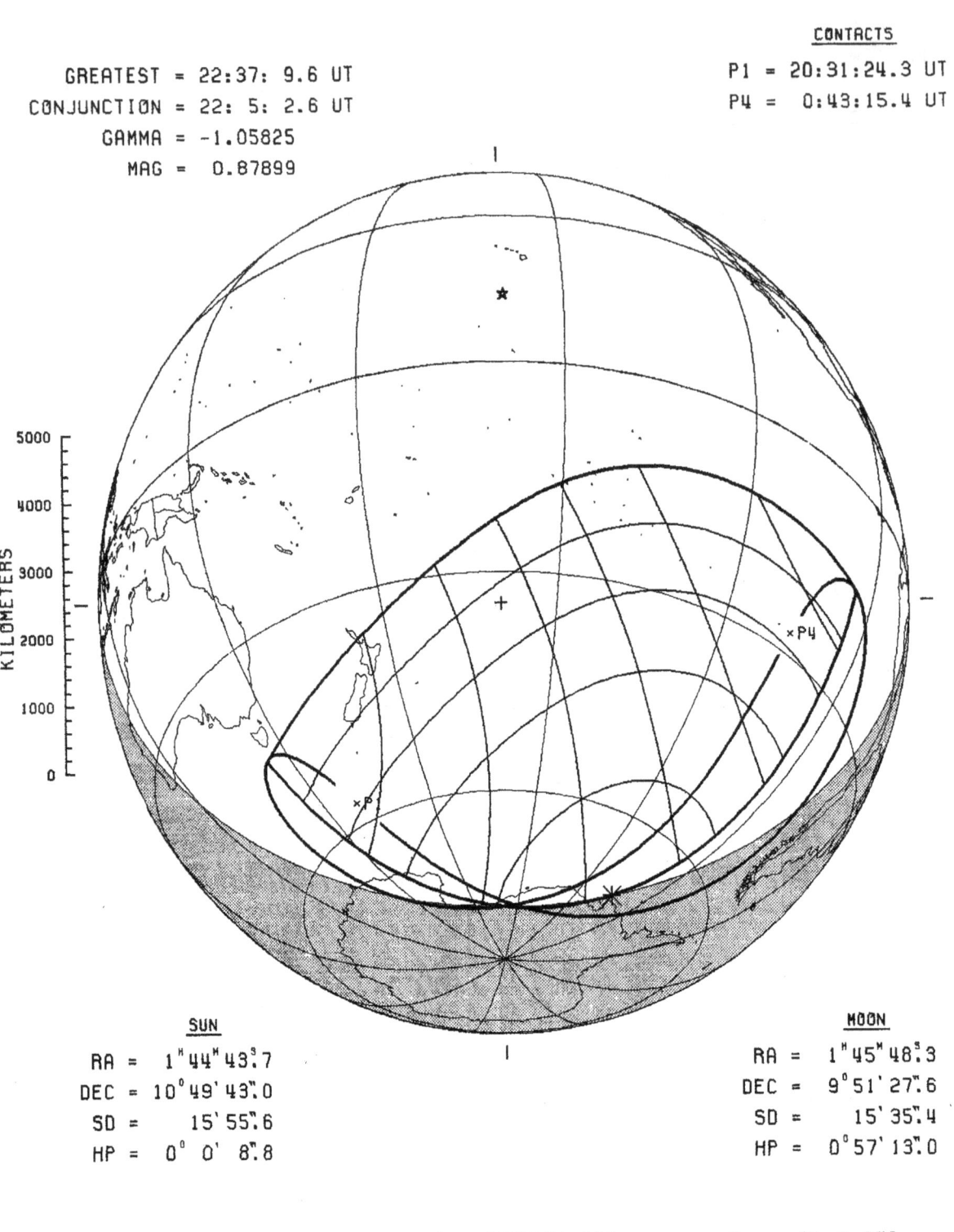

KILOMETERS
5000
4000
3000
2000
1000
0

SUN
RA = 1ʰ44ᵐ43ˢ.7
DEC = 10°49'43".0
SD = 15'55".6
HP = 0°0'8".8

MOON
RA = 1ʰ45ᵐ48ˢ.3
DEC = 9°51'27".6
SD = 15'35".4
HP = 0°57'13".0

SAROS 148 (20/75)   JD = 2450191.443   ΔT = 61.9 SEC

Figure 52

164

PARTIAL SOLAR ECLIPSE —   12 OCT 1996

GREATEST = 14: 2: 0.3 UT
CONJUNCTION = 13:23:49.5 UT
GAMMA = 1.12255
MAG = 0.75723

CONTACTS

P1 = 11:59:27.0 UT
P4 = 16: 4:46.2 UT

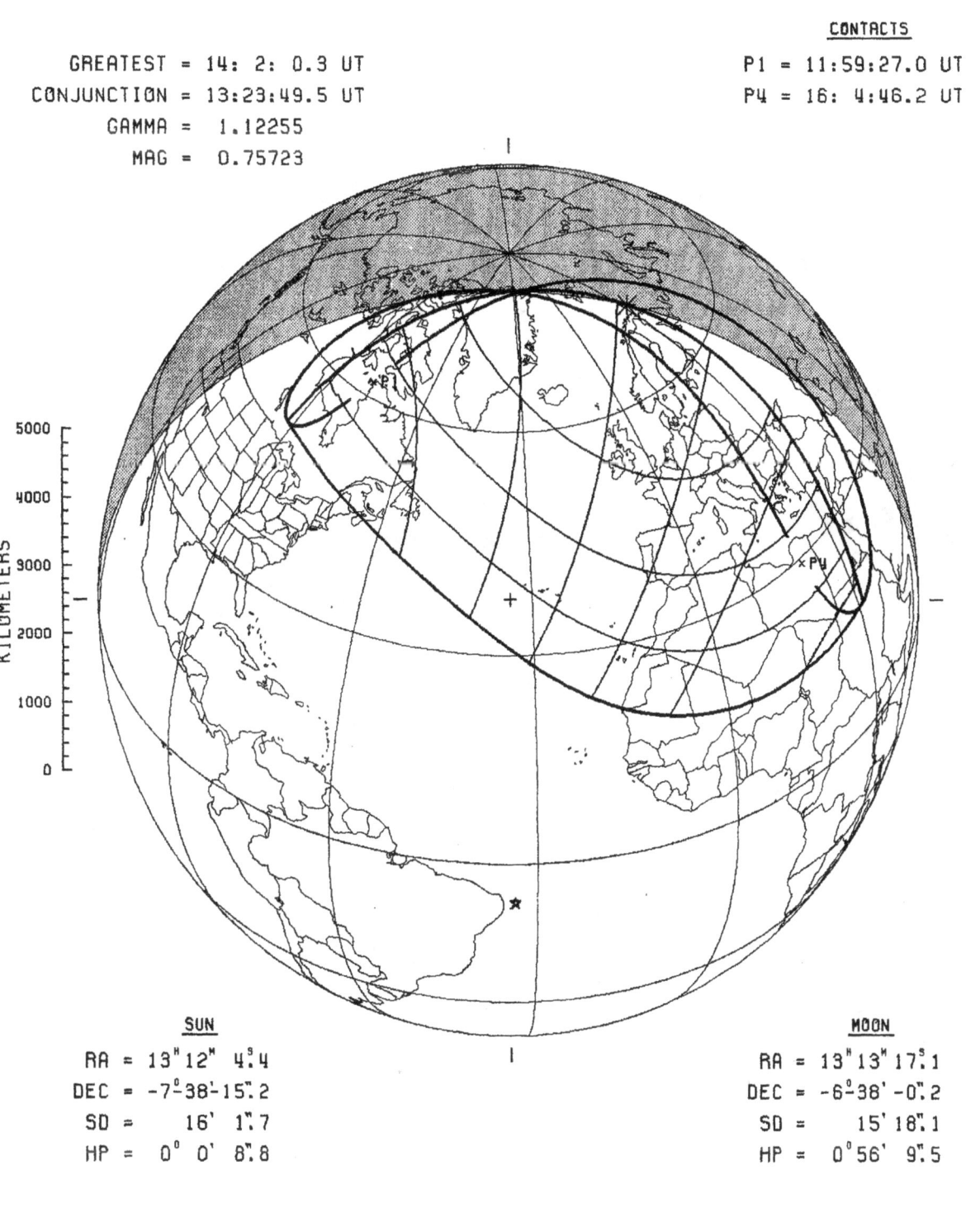

SUN

RA = $13^h 12^m 4^s.4$
DEC = $-7°38'-15".2$
SD = $16' 1".7$
HP = $0° 0' 8".8$

MOON

RA = $13^h 13^m 17^s.1$
DEC = $-6°38'-0".2$
SD = $15' 18".1$
HP = $0° 56' 9".5$

SAROS 153 (8/70)   JD = 2450369.085   ΔT =   62.3 SEC

Figure 53

165

TOTAL SOLAR ECLIPSE – 9 MAR 1997

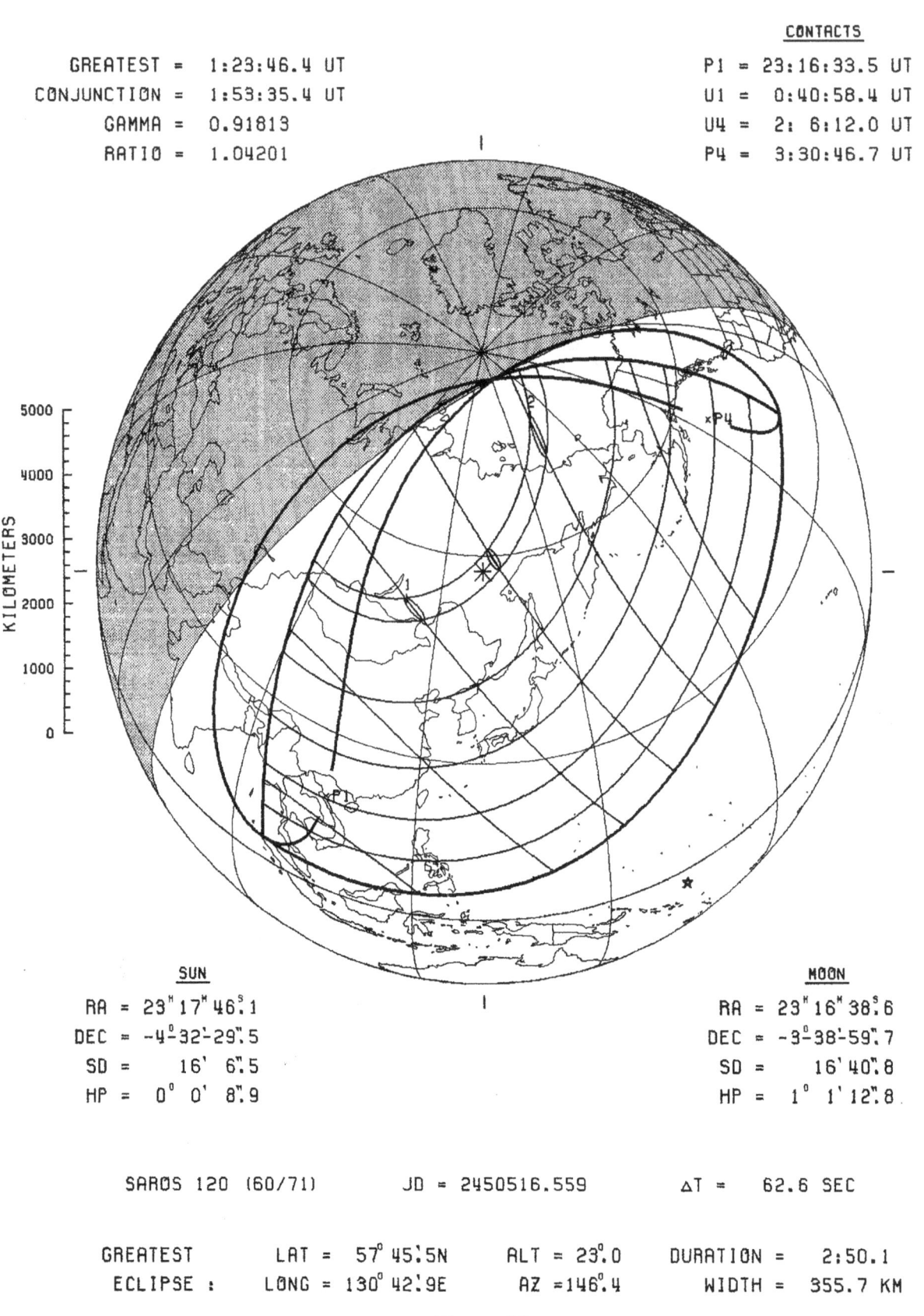

GREATEST = 1:23:46.4 UT

CONJUNCTION = 1:53:35.4 UT

GAMMA = 0.91813

RATIO = 1.04201

CONTACTS

P1 = 23:16:33.5 UT

U1 = 0:40:58.4 UT

U4 = 2: 6:12.0 UT

P4 = 3:30:46.7 UT

KILOMETERS

5000
4000
3000
2000
1000
0

SUN

RA = $23^H 17^M 46\overset{s}{.}1$

DEC = $-4^\circ 32' 29\overset{"}{.}5$

SD = $16' 6\overset{"}{.}5$

HP = $0^\circ 0' 8\overset{"}{.}9$

MOON

RA = $23^H 16^M 38\overset{s}{.}6$

DEC = $-3^\circ 38' 59\overset{"}{.}7$

SD = $16' 40\overset{"}{.}8$

HP = $1^\circ 1' 12\overset{"}{.}8$

SAROS 120 (60/71)   JD = 2450516.559   ΔT = 62.6 SEC

GREATEST   LAT = $57^\circ 45\overset{.}{.}5N$ ALT = $23\overset{\circ}{.}0$ DURATION = 2:50.1

ECLIPSE :   LONG = $130^\circ 42\overset{.}{.}9E$ AZ = $146\overset{\circ}{.}4$ WIDTH = 355.7 KM

Figure 54
166

PARTIAL SOLAR ECLIPSE – 2 SEP 1997

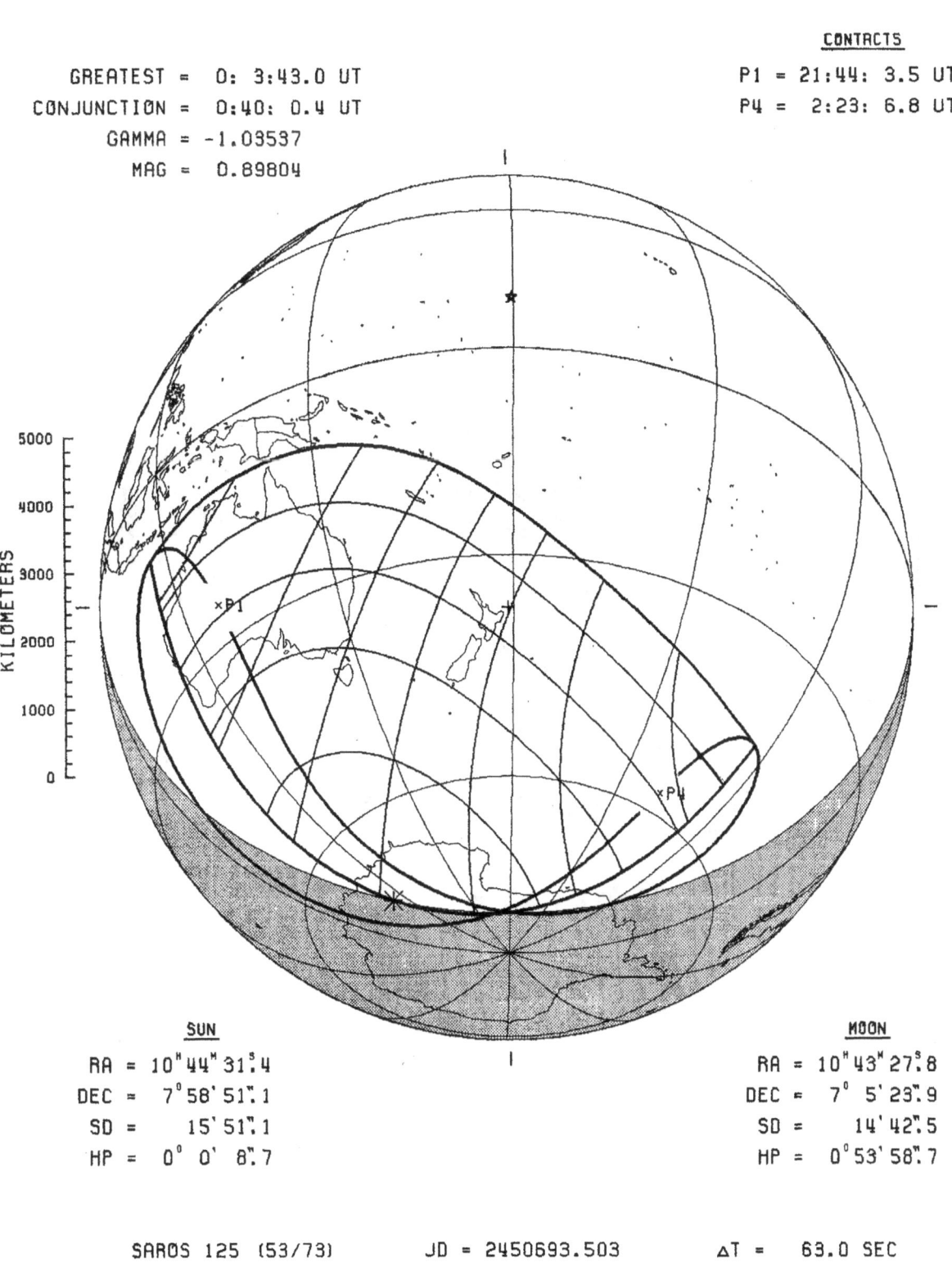

CONTACTS

GREATEST = 0: 3:43.0 UT
CONJUNCTION = 0:40: 0.4 UT
GAMMA = -1.03537
MAG = 0.89804

P1 = 21:44: 3.5 UT
P4 = 2:23: 6.8 UT

KILOMETERS

5000
4000
3000
2000
1000
0

×P1

×P4

SUN

RA = 10ʰ 44ᴹ 31ˢ.4
DEC = 7° 58' 51".1
SD = 15' 51".1
HP = 0° 0' 8".7

MOON

RA = 10ʰ 43ᴹ 27ˢ.8
DEC = 7° 5' 23".9
SD = 14' 42".5
HP = 0° 53' 58".7

SAROS 125 (53/73)   JD = 2450693.503   ΔT =   63.0 SEC

Figure 55

167

TOTAL SOLAR ECLIPSE — 26 FEB 1998

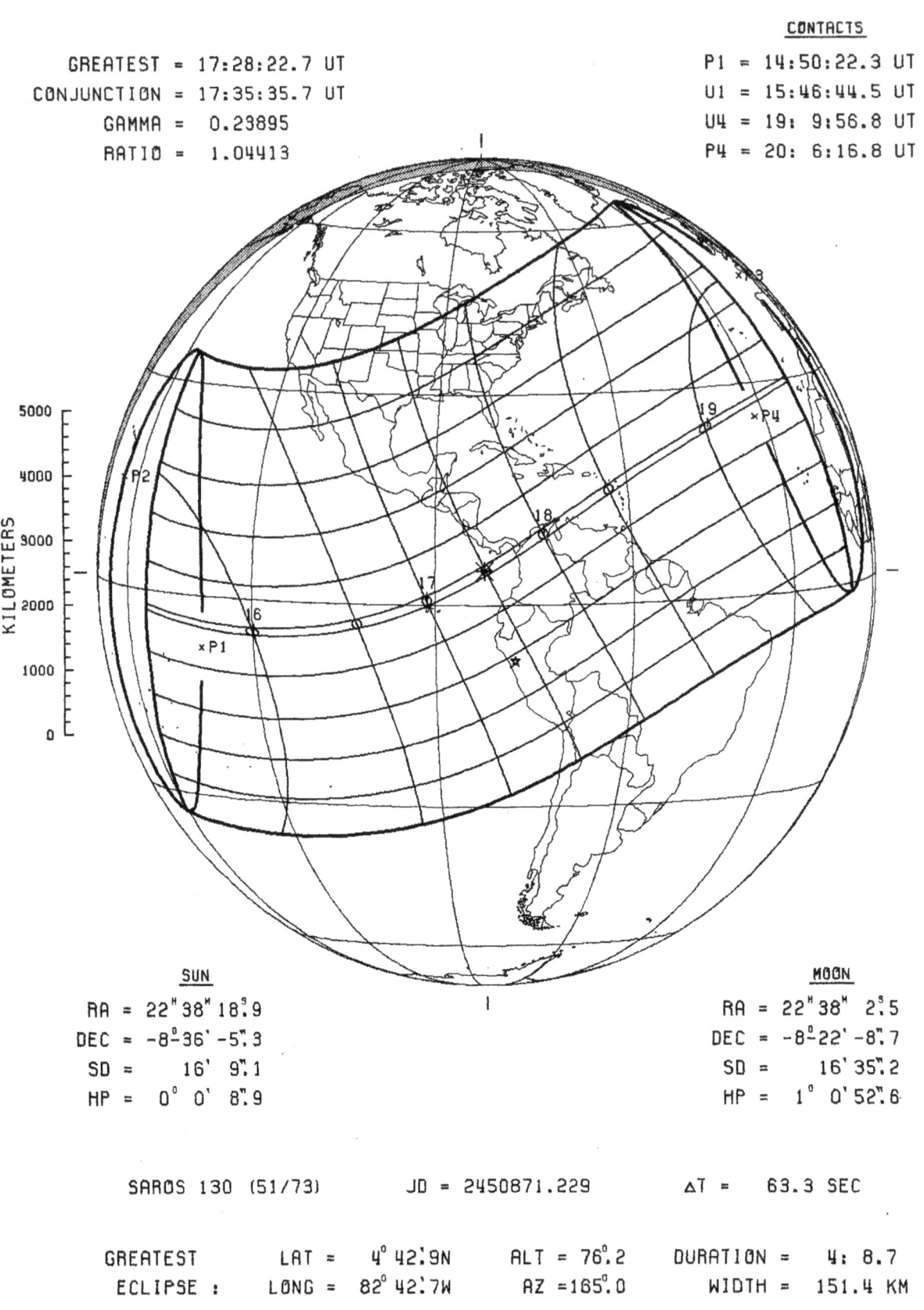

GREATEST = 17:28:22.7 UT
CONJUNCTION = 17:35:35.7 UT
GAMMA = 0.23895
RATIO = 1.04413

CONTACTS

P1 = 14:50:22.3 UT
U1 = 15:46:44.5 UT
U4 = 19: 9:56.8 UT
P4 = 20: 6:16.8 UT

KILOMETERS

5000
4000
3000
2000
1000
0

SUN

RA = 22^{h}38^{m}18^{s}9
DEC = -8^{0}36' -5".3
SD = 16' 9".1
HP = 0^0 0' 8".9

MOON

RA = 22^{h}38^m 2^{s}5
DEC = -8^{0}22' -8".7
SD = 16' 35".2
HP = 1^0 0' 52".6

SAROS 130 (51/73)   JD = 2450871.229   ΔT =   63.3 SEC

GREATEST   LAT = 4^0 42'9N   ALT = 76^0.2   DURATION =   4: 8.7
ECLIPSE :   LONG = 82^0 42'7W   AZ =165^0.0   WIDTH =   151.4 KM

Figure 56
168

ANNULAR SOLAR ECLIPSE -   22 AUG 1998

GREATEST = 2: 6: 5.8 UT
CONJUNCTION = 2:14: 0.9 UT
GAMMA = -0.26459
RATIO = 0.97335

CONTACTS

P1 = 23:10:15.4 UT
U1 = 0:14:14.8 UT
U4 = 3:57:54.7 UT
P4 = 5: 1:58.6 UT

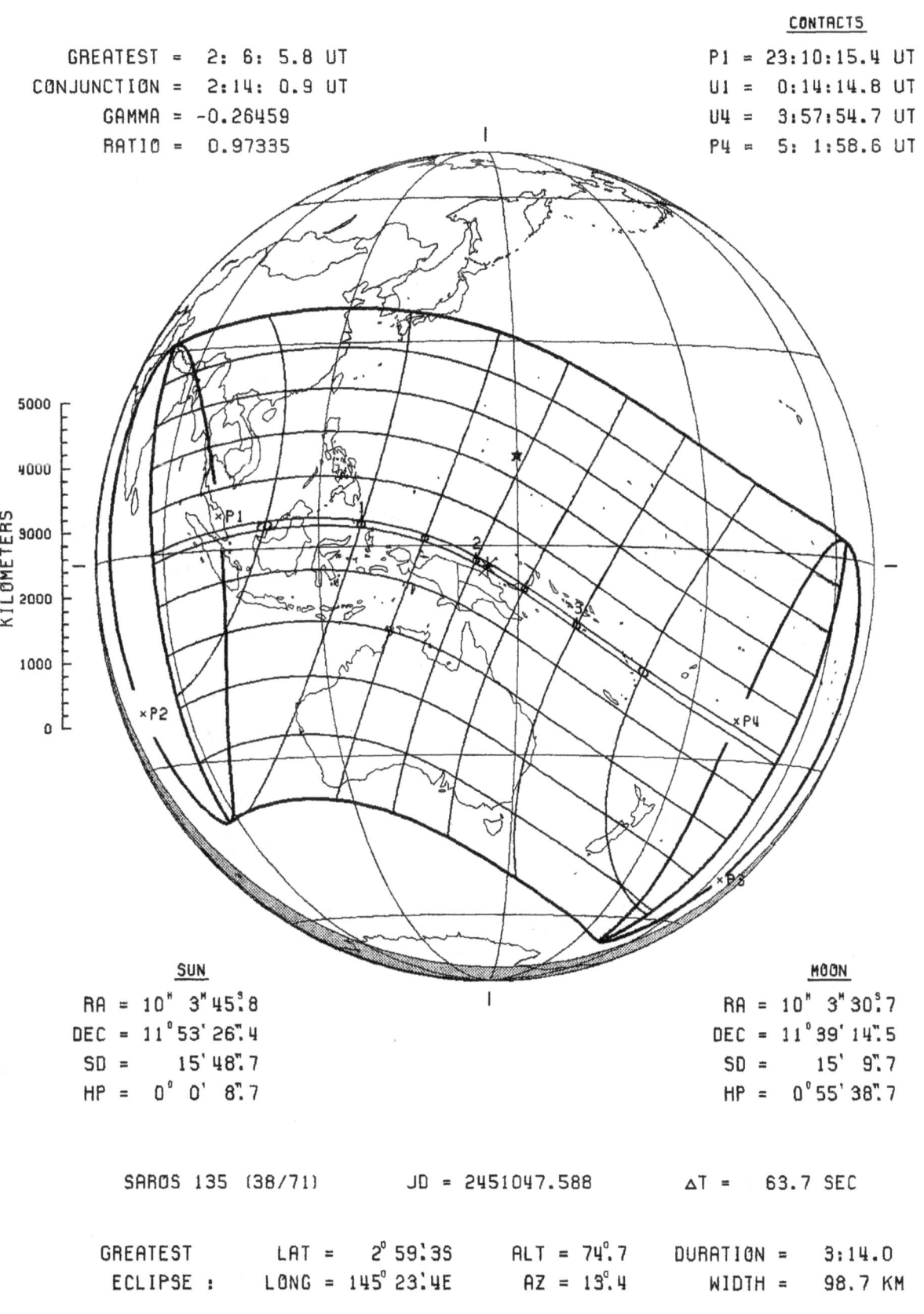

KILOMETERS

5000
4000
9000
2000
1000
0

×P1

×P2

×P4

×P3

SUN

RA = 10ʰ 3ᵐ 45ˢ.8
DEC = 11° 53' 26".4
SD = 15' 48".7
HP = 0° 0' 8".7

MOON

RA = 10ʰ 3ᵐ 30ˢ.7
DEC = 11° 39' 14".5
SD = 15' 9".7
HP = 0° 55' 38".7

SAROS 135 (38/71)   JD = 2451047.588   ΔT = 63.7 SEC

GREATEST   LAT = 2° 59'.3S   ALT = 74°.7   DURATION = 3:14.0
ECLIPSE :   LONG = 145° 23'.4E   AZ = 19°.4   WIDTH = 98.7 KM

Figure 57
169

ANNULAR SOLAR ECLIPSE - 16 FEB 1999

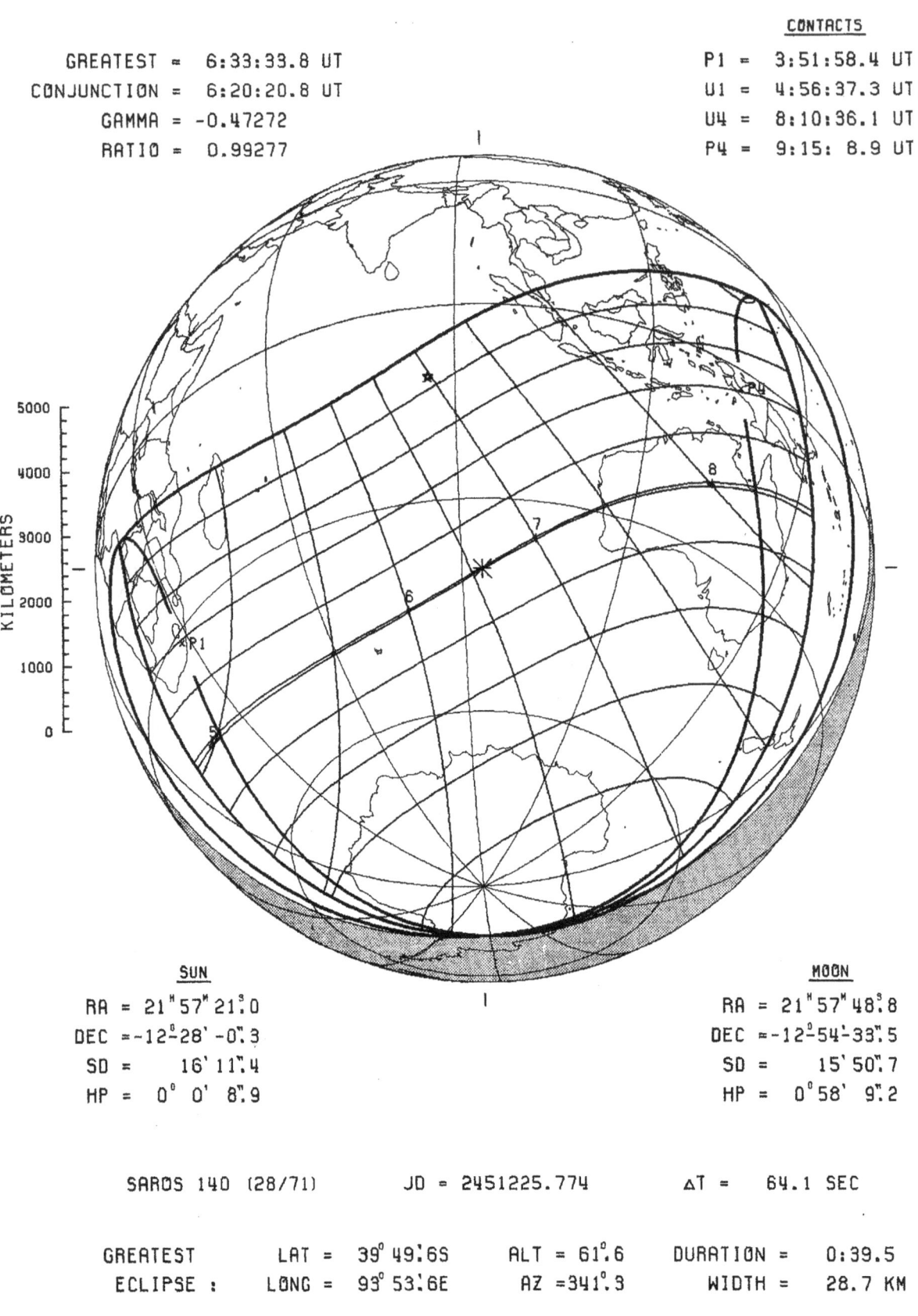

GREATEST = 6:33:33.8 UT
CONJUNCTION = 6:20:20.8 UT
GAMMA = -0.47272
RATIO = 0.99277

CONTACTS

P1 = 3:51:58.4 UT
U1 = 4:56:37.3 UT
U4 = 8:10:36.1 UT
P4 = 9:15: 8.9 UT

KILOMETERS

5000
4000
3000
2000
1000
0

SUN

RA = 21ʰ57ᵐ21ˢ0
DEC = -12°28'-0".3
SD =   16'11".4
HP = 0° 0' 8".9

MOON

RA = 21ʰ57ᵐ48ˢ8
DEC = -12°54'-33".5
SD =   15'50".7
HP = 0°58' 9".2

SAROS 140 (28/71)   JD = 2451225.774   ΔT =   64.1 SEC

GREATEST   LAT = 39° 49.6S   ALT = 61°.6   DURATION =   0:39.5
ECLIPSE :   LONG = 99° 53.6E   AZ = 341°.3   WIDTH =   28.7 KM

Figure 58
170

TOTAL SOLAR ECLIPSE — 11 AUG 1999

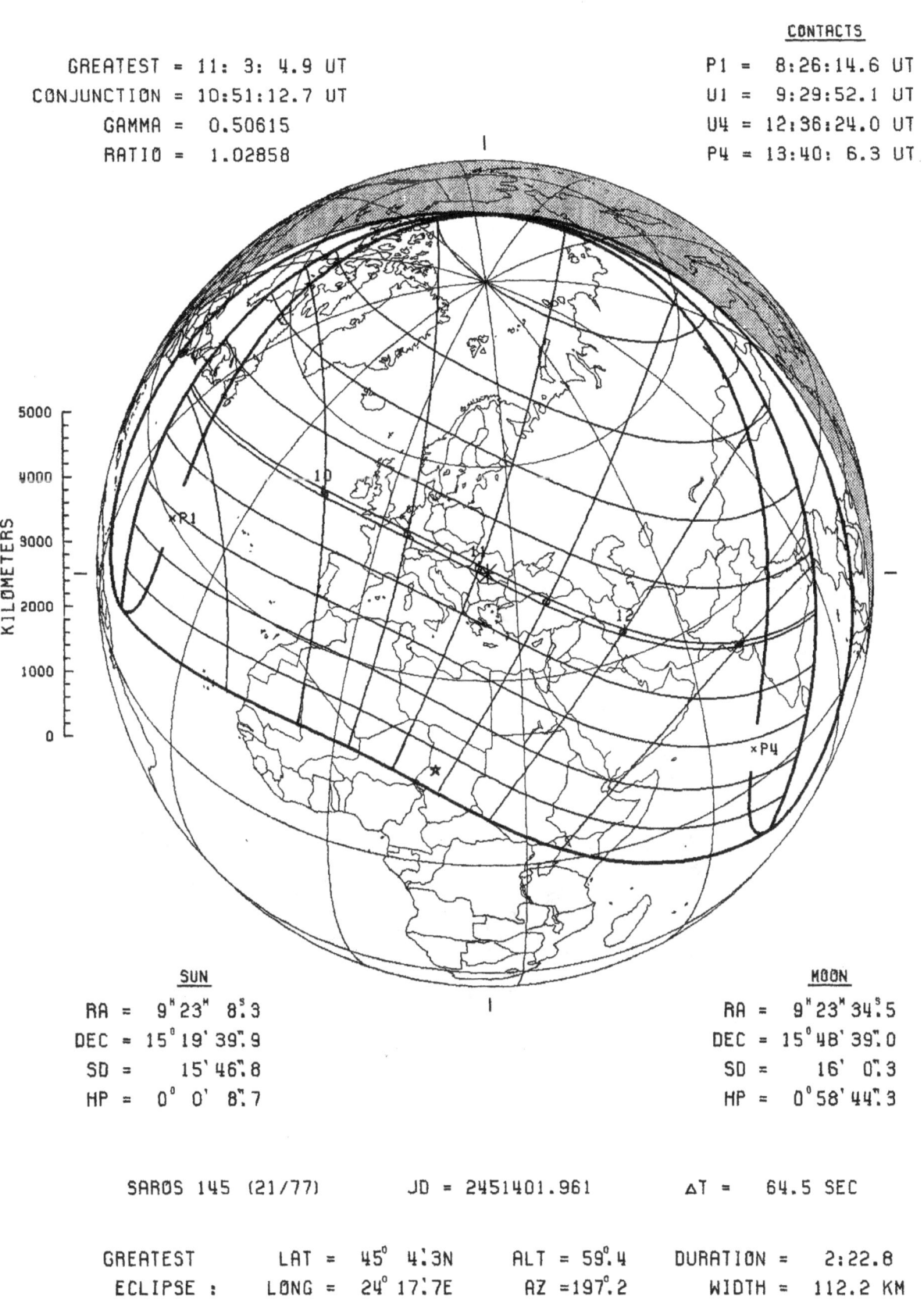

GREATEST = 11: 3: 4.9 UT
CONJUNCTION = 10:51:12.7 UT
GAMMA = 0.50615
RATIO = 1.02858

<u>CONTACTS</u>

P1 = 8:26:14.6 UT
U1 = 9:29:52.1 UT
U4 = 12:36:24.0 UT
P4 = 13:40: 6.3 UT

<u>SUN</u>

RA = 9ʰ23ᵐ 8ˢ3
DEC = 15°19' 39".9
SD = 15' 46".8
HP = 0° 0' 8".7

<u>MOON</u>

RA = 9ʰ23ᵐ34ˢ5
DEC = 15°48' 39".0
SD = 16' 0".3
HP = 0°58' 44".3

SAROS 145 (21/77)   JD = 2451401.961   ΔT =   64.5 SEC

GREATEST   LAT = 45° 4.3N   ALT = 59°.4   DURATION =   2:22.8
ECLIPSE :   LONG = 24° 17.7E   AZ =197°.2   WIDTH = 112.2 KM

Figure 59
171

PARTIAL SOLAR ECLIPSE - 5 FEB 2000

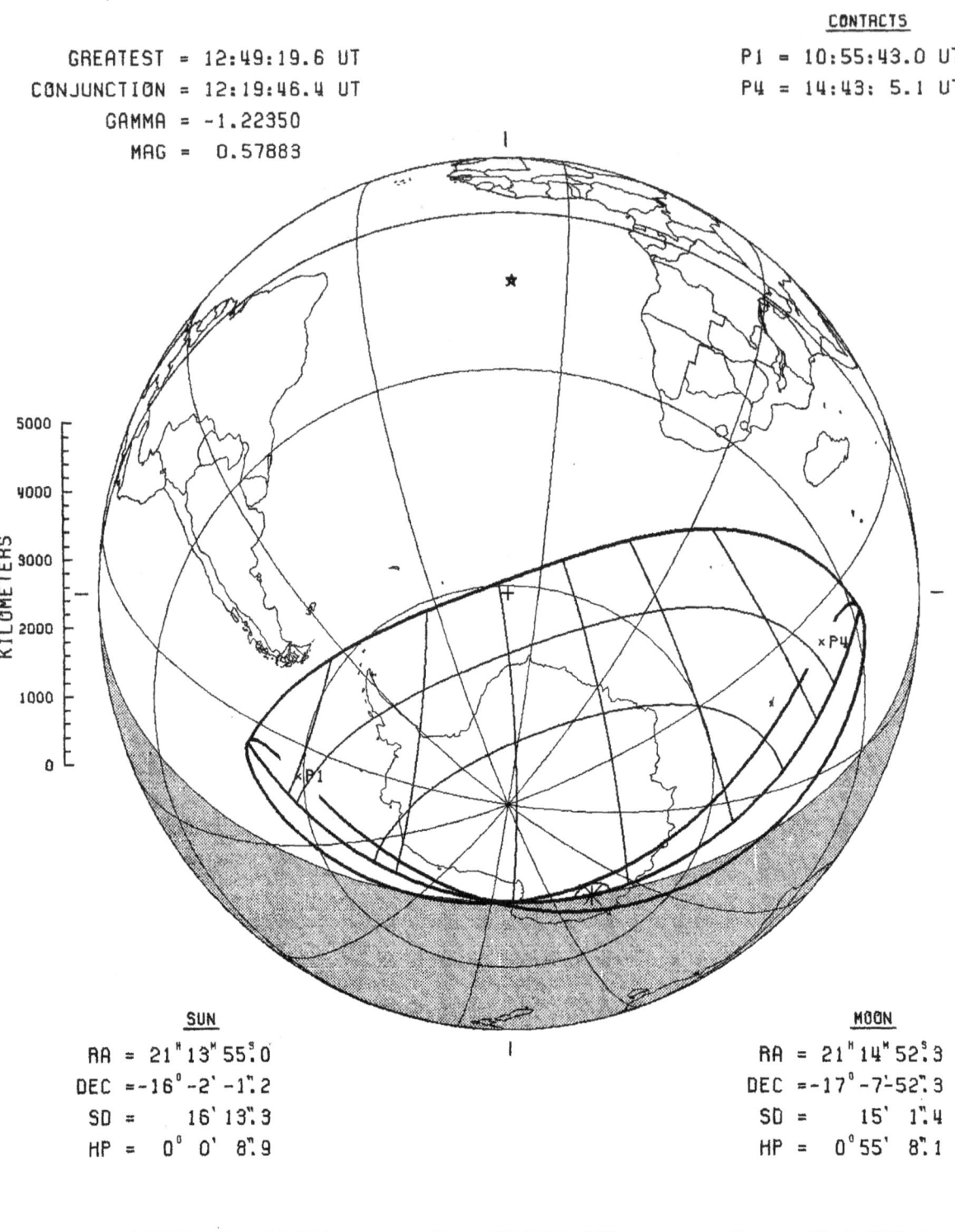

GREATEST = 12:49:19.6 UT
CONJUNCTION = 12:19:46.4 UT
GAMMA = -1.22350
MAG = 0.57883

<u>CONTACTS</u>
P1 = 10:55:43.0 UT
P4 = 14:43: 5.1 UT

KILOMETERS

<u>SUN</u>
RA = 21ʰ13ᵐ55ˢ0
DEC =-16°-2'-1".2
SD =   16'13".3
HP = 0° 0' 8".9

<u>MOON</u>
RA = 21ʰ14ᵐ52ˢ3
DEC =-17°-7'-52".3
SD =   15' 1".4
HP = 0°55' 8".1

SAROS 150 (16/71)   JD = 2451580.035   ΔT =   64.8 SEC

Figure 60

172

PARTIAL SOLAR ECLIPSE – 1 JUL 2000

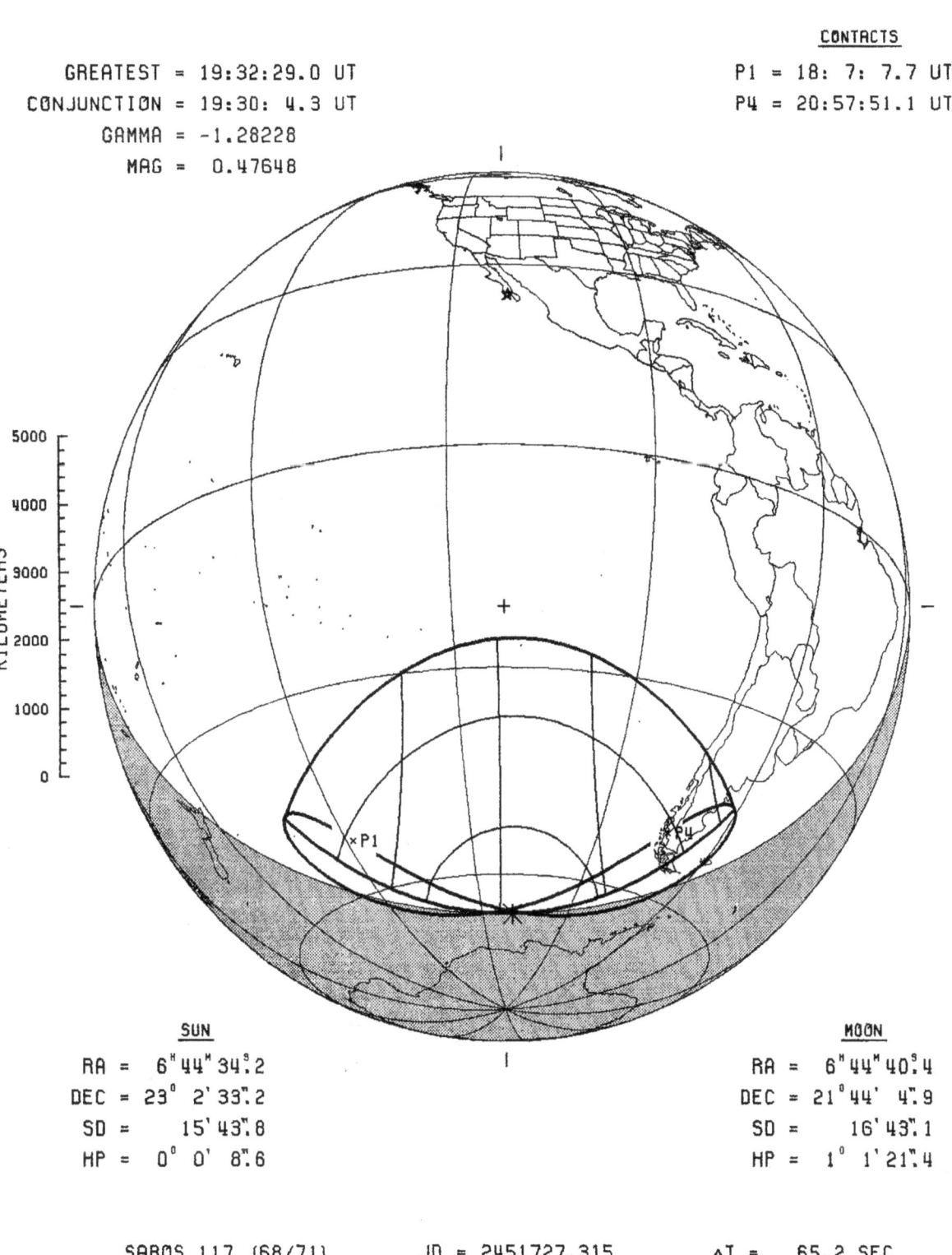

GREATEST = 19:32:29.0 UT
CONJUNCTION = 19:30: 4.3 UT
GAMMA = -1.28228
MAG = 0.47648

CONTACTS
P1 = 18: 7: 7.7 UT
P4 = 20:57:51.1 UT

KILOMETERS

5000
4000
3000
2000
1000
0

×P1

×P4

SUN
RA = 6ʰ44ᵐ34ˢ.2
DEC = 23° 2' 33".2
SD = 15' 43".8
HP = 0° 0' 8".6

MOON
RA = 6ʰ44ᵐ40ˢ.4
DEC = 21°44' 4".9
SD = 16' 43".1
HP = 1° 1' 21".4

SAROS 117 (68/71)   JD = 2451727.315   ΔT = 65.2 SEC

Figure 61

173

PARTIAL SOLAR ECLIPSE - 31 JUL 2000

GREATEST = 2:13: 3.7 UT
CONJUNCTION = 1:52: 3.5 UT
GAMMA = 1.21660
MAG = 0.60327

P1 = 0:37:27.1 UT
P4 = 3:48:51.8 UT

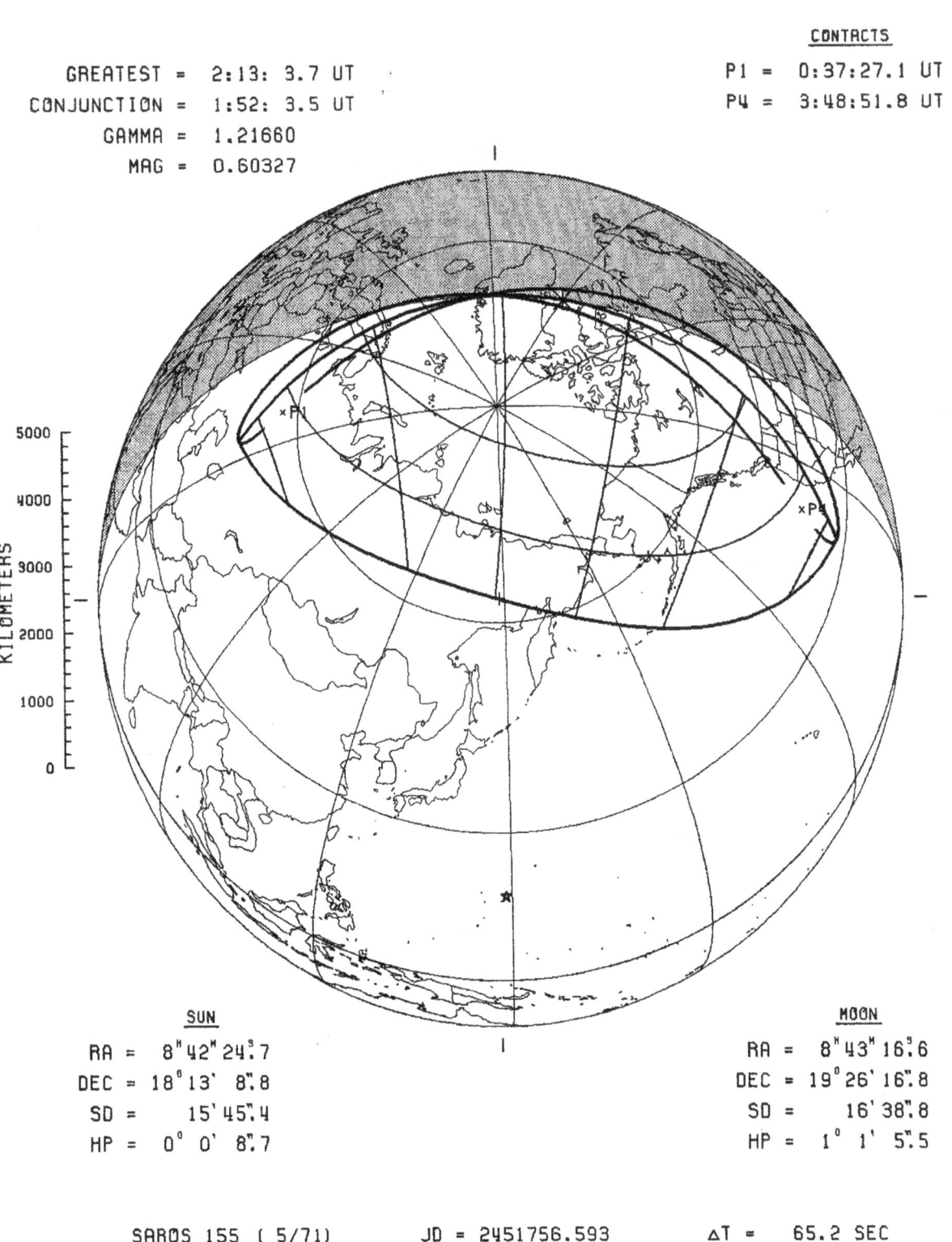

KILOMETERS

5000
4000
3000
2000
1000
0

SUN

RA = 8ʰ42ᴹ24ˢ7
DEC = 18°13' 8".8
SD = 15'45".4
HP = 0° 0' 8".7

MOON

RA = 8ʰ43ᴹ16ˢ6
DEC = 19°26'16".8
SD = 16'38".8
HP = 1° 1' 5".5

SAROS 155 (5/71)   JD = 2451756.593   ΔT =   65.2 SEC

Figure 62

174

PARTIAL SOLAR ECLIPSE - 25 DEC 2000

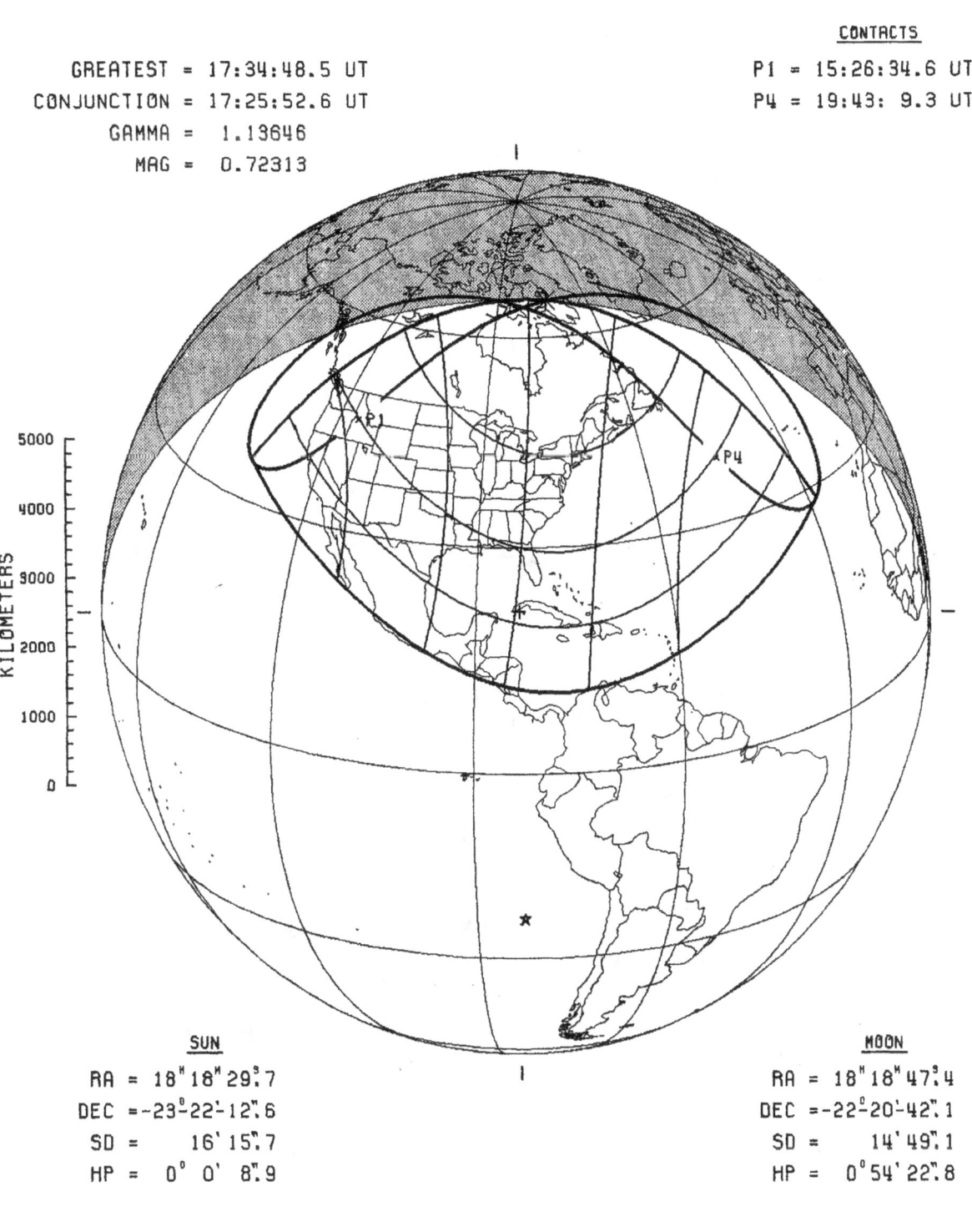

GREATEST = 17:34:48.5 UT
CONJUNCTION = 17:25:52.6 UT
GAMMA = 1.13646
MAG = 0.72313

CONTACTS
P1 = 15:26:34.6 UT
P4 = 19:43: 9.3 UT

KILOMETERS
5000
4000
3000
2000
1000
0

SUN
RA = 18ʰ18ᵐ29ˢ.7
DEC = -23°22'12".6
SD = 16'15".7
HP = 0° 0' 8".9

MOON
RA = 18ʰ18ᵐ47ˢ.4
DEC = -22°20'42".1
SD = 14'49".1
HP = 0°54'22".8

SAROS 122 (57/70)   JD = 2451904.233   ΔT = 65.5 SEC

Figure 63

175

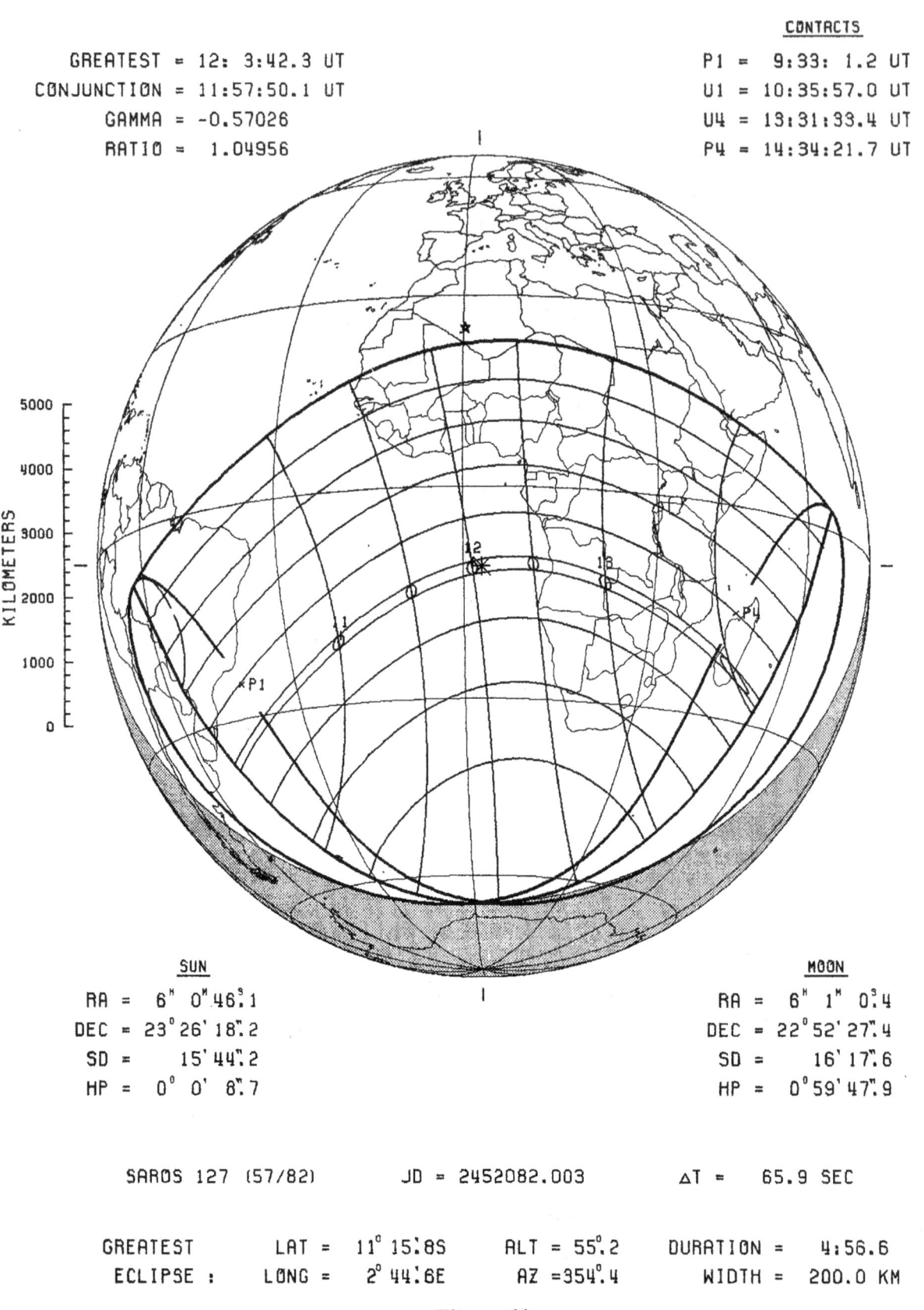

TOTAL SOLAR ECLIPSE -   21 JUN 2001

GREATEST = 12: 3:42.3 UT
CONJUNCTION = 11:57:50.1 UT
GAMMA = -0.57026
RATIO = 1.04956

CONTACTS

P1 = 9:33: 1.2 UT
U1 = 10:35:57.0 UT
U4 = 13:31:33.4 UT
P4 = 14:34:21.7 UT

KILOMETERS

5000
4000
3000
2000
1000
0

SUN

RA = 6ʰ 0ᵐ 46ˢ.1
DEC = 23° 26' 18".2
SD =   15' 44".2
HP = 0° 0' 8".7

MOON

RA = 6ʰ 1ᵐ 0ˢ.4
DEC = 22° 52' 27".4
SD =   16' 17".6
HP = 0° 59' 47".9

SAROS 127 (57/82)   JD = 2452082.003   ΔT =   65.9 SEC

GREATEST   LAT = 11° 15.8S   ALT = 55°.2   DURATION =   4:56.6
ECLIPSE :   LONG = 2° 44.6E   AZ =354°.4   WIDTH = 200.0 KM

Figure 64
176

ANNULAR SOLAR ECLIPSE — 14 DEC 2001

GREATEST = 20:51:52.6 UT
CONJUNCTION = 20:44:49.9 UT
GAMMA = 0.40865
RATIO = 0.96811

CONTACTS

P1 = 18: 3:16.5 UT
U1 = 19: 8: 4.2 UT
U4 = 22:35:49.5 UT
P4 = 23:40:38.4 UT

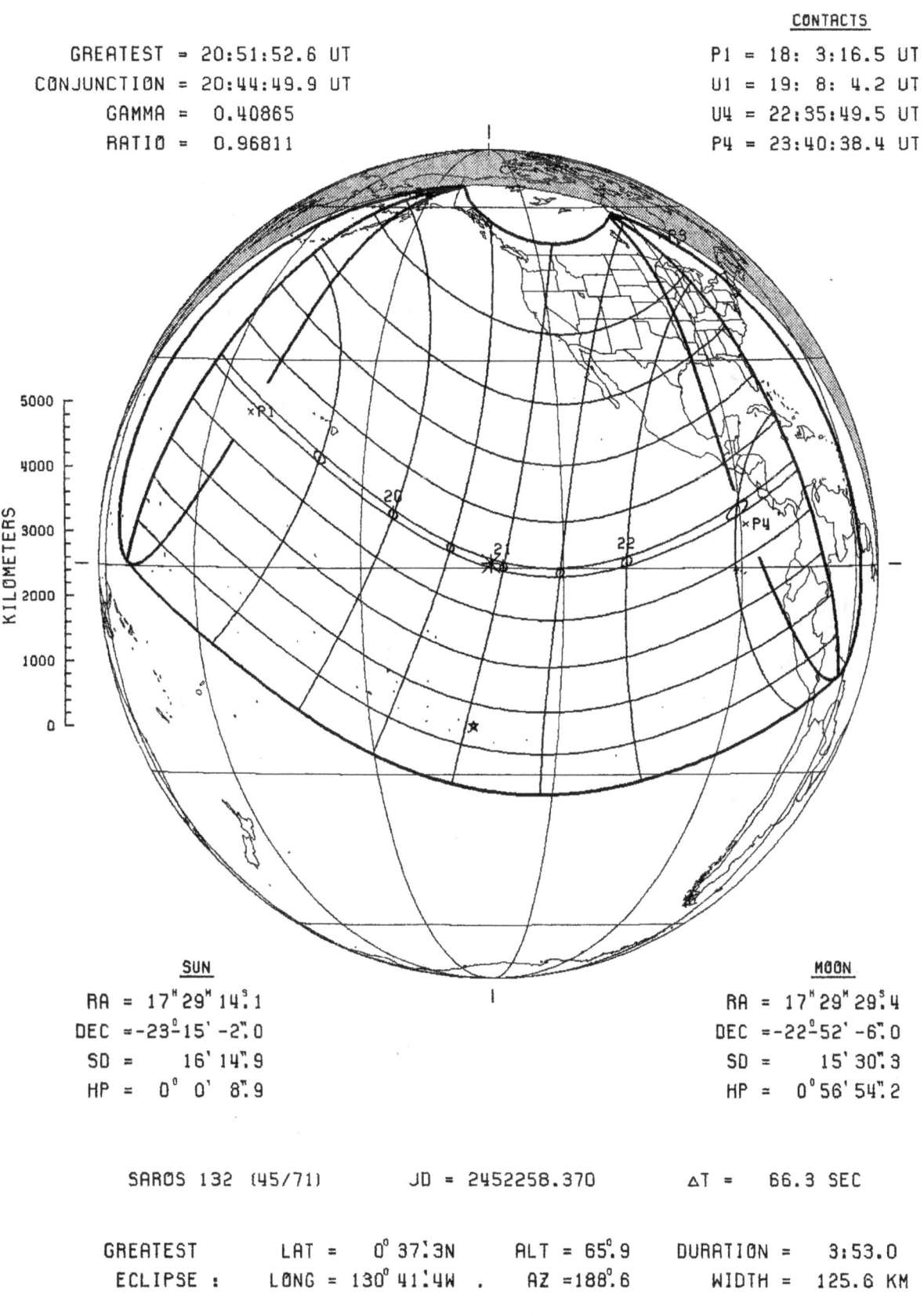

SUN

RA = 17ʰ 29ᵐ 14ˢ1
DEC = -23°15' -2".0
SD = 16' 14".9
HP = 0° 0' 8".9

MOON

RA = 17ʰ 29ᵐ 29ˢ4
DEC = -22°52' -6".0
SD = 15' 30".3
HP = 0°56' 54".2

SAROS 132 (45/71)   JD = 2452258.370   ΔT = 66.3 SEC

GREATEST   LAT = 0° 37.3N   ALT = 65°.9   DURATION = 3:53.0
ECLIPSE :   LONG = 130° 41.4W .   AZ = 188°.6   WIDTH = 125.6 KM

Figure 65
177

ANNULAR SOLAR ECLIPSE — 10 JUN 2002

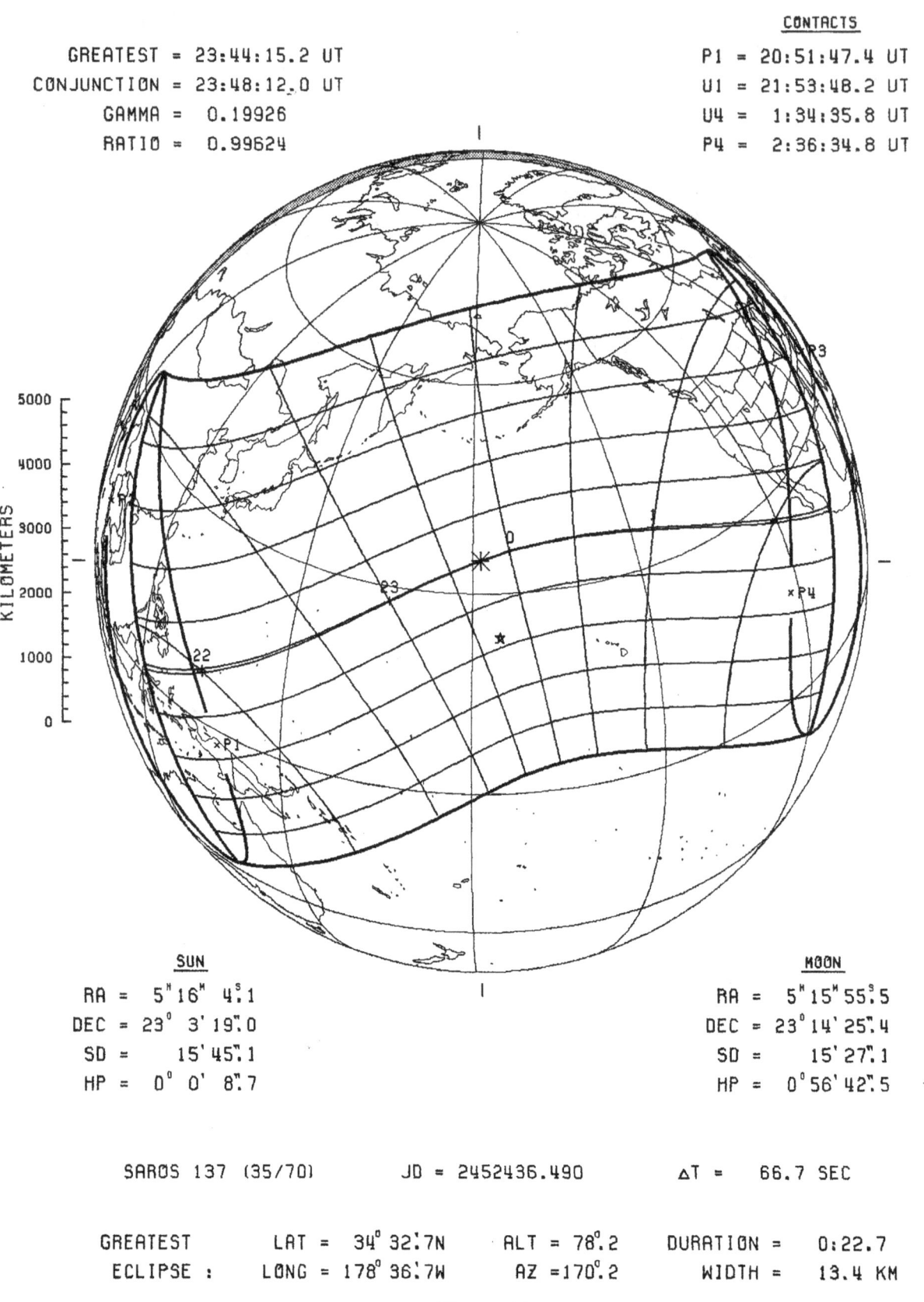

GREATEST = 23:44:15.2 UT
CONJUNCTION = 23:48:12.0 UT
GAMMA = 0.19926
RATIO = 0.99624

CONTACTS

P1 = 20:51:47.4 UT
U1 = 21:53:48.2 UT
U4 = 1:34:35.8 UT
P4 = 2:36:34.8 UT

KILOMETERS
5000
4000
3000
2000
1000
0

SUN

RA = $5^h 16^m 4^s.1$
DEC = $23° 3' 19".0$
SD = $15' 45".1$
HP = $0° 0' 8".7$

MOON

RA = $5^h 15^m 55^s.5$
DEC = $23° 14' 25".4$
SD = $15' 27".1$
HP = $0° 56' 42".5$

SAROS 137 (35/70)   JD = 2452436.490   ΔT =   66.7 SEC

GREATEST   LAT = $34° 32".7N$ ALT = $78°.2$ DURATION =   0:22.7
ECLIPSE :   LONG = $178° 36".7W$ AZ = $170°.2$ WIDTH =   13.4 KM

Figure 66
178

TOTAL SOLAR ECLIPSE — 4 DEC 2002

GREATEST = 7:31: 7.5 UT
CONJUNCTION = 7:38:41.0 UT
GAMMA = -0.30228
RATIO = 1.02436

P1 = 4:51:19.6 UT
U1 = 5:50:16.3 UT
U4 = 9:11:52.2 UT
P4 = 10:10:56.9 UT

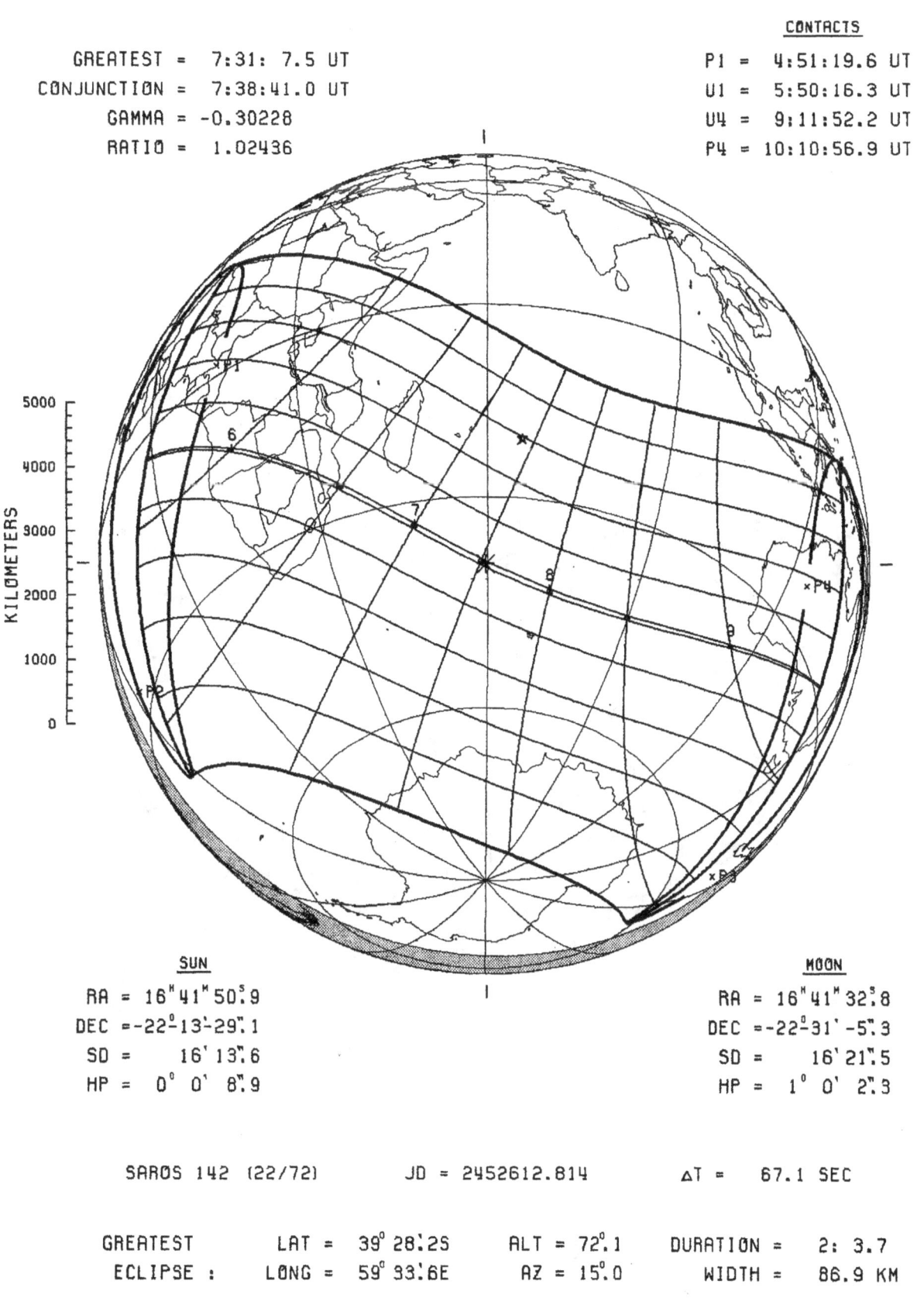

KILOMETERS

5000
4000
3000
2000
1000
0

SUN
RA = 16ʰ41ᵐ50ˢ9
DEC = -22°13'29".1
SD = 16'13".6
HP = 0° 0' 8".9

MOON
RA = 16ʰ41ᵐ32ˢ8
DEC = -22°31'-5".3
SD = 16'21".5
HP = 1° 0' 2".3

SAROS 142 (22/72)   JD = 2452612.814   ΔT = 67.1 SEC

GREATEST   LAT = 39° 28'.2S   ALT = 72°.1   DURATION = 2: 3.7
ECLIPSE :   LONG = 59° 33'.6E   AZ = 15°.0   WIDTH = 86.9 KM

Figure 67
179

ANNULAR SOLAR ECLIPSE - 31 MAY 2003

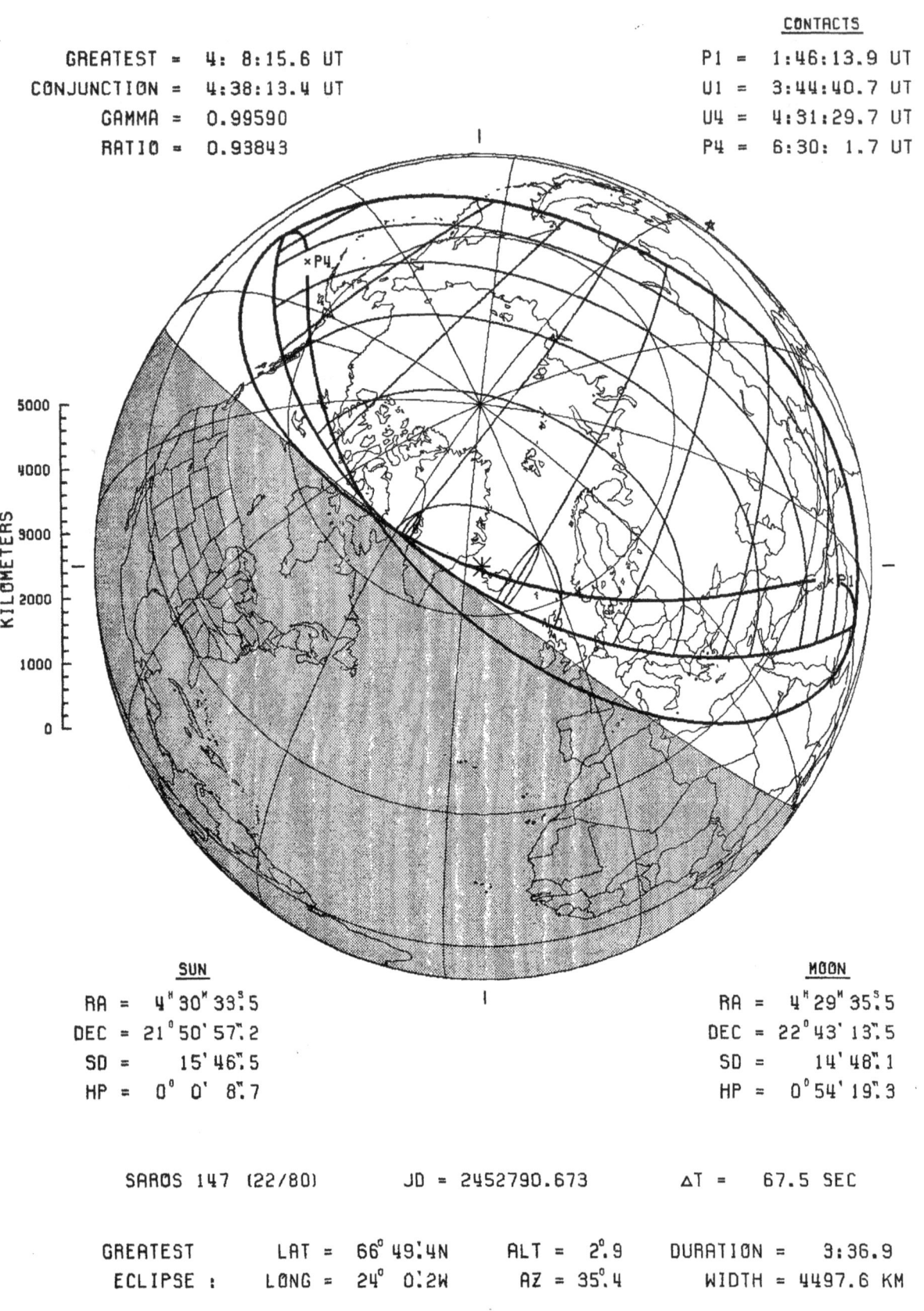

GREATEST = 4: 8:15.6 UT
CONJUNCTION = 4:38:13.4 UT
GAMMA = 0.99590
RATIO = 0.93843

CONTACTS

P1 = 1:46:13.9 UT
U1 = 3:44:40.7 UT
U4 = 4:31:29.7 UT
P4 = 6:30: 1.7 UT

KILOMETERS
5000
4000
3000
2000
1000
0

SUN

RA = 4ʰ30ᴹ33ˢ.5
DEC = 21°50'57".2
SD = 15'46".5
HP = 0° 0' 8".7

MOON

RA = 4ʰ29ᴹ35ˢ.5
DEC = 22°43'13".5
SD = 14'48".1
HP = 0°54'19".3

SAROS 147 (22/80)   JD = 2452790.673   ΔT = 67.5 SEC

GREATEST   LAT = 66°49'.4N   ALT = 2°.9   DURATION = 3:36.9
ECLIPSE :   LONG = 24° 0'.2W   AZ = 35°.4   WIDTH = 4497.6 KM

Figure 68
180

TOTAL SOLAR ECLIPSE – 23 NOV 2003

GREATEST = 22:49:15.2 UT
CONJUNCTION = 23:20:13.5 UT
GAMMA = -0.96394
RATIO = 1.03786

CONTACTS

P1 = 20:46: 4.3 UT
U1 = 22:19:22.1 UT
U4 = 23:18:48.4 UT
P4 = 0:52:13.1 UT

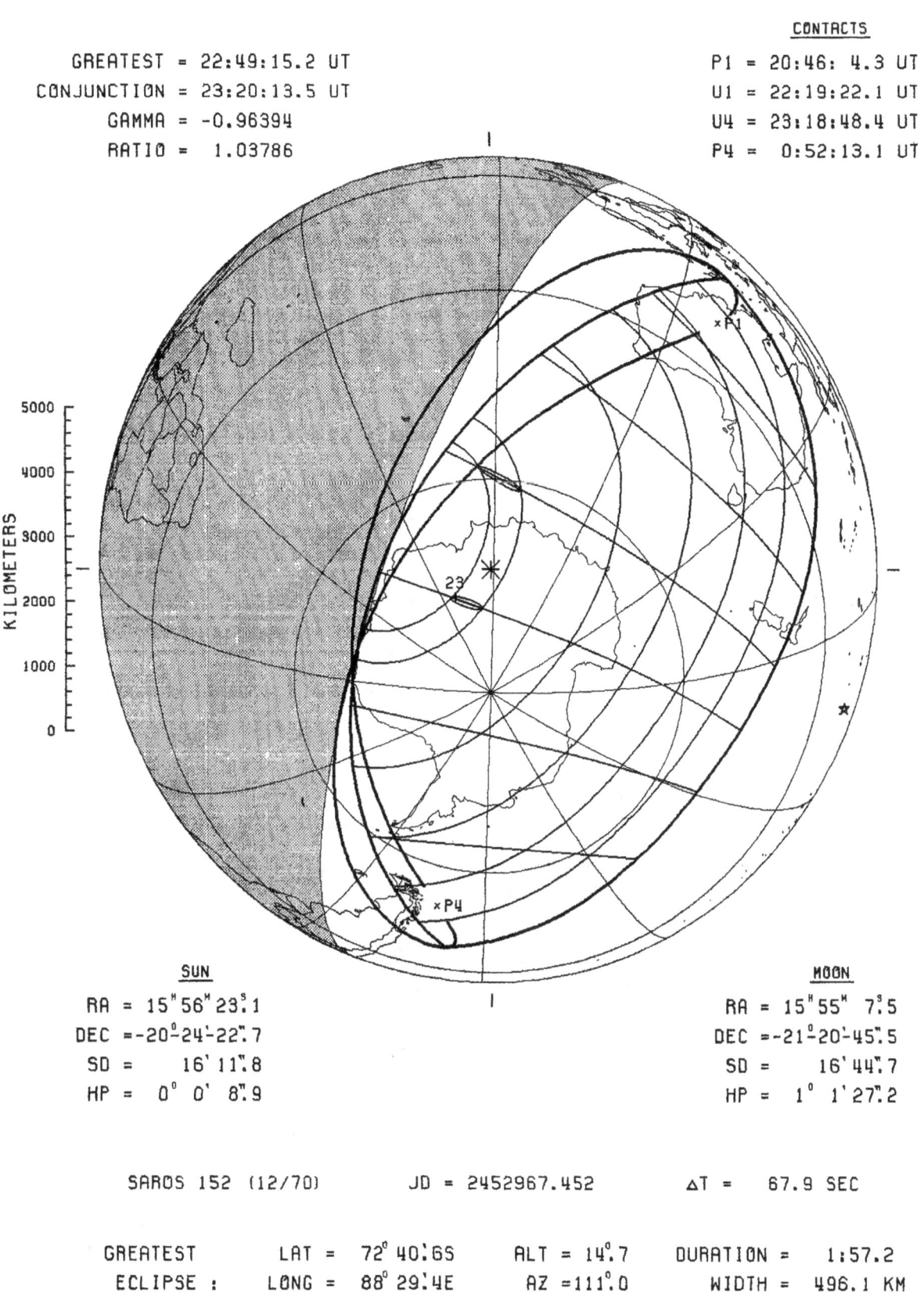

SUN

RA = 15ʰ56ᵐ23ˢ1
DEC = -20°24'-22".7
SD = 16' 11".8
HP = 0° 0' 8".9

MOON

RA = 15ʰ55ᵐ 7ˢ5
DEC = -21°20'-45".5
SD = 16' 44".7
HP = 1° 1' 27".2

SAROS 152 (12/70)   JD = 2452967.452   ΔT = 67.9 SEC

GREATEST   LAT = 72° 40.'6S   ALT = 14°.7   DURATION = 1:57.2
ECLIPSE :   LONG = 88° 29.'4E   AZ = 111°.0   WIDTH = 496.1 KM

Figure 69
181

PARTIAL SØLAR ECLIPSE – 19 APR 2004

GREATEST = 13:33:57.9 UT
CØNJUNCTIØN = 12:29:21.4 UT
GAMMA = -1.13363
MAG = 0.73535

CONTACTS
P1 = 11:29:53.6 UT
P4 = 15:38:32.3 UT

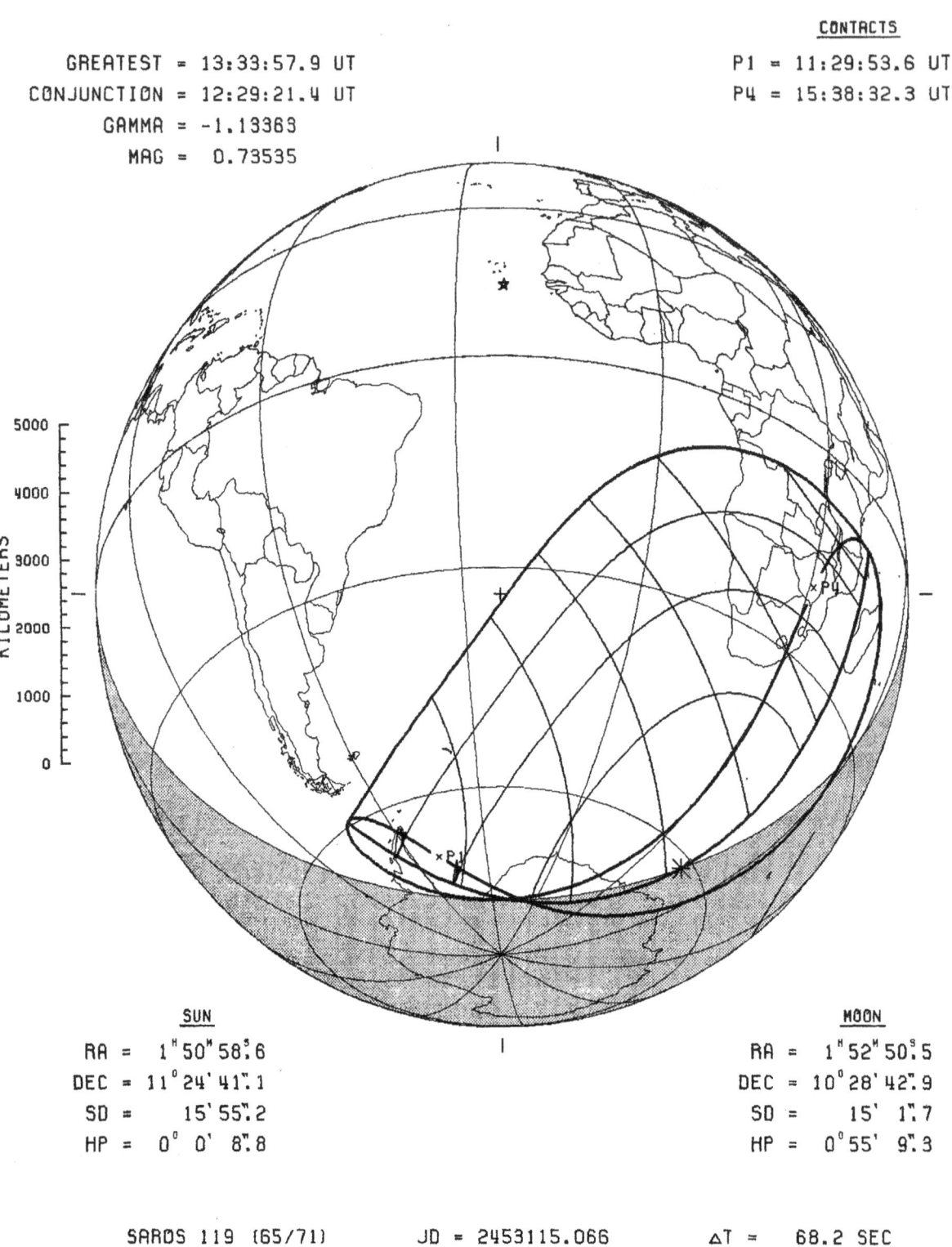

SUN
RA = 1ʰ50ᴹ58ˢ6
DEC = 11°24'41".1
SD = 15'55".2
HP = 0° 0' 8".8

MOON
RA = 1ʰ52ᴹ50ˢ5
DEC = 10°28'42".9
SD = 15' 1".7
HP = 0°55' 9".3

SARØS 119 (65/71)   JD = 2453115.066   ΔT = 68.2 SEC

Figure 70

182

PARTIAL SOLAR ECLIPSE   —   14 OCT 2004

GREATEST = 2:59:13.7 UT
CONJUNCTION = 2: 0:23.6 UT
GAMMA = 1.03451
MAG = 0.92753

CONTACTS
P1 = 0:54:31.5 UT
P4 = 5: 4:14.3 UT

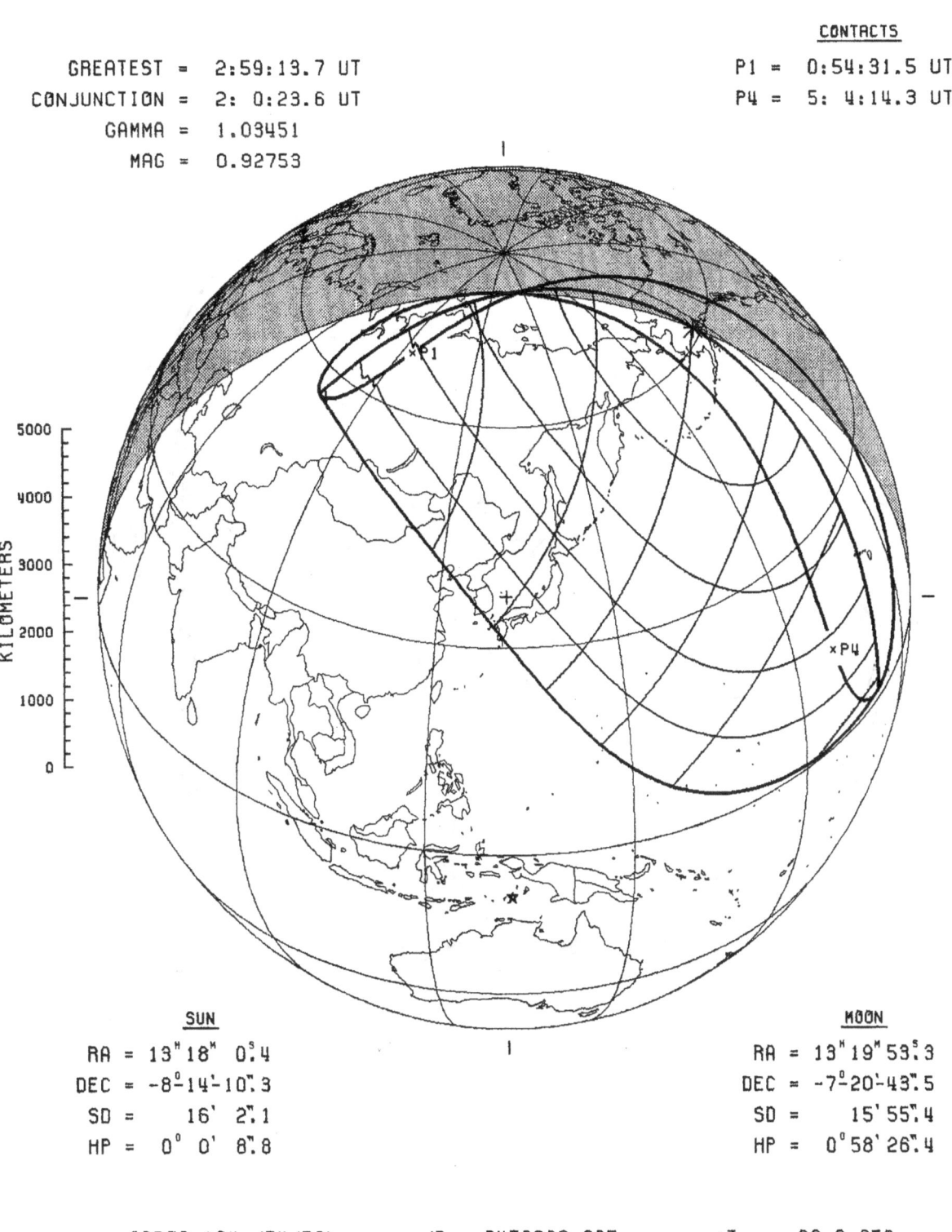

SUN
RA = 13ʰ18ᴹ 0ˢ4
DEC = -8°14'10".3
SD = 16' 2".1
HP = 0° 0' 8".8

MOON
RA = 13ʰ19ᴹ 53ˢ3
DEC = -7°20'43".5
SD = 15'55".4
HP = 0°58'26".4

SAROS 124 (54/73)   JD = 2453292.625   ΔT = 68.6 SEC

Figure 71

183

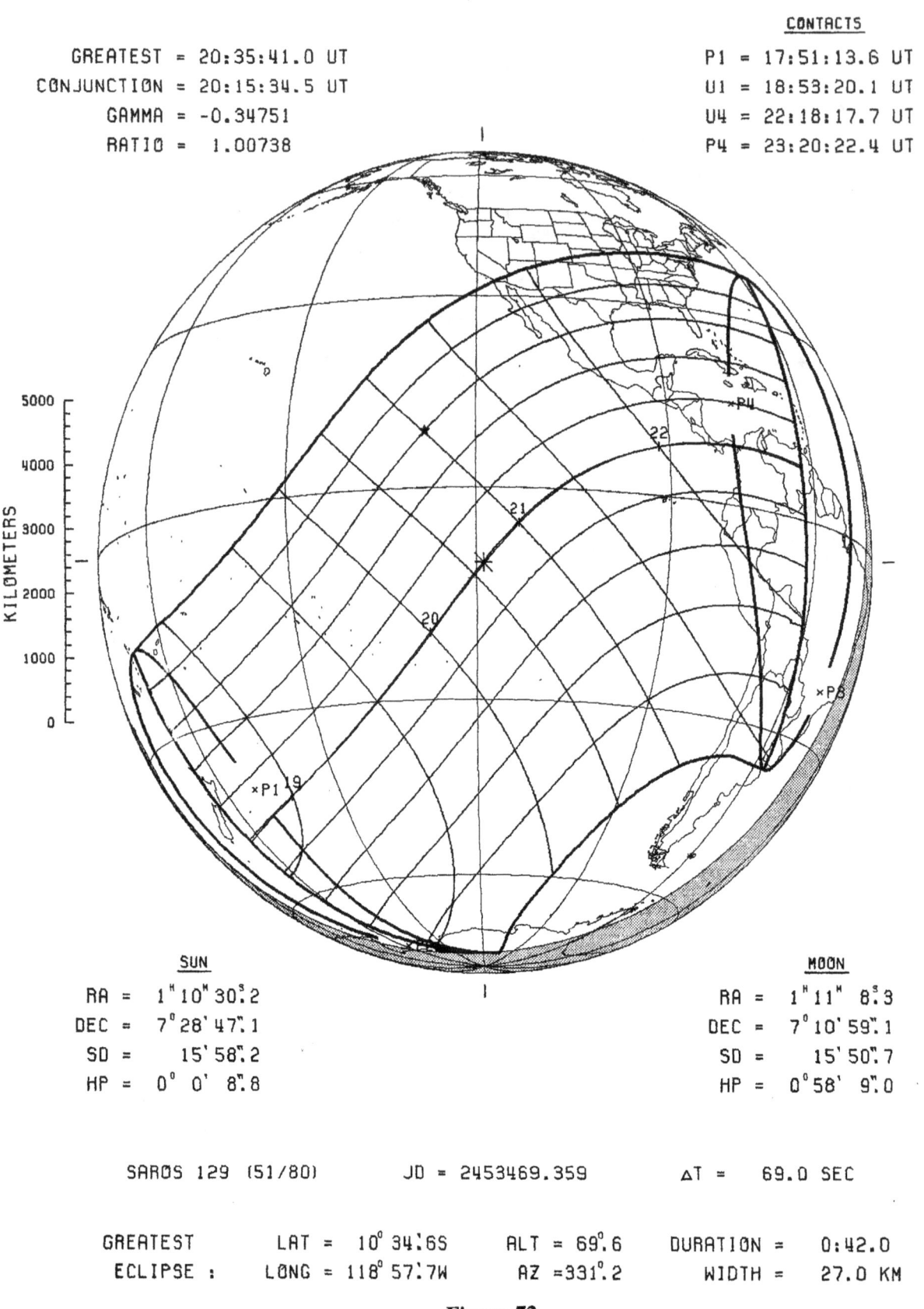

ANN/TOT SOLAR ECLIPSE - 8 APR 2005

GREATEST = 20:35:41.0 UT
CONJUNCTION = 20:15:34.5 UT
GAMMA = -0.34751
RATIO = 1.00738

CONTACTS

P1 = 17:51:13.6 UT
U1 = 18:53:20.1 UT
U4 = 22:18:17.7 UT
P4 = 23:20:22.4 UT

KILOMETERS

5000
4000
3000
2000
1000
0

SUN

RA = 1ʰ10ᴹ30ˢ2
DEC = 7°28'47".1
SD = 15'58".2
HP = 0° 0' 8".8

MOON

RA = 1ʰ11ᴹ 8ˢ3
DEC = 7°10'59".1
SD = 15'50".7
HP = 0°58' 9".0

SAROS 129 (51/80)   JD = 2453469.359   ΔT =   69.0 SEC

GREATEST   LAT = 10°34'.6S   ALT = 69°.6   DURATION =   0:42.0
ECLIPSE :   LONG = 118°57'.7W   AZ = 331°.2   WIDTH =   27.0 KM

Figure 72
184

ANNULAR SOLAR ECLIPSE – 3 OCT 2005

GREATEST = 10:31:36.5 UT
CONJUNCTION = 10:10:37.3 UT
GAMMA = 0.33027
RATIO = 0.95761

CONTACTS
P1 = 7:35:28.5 UT
U1 = 8:40:52.6 UT
U4 = 12:22:30.0 UT
P4 = 13:27:47.2 UT

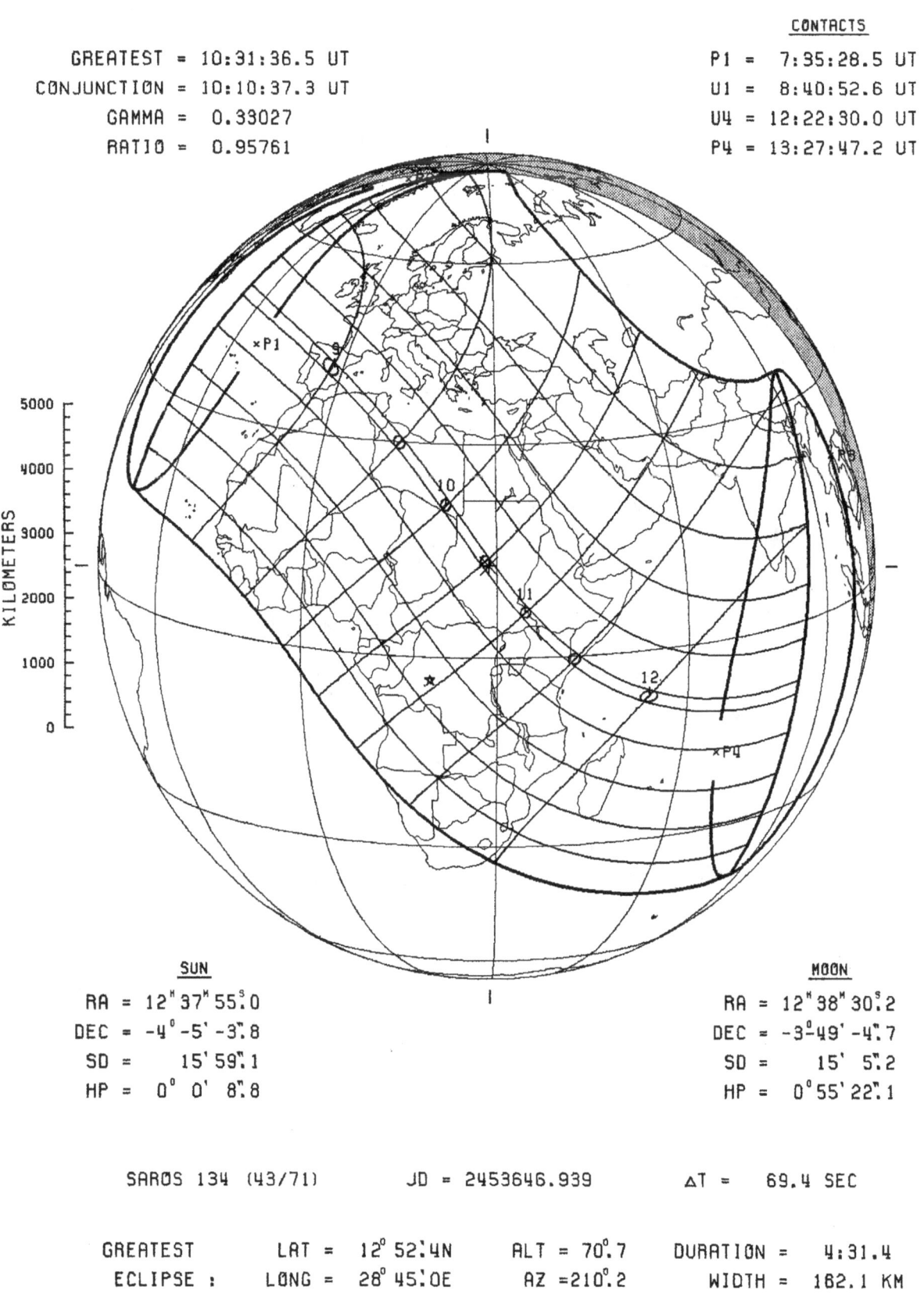

KILOMETERS

5000
4000
3000
2000
1000
0

SUN

RA = 12ʰ 37ᵐ 55ˢ.0
DEC = -4° -5' -3".8
SD =   15' 59".1
HP = 0° 0' 8".8

MOON

RA = 12ʰ 38ᵐ 30ˢ.2
DEC = -3° 49' -4".7
SD =   15' 5".2
HP = 0° 55' 22".1

SAROS 134 (43/71)   JD = 2453646.939   ΔT = 69.4 SEC

GREATEST   LAT = 12° 52".4N   ALT = 70°.7   DURATION =   4:31.4
ECLIPSE :   LONG = 28° 45".0E   AZ =210°.2   WIDTH = 162.1 KM

Figure 73
185

TOTAL SOLAR ECLIPSE – 29 MAR 2006

GREATEST = 10:11:15.0 UT
CONJUNCTION = 10:33:14.2 UT
GAMMA = 0.38417
RATIO = 1.05151

P1 = 7:36:45.8 UT
U1 = 8:34:21.3 UT
U4 = 11:47:54.1 UT
P4 = 12:45:38.0 UT

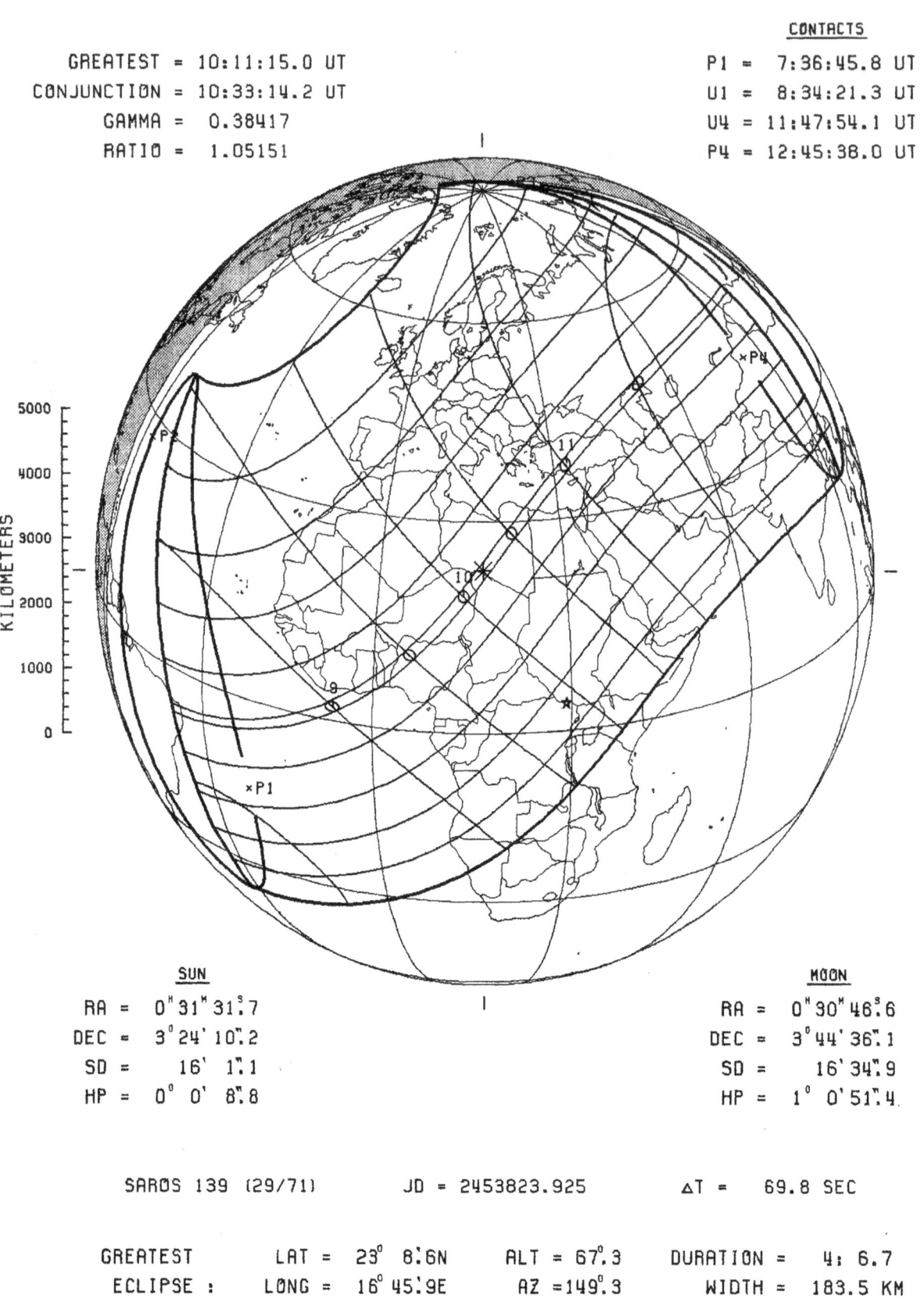

KILOMETERS
5000
4000
3000
2000
1000
0

SUN

RA = $0^h 31^m 31^s.7$
DEC = $3^\circ 24' 10''.2$
SD = $16' 1''.1$
HP = $0^\circ 0' 8''.8$

MOON

RA = $0^h 30^m 46^s.6$
DEC = $3^\circ 44' 36''.1$
SD = $16' 34''.9$
HP = $1^\circ 0' 51''.4$

SAROS 139 (29/71)   JD = 2453823.925   ΔT = 69.8 SEC

GREATEST   LAT = $23^\circ 8'.6N$ ALT = $67^\circ.3$ DURATION = 4: 6.7
ECLIPSE :   LONG = $16^\circ 45'.9E$ AZ = $149^\circ.3$ WIDTH = 183.5 KM

Figure 74
186

ANNULAR SOLAR ECLIPSE - 22 SEP 2006

GREATEST = 11:40: 3.7 UT
CONJUNCTION = 12: 7: 4.4 UT
GAMMA = -0.40659
RATIO = 0.93516

CONTACTS

P1 = 8:39:50.2 UT
U1 = 9:48:25.4 UT
U4 = 13:31:25.7 UT
P4 = 14:40: 6.5 UT

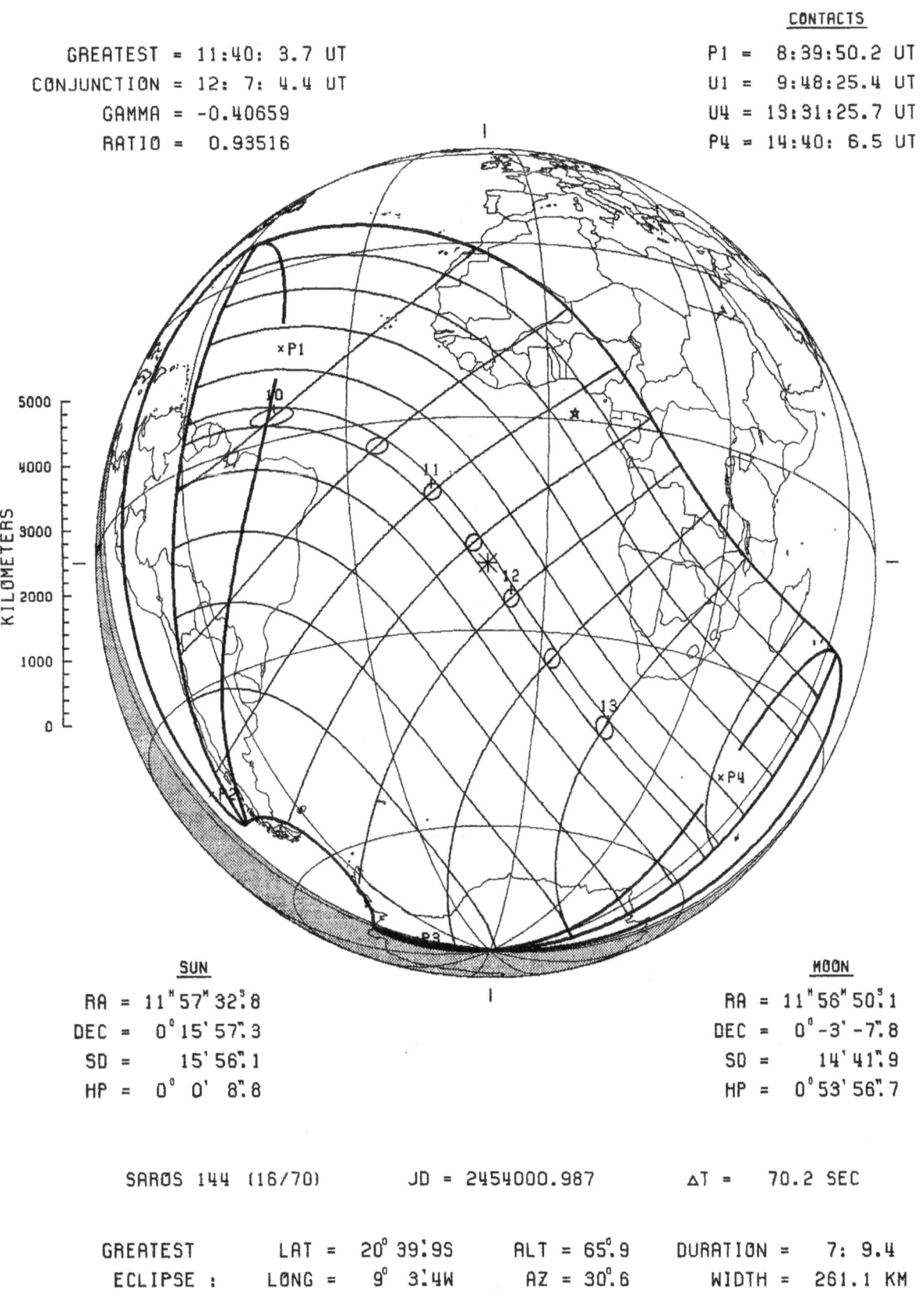

KILOMETERS

5000
4000
3000
2000
1000
0

SUN

RA = 11ʰ57ᵐ32ˢ.8
DEC = 0°15'57".3
SD = 15'56".1
HP = 0° 0' 8".8

MOON

RA = 11ʰ56ᵐ50ˢ.1
DEC = 0°-3'-7".8
SD = 14'41".9
HP = 0°53'56".7

SAROS 144 (16/70)   JD = 2454000.987   ΔT = 70.2 SEC

GREATEST   LAT = 20° 39'.9S   ALT = 65°.9   DURATION = 7: 9.4
ECLIPSE :   LONG = 9° 3'.4W   AZ = 30°.6   WIDTH = 261.1 KM

Figure 75
187

PARTIAL SOLAR ECLIPSE - 19 MAR 2007

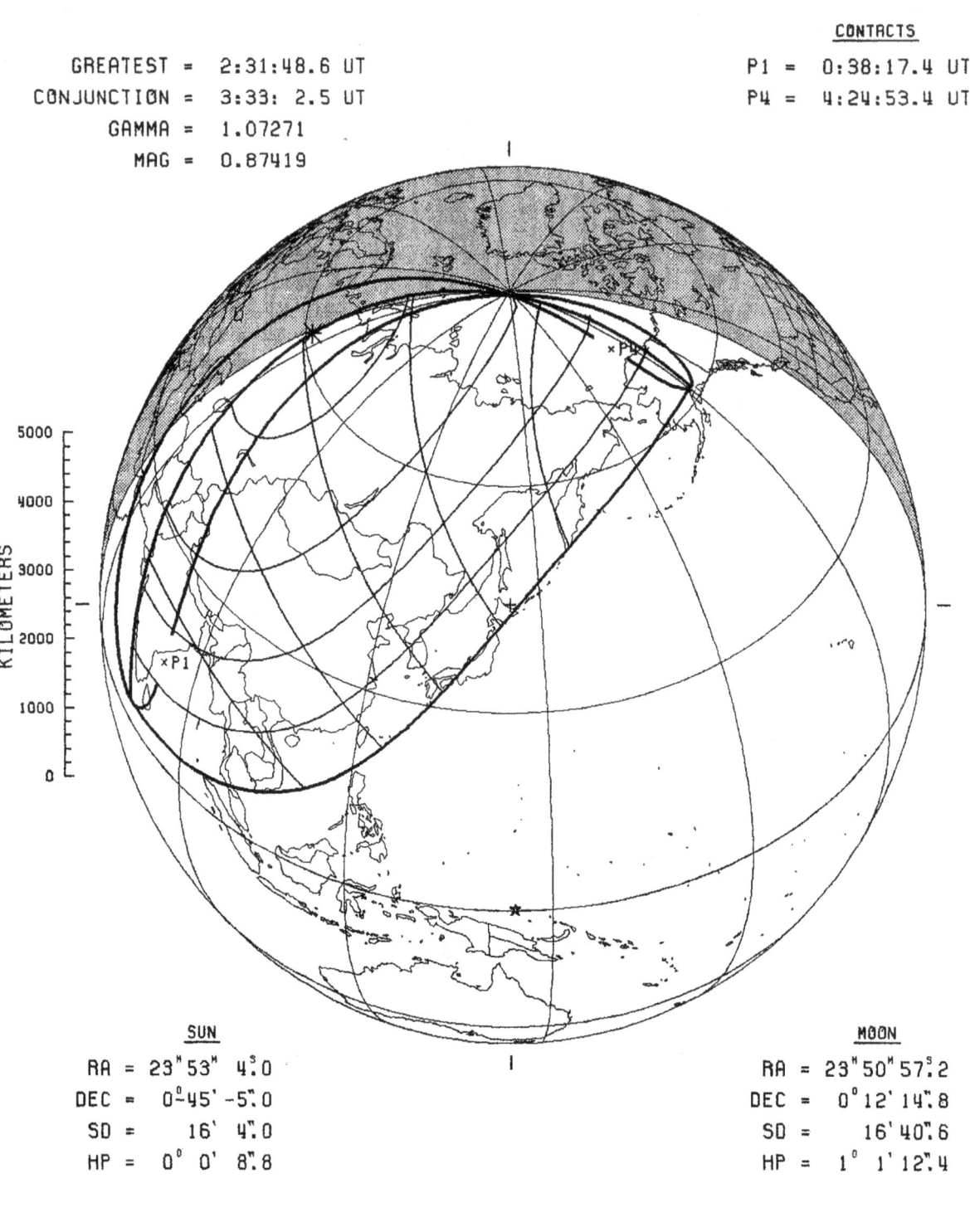

GREATEST = 2:31:48.6 UT
CONJUNCTION = 3:33: 2.5 UT
GAMMA = 1.07271
MAG = 0.87419

P1 = 0:38:17.4 UT
P4 = 4:24:53.4 UT

5000
4000
3000
2000
1000
0

KILOMETERS

×P1

×P4

SUN
RA = 23ʰ53ᴹ 4ˢ0
DEC = 0°45'-5".0
SD = 16' 4".0
HP = 0° 0' 8".8

MOON
RA = 23ʰ50ᴹ57ˢ2
DEC = 0°12'14".8
SD = 16'40".6
HP = 1° 1'12".4

SAROS 149 (20/71)   JD = 2454178.606   ΔT =   70.6 SEC

Figure 76

188

PARTIAL SOLAR ECLIPSE - 11 SEP 2007

GREATEST = 12:31:13.0 UT
CONJUNCTION = 13:42:35.9 UT
GAMMA = -1.12582
MAG = 0.74886

P1 = 10:25:38.8 UT
P4 = 14:36:23.8 UT

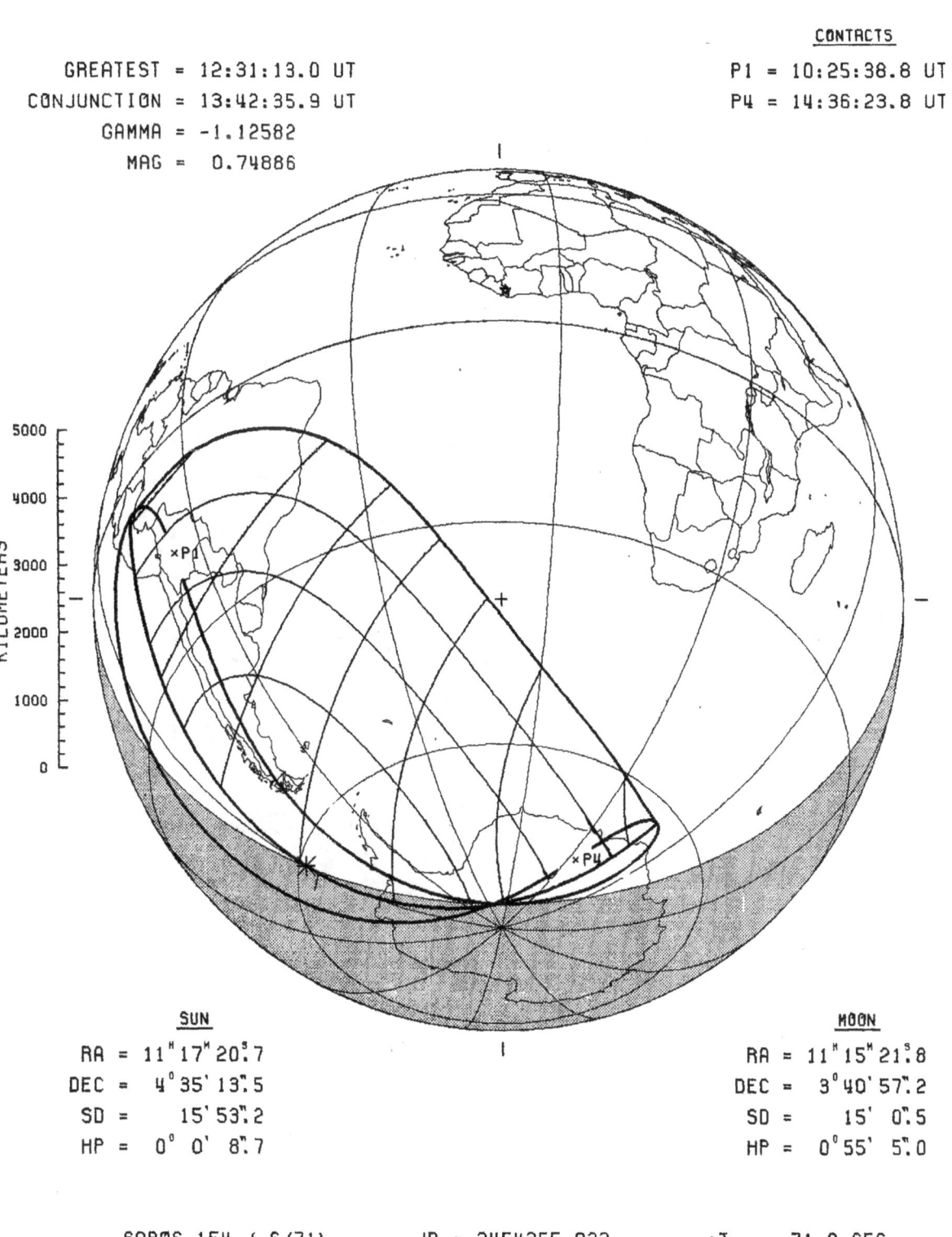

KILOMETERS

5000
4000
3000
2000
1000
0

×P1

×P4

SUN

RA = 11ʰ17ᵐ20ˢ7
DEC = 4°35'13".5
SD = 15'53".2
HP = 0°0'8".7

MOON

RA = 11ʰ15ᵐ21ˢ8
DEC = 3°40'57".2
SD = 15'0".5
HP = 0°55'5".0

SAROS 154 (6/71)   JD = 2454355.023   ΔT = 71.0 SEC

Figure 77

189

ANNULAR SOLAR ECLIPSE   –   7 FEB 2008

GREATEST = 3:54:57.4 UT

CONJUNCTION = 3: 8:43.3 UT

GAMMA = -0.95704

RATIO = 0.96499

P1 = 1:38:22.1 UT

U1 = 3:19:37.8 UT

U4 = 4:30:42.8 UT

P4 = 6:11:45.5 UT

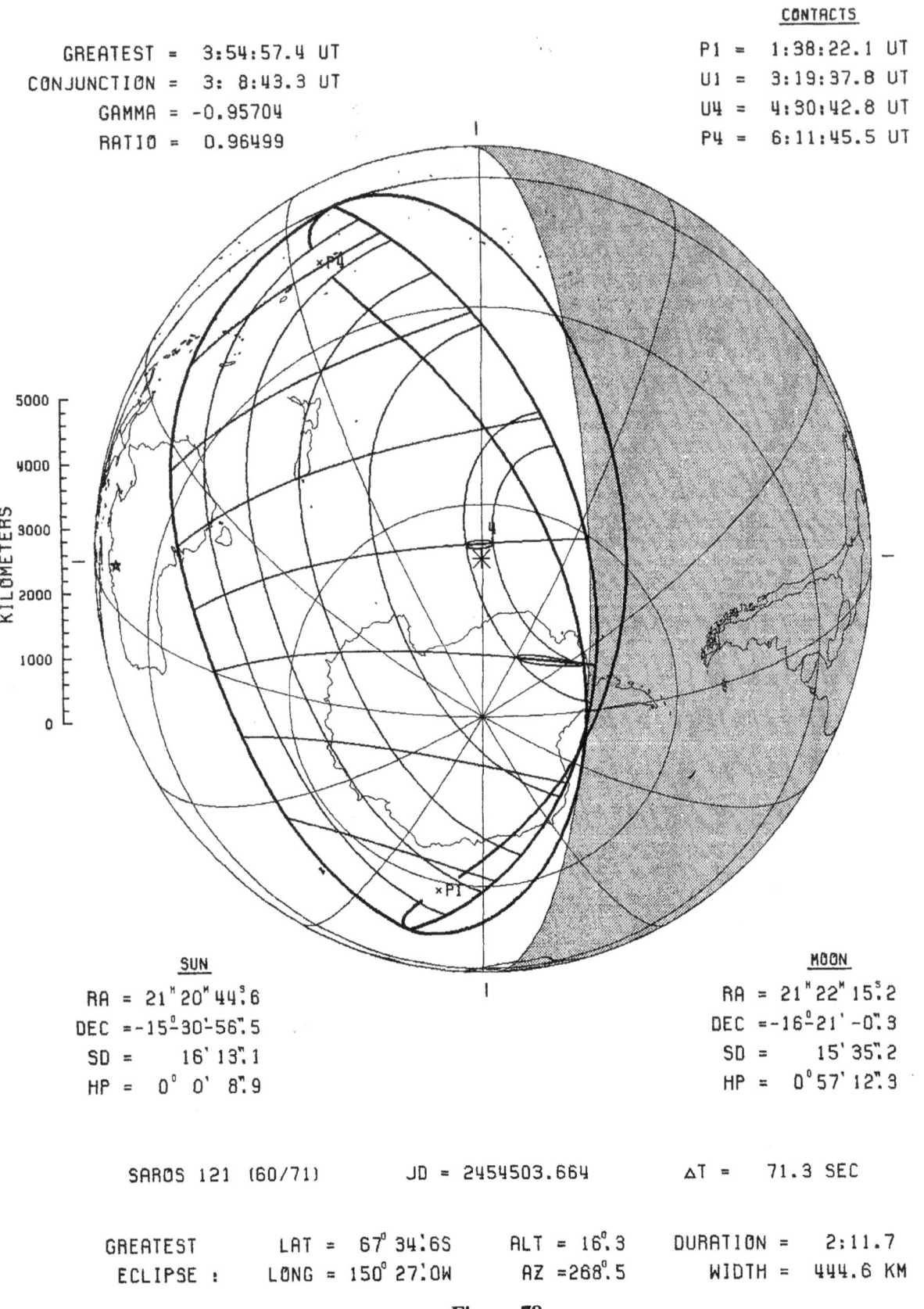

SUN

RA = $21^H 20^M 44^S.6$

DEC = $-15^\circ 30' 56".5$

SD = $16' 13".1$

HP = $0^\circ 0' 8".9$

MOON

RA = $21^H 22^M 15^S.2$

DEC = $-16^\circ 21' -0".3$

SD = $15' 35".2$

HP = $0^\circ 57' 12".3$

SAROS 121 (60/71)   JD = 2454503.664   ΔT =   71.3 SEC

GREATEST   LAT = $67^\circ 34'.6S$ ALT = $16^\circ.3$ DURATION =   2:11.7

ECLIPSE :   LONG = $150^\circ 27'.0W$ AZ = $268^\circ.5$ WIDTH =   444.6 KM

Figure 78
190

TOTAL SOLAR ECLIPSE — 1 AUG 2008

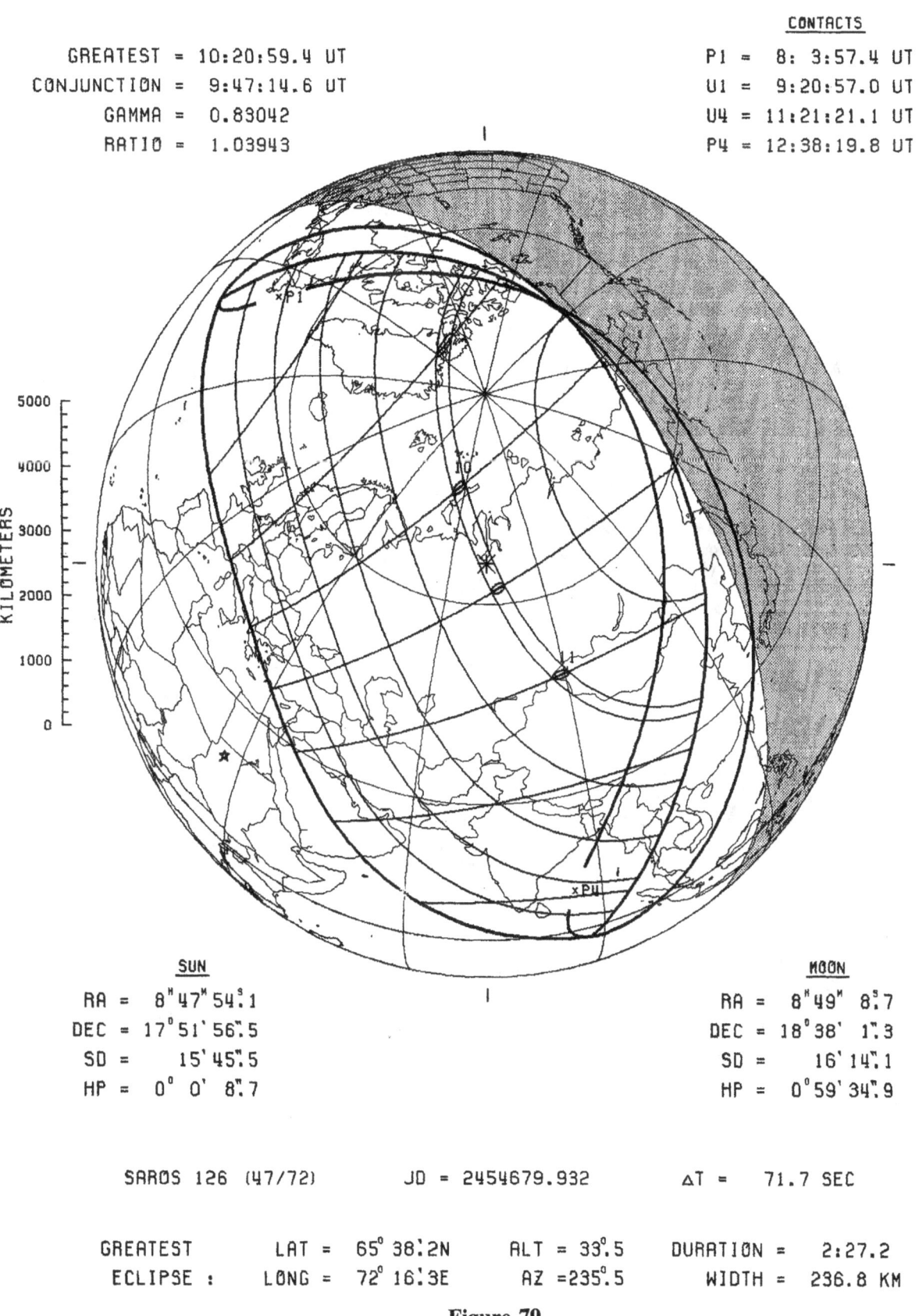

GREATEST = 10:20:59.4 UT
CONJUNCTION = 9:47:14.6 UT
GAMMA = 0.83042
RATIO = 1.03943

CONTACTS

P1 = 8: 3:57.4 UT
U1 = 9:20:57.0 UT
U4 = 11:21:21.1 UT
P4 = 12:38:19.8 UT

SUN

RA = 8ʰ47ᵐ54ˢ.1
DEC = 17°51' 56".5
SD =   15' 45".5
HP = 0° 0' 8".7

MOON

RA = 8ʰ49ᵐ 8ˢ.7
DEC = 18°38' 1".3
SD =   16' 14".1
HP = 0°59' 34".9

SAROS 126 (47/72)   JD = 2454679.932   ΔT = 71.7 SEC

GREATEST   LAT = 65° 38'.2N   ALT = 33°.5   DURATION = 2:27.2
ECLIPSE :   LONG = 72° 16'.3E   AZ = 235°.5   WIDTH = 236.8 KM

Figure 79
191

ANNULAR SOLAR ECLIPSE – 26 JAN 2009

GREATEST = 7:58:30.3 UT
CONJUNCTION = 7:46:15.8 UT
GAMMA = -0.28204
RATIO = 0.92825

CONTACTS
P1 = 4:56:28.9 UT
U1 = 6: 2:30.6 UT
U4 = 9:54:35.1 UT
P4 = 11: 0:32.2 UT

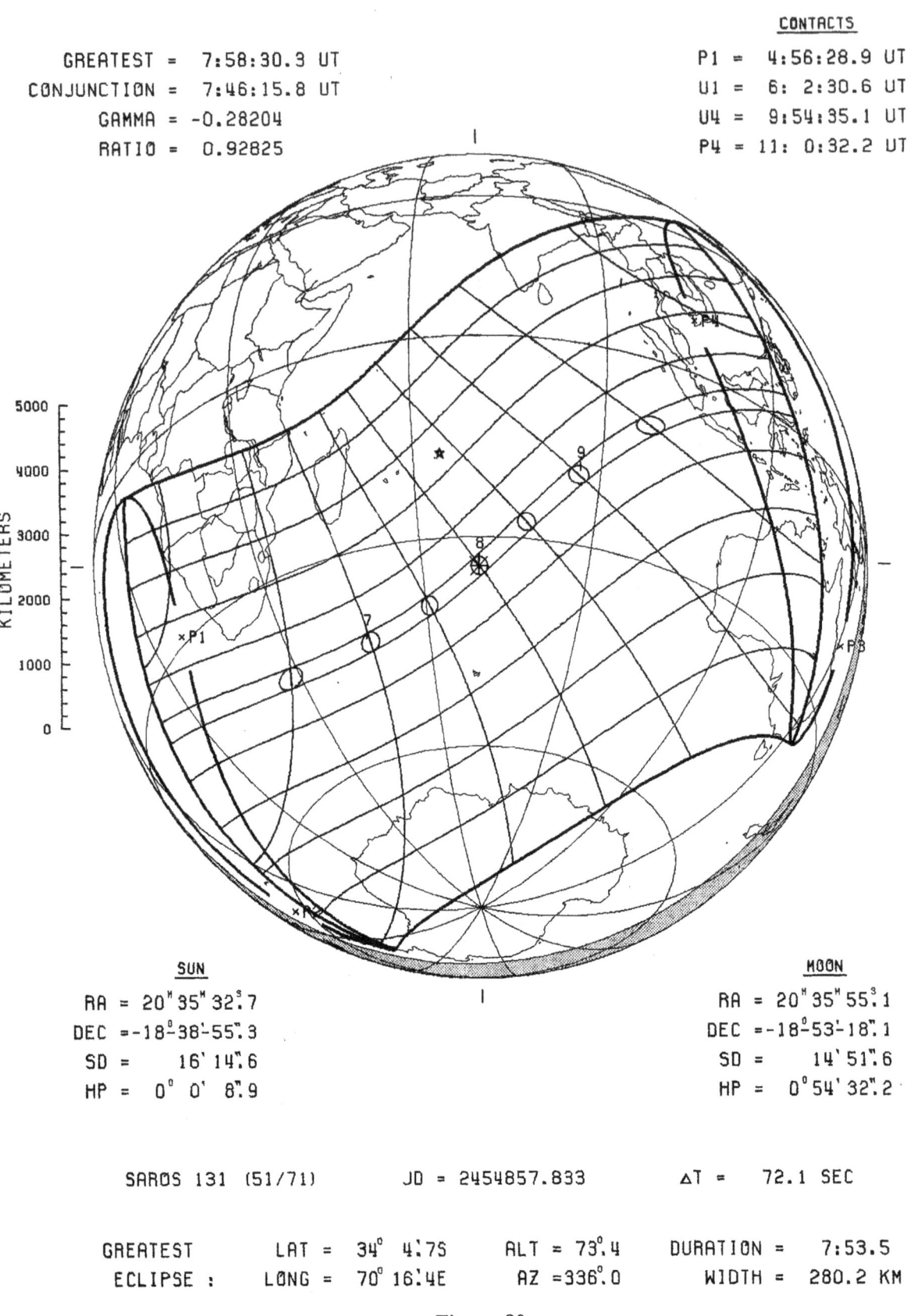

KILOMETERS

5000
4000
3000
2000
1000
0

SUN

RA = 20ʰ35ᵐ32ˢ.7
DEC = -18°38'55".3
SD =   16' 14".6
HP = 0° 0' 8".9

MOON

RA = 20ʰ35ᵐ55ˢ.1
DEC = -18°53'18".1
SD =   14' 51".6
HP = 0°54' 32".2

SAROS 131 (51/71)   JD = 2454857.833   ΔT =   72.1 SEC

GREATEST   LAT = 34° 4.7S   ALT = 73°.4   DURATION =   7:53.5
ECLIPSE :   LONG = 70° 16.4E   AZ =336°.0   WIDTH = 280.2 KM

Figure 80
192

TOTAL SOLAR ECLIPSE -   22 JUL 2009

GREATEST = 2:35:12.4 UT

CONJUNCTION = 2:32:56.0 UT

GAMMA = 0.06942

RATIO = 1.07991

P1 = 23:58:10.0 UT

U1 = 0:51: 8.0 UT

U4 = 4:19:17.9 UT

P4 = 5:12:16.4 UT

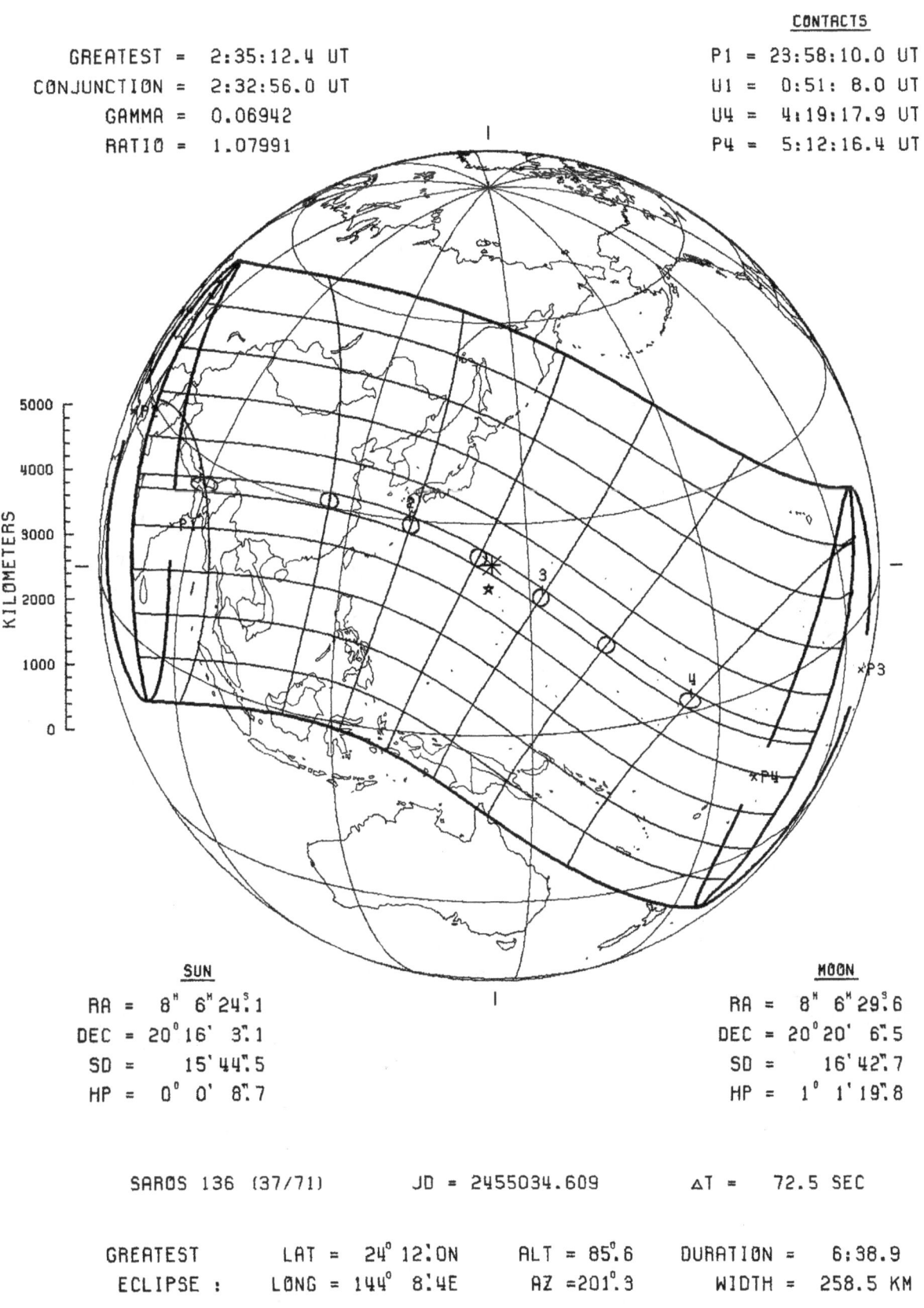

KILOMETERS

SUN

RA = 8^h 6^m $24^s.1$

DEC = $20°16'$ $3".1$

SD = $15'44".5$

HP = $0°$ $0'$ $8".7$

MOON

RA = 8^h 6^m $29^s.6$

DEC = $20°20'$ $6".5$

SD = $16'42".7$

HP = $1°$ $1'19".8$

SAROS 136 (37/71)   JD = 2455034.609   ΔT =   72.5 SEC

GREATEST   LAT = $24°$ $12'.0$N   ALT = $85°.6$ DURATION =   6:38.9

ECLIPSE :   LONG = $144°$ $8'.4$E   AZ = $201°.3$ WIDTH =   258.5 KM

Figure 81

ANNULAR SOLAR ECLIPSE - 15 JAN 2010

GREATEST = 7: 6:22.9 UT
CONJUNCTION = 7:20:11.1 UT
GAMMA = 0.40013
RATIO = 0.91902

CONTACTS
P1 = 4: 5:17.5 UT
U1 = 5:13:44.7 UT
U4 = 8:58:53.7 UT
P4 = 10: 7:24.8 UT

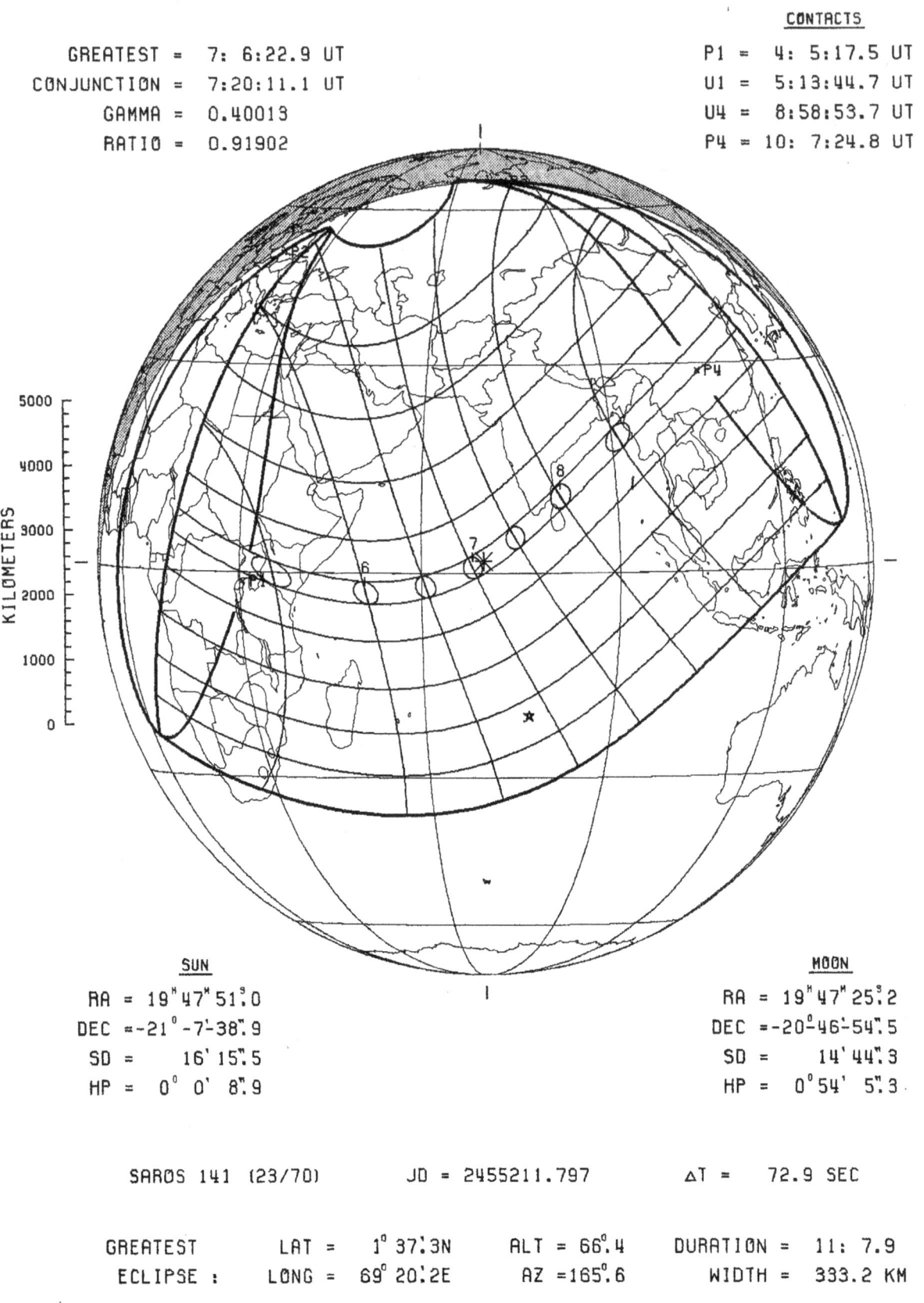

KILOMETERS
5000
4000
3000
2000
1000
0

SUN
RA = 19ʰ47ᵐ51ˢ0
DEC = -21°-7'-38".9
SD = 16' 15".5
HP = 0° 0' 8".9

MOON
RA = 19ʰ47ᵐ25ˢ2
DEC = -20°-46'-54".5
SD = 14' 44".3
HP = 0°54' 5".3

SAROS 141 (23/70)   JD = 2455211.797   ΔT =   72.9 SEC

GREATEST   LAT =   1° 37'.3N   ALT = 66°.4   DURATION = 11: 7.9
ECLIPSE :   LONG = 69° 20'.2E   AZ =165°.6   WIDTH = 333.2 KM

Figure 82
194

TOTAL SOLAR ECLIPSE — 11 JUL 2010

GREATEST = 19:33:25.2 UT
CONJUNCTION = 19:50:49.4 UT
GAMMA = -0.67913
RATIO = 1.05804

CONTACTS

P1 = 17: 9:32.6 UT
U1 = 18:15: 7.9 UT
U4 = 20:51:32.7 UT
P4 = 21:57: 6.8 UT

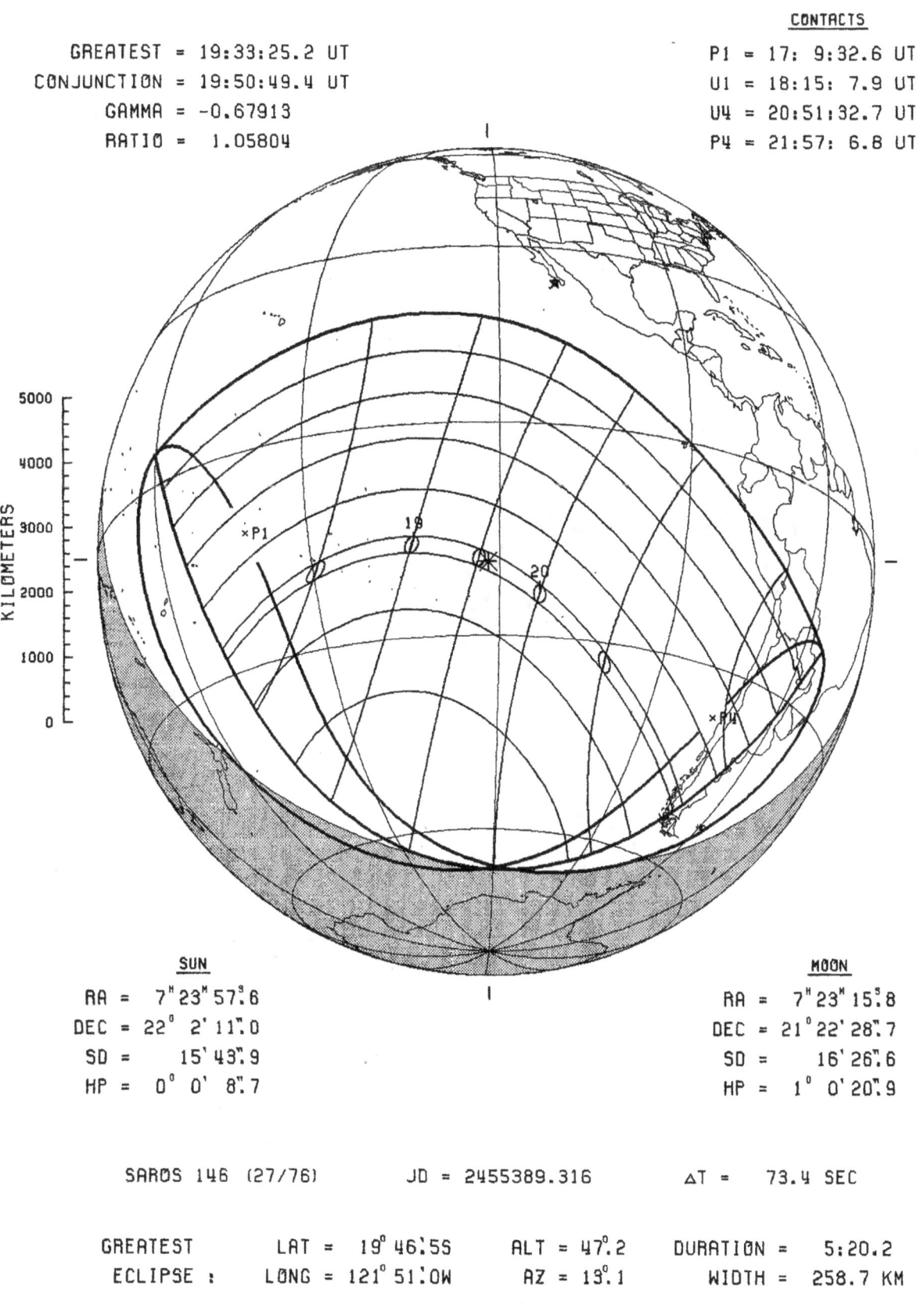

KILOMETERS

5000
4000
3000
2000
1000
0

×P1

19

20

×P4

SUN

RA = 7ʰ23ᵐ57ˢ6
DEC = 22° 2' 11".0
SD = 15' 43".9
HP = 0° 0' 8".7

MOON

RA = 7ʰ23ᵐ15ˢ8
DEC = 21° 22' 28".7
SD = 16' 26".6
HP = 1° 0' 20".9

SAROS 146 (27/76)   JD = 2455389.316   ΔT = 73.4 SEC

GREATEST   LAT = 19° 46'.5S   ALT = 47°.2   DURATION = 5:20.2
ECLIPSE :   LONG = 121° 51'.0W   AZ = 19°.1   WIDTH = 258.7 KM

Figure 83
195

PARTIAL SOLAR ECLIPSE – 4 JAN 2011

GREATEST = 8:50:25.7 UT

CONJUNCTION = 9:15: 3.8 UT

GAMMA = 1.06257

MAG = 0.85752

P1 = 6:40: 1.6 UT

P4 = 11: 0:45.0 UT

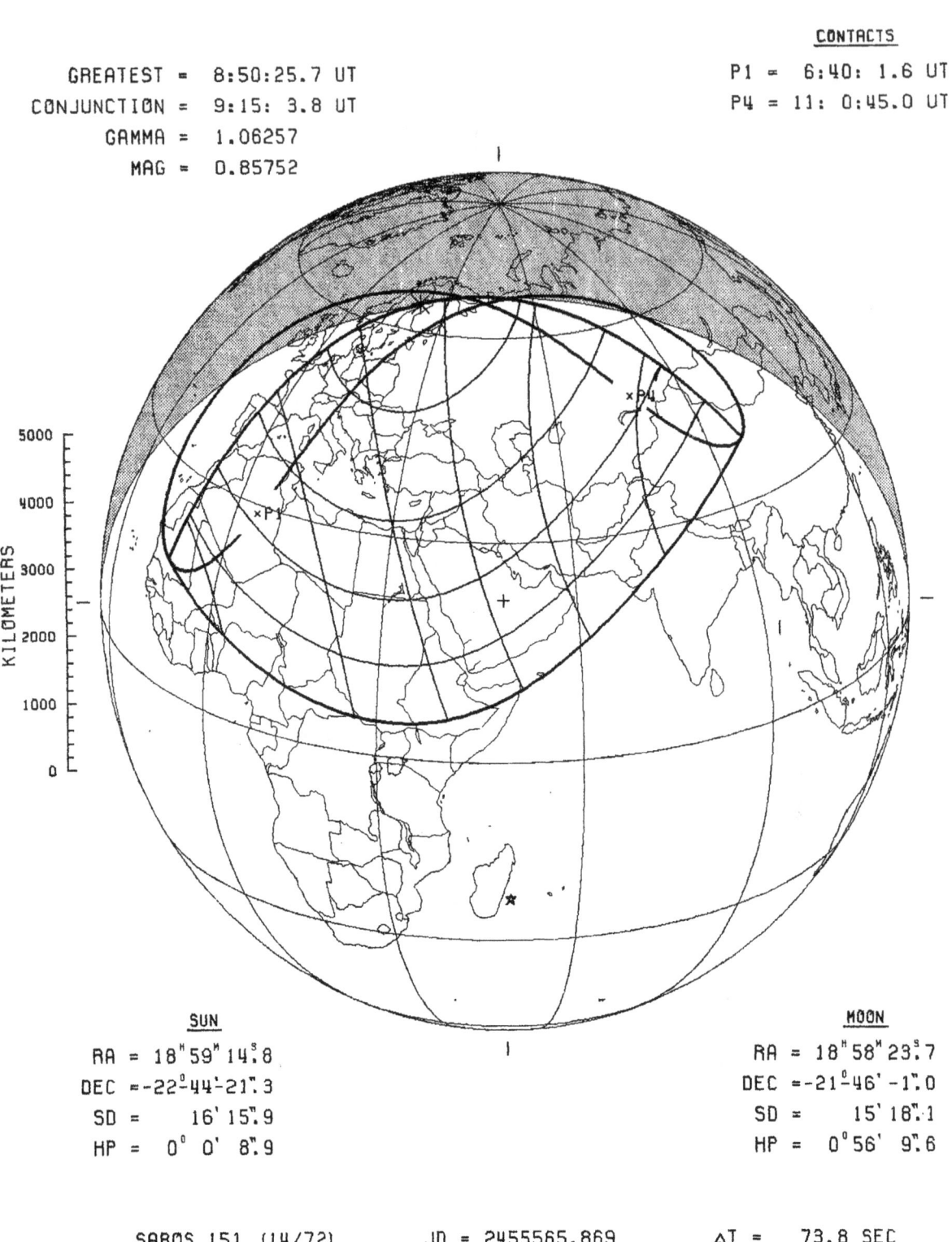

KILOMETERS

5000
4000
3000
2000
1000
0

SUN

RA = $18^h59^m14\overset{s}{.}8$

DEC = $-22°44'21".3$

SD = $16'15".9$

HP = $0°0'8".9$

MOON

RA = $18^h58^m23\overset{s}{.}7$

DEC = $-21°46'-1".0$

SD = $15'18".1$

HP = $0°56'9".6$

SAROS 151 (14/72)   JD = 2455565.869   ΔT = 73.8 SEC

Figure 84

196

PARTIAL SOLAR ECLIPSE - 1 JUN 2011

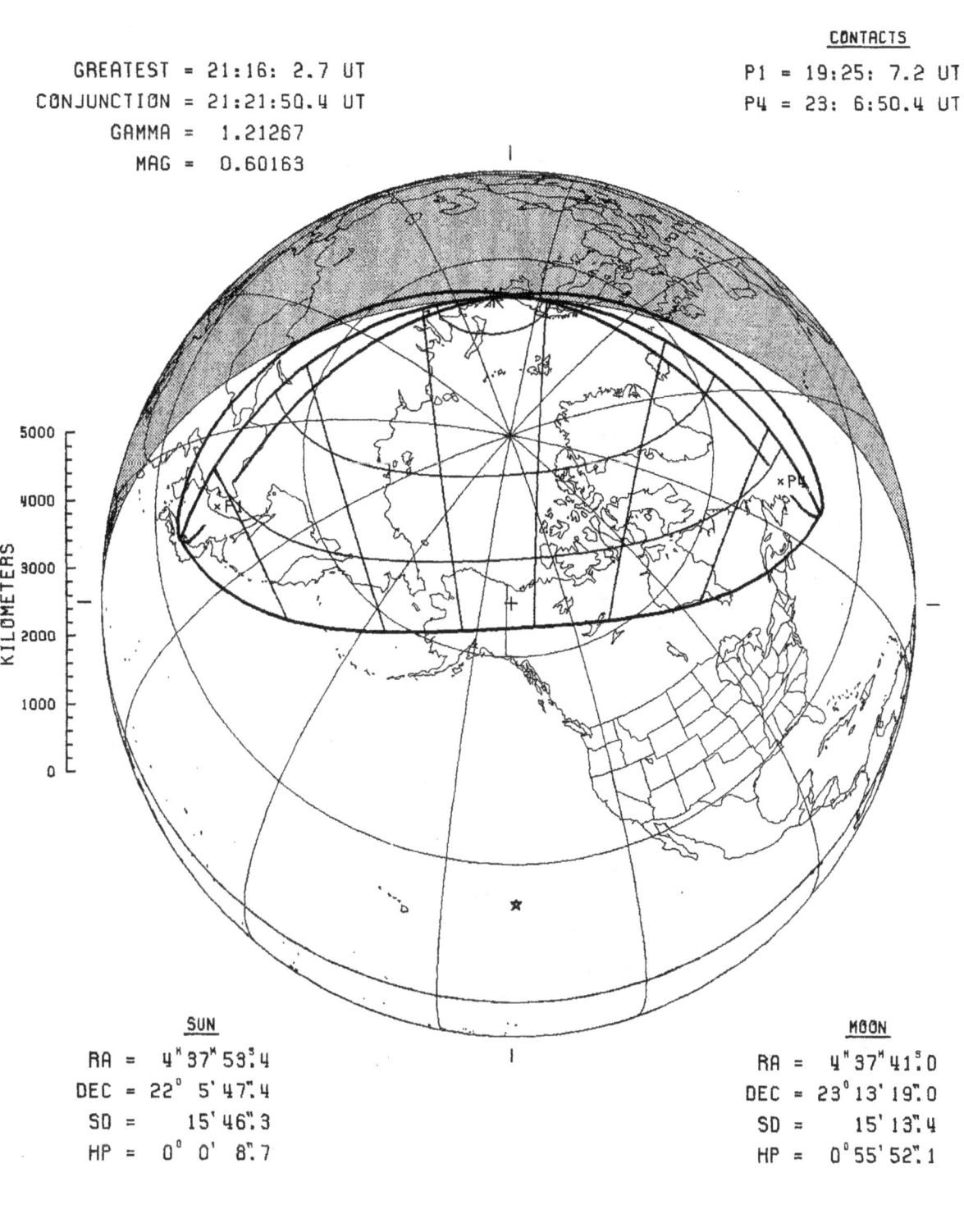

GREATEST = 21:16: 2.7 UT
CONJUNCTION = 21:21:50.4 UT
GAMMA = 1.21267
MAG = 0.60163

P1 = 19:25: 7.2 UT
P4 = 23: 6:50.4 UT

KILOMETERS

5000
4000
3000
2000
1000
0

SUN
RA = 4ʰ37ᴹ53ˢ4
DEC = 22° 5' 47".4
SD = 15' 46".3
HP = 0° 0' 8".7

MOON
RA = 4ʰ37ᴹ41ˢ0
DEC = 23°13' 19".0
SD = 15' 13".4
HP = 0°55' 52".1

SAROS 118 (68/72)   JD = 2455714.387   ΔT =   74.1 SEC

Figure 85

197

PARTIAL SOLAR ECLIPSE - 1 JUL 2011

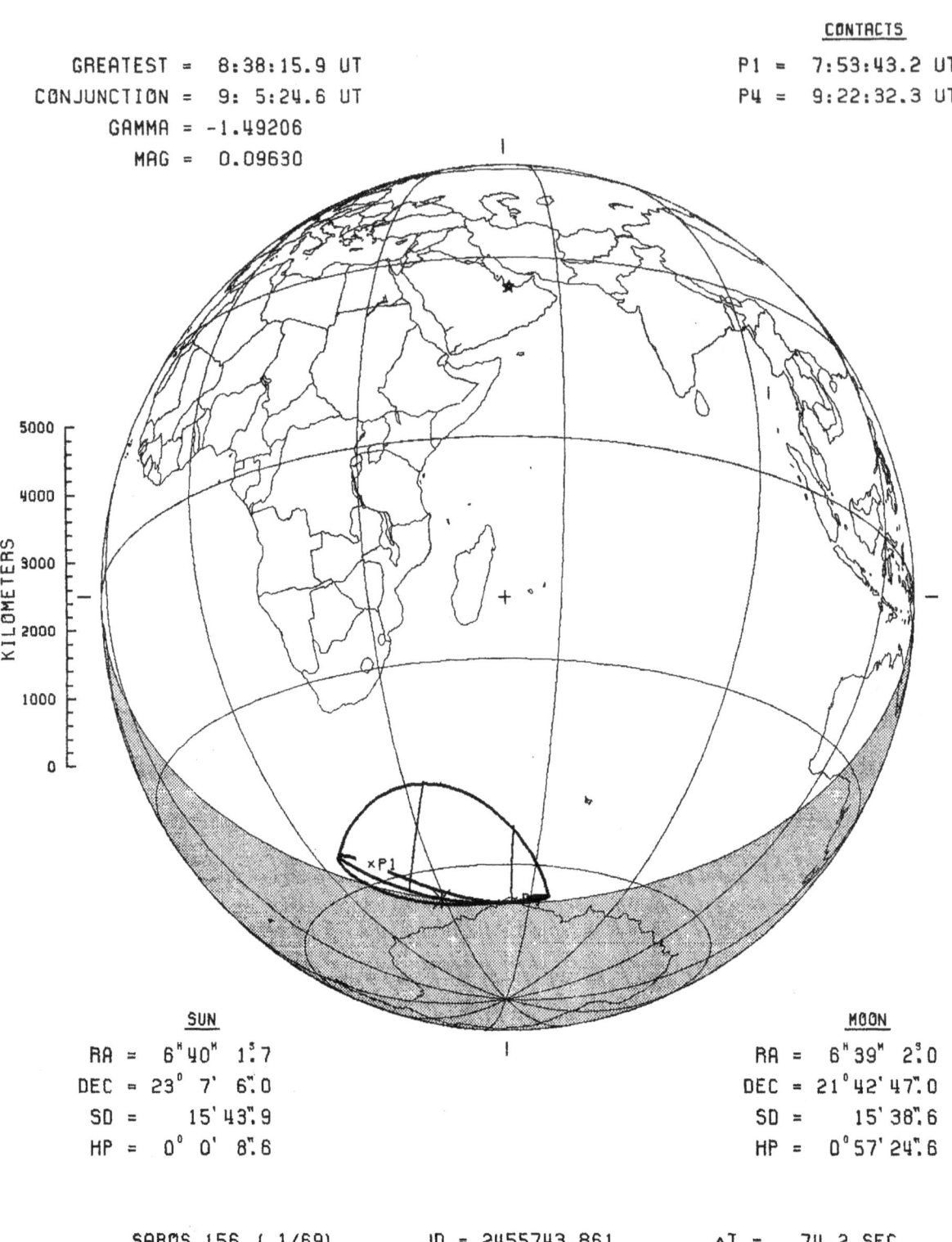

GREATEST = 8:38:15.9 UT
CONJUNCTION = 9: 5:24.6 UT
GAMMA = -1.49206
MAG = 0.09630

<u>CONTACTS</u>
P1 = 7:53:43.2 UT
P4 = 9:22:32.3 UT

KILOMETERS

5000
4000
3000
2000
1000
0

×P1

<u>SUN</u>
RA = 6ʰ40ᵐ 1ˢ7
DEC = 23° 7' 6".0
SD = 15' 43".9
HP = 0° 0' 8".6

<u>MOON</u>
RA = 6ʰ39ᵐ 2ˢ0
DEC = 21° 42' 47".0
SD = 15' 38".6
HP = 0° 57' 24".6

SAROS 156 (1/69)   JD = 2455743.861   ΔT =   74.2 SEC

Figure 86

PARTIAL SOLAR ECLIPSE — 25 NOV 2011

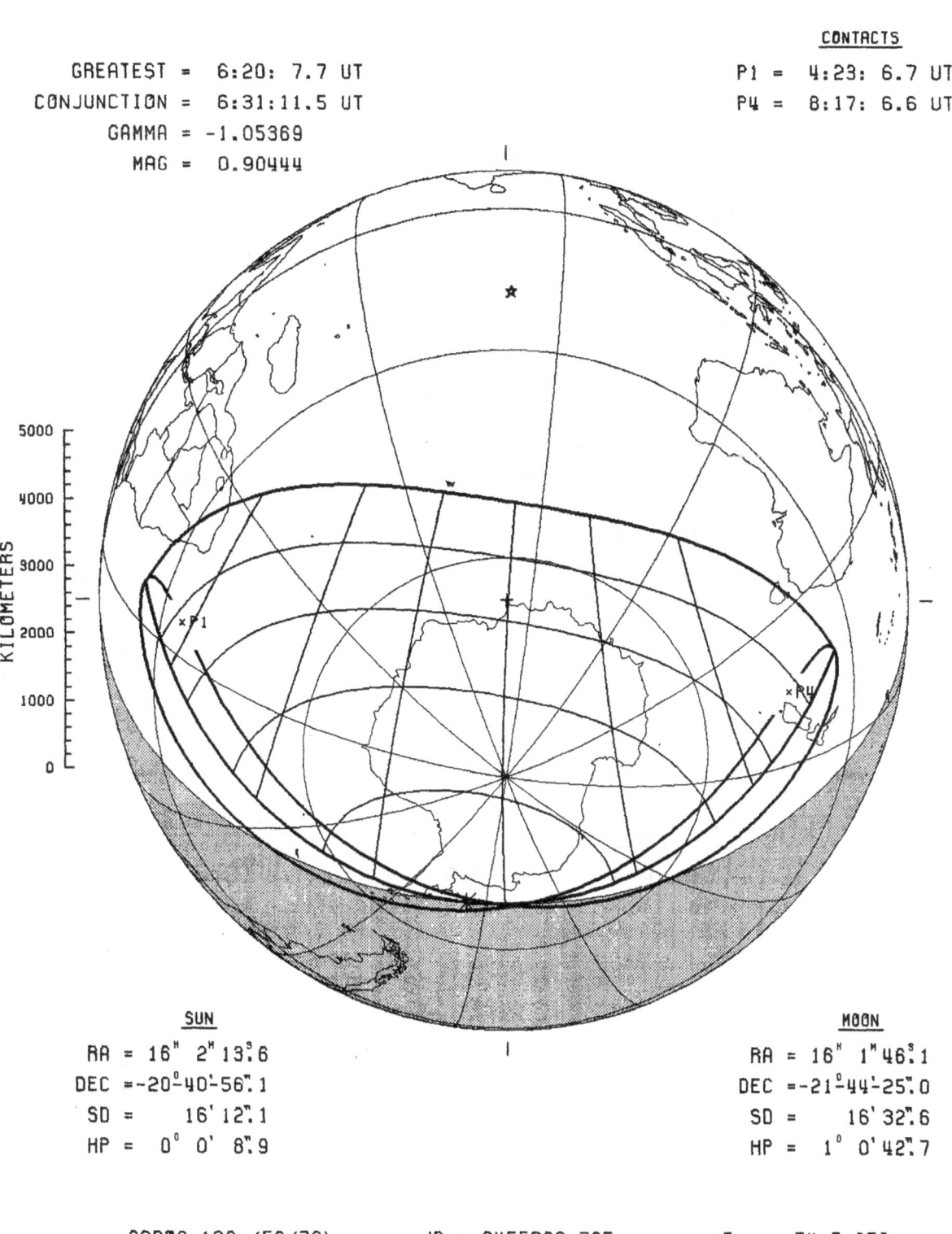

GREATEST = 6:20: 7.7 UT
CONJUNCTION = 6:31:11.5 UT
GAMMA = -1.05369
MAG = 0.90444

CONTACTS

P1 = 4:23: 6.7 UT
P4 = 8:17: 6.6 UT

KILOMETERS
5000
4000
3000
2000
1000
0

×P1

×P4

SUN
RA = 16ʰ 2ᵐ 13ˢ.6
DEC = -20°40'-56".1
SD = 16'12".1
HP = 0° 0' 8".9

MOON
RA = 16ʰ 1ᵐ 46ˢ.1
DEC = -21°44'-25".0
SD = 16'32".6
HP = 1° 0'42".7

SAROS 123 (53/70)   JD = 2455890.765   ΔT = 74.5 SEC

Figure 87

199

ANNULAR SOLAR ECLIPSE – 20 MAY 2012

GREATEST = 23:52:38.2 UT
CONJUNCTION = 0: 0:×××× UT
GAMMA = 0.48252
RATIO = 0.94389

CONTACTS

P1 = 20:55:58.2 UT
U1 = 22: 6: 7.5 UT
U4 = 1:39: 3.4 UT
P4 = 2:49:13.5 UT

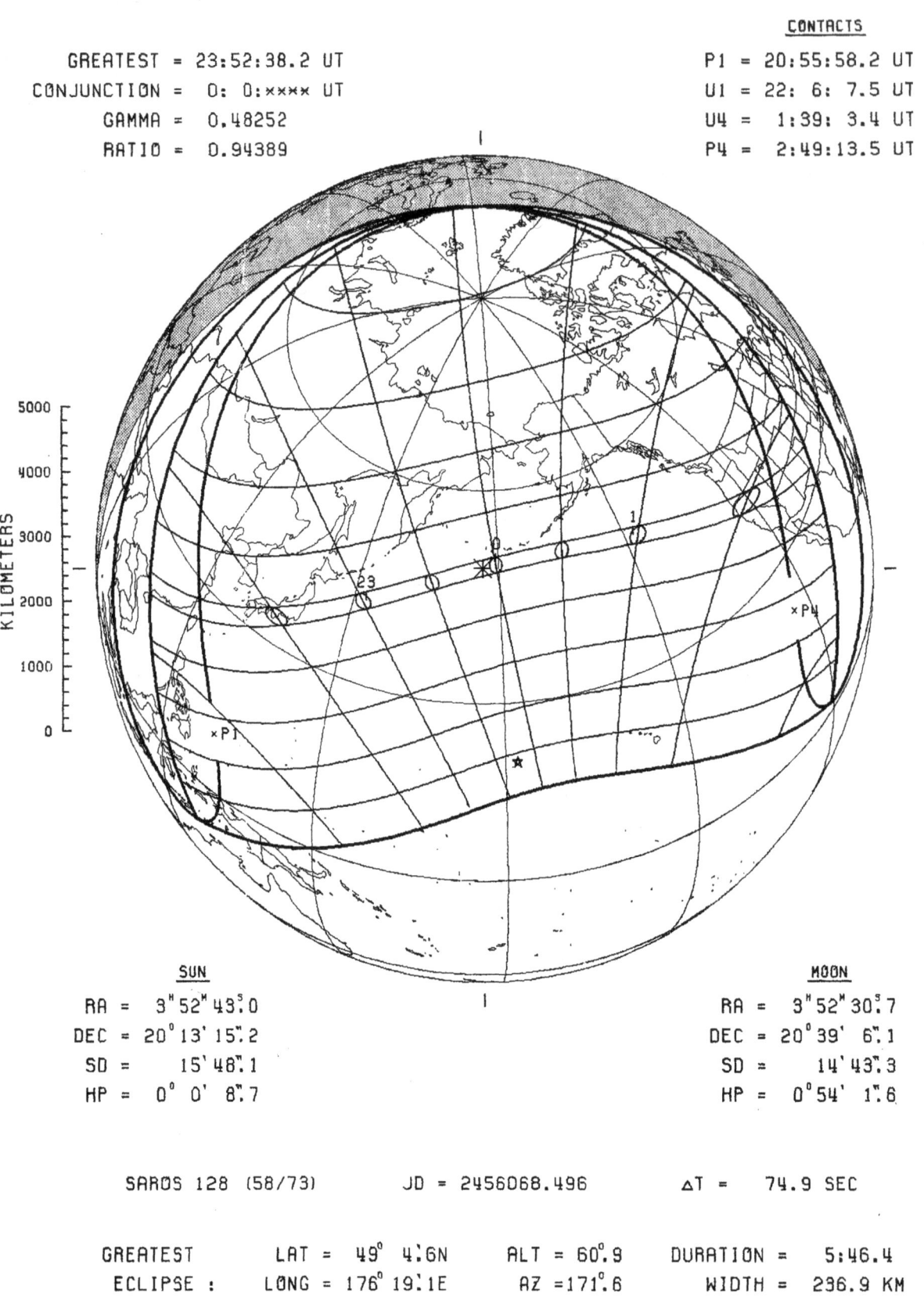

KILOMETERS

5000
4000
3000
2000
1000
0

SUN

RA = 3ʰ52ᴹ43ˢ0
DEC = 20°13' 15".2
SD = 15' 48".1
HP = 0° 0' 8".7

MOON

RA = 3ʰ52ᴹ30ˢ7
DEC = 20°39' 6".1
SD = 14' 43".3
HP = 0°54' 1".6

SAROS 128 (58/73)   JD = 2456068.496   ΔT =   74.9 SEC

GREATEST   LAT = 49° 4.6N   ALT = 60°.9   DURATION =   5:46.4
ECLIPSE :   LONG = 176° 19.1E   AZ = 171°.6   WIDTH =   236.9 KM

Figure 88
200

TOTAL SOLAR ECLIPSE - 13 NOV 2012

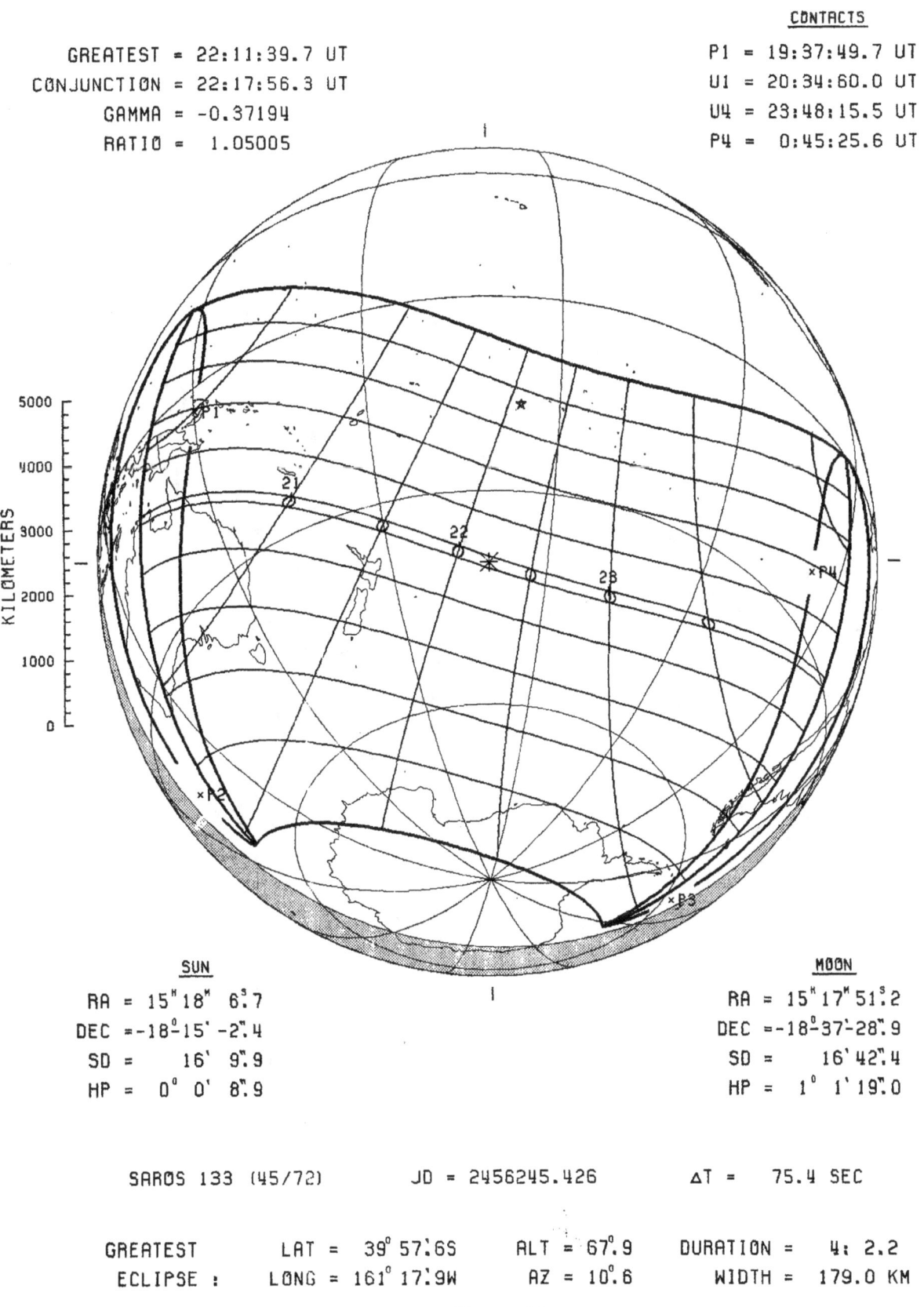

GREATEST = 22:11:39.7 UT
CONJUNCTION = 22:17:56.3 UT
GAMMA = -0.37194
RATIO = 1.05005

P1 = 19:37:49.7 UT
U1 = 20:34:60.0 UT
U4 = 23:48:15.5 UT
P4 = 0:45:25.6 UT

KILOMETERS

5000
4000
3000
2000
1000
0

×P2

×P3

×P4

SUN

RA = $15^h 18^m 6^s.7$
DEC = $-18^o 15' -2".4$
SD = $16' 9".9$
HP = $0^o 0' 8".9$

MOON

RA = $15^h 17^m 51^s.2$
DEC = $-18^o 37' -28".9$
SD = $16' 42".4$
HP = $1^o 1' 19".0$

SAROS 133 (45/72)   JD = 2456245.426   ΔT = 75.4 SEC

GREATEST   LAT = $39^o 57".6S$ ALT = $67^o.9$ DURATION = 4: 2.2
ECLIPSE :   LONG = $161^o 17".9W$ AZ = $10^o.6$ WIDTH = 179.0 KM

Figure 89
201

ANNULAR SOLAR ECLIPSE — 10 MAY 2013

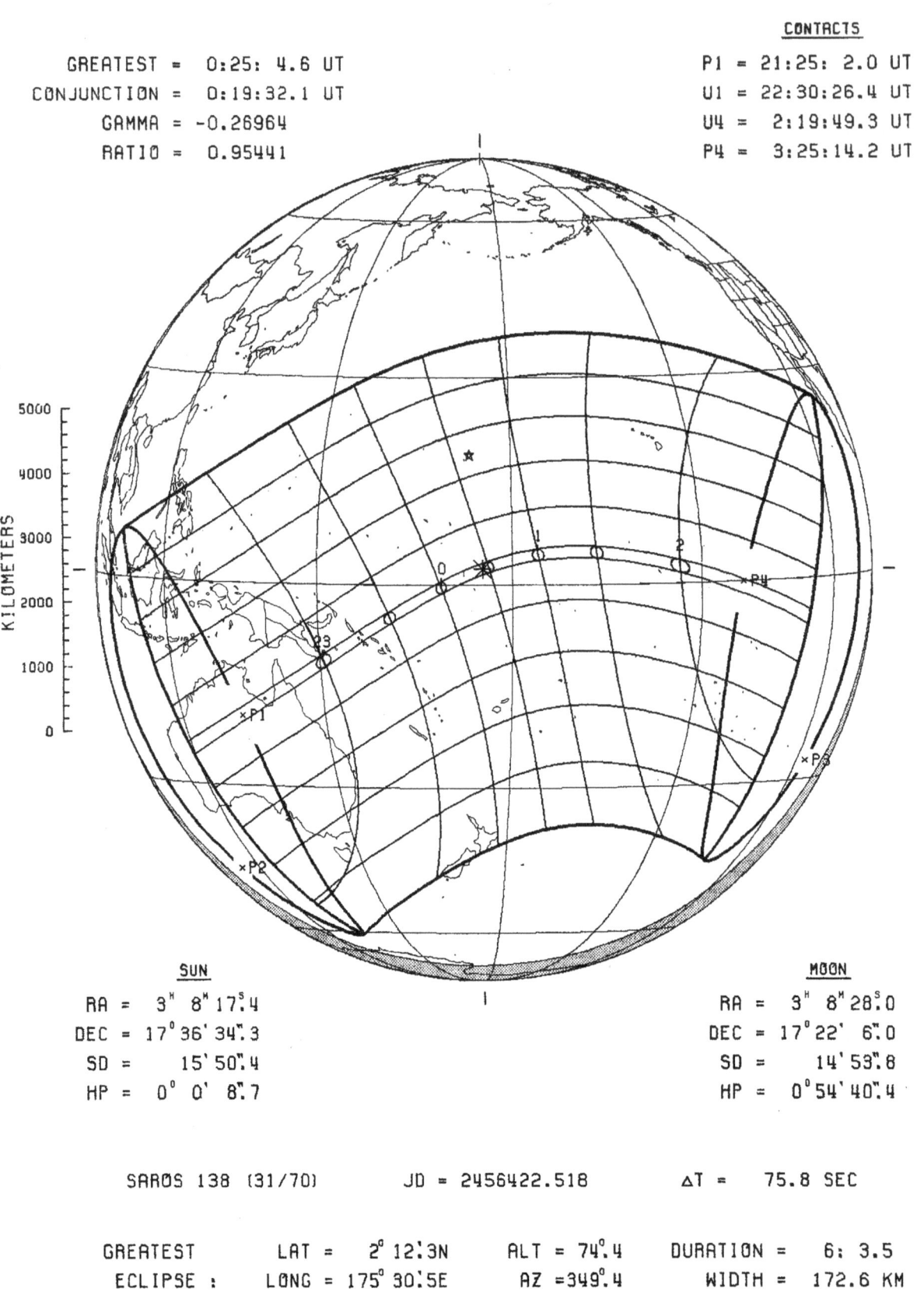

GREATEST = 0:25: 4.6 UT
CONJUNCTION = 0:19:32.1 UT
GAMMA = -0.26964
RATIO = 0.95441

CONTACTS

P1 = 21:25: 2.0 UT
U1 = 22:30:26.4 UT
U4 = 2:19:49.3 UT
P4 = 3:25:14.2 UT

KILOMETERS

SUN

RA = 3ʰ 8ᴹ 17ˢ.4
DEC = 17° 36' 34".3
SD = 15' 50".4
HP = 0° 0' 8".7

MOON

RA = 3ʰ 8ᴹ 28ˢ.0
DEC = 17° 22' 6".0
SD = 14' 53".8
HP = 0° 54' 40".4

SAROS 138 (31/70)   JD = 2456422.518   ΔT =   75.8 SEC

GREATEST   LAT = 2° 12'.3N   ALT = 74°.4   DURATION =   6: 3.5
ECLIPSE :   LONG = 175° 30'.5E   AZ =349°.4   WIDTH = 172.6 KM

Figure 90
202

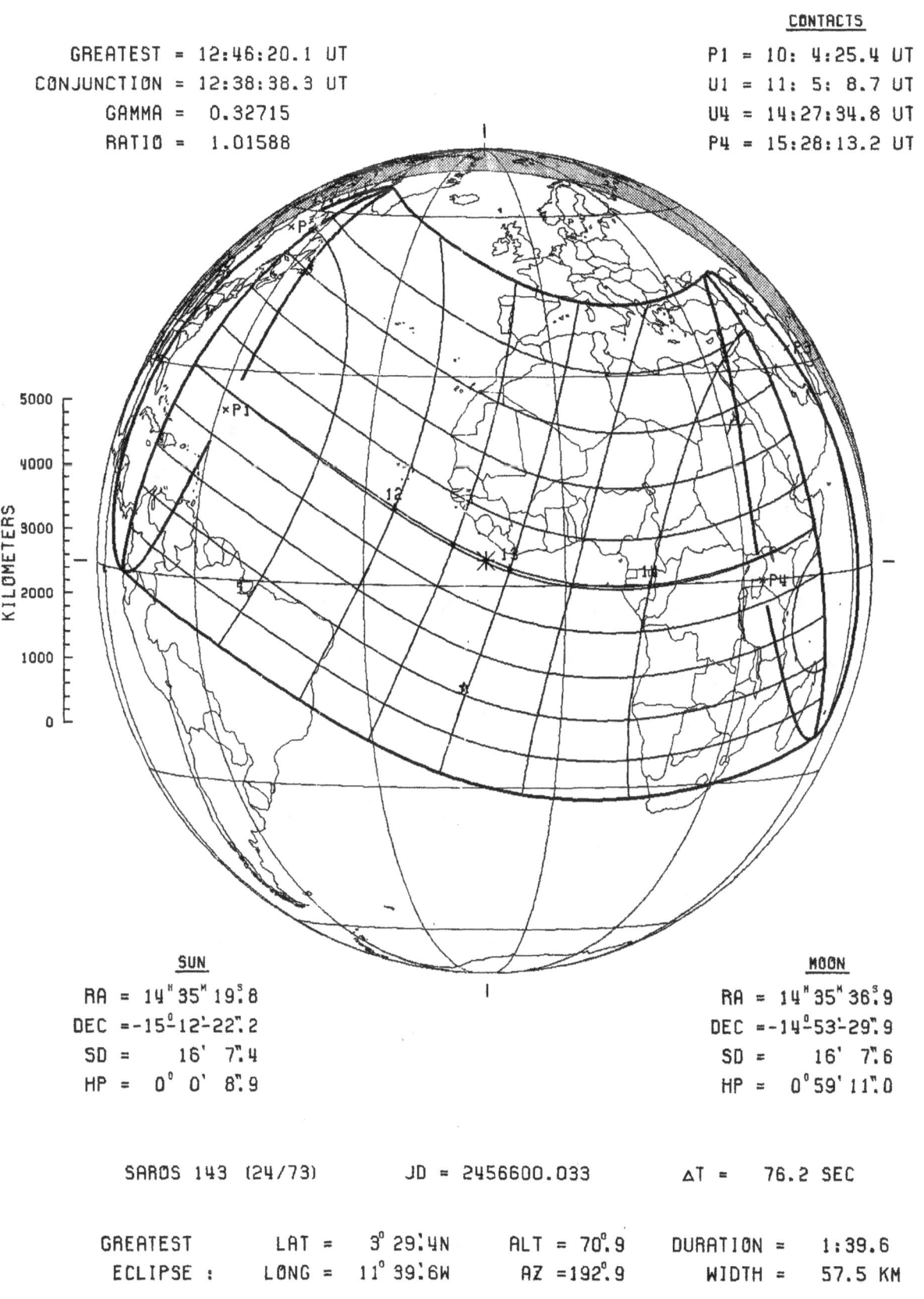

GREATEST = 12:46:20.1 UT
CONJUNCTION = 12:38:38.3 UT
GAMMA = 0.32715
RATIO = 1.01588

CONTACTS

P1 = 10: 4:25.4 UT
U1 = 11: 5: 8.7 UT
U4 = 14:27:34.8 UT
P4 = 15:28:13.2 UT

SUN

RA = 14ʰ35ᵐ19ˢ.8
DEC = -15°12'22".2
SD = 16' 7".4
HP = 0° 0' 8".9

MOON

RA = 14ʰ35ᵐ36ˢ.9
DEC = -14°53'29".9
SD = 16' 7".6
HP = 0°59'11".0

SAROS 143 (24/73)   JD = 2456600.033   ΔT = 76.2 SEC

GREATEST   LAT = 3° 29'.4N   ALT = 70°.9   DURATION = 1:39.6
ECLIPSE :   LONG = 11° 39'.6W   AZ = 192°.9   WIDTH = 57.5 KM

Figure 91
203

ANNULAR SOLAR ECLIPSE — 29 APR 2014

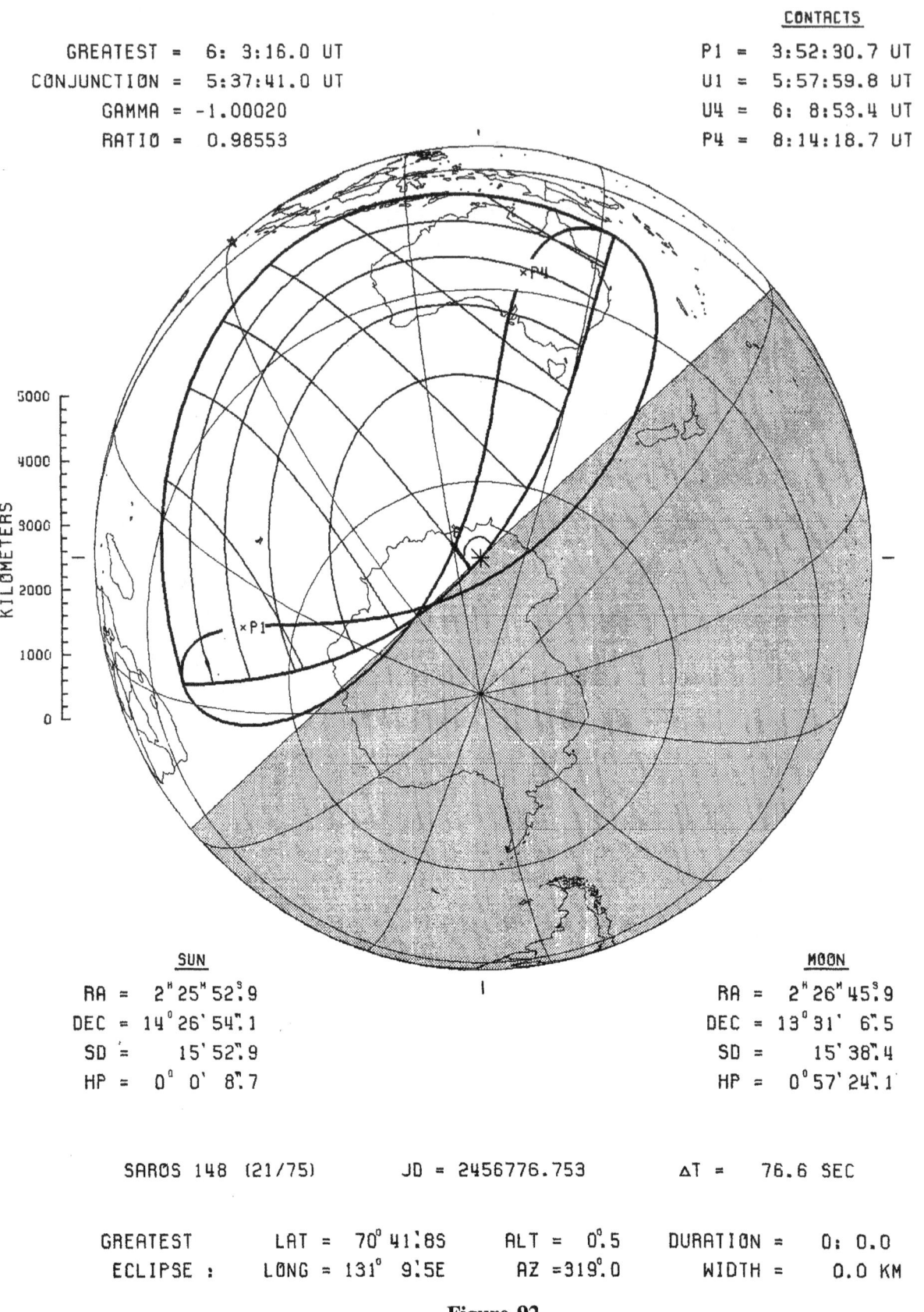

Figure 92
204

PARTIAL SOLAR ECLIPSE   -   23 OCT 2014

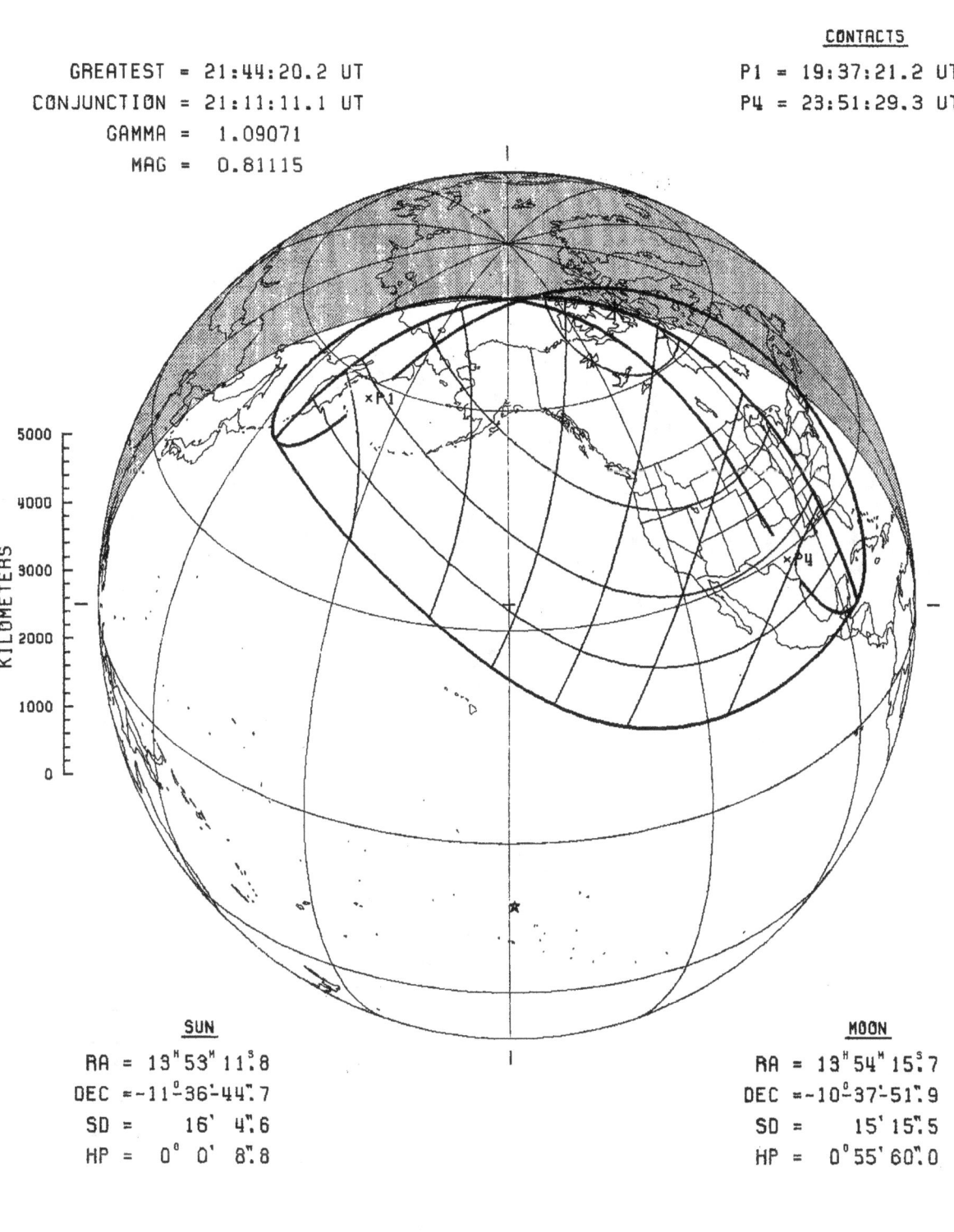

Figure 93

TOTAL SOLAR ECLIPSE — 20 MAR 2015

GREATEST = 9:45:29.5 UT
CONJUNCTION = 10:16:56.5 UT
GAMMA = 0.94524
RATIO = 1.04455

CONTACTS
P1 = 7:40:41.8 UT
U1 = 9: 9:20.9 UT
U4 = 10:21:14.6 UT
P4 = 11:50: 3.5 UT

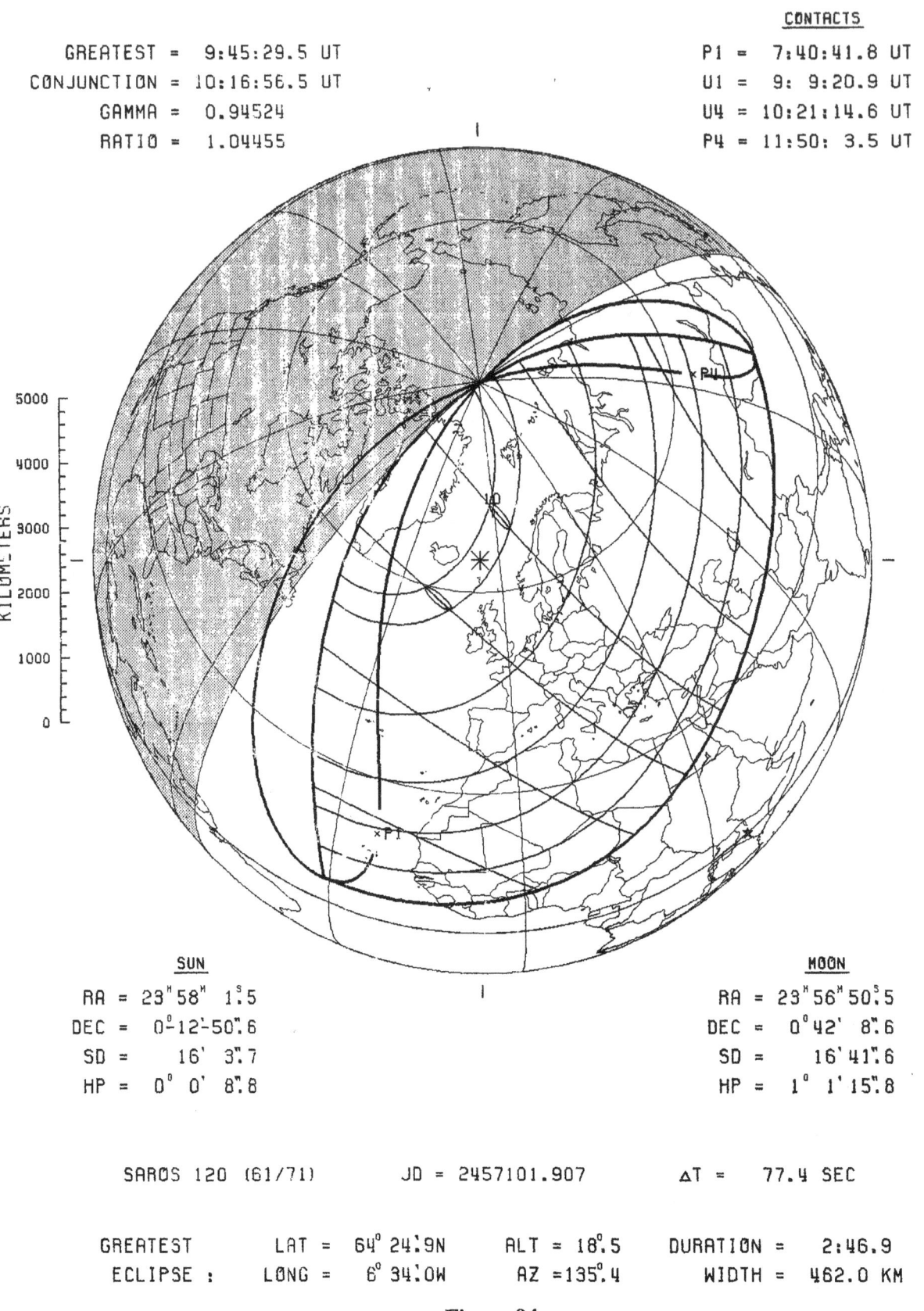

SUN
RA = $23^h58^m 1\overset{s}{.}5$
DEC = $0°-12'-50".6$
SD = $16' 3".7$
HP = $0° 0' 8".8$

MOON
RA = $23^h56^m 50\overset{s}{.}5$
DEC = $0° 42' 8".6$
SD = $16' 41".6$
HP = $1° 1' 15".8$

SAROS 120 (61/71) JD = 2457101.907 ΔT = 77.4 SEC

GREATEST   LAT = 64° 24'.9N   ALT = 18°.5   DURATION = 2:46.9
ECLIPSE :   LONG = 6° 34'.0W   AZ = 135°.4   WIDTH = 462.0 KM

Figure 94
206

PARTIAL SOLAR ECLIPSE - 13 SEP 2015

GREATEST = 6:53:58.8 UT
CONJUNCTION = 7:35: 6.8 UT
GAMMA = -1.10052
MAG = 0.78675

P1 = 4:41:28.8 UT
P4 = 9: 6:11.1 UT

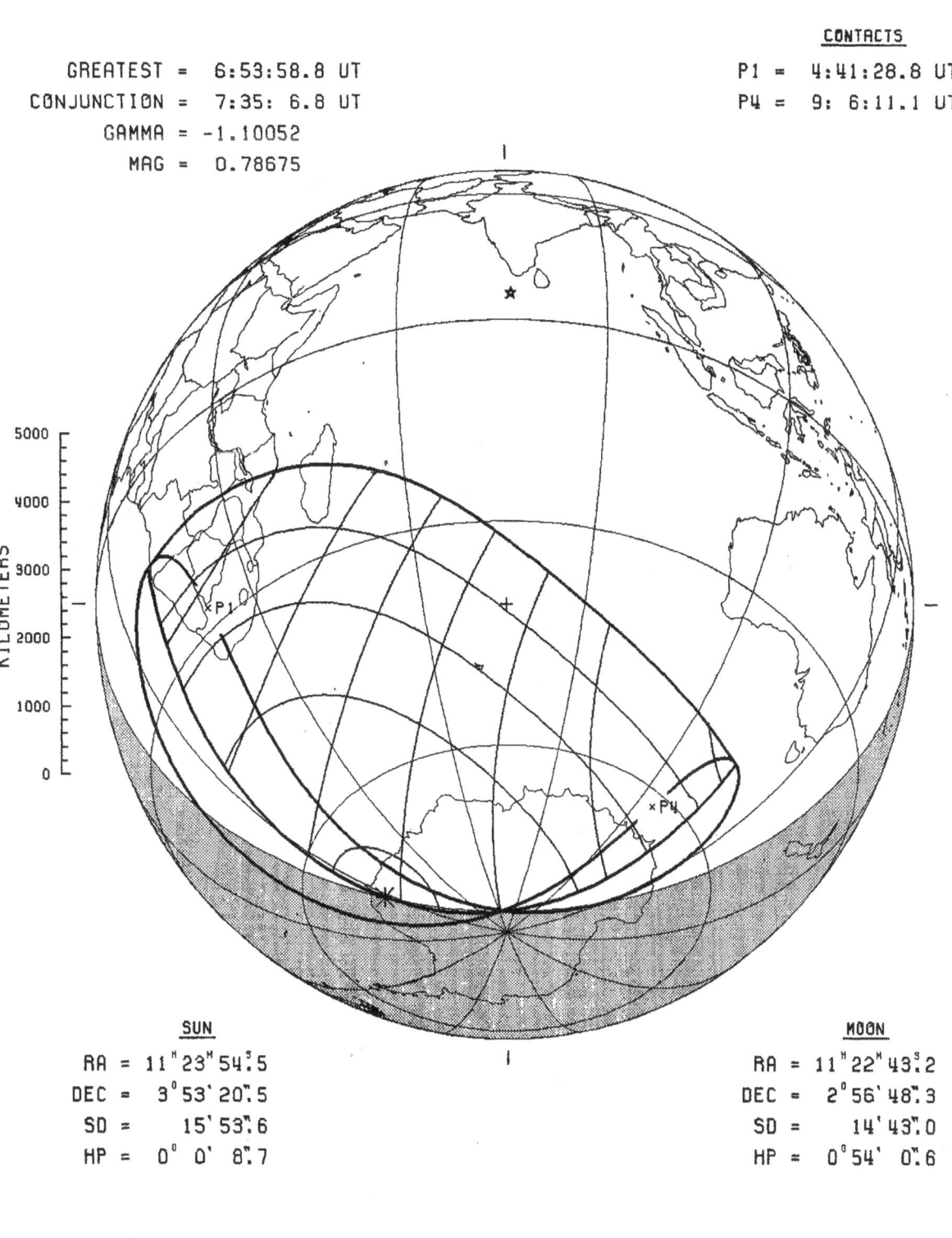

5000
4000
3000
2000
1000
0

KILOMETERS

SUN
RA = 11ʰ23ᵐ54ˢ.5
DEC = 3°53'20".5
SD = 15'53".6
HP = 0° 0' 8".7

MOON
RA = 11ʰ22ᵐ43ˢ.2
DEC = 2°56'48".3
SD = 14'43".0
HP = 0°54' 0".6

SAROS 125 (54/73)   JD = 2457278.788   ΔT =   77.8 SEC

Figure 95

207

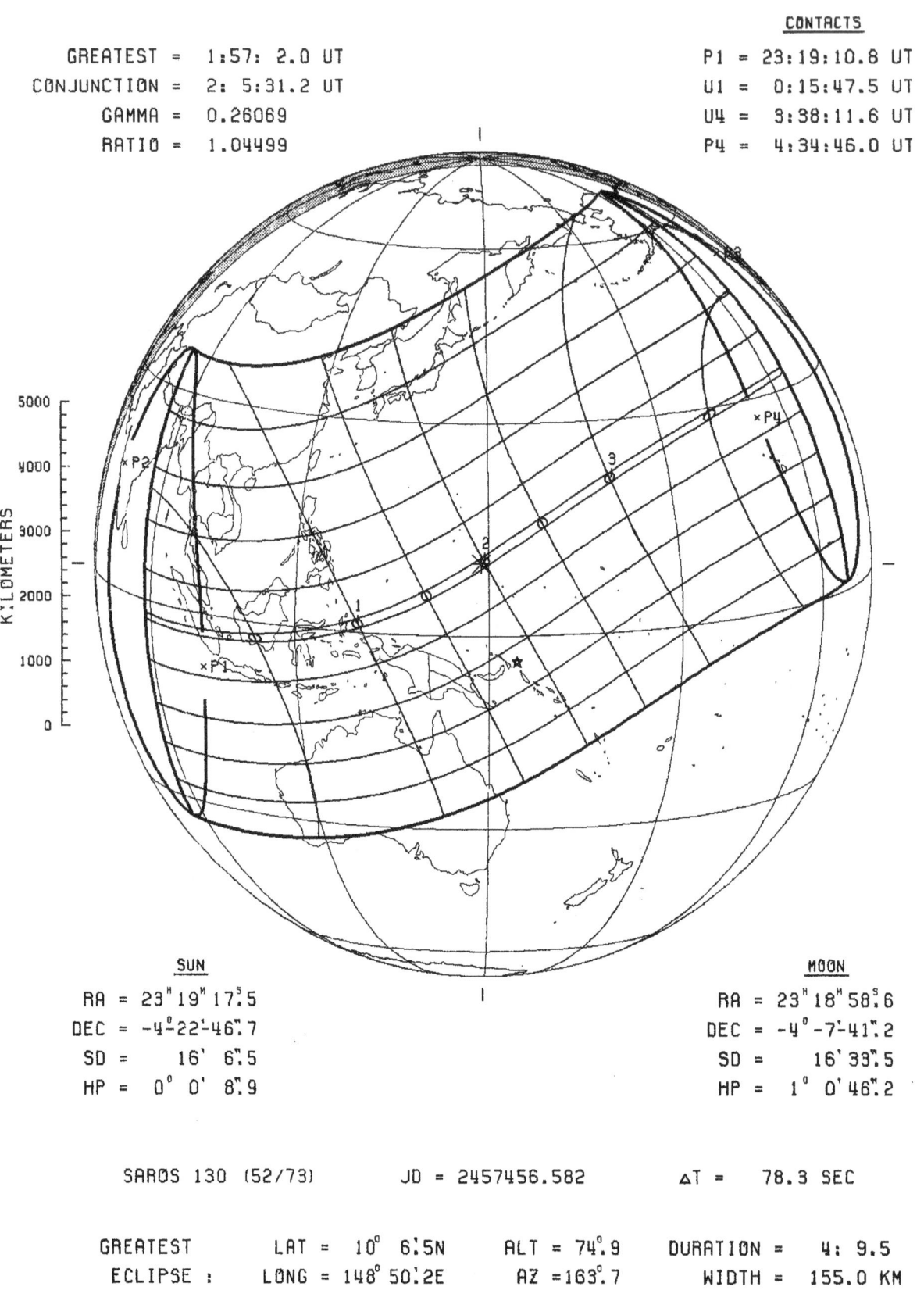

TOTAL SOLAR ECLIPSE – 9 MAR 2016

GREATEST = 1:57: 2.0 UT
CONJUNCTION = 2: 5:31.2 UT
GAMMA = 0.26069
RATIO = 1.04499

P1 = 23:19:10.8 UT
U1 = 0:15:47.5 UT
U4 = 3:38:11.6 UT
P4 = 4:34:46.0 UT

KILOMETERS
5000
4000
3000
2000
1000
0

×P2
×P3

SUN

RA = $23^H 19^M 17^S.5$
DEC = $-4°22'46".7$
SD = $16' 6".5$
HP = $0° 0' 8".9$

MOON

RA = $23^H 18^M 58^S.6$
DEC = $-4°-7'-41".2$
SD = $16' 33".5$
HP = $1° 0' 46".2$

SAROS 130 (52/73)   JD = 2457456.582   ΔT = 78.3 SEC

GREATEST   LAT = $10° 6'.5N$ ALT = $74°.9$ DURATION = 4: 9.5
ECLIPSE :   LONG = $148° 50'.2E$ AZ = $163°.7$ WIDTH = 155.0 KM

Figure 96
208

ANNULAR SOLAR ECLIPSE — 1 SEP 2016

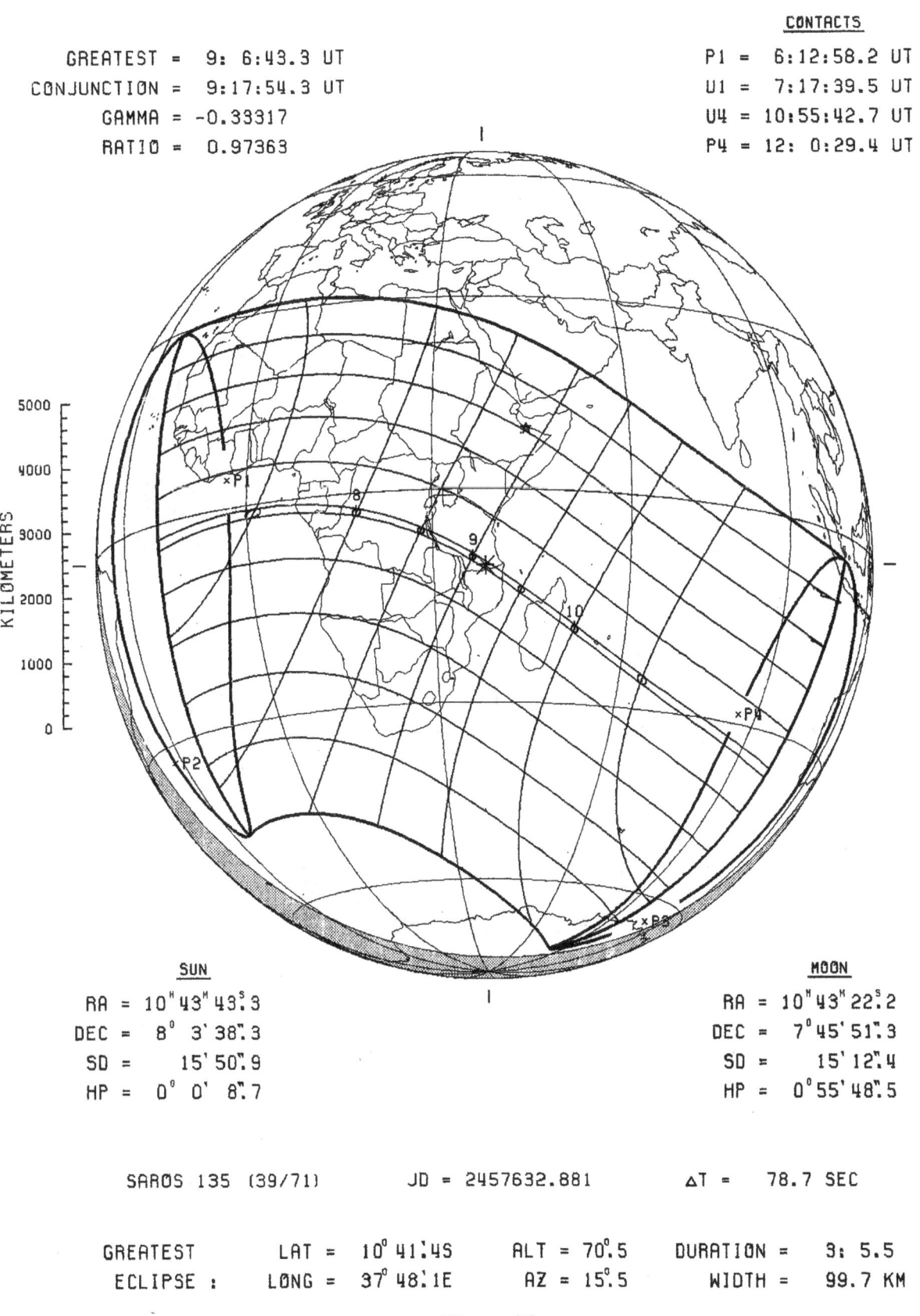

GREATEST = 9: 6:43.3 UT
CONJUNCTION = 9:17:54.3 UT
GAMMA = -0.33317
RATIO = 0.97363

CONTACTS

P1 = 6:12:58.2 UT
U1 = 7:17:39.5 UT
U4 = 10:55:42.7 UT
P4 = 12: 0:29.4 UT

KILOMETERS

5000
4000
3000
2000
1000
0

SUN

RA = $10^h 43^m 43^s.3$
DEC = $8° 3' 38".3$
SD = $15' 50".9$
HP = $0° 0' 8".7$

MOON

RA = $10^h 43^m 22^s.2$
DEC = $7° 45' 51".3$
SD = $15' 12".4$
HP = $0° 55' 48".5$

SAROS 135 (39/71)   JD = 2457632.881   ΔT =   78.7 SEC

GREATEST   LAT = $10° 41'.45$ ALT = $70°.5$ DURATION =   3: 5.5
ECLIPSE :   LONG = $37° 48'.1E$ AZ = $15°.5$ WIDTH =   99.7 KM

Figure 97
209

ANNULAR SOLAR ECLIPSE — 26 FEB 2017

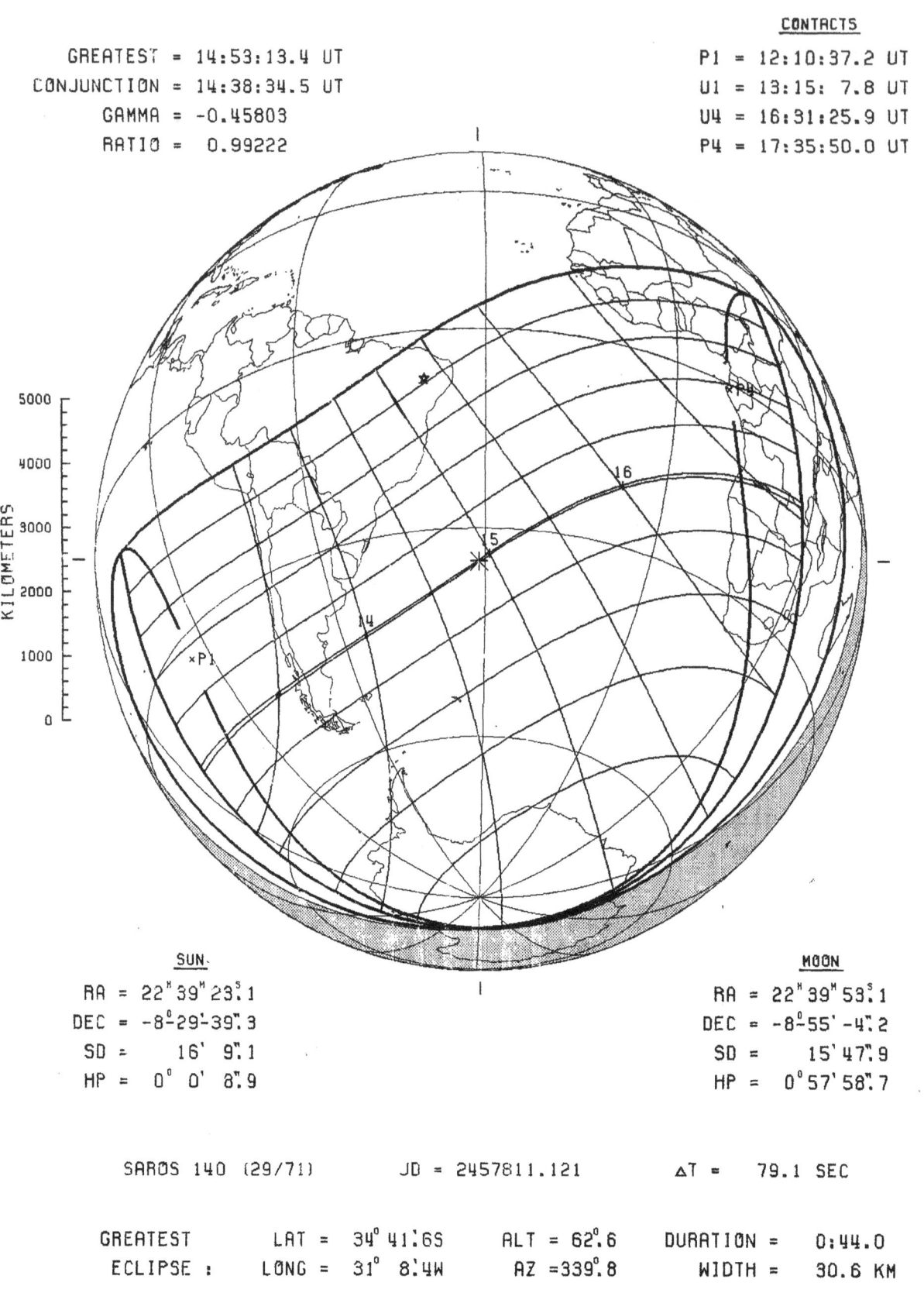

GREATEST = 14:53:13.4 UT
CONJUNCTION = 14:38:34.5 UT
GAMMA = -0.45803
RATIO = 0.99222

CONTACTS
P1 = 12:10:37.2 UT
U1 = 13:15: 7.8 UT
U4 = 16:31:25.9 UT
P4 = 17:35:50.0 UT

KILOMETERS
5000
4000
3000
2000
1000
0

SUN
RA = $22^h 39^m 23^s.1$
DEC = $-8^{\circ} 29' -39".3$
SD = $16' 9".1$
HP = $0^{\circ} 0' 8".9$

MOON
RA = $22^h 39^m 53^s.1$
DEC = $-8^{\circ} 55' -4".2$
SD = $15' 47".9$
HP = $0^{\circ} 57' 58".7$

SAROS 140 (29/71)   JD = 2457811.121   ΔT = 79.1 SEC

GREATEST   LAT = $34^{\circ} 41'.6S$ ALT = $62^{\circ}.6$ DURATION = 0:44.0
ECLIPSE :   LONG = $31^{\circ} 8'.4W$ AZ = $339^{\circ}.8$ WIDTH = 30.6 KM

Figure 98
210

TOTAL SOLAR ECLIPSE — 21 AUG 2017

GREATEST = 18:25:20.7 UT
CONJUNCTION = 18:13: 3.3 UT
GAMMA = 0.43670
RATIO = 1.03059

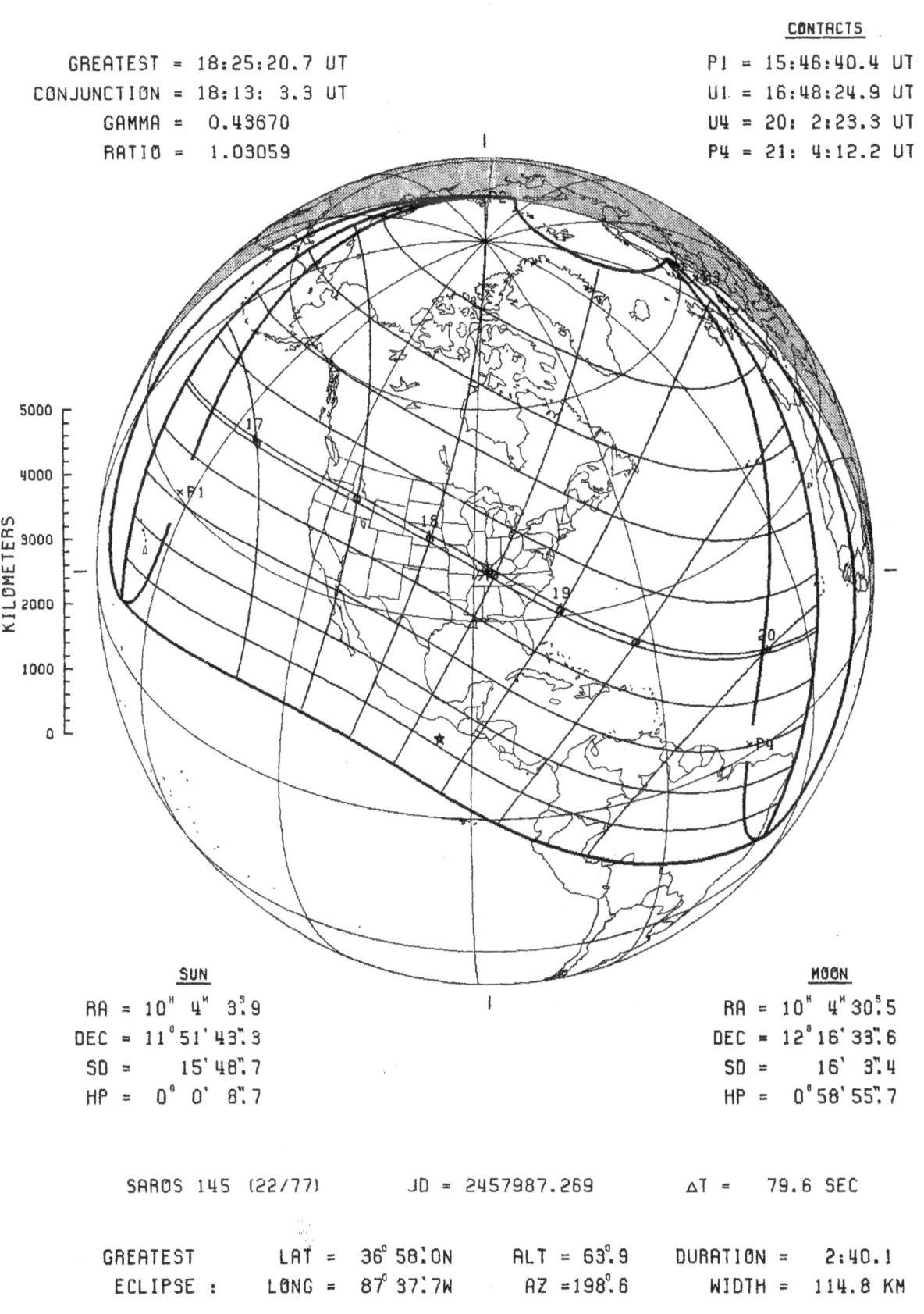

SUN

RA = 10^h 4^m $3^s.9$
DEC = $11^\circ 51' 43".3$
SD = $15' 48".7$
HP = $0^\circ 0' 8".7$

MOON

RA = 10^h 4^m $30^s.5$
DEC = $12^\circ 16' 33".6$
SD = $16' 3".4$
HP = $0^\circ 58' 55".7$

SAROS 145 (22/77)   JD = 2457987.269   ΔT = 79.6 SEC

GREATEST   LAT = $36^\circ 58'.0$N   ALT = $63^\circ.9$ DURATION = 2:40.1
ECLIPSE :   LONG = $87^\circ 37'.7$W   AZ = $198^\circ.6$ WIDTH = 114.8 KM

Figure 99
211

PARTIAL SOLAR ECLIPSE – 15 FEB 2018

GREATEST = 20:51:10.7 UT
CONJUNCTION = 20:14:54.0 UT
GAMMA = -1.21194
MAG = 0.59823

P1 = 18:55:39.8 UT
P4 = 22:46:53.9 UT

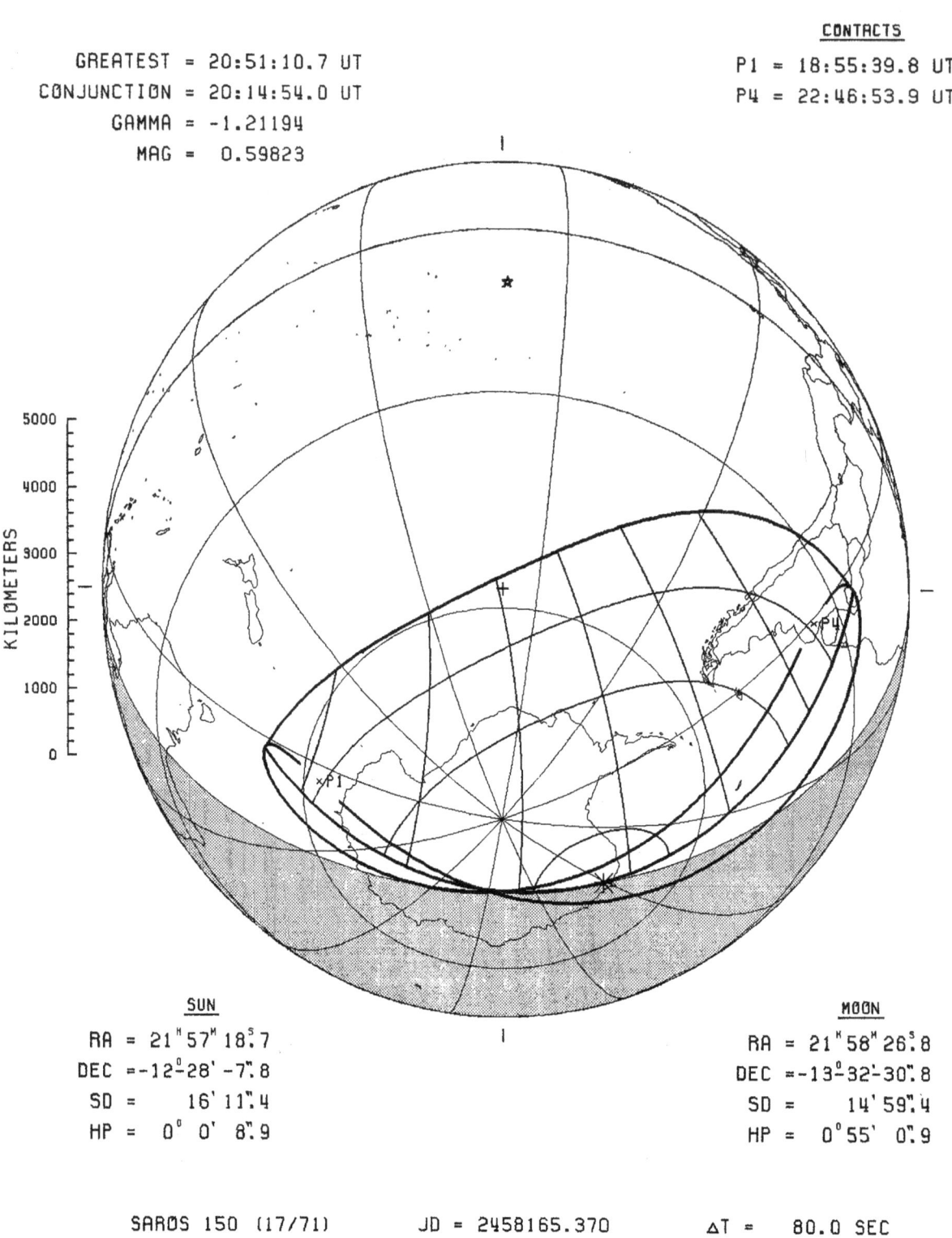

SUN
RA = 21^{h}57^{m}18^s.7
DEC = -12^{0}28'-7".8
SD = 16'11".4
HP = 0^{0}0'8".9

MOON
RA = 21^{h}58^{m}26^s.8
DEC = -13^{0}32'-30".8
SD = 14'59".4
HP = 0^{0}55'0".9

SAROS 150 (17/71)   JD = 2458165.370   ΔT = 80.0 SEC

Figure 100

212

PARTIAL SOLAR ECLIPSE - 13 JUL 2018

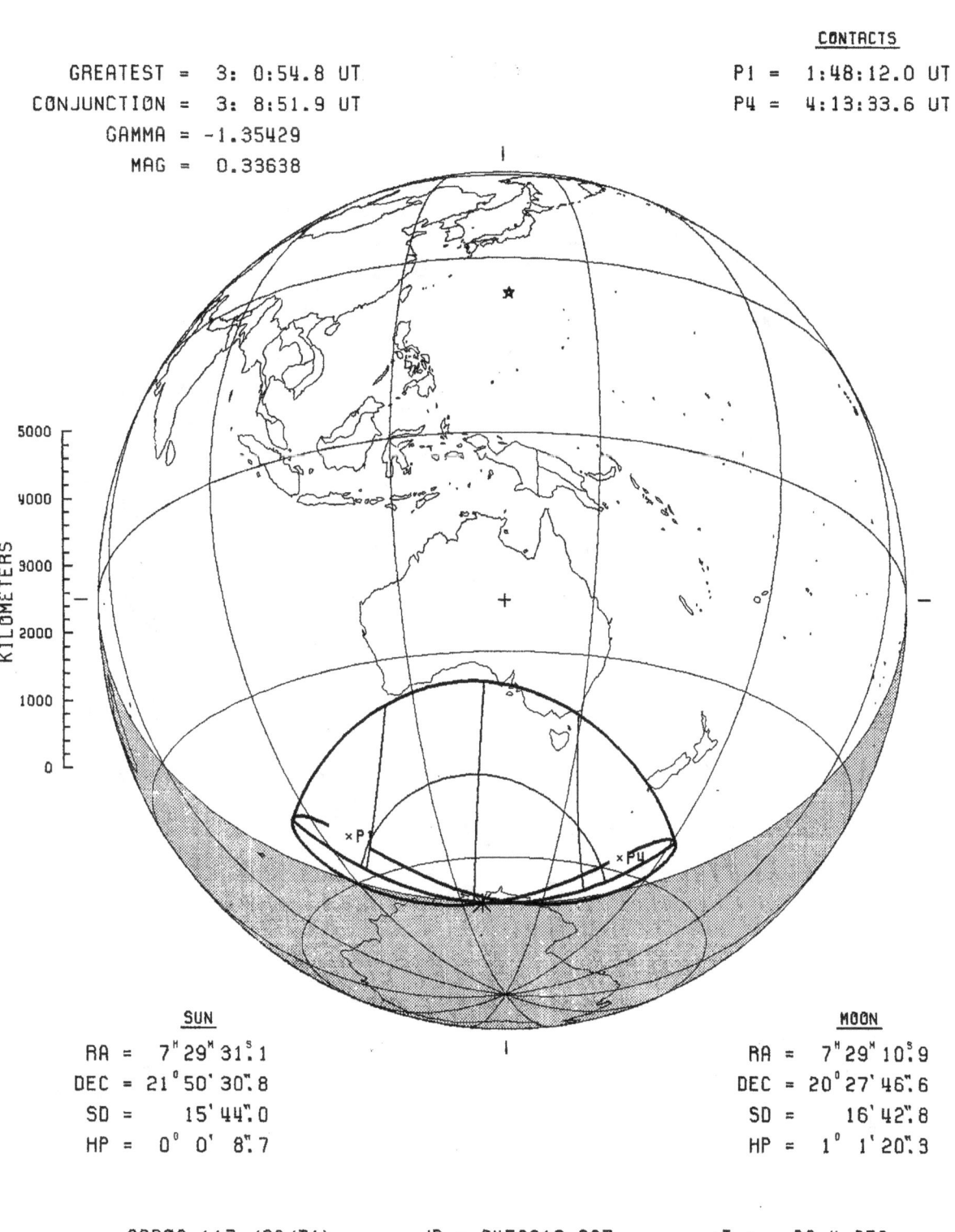

GREATEST = 3: 0:54.8 UT
CONJUNCTION = 3: 8:51.9 UT
GAMMA = -1.35429
MAG = 0.33638

CONTACTS
P1 = 1:48:12.0 UT
P4 = 4:13:33.6 UT

KILOMETERS

5000
4000
3000
2000
1000
0

×P1
×P4

SUN
RA = 7^{H}29^{M}31^{S}1
DEC = 21°50'30".8
SD = 15'44".0
HP = 0° 0' 8".7

MOON
RA = 7^{H}29^{M}10^{S}9
DEC = 20°27'46".6
SD = 16'42".8
HP = 1° 1'20".3

SAROS 117 (69/71)   JD = 2458312.627   ΔT = 80.4 SEC

Figure 101

213

PARTIAL SOLAR ECLIPSE –   11 AUG 2018

GREATEST = 9:46: 7.4 UT

CONJUNCTION = 9:19:52.3 UT

GAMMA = 1.14765

MAG = 0.73633

CONTACTS

P1 =   8: 1:56.4 UT

P4 = 11:30:32.2 UT

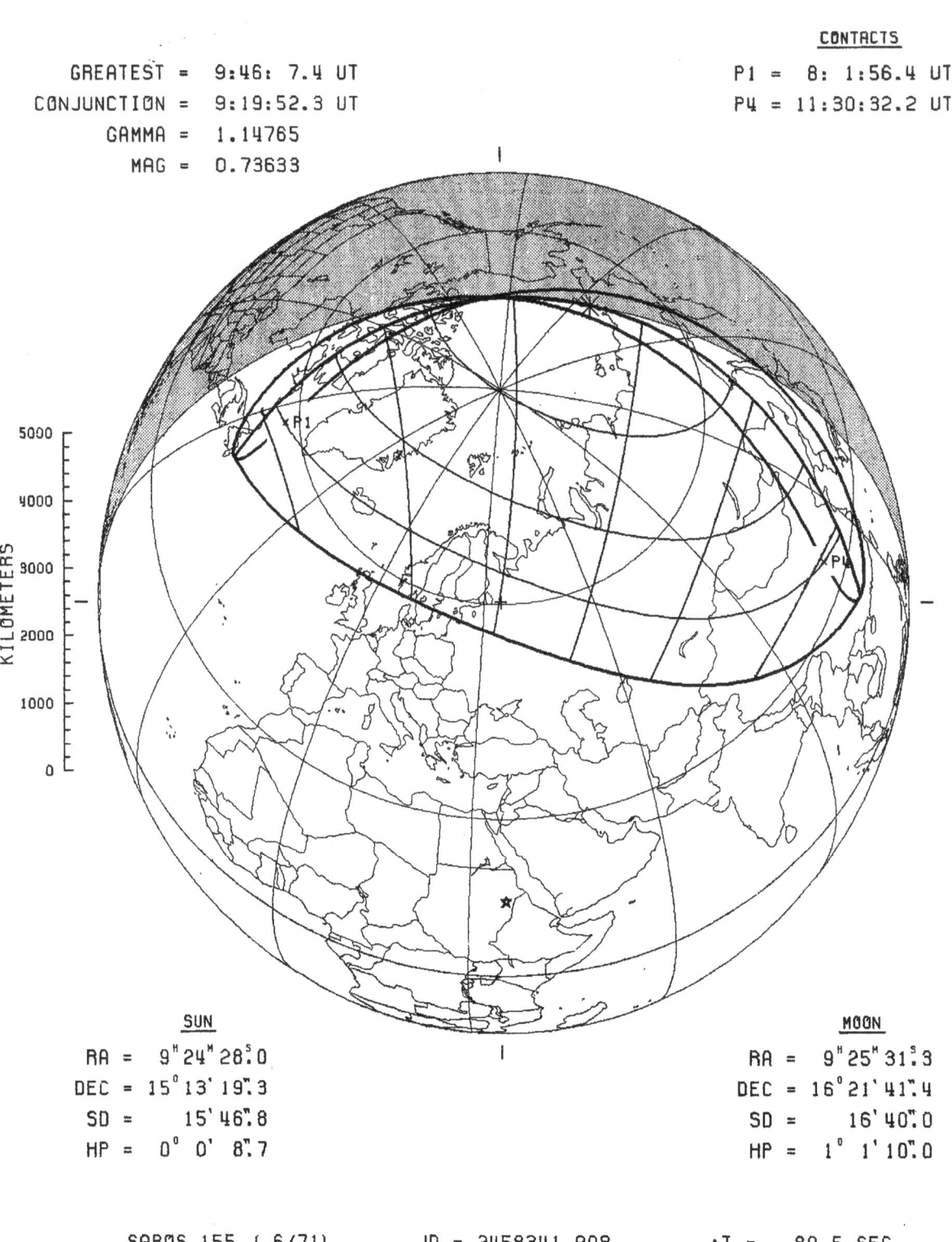

SUN

RA = 9ʰ24ᴹ28ˢ0

DEC = 15°13'19".3

SD =   15'46".8

HP =   0° 0' 8".7

MOON

RA = 9ʰ25ᴹ31ˢ3

DEC = 16°21'41".4

SD =   16'40".0

HP =   1° 1'10".0

SAROS 155 (6/71)   JD = 2458341.908   ΔT =   80.5 SEC

Figure 102

PARTIAL SOLAR ECLIPSE — 6 JAN 2019

GREATEST = 1:41:13.7 UT
CONJUNCTION = 1:43:26.7 UT
GAMMA = 1.14150
MAG = 0.71496

CONTACTS

P1 = 23:33:52.5 UT
P4 = 3:48:37.3 UT

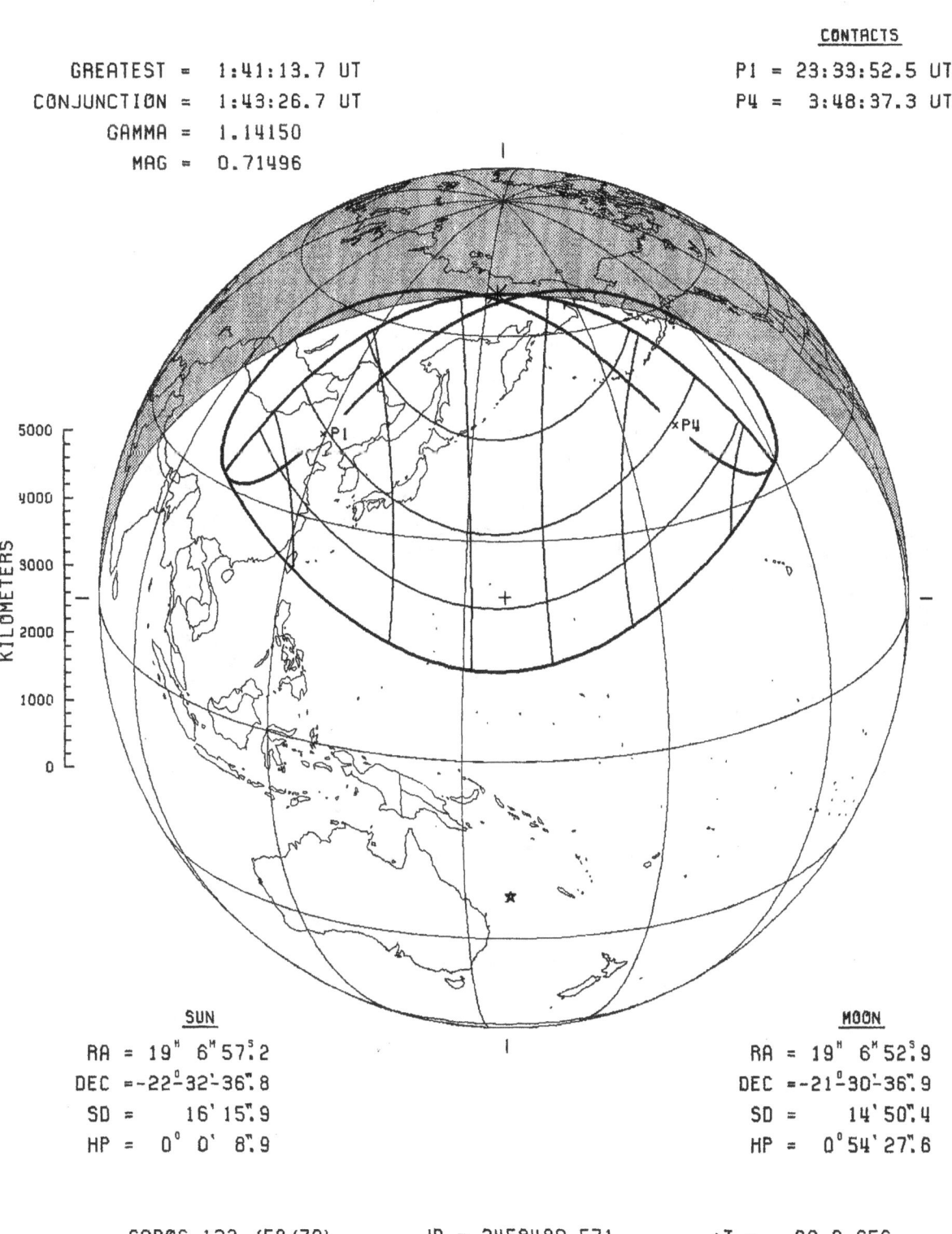

KILOMETERS

5000
4000
3000
2000
1000
0

SUN

RA = 19^h 6^{m}57^s.2
DEC = -22$^°$32$'$36$''$.8
SD =   16$'$ 15$''$.9
HP = 0$^°$ 0$'$ 8$''$.9

MOON

RA = 19^h 6^{m}52^s.9
DEC = -21$^°$30$'$36$''$.9
SD =   14$'$ 50$''$.4
HP = 0$^°$54$'$ 27$''$.6

SAROS 122 (58/70)   JD = 2458489.571   ΔT =   80.8 SEC

Figure 103

215

TOTAL SOLAR ECLIPSE – 2 JUL 2019

GREATEST = 19:22:45.6 UT
CONJUNCTION = 19:21:29.1 UT
GAMMA = -0.64660
RATIO = 1.04592

CONTACTS

P1 = 16:55: 1.3 UT
U1 = 18: 0:57.9 UT
U4 = 20:44:36.0 UT
P4 = 21:50:26.0 UT

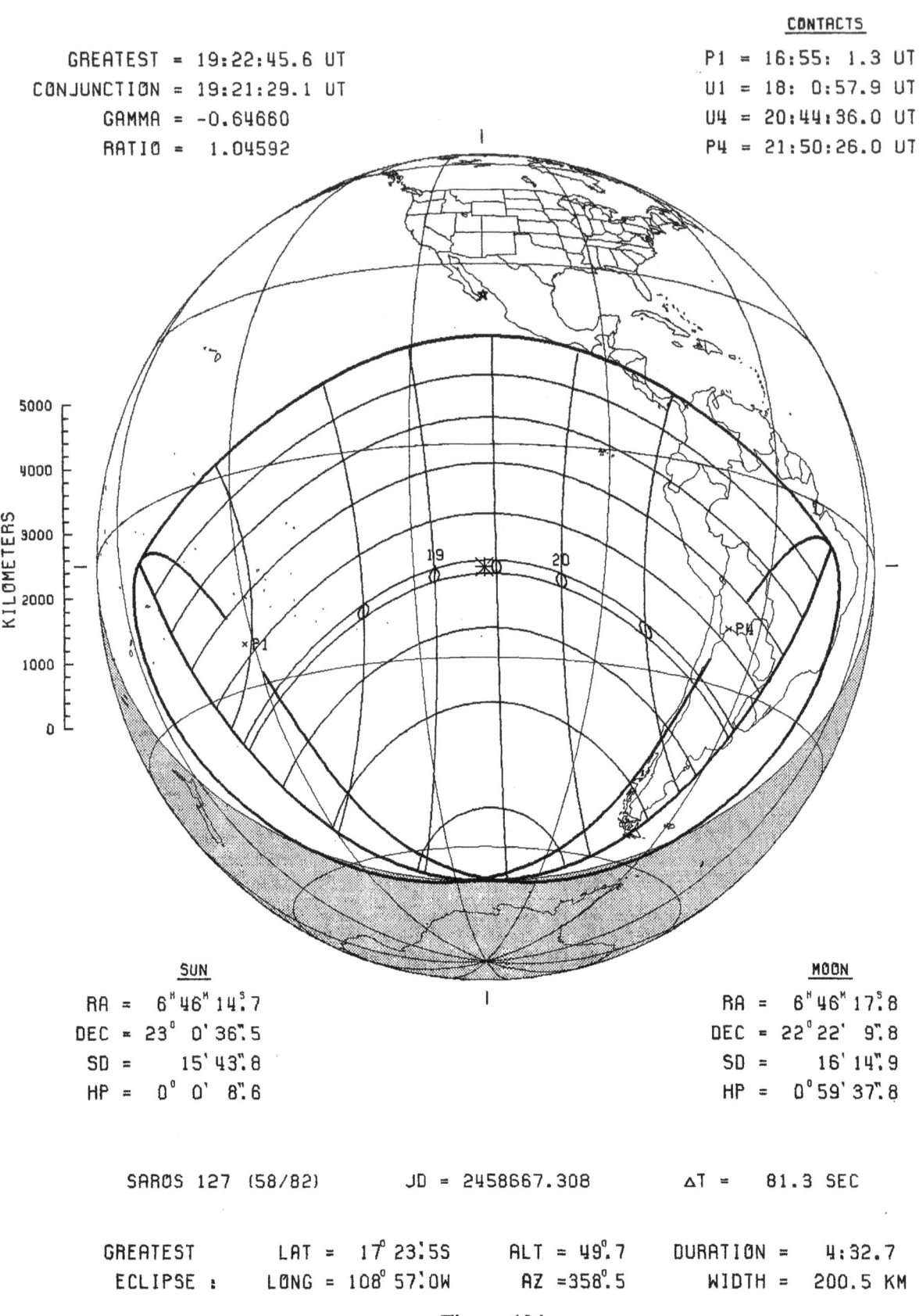

KILOMETERS
5000
4000
3000
2000
1000
0

SUN

RA =   6^{h}46^{m}14^s.7
DEC = 23° 0'36".5
SD =   15'43".8
HP =   0° 0' 8".6

MOON

RA =   6^{h}46^{m}17^s.8
DEC = 22°22' 9".8
SD =   16'14".9
HP =   0°59'37".8

SAROS 127 (58/82)   JD = 2458667.308   ΔT =   81.3 SEC

GREATEST   LAT = 17°23'.5S   ALT = 49°.7   DURATION =   4:32.7
ECLIPSE :   LONG = 108°57'.0W   AZ =358°.5   WIDTH = 200.5 KM

Figure 104
216

ANNULAR SOLAR ECLIPSE   —   26 DEC 2019

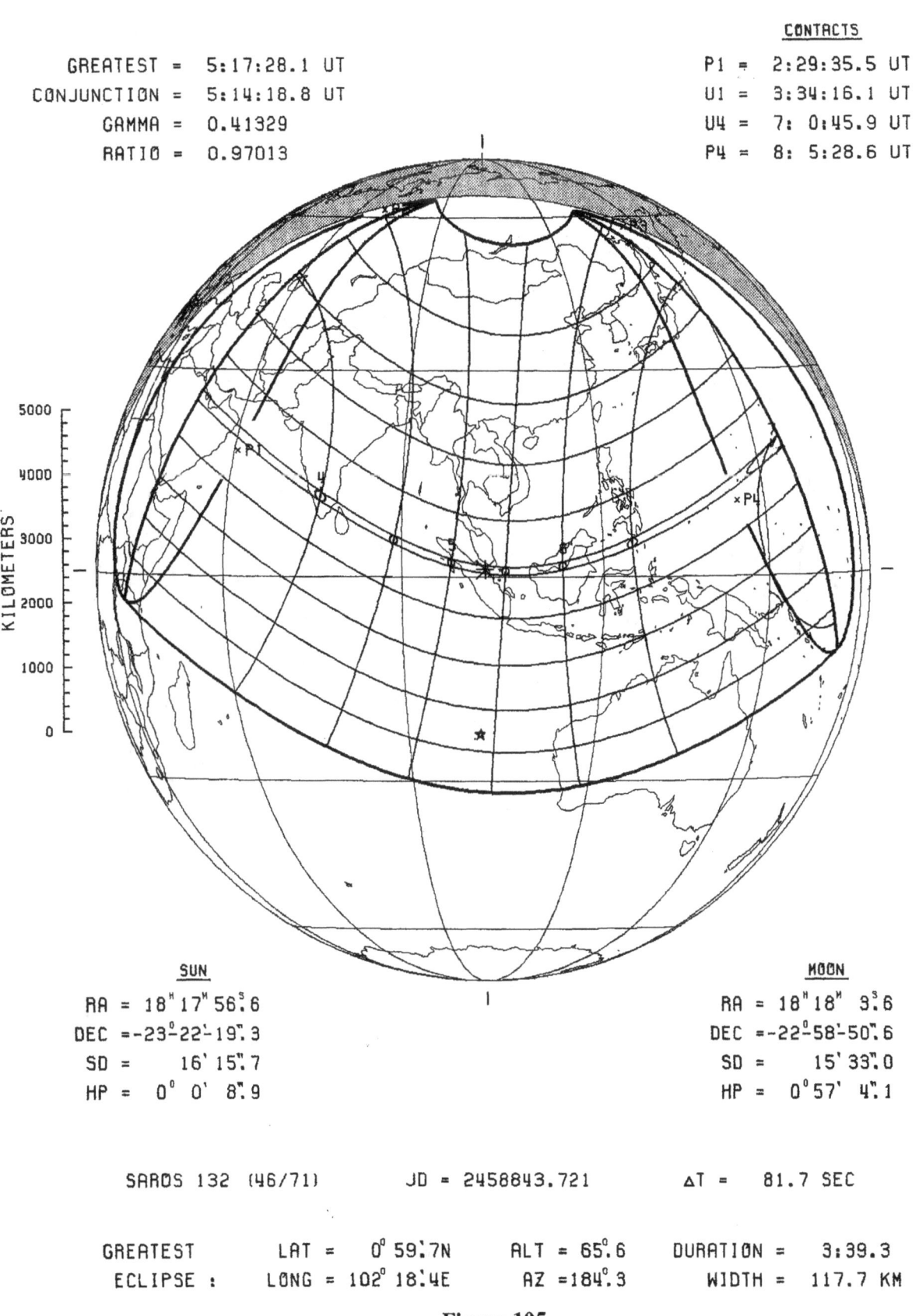

GREATEST = 5:17:28.1 UT
CONJUNCTION = 5:14:18.8 UT
GAMMA = 0.41329
RATIO = 0.97013

CONTACTS

P1 = 2:29:35.5 UT
U1 = 3:34:16.1 UT
U4 = 7: 0:45.9 UT
P4 = 8: 5:28.6 UT

SUN

RA = $18^h 17^m 56^s.6$
DEC = $-23°22'19".3$
SD = $16'15".7$
HP = $0°0'8".9$

MOON

RA = $18^h 18^m 9^s.6$
DEC = $-22°58'50".6$
SD = $15'33".0$
HP = $0°57'4".1$

SAROS 132 (46/71)   JD = 2458843.721   ΔT = 81.7 SEC

GREATEST   LAT = 0°59'.7N   ALT = 65°.6   DURATION = 3:39.3
ECLIPSE :   LONG = 102°18'.4E   AZ = 184°.3   WIDTH = 117.7 KM

Figure 105
217

ANNULAR SOLAR ECLIPSE — 21 JUN 2020

GREATEST = 6:39:51.8 UT

CONJUNCTION = 6:41:10.9 UT

GAMMA = 0.12086

RATIO = 0.99400

P1 = 3:45:46.4 UT

U1 = 4:47:30.5 UT

U4 = 8:32: 8.6 UT

P4 = 9:33:50.2 UT

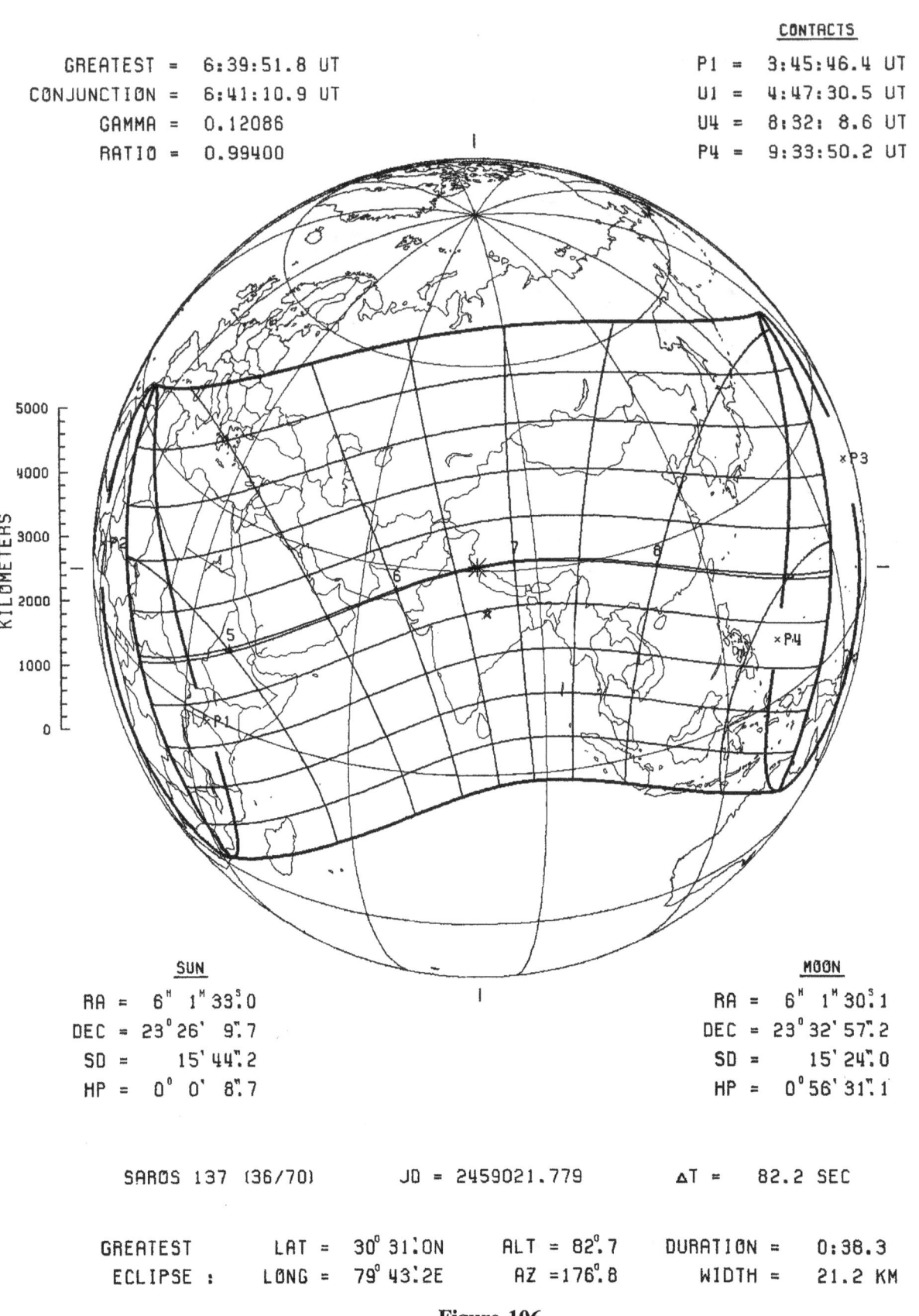

SUN

RA = $6^H 1^M 33^S.0$

DEC = $23° 26' 9".7$

SD = $15' 44".2$

HP = $0° 0' 8".7$

MOON

RA = $6^H 1^M 30^S.1$

DEC = $23° 32' 57".2$

SD = $15' 24".0$

HP = $0° 56' 31".1$

SAROS 137 (36/70)   JD = 2459021.779   ΔT = 82.2 SEC

GREATEST   LAT = 30° 31'.0N   ALT = 82°.7   DURATION = 0:38.3

ECLIPSE :   LONG = 79° 43'.2E   AZ = 176°.8   WIDTH = 21.2 KM

Figure 106
218

TOTAL SOLAR ECLIPSE –   14 DEC 2020

GREATEST = 16:13:15.1 UT
CONJUNCTION = 16:17:57.9 UT
GAMMA = -0.29421
RATIO = 1.02536

CONTACTS
P1 = 13:33:40.4 UT
U1 = 14:32:20.5 UT
U4 = 17:54: 5.0 UT
P4 = 18:52:52.0 UT

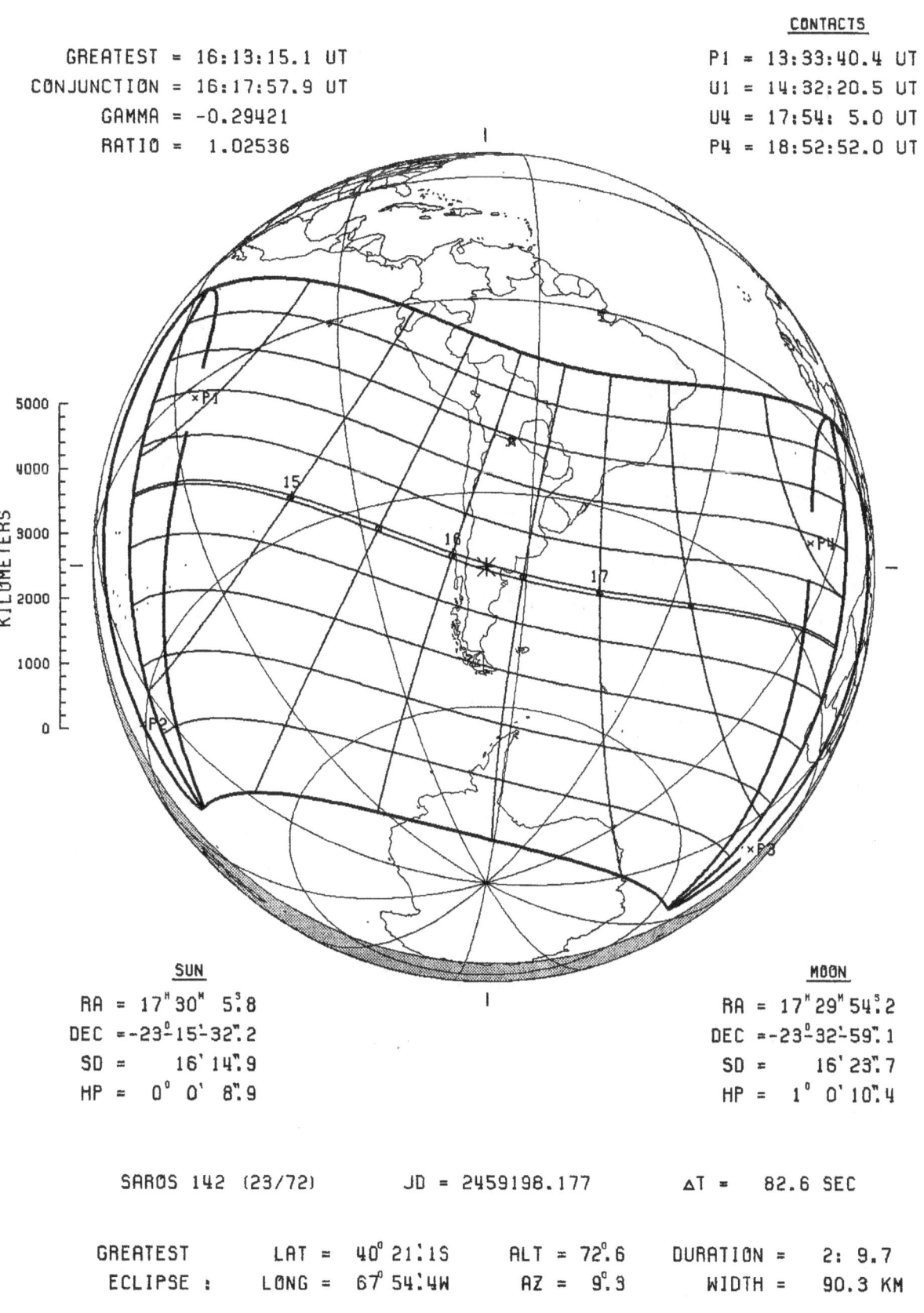

KILOMETERS
5000
4000
9000
2000
1000
0

SUN
RA = 17^{h}30^m 5^s.8
DEC =-23°15'32".2
SD =   16'14".9
HP = 0° 0' 8".9

MOON
RA = 17^{h}29^{m}54^s.2
DEC =-23°32'59".1
SD =   16'23".7
HP = 1° 0'10".4

SAROS 142 (23/72)   JD = 2459198.177   ΔT =   82.6 SEC

| GREATEST | LAT = 40° 21'.1S | ALT = 72°.6 | DURATION = 2: 9.7 |
| ECLIPSE : | LONG = 67° 54'.4W | AZ = 9°.3 | WIDTH = 90.3 KM |

Figure 107
219

ANNULAR SOLAR ECLIPSE   -   10 JUN 2021

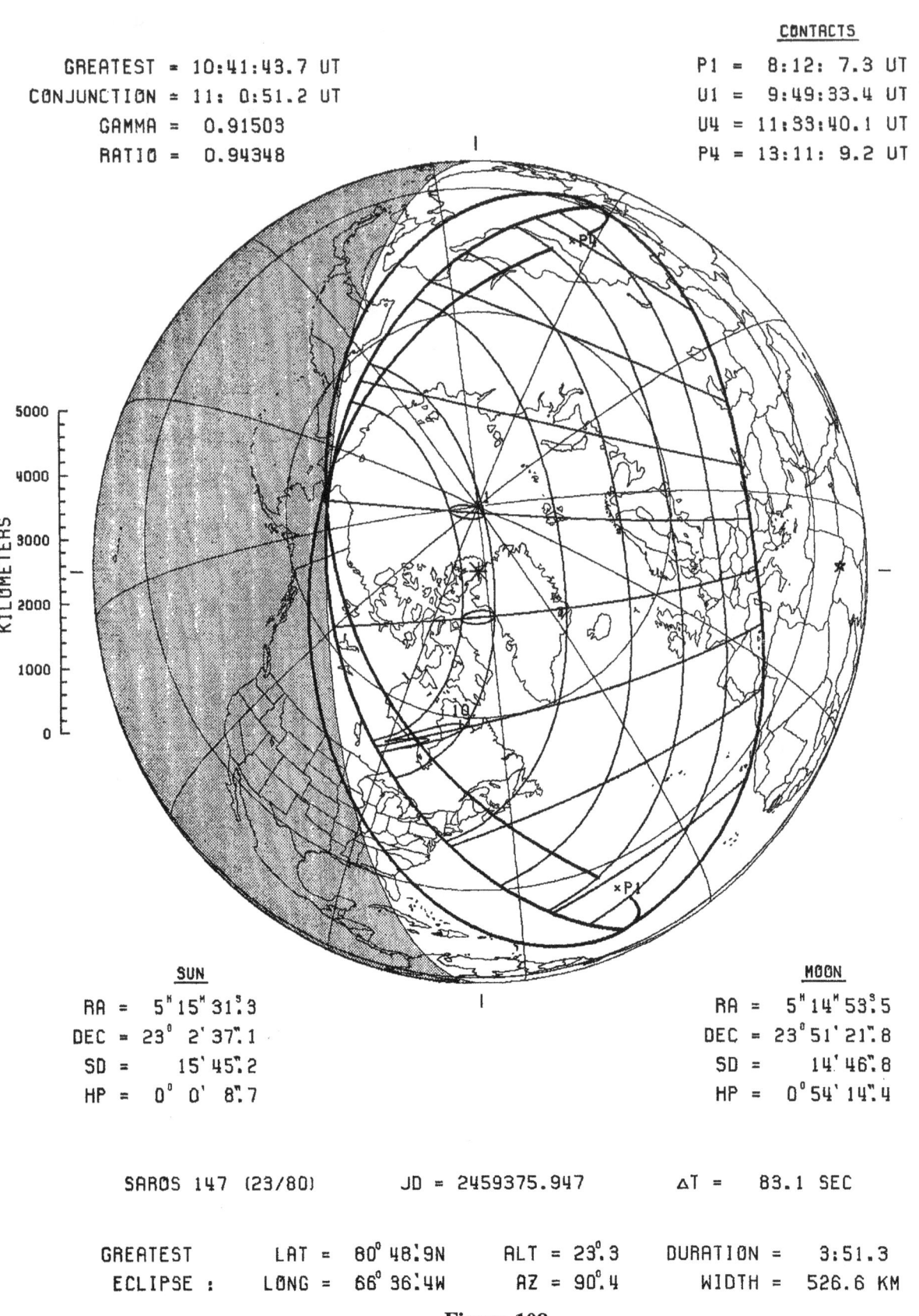

GREATEST = 10:41:43.7 UT
CONJUNCTION = 11: 0:51.2 UT
GAMMA = 0.91509
RATIO = 0.94348

CONTACTS

P1 = 8:12: 7.3 UT
U1 = 9:49:33.4 UT
U4 = 11:33:40.1 UT
P4 = 13:11: 9.2 UT

SUN

RA = 5^{H}15^{M}31^{S}3
DEC = 23^0 2' 37".1
SD =   15' 45".2
HP = 0^0 0' 8".7

MOON

RA = 5^{H}14^{M}53^{S}5
DEC = 23^{0}51' 21".8
SD =   14' 46".8
HP = 0^{0}54' 14".4

SAROS 147 (23/80)   JD = 2459375.947   ΔT =   83.1 SEC

GREATEST   LAT = 80^0 48'.9N   ALT = 23^0.3   DURATION =   3:51.3
ECLIPSE :   LONG = 66^0 36'.4W   AZ = 90^0.4   WIDTH =   526.6 KM

Figure 108
220

TOTAL SOLAR ECLIPSE — 4 DEC 2021

GREATEST = 7:33:15.0 UT
CONJUNCTION = 7:55:57.7 UT
GAMMA = -0.95276
RATIO = 1.03669

CONTACTS

P1 = 5:29: 4.5 UT
U1 = 6:59:56.7 UT
U4 = 8: 6:18.5 UT
P4 = 9:37:15.6 UT

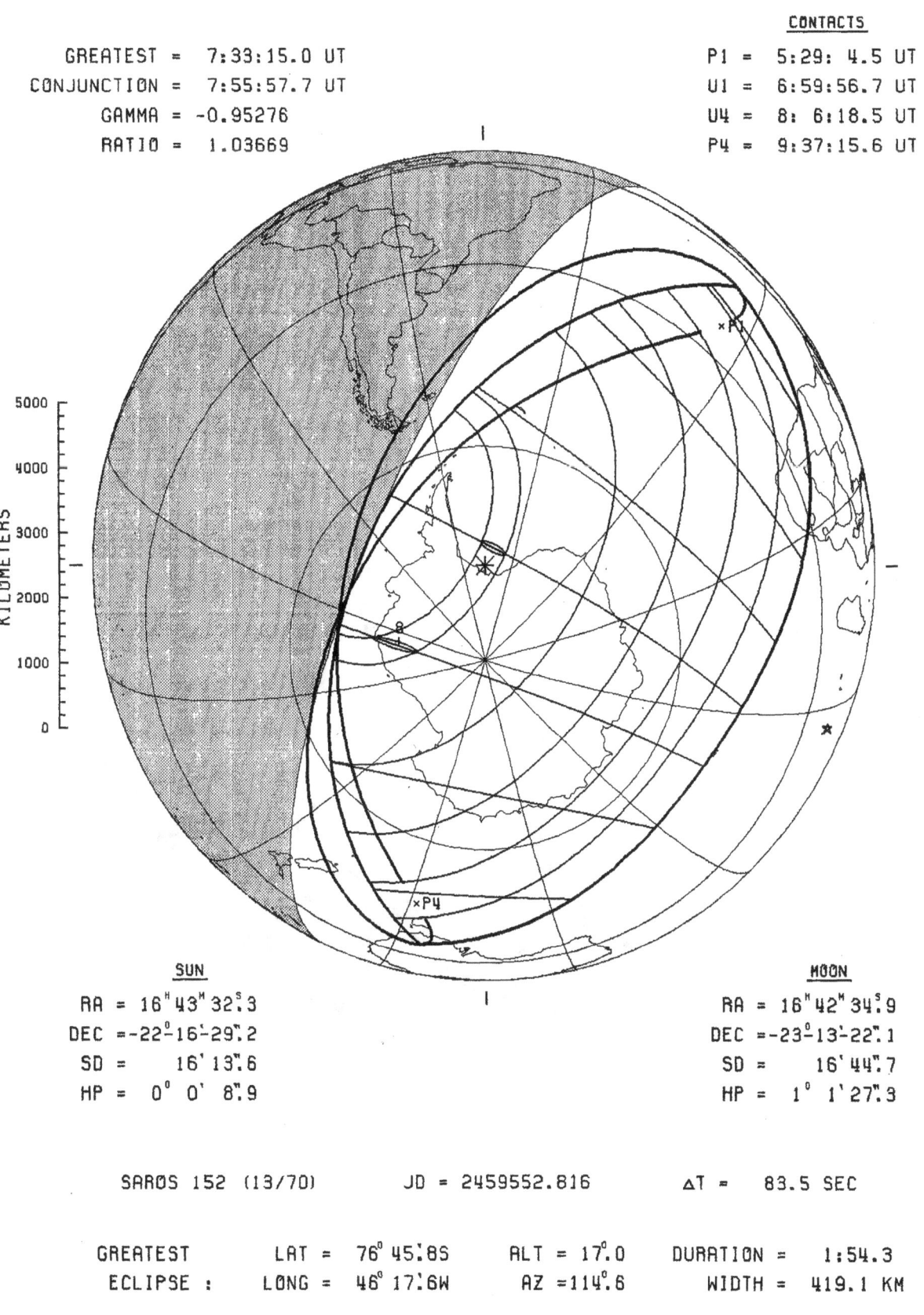

KILOMETERS

5000
4000
3000
2000
1000
0

×P1
×P4

SUN

RA = $16^h 43^m 32^s.3$
DEC = $-22^\circ 16'29''.2$
SD = $16'13''.6$
HP = $0^\circ 0' 8''.9$

MOON

RA = $16^h 42^m 34^s.9$
DEC = $-23^\circ 13'22''.1$
SD = $16'44''.7$
HP = $1^\circ 1'27''.3$

SAROS 152 (13/70)   JD = 2459552.816   ΔT = 83.5 SEC

GREATEST   LAT = $76^\circ 45'.8S$ ALT = $17^\circ.0$ DURATION = 1:54.3
ECLIPSE :   LONG = $46^\circ 17'.6W$ AZ = $114^\circ.6$ WIDTH = 419.1 KM

Figure 109
221

PARTIAL SOLAR ECLIPSE – 30 APR 2022

GREATEST = 20:41:12.8 UT
CONJUNCTION = 19:40:34.5 UT
GAMMA = -1.19022
MAG = 0.63850

P1 = 18:45: 7.7 UT
P4 = 22:37:46.8 UT

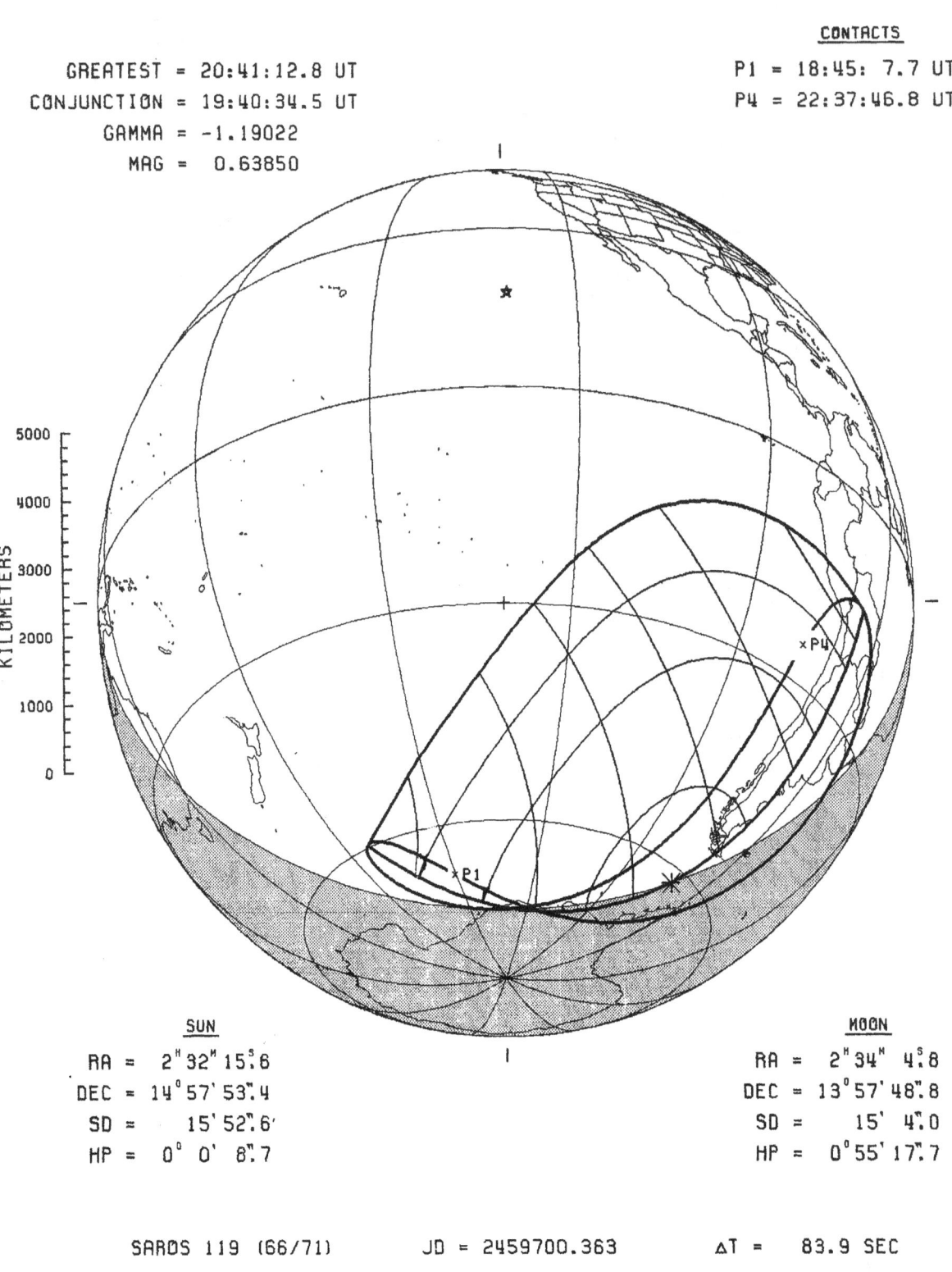

KILOMETERS

5000
4000
3000
2000
1000
0

×P4

×P1

SUN

RA = 2ʰ32ᵐ15ˢ.6
DEC = 14°57'53".4
SD = 15'52".6
HP = 0°0'8".7

MOON

RA = 2ʰ34ᵐ4ˢ.8
DEC = 13°57'48".8
SD = 15'4".0
HP = 0°55'17".7

SAROS 119 (66/71)   JD = 2459700.363   ΔT = 83.9 SEC

Figure 110

222

PARTIAL SOLAR ECLIPSE — 25 OCT 2022

GREATEST = 10:59:52.7 UT

CONJUNCTION = 10: 3:29.6 UT

GAMMA = 1.06984

MAG = 0.86133

P1 = 8:58: 1.9 UT

P4 = 13: 2: 1.1 UT

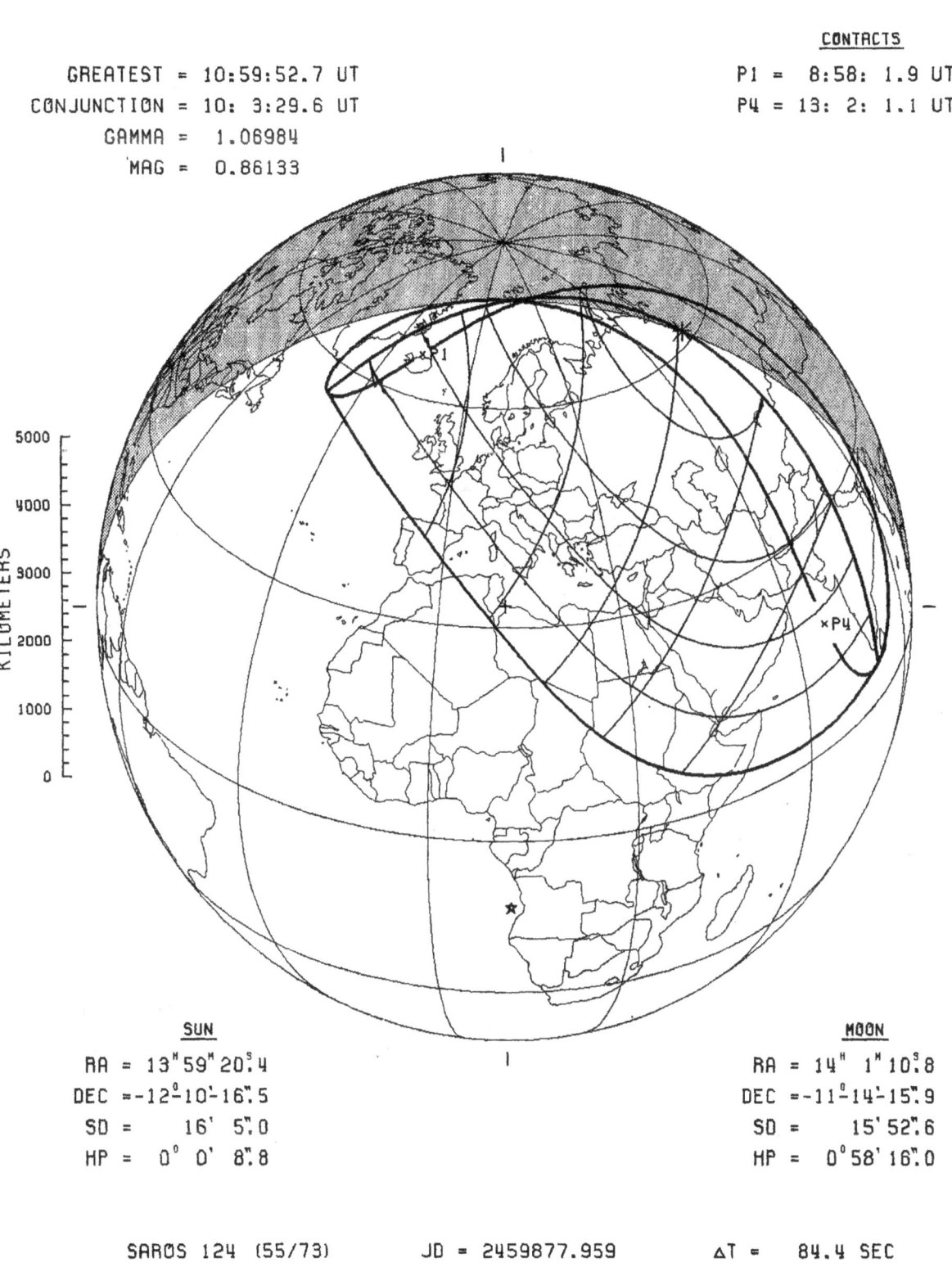

KILOMETERS

5000
4000
3000
2000
1000
0

SUN

RA = $13^h59^m20^s.4$

DEC = $-12^0 10' 16''.5$

SD = $16' 5''.0$

HP = $0^0 0' 8''.8$

MOON

RA = $14^h 1^m 10^s.8$

DEC = $-11^0 14' 15''.9$

SD = $15' 52''.6$

HP = $0^0 58' 16''.0$

SAROS 124 (55/73)   JD = 2459877.959   ΔT =   84.4 SEC

Figure 111

223

ANN/TOT SOLAR ECLIPSE – 20 APR 2023

GREATEST = 4:16:30.3 UT
CONJUNCTION = 3:55:18.7 UT
GAMMA = -0.39526
RATIO = 1.01323

P1 = 1:34: 9.0 UT
U1 = 2:36:49.6 UT
U4 = 5:56:27.4 UT
P4 = 6:59: 5.8 UT

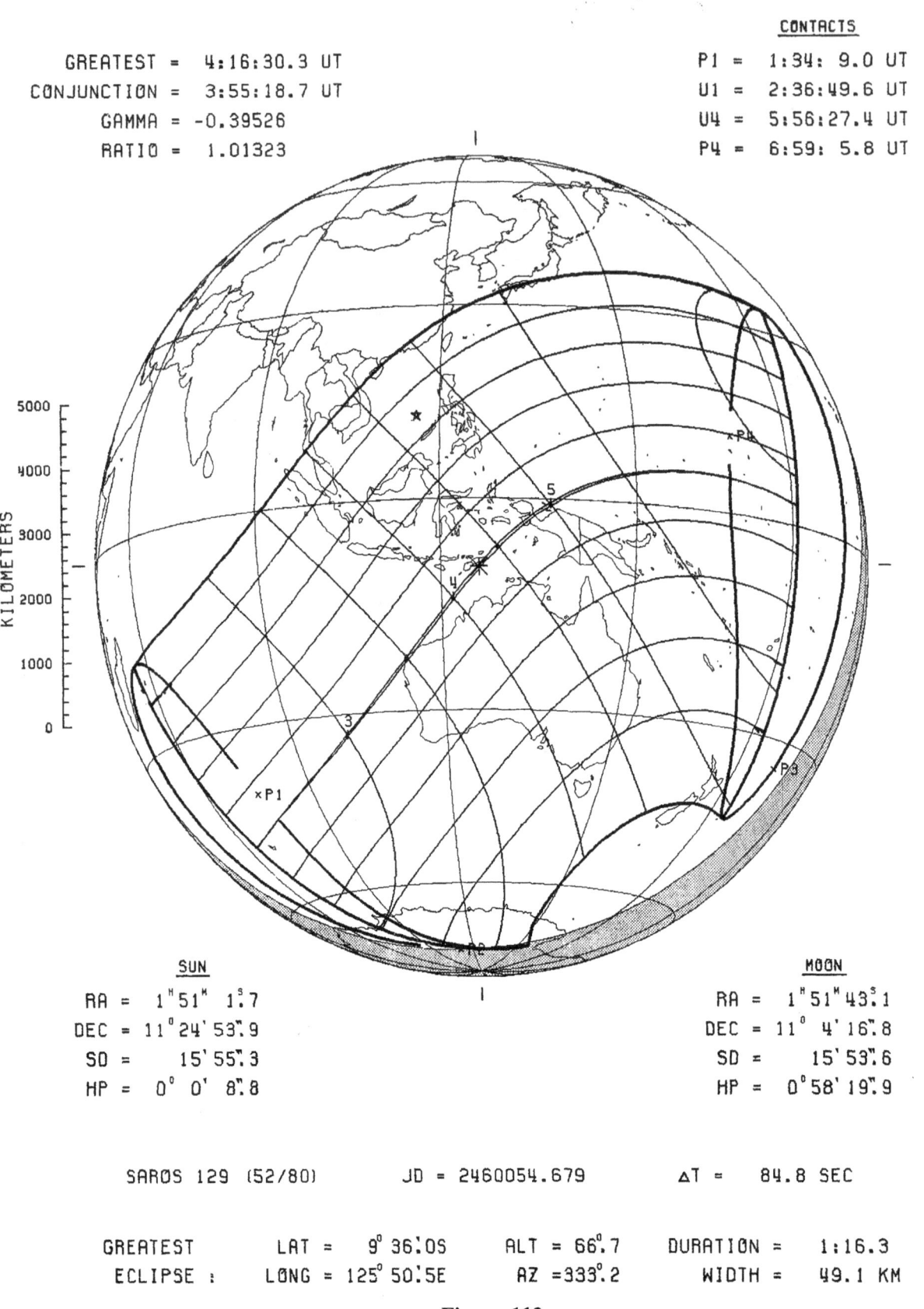

KILOMETERS
5000
4000
3000
2000
1000
0

SUN

RA = $1^h51^m 1^s.7$
DEC = $11°24'53".9$
SD = $15'55".3$
HP = $0°0' 8".8$

MOON

RA = $1^h51^m43^s.1$
DEC = $11° 4'16".8$
SD = $15'53".6$
HP = $0°58'19".9$

SAROS 129 (52/80)   JD = 2460054.679   ΔT = 84.8 SEC

GREATEST   LAT = $9°36'.0S$ ALT = $66°.7$ DURATION = 1:16.3
ECLIPSE :   LONG = $125°50'.5E$ AZ = $333°.2$ WIDTH = 49.1 KM

Figure 112
224

ANNULAR SOLAR ECLIPSE - 14 OCT 2023

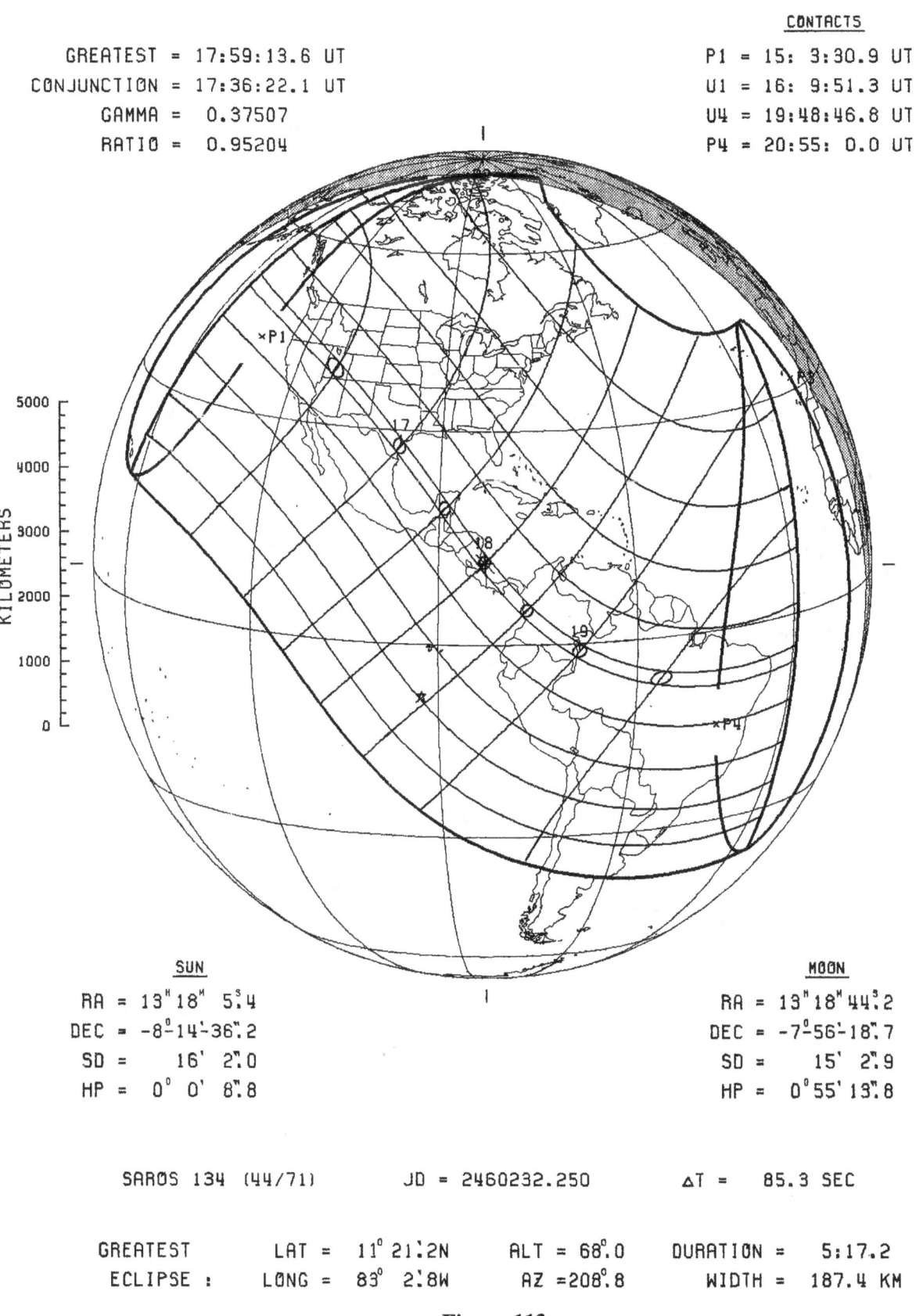

GREATEST = 17:59:13.6 UT
CONJUNCTION = 17:36:22.1 UT
GAMMA = 0.37507
RATIO = 0.95204

CONTACTS

P1 = 15: 3:30.9 UT
U1 = 16: 9:51.3 UT
U4 = 19:48:46.8 UT
P4 = 20:55: 0.0 UT

KILOMETERS

5000
4000
3000
2000
1000
0

SUN

RA = 13ʰ18ᴹ 5ˢ4
DEC = -8°14'-36".2
SD = 16' 2".0
HP = 0° 0' 8".8

MOON

RA = 13ʰ18ᴹ44ˢ2
DEC = -7°56'-18".7
SD = 15' 2".9
HP = 0°55' 13".8

SAROS 134 (44/71)   JD = 2460232.250   ΔT = 85.3 SEC

GREATEST   LAT = 11° 21'.2N   ALT = 68°.0   DURATION = 5:17.2
ECLIPSE :   LONG = 83° 2'.8W   AZ =208°.8   WIDTH = 187.4 KM

Figure 113
225

TOTAL SOLAR ECLIPSE – 8 APR 2024

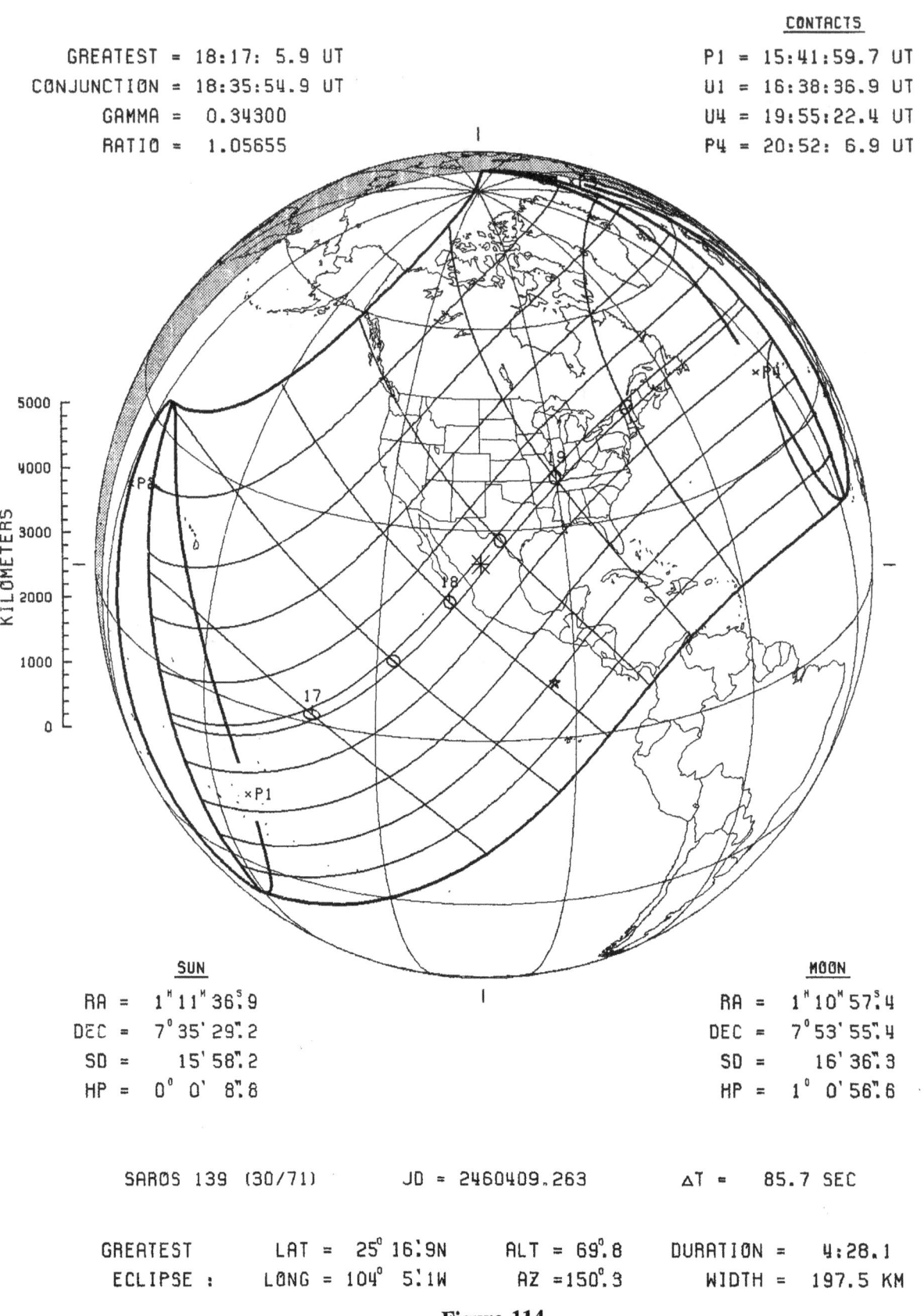

GREATEST = 18:17: 5.9 UT
CONJUNCTION = 18:35:54.9 UT
GAMMA = 0.34300
RATIO = 1.05655

CONTACTS
P1 = 15:41:59.7 UT
U1 = 16:38:36.9 UT
U4 = 19:55:22.4 UT
P4 = 20:52: 6.9 UT

SUN
RA = 1ʰ11ᴹ36ˢ.9
DEC = 7°35'29".2
SD =   15'58".2
HP = 0° 0' 8".8

MOON
RA = 1ʰ10ᴹ57ˢ.4
DEC = 7°53'55".4
SD =   16'36".3
HP = 1° 0'56".6

SAROS 139 (30/71)   JD = 2460409.263   ΔT =   85.7 SEC

GREATEST   LAT = 25° 16'.9N   ALT = 69°.8   DURATION =   4:28.1
ECLIPSE :   LONG = 104° 5'.1W   AZ = 150°.3   WIDTH = 197.5 KM

Figure 114
226

ANNULAR SOLAR ECLIPSE — 2 OCT 2024

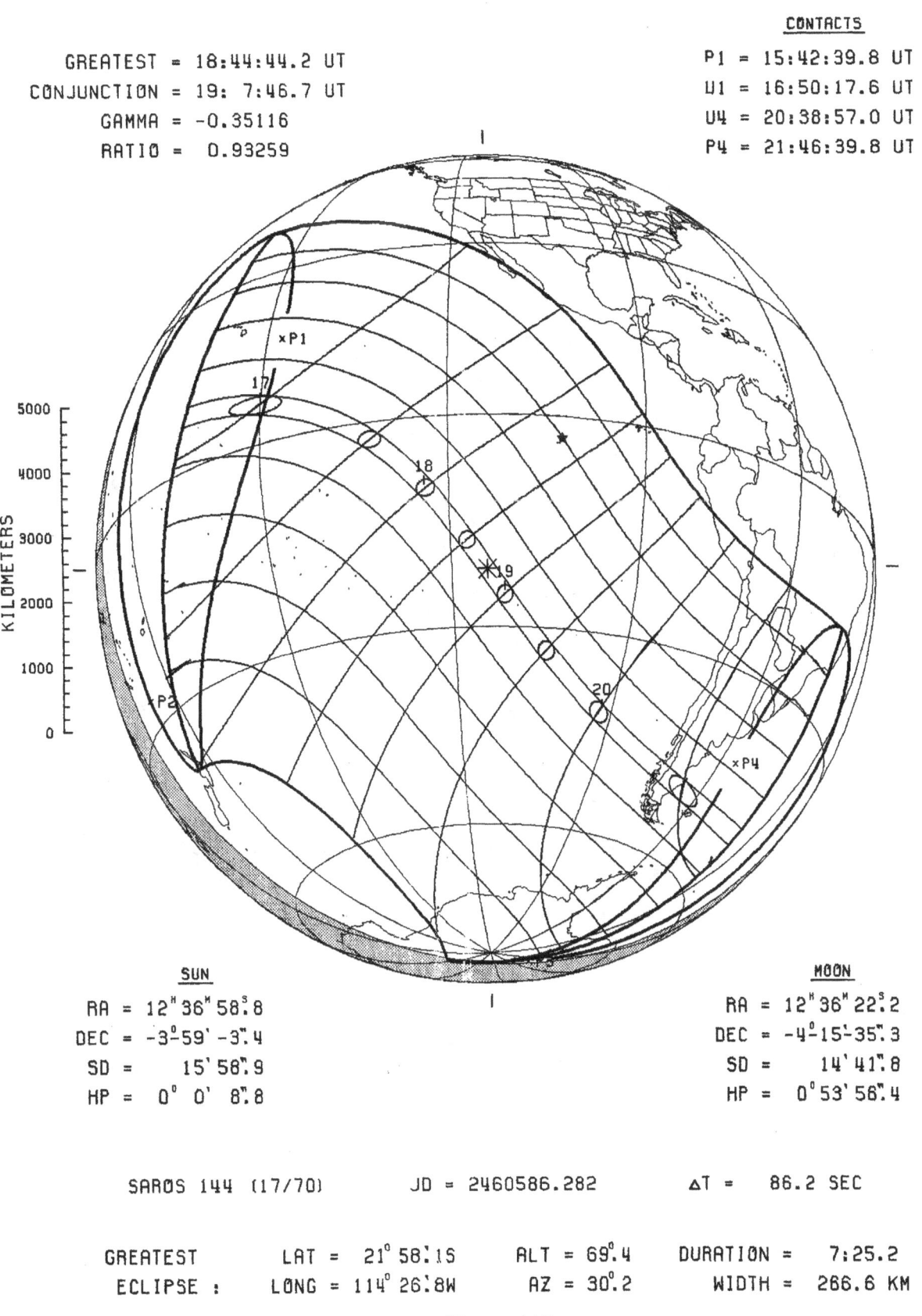

GREATEST = 18:44:44.2 UT
CONJUNCTION = 19: 7:46.7 UT
GAMMA = -0.35116
RATIO = 0.93259

CONTACTS

P1 = 15:42:39.8 UT
U1 = 16:50:17.6 UT
U4 = 20:38:57.0 UT
P4 = 21:46:39.8 UT

SUN

RA = $12^h 36^m 58^s.8$
DEC = $-3° 59' -3".4$
SD = $15' 58".9$
HP = $0° 0' 8".8$

MOON

RA = $12^h 36^m 22^s.2$
DEC = $-4° 15' -35".3$
SD = $14' 41".8$
HP = $0° 53' 56".4$

SAROS 144 (17/70)   JD = 2460586.282   ΔT = 86.2 SEC

GREATEST   LAT = $21° 58'.1S$ ALT = $69°.4$ DURATION = 7:25.2
ECLIPSE :   LONG = $114° 26'.8W$ AZ = $30°.2$ WIDTH = 266.6 KM

Figure 115
227

PARTIAL SOLAR ECLIPSE – 29 MAR 2025

GREATEST = 10:47:11.4 UT
CONJUNCTION = 11:46: 1.7 UT
GAMMA = 1.04035
MAG = 0.93647

P1 = 8:50:27.0 UT
P4 = 12:43:30.3 UT

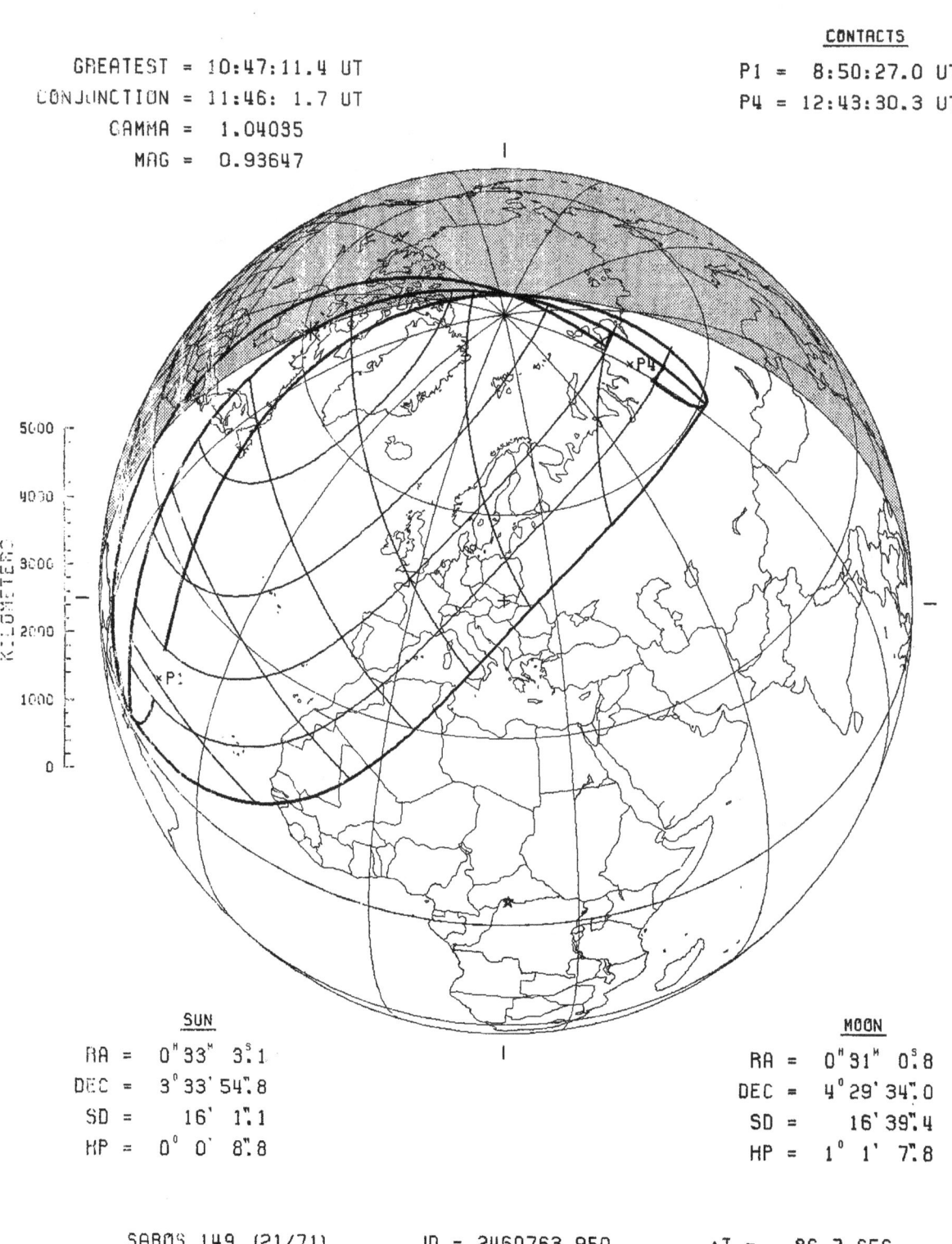

SUN

RA = 0^{h}33^m 3^{s}1
DEC = 3°33' 54".8
SD =   16' 1".1
HP = 0° 0' 8".8

MOON

RA = 0^{h}31^m 0^{s}8
DEC = 4°29' 34".0
SD =   16' 39".4
HP = 1° 1' 7".8

SAROS 149 (21/71)   JD = 2460763.950   ΔT =   86.7 SEC

Figure 116

228

PARTIAL SOLAR ECLIPSE – 21 SEP 2025

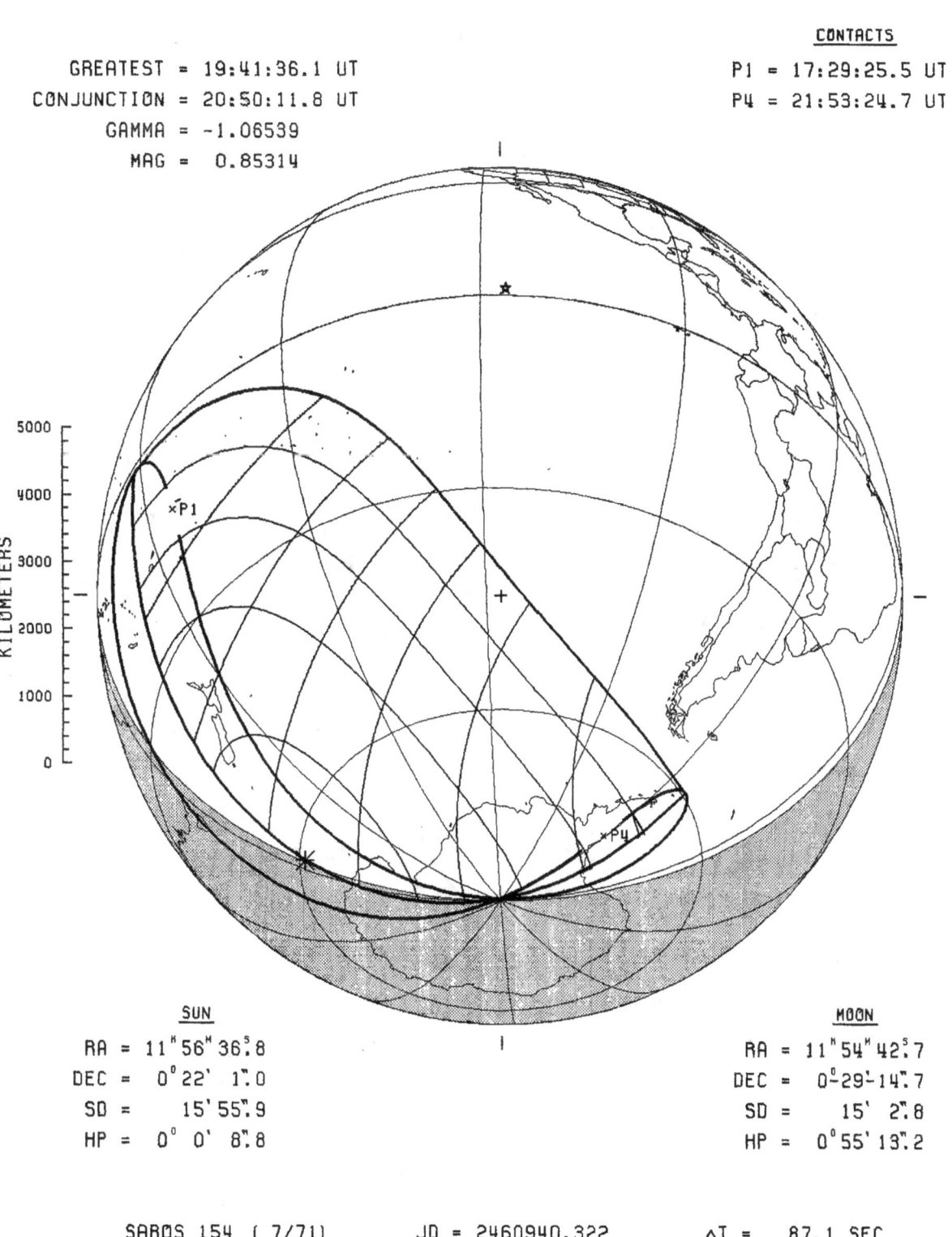

SUN
RA = 11ʰ56ᴹ36ˢ.8
DEC = 0°22' 1".0
SD = 15' 55".9
HP = 0° 0' 8".8

MOON
RA = 11ʰ54ᴹ42ˢ.7
DEC = 0°29'14".7
SD = 15' 2".8
HP = 0°55'13".2

SAROS 154 (7/71)   JD = 2460940.322   ΔT =   87.1 SEC

Figure 117

229

ANNULAR SOLAR ECLIPSE - 17 FEB 2026

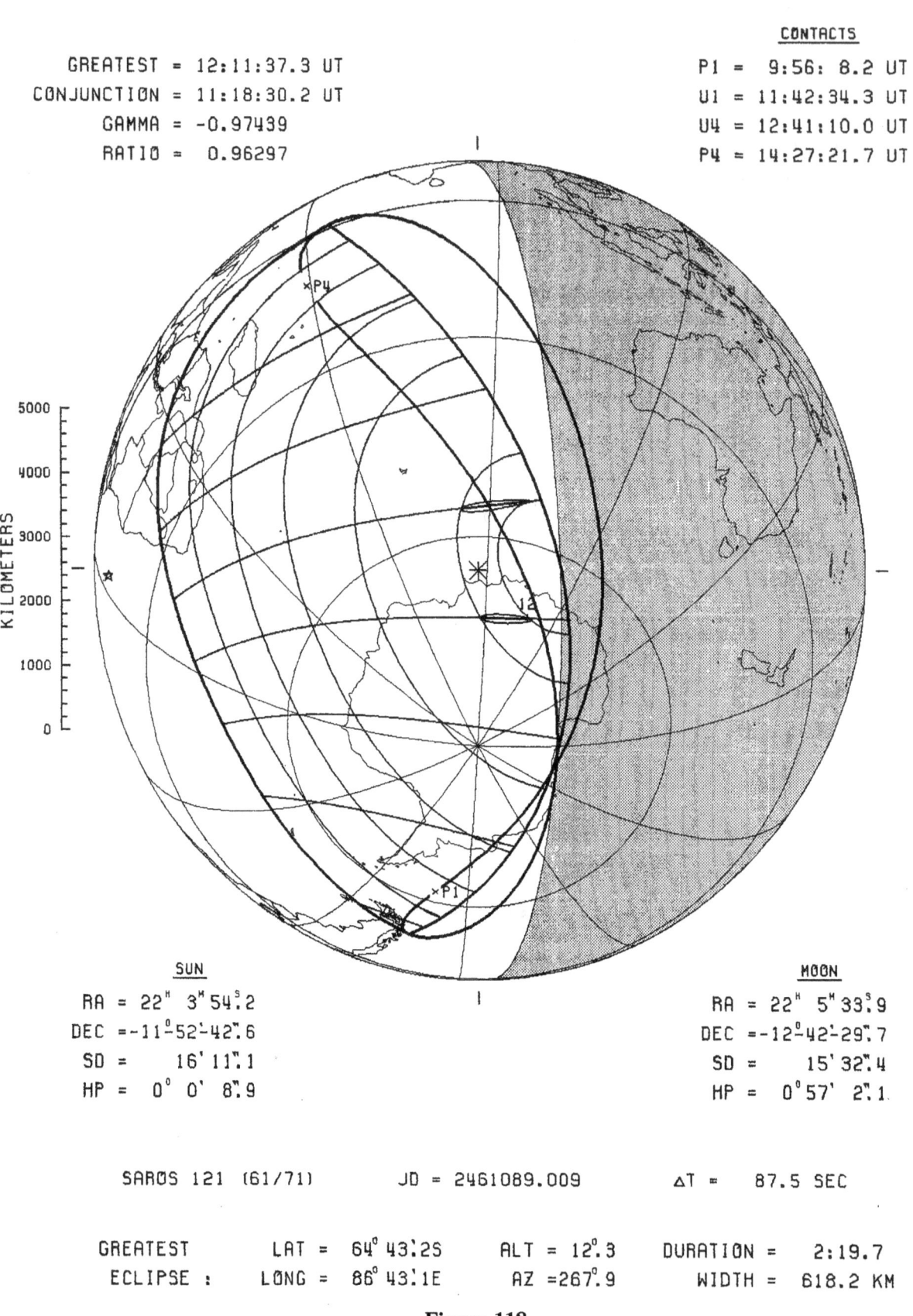

GREATEST = 12:11:37.3 UT
CONJUNCTION = 11:18:30.2 UT
GAMMA = -0.97439
RATIO = 0.96297

CONTACTS
P1 = 9:56: 8.2 UT
U1 = 11:42:34.3 UT
U4 = 12:41:10.0 UT
P4 = 14:27:21.7 UT

SUN
RA = 22ʰ 3ᵐ 54ˢ.2
DEC = -11°52'42".6
SD =   16' 11".1
HP = 0° 0' 8".9

MOON
RA = 22ʰ 5ᵐ 33ˢ.9
DEC = -12°42'29".7
SD =   15' 32".4
HP = 0°57' 2".1

SAROS 121 (61/71)   JD = 2461089.009   ΔT =   87.5 SEC

GREATEST   LAT = 64° 43'.2S   ALT = 12°.3   DURATION =   2:19.7
ECLIPSE :   LONG = 86° 43'.1E   AZ = 267°.9   WIDTH = 618.2 KM

Figure 118
230

TOTAL SOLAR ECLIPSE — 12 AUG 2026

GREATEST = 17:45:36.7 UT
CONJUNCTION = 17: 3:33.4 UT
GAMMA = 0.89741
RATIO = 1.03867

P1 = 15:33:53.4 UT
U1 = 16:57:44.6 UT
U4 = 18:33:52.9 UT
P4 = 19:57:41.5 UT

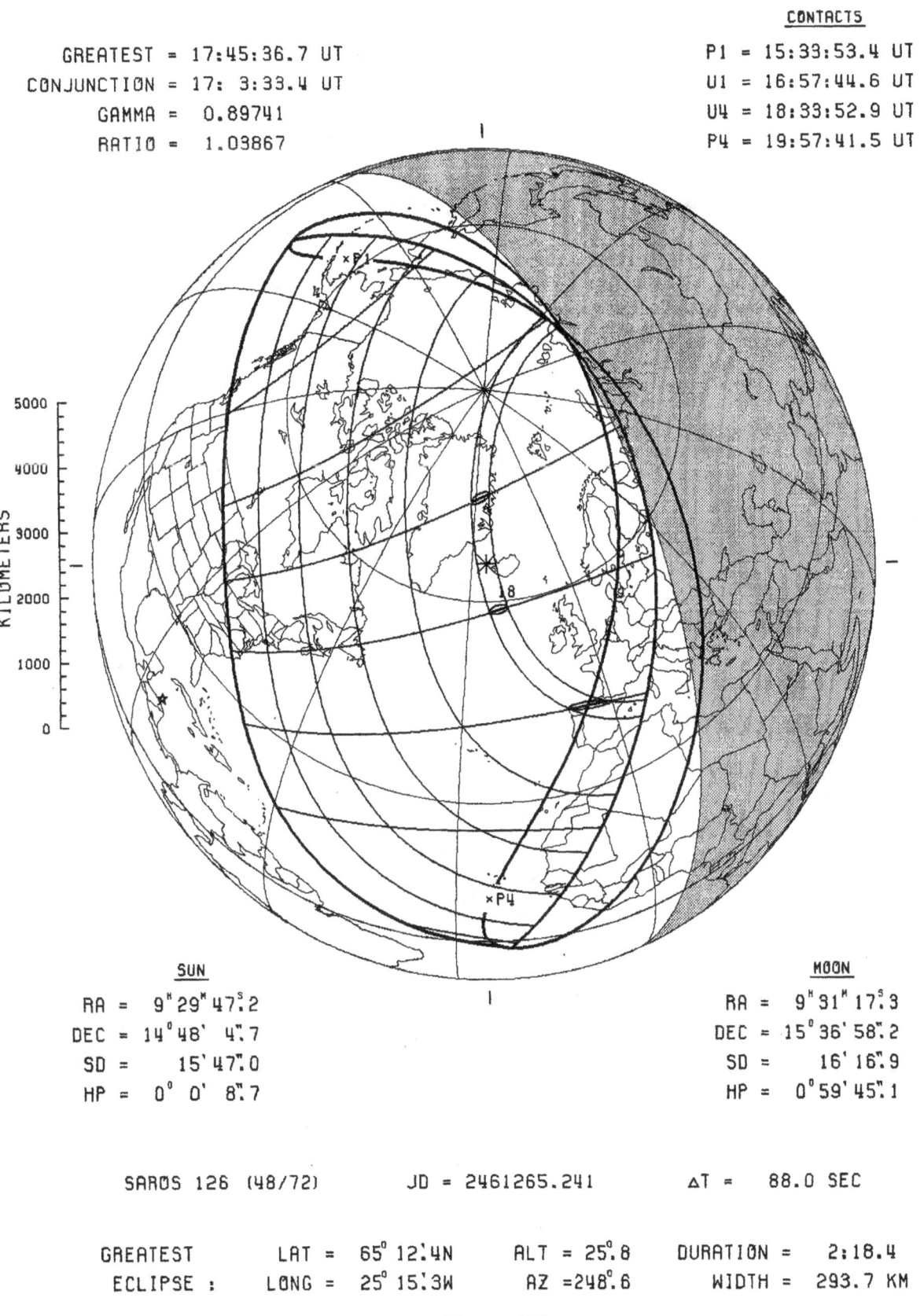

5000
4000
3000
2000
1000
0

KILOMETERS

SUN

RA = $9^h 29^m 47^s.2$
DEC = $14^\circ 48' 4".7$
SD = $15' 47".0$
HP = $0^\circ 0' 8".7$

MOON

RA = $9^h 31^m 17^s.3$
DEC = $15^\circ 36' 58".2$
SD = $16' 16".9$
HP = $0^\circ 59' 45".1$

SAROS 126 (48/72)   JD = 2461265.241   ∆T = 88.0 SEC

GREATEST   LAT = $65^\circ 12'.4$N   ALT = $25^\circ.8$ DURATION = 2:18.4
ECLIPSE :   LONG = $25^\circ 15'.3$W   AZ = $248^\circ.6$ WIDTH = 293.7 KM

Figure 119
231

ANNULAR SOLAR ECLIPSE – 6 FEB 2027

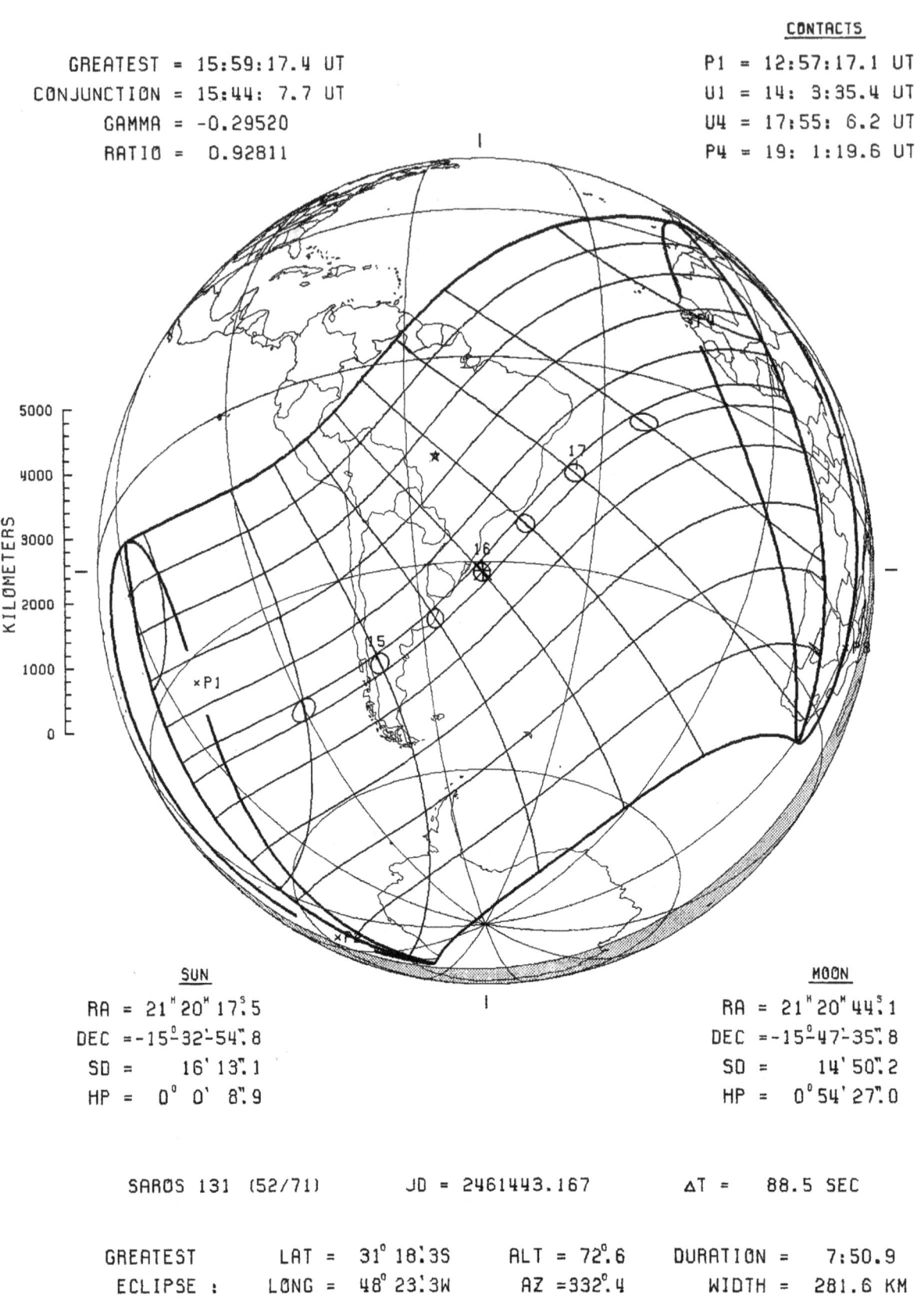

GREATEST = 15:59:17.4 UT
CONJUNCTION = 15:44: 7.7 UT
GAMMA = -0.29520
RATIO = 0.92811

CONTACTS

P1 = 12:57:17.1 UT
U1 = 14: 3:35.4 UT
U4 = 17:55: 6.2 UT
P4 = 19: 1:19.6 UT

SUN

RA = 21ʰ20ᴹ17ˢ5
DEC = -15°32'54".8
SD = 16'13".1
HP = 0° 0' 8".9

MOON

RA = 21ʰ20ᴹ44ˢ1
DEC = -15°47'35".8
SD = 14'50".2
HP = 0°54'27".0

SAROS 131 (52/71)   JD = 2461443.167   ΔT = 88.5 SEC

GREATEST   LAT = 31° 18.3S   ALT = 72°.6   DURATION = 7:50.9
ECLIPSE :   LONG = 48° 23.3W   AZ = 332°.4   WIDTH = 281.6 KM

Figure 120
232

TOTAL SOLAR ECLIPSE – 2 AUG 2027

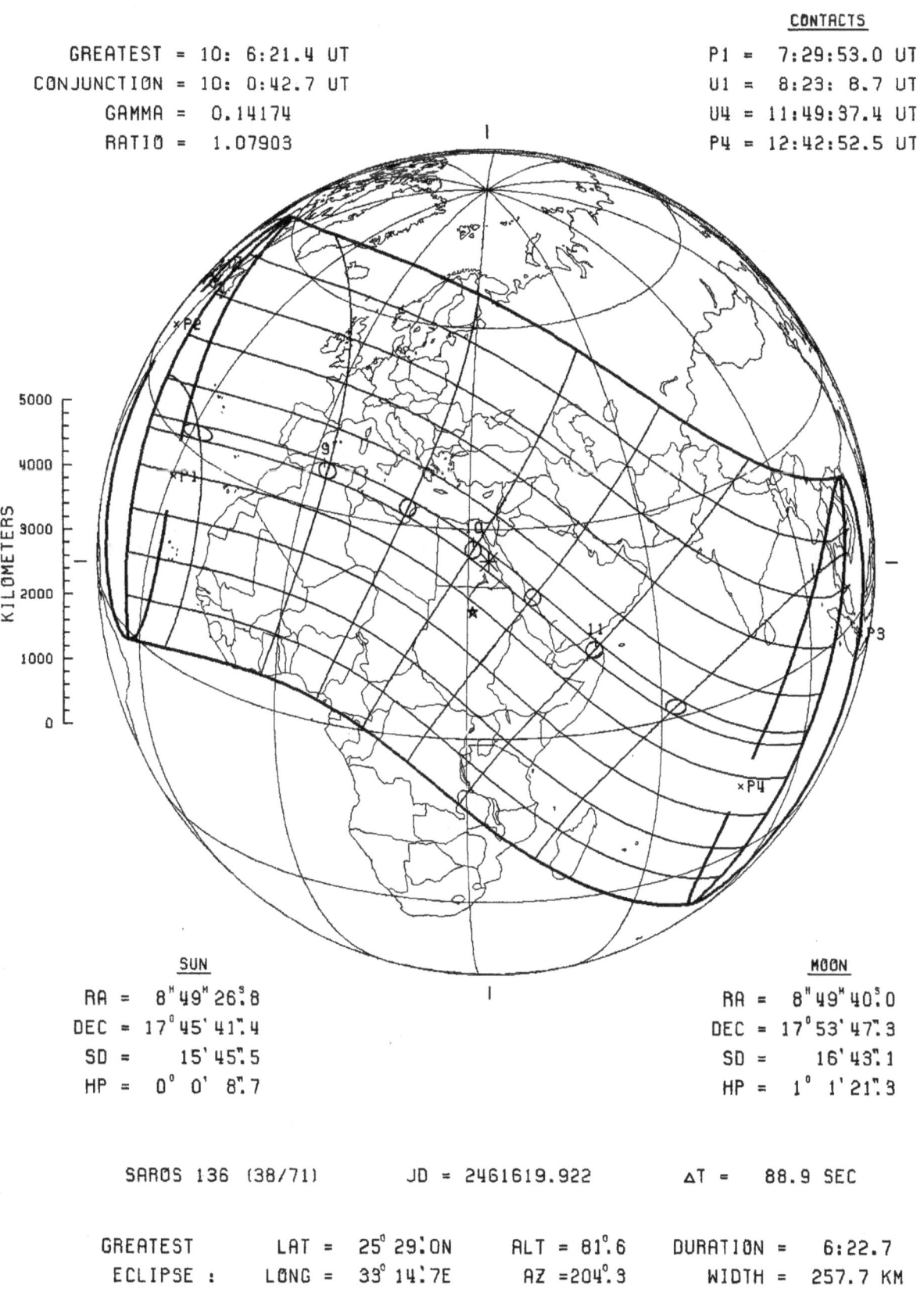

GREATEST = 10: 6:21.4 UT
CONJUNCTION = 10: 0:42.7 UT
GAMMA = 0.14174
RATIO = 1.07903

CONTACTS

P1 = 7:29:53.0 UT
U1 = 8:23: 8.7 UT
U4 = 11:49:37.4 UT
P4 = 12:42:52.5 UT

SUN

RA = $8^h 49^m 26^s.8$
DEC = $17^\circ 45' 41''.4$
SD = $15' 45''.5$
HP = $0^\circ 0' 8''.7$

MOON

RA = $8^h 49^m 40^s.0$
DEC = $17^\circ 53' 47''.3$
SD = $16' 43''.1$
HP = $1^\circ 1' 21''.3$

SAROS 136 (38/71)   JD = 2461619.922   ΔT = 88.9 SEC

GREATEST   LAT = $25^\circ 29'.0$N   ALT = $81^\circ.6$ DURATION = 6:22.7
ECLIPSE :   LONG = $39^\circ 14'.7$E   AZ = $204^\circ.3$ WIDTH = 257.7 KM

Figure 121
233

ANNULAR SOLAR ECLIPSE — 26 JAN 2028

GREATEST = 15: 7:26.7 UT
CONJUNCTION = 15:24:26.3 UT
GAMMA = 0.39018
RATIO = 0.92079

CONTACTS

P1 = 12: 6:20.8 UT
U1 = 13:14:31.4 UT
U4 = 17: 0:12.8 UT
P4 = 18: 8:28.1 UT

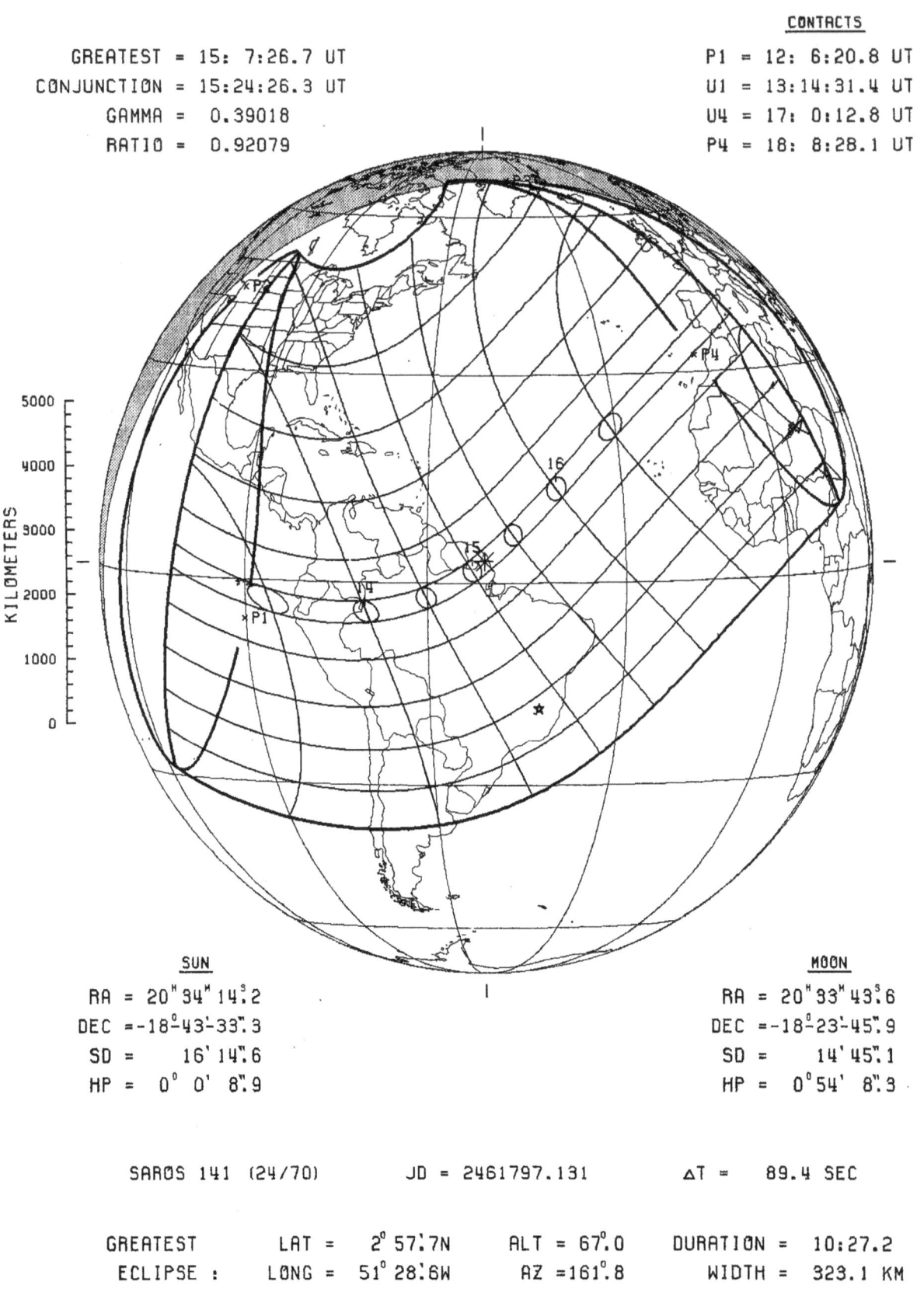

KILOMETERS

5000
4000
3000
2000
1000
0

SUN

RA = $20^h 34^m 14^s.2$
DEC = $-18^{\circ} 43' 33''.3$
SD = $16' 14''.6$
HP = $0^{\circ} 0' 8''.9$

MOON

RA = $20^h 33^m 43^s.6$
DEC = $-18^{\circ} 23' 45''.9$
SD = $14' 45''.1$
HP = $0^{\circ} 54' 8''.3$

SAROS 141 (24/70)   JD = 2461797.131   ΔT =   89.4 SEC

GREATEST   LAT = $2^{\circ} 57'.7N$ ALT = $67^{\circ}.0$ DURATION = 10:27.2
ECLIPSE :   LONG = $51^{\circ} 28'.6W$ AZ = $161^{\circ}.8$ WIDTH = 323.1 KM

Figure 122
234

TOTAL SOLAR ECLIPSE   -   22 JUL 2028

GREATEST = 2:55:10.7 UT
CONJUNCTION = 3:15:31.7 UT
GAMMA = -0.60588
RATIO = 1.05602

CONTACTS
P1 = 0:27:16.5 UT
U1 = 1:30:24.5 UT
U4 = 4:19:45.4 UT
P4 = 5:22:52.6 UT

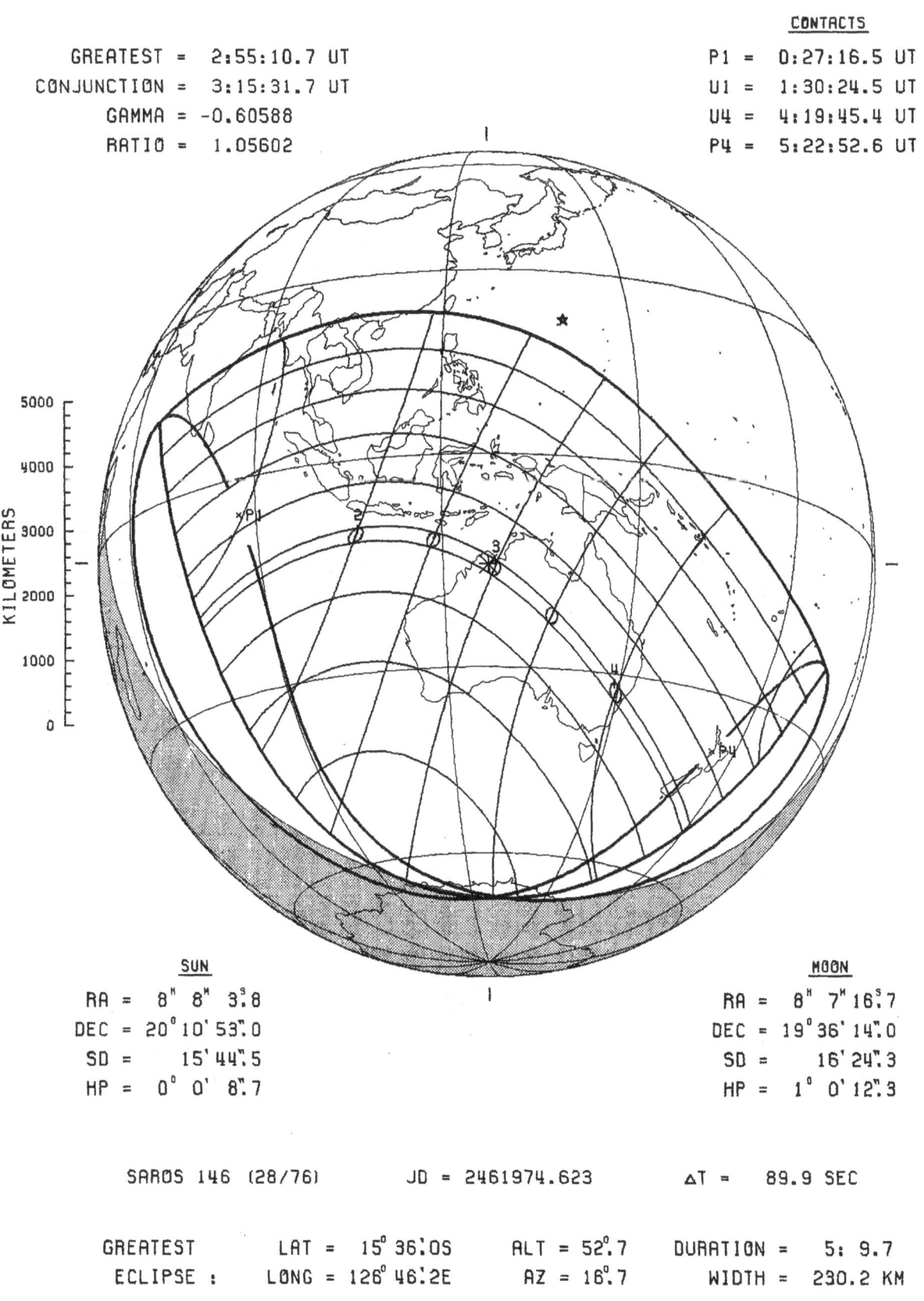

KILOMETERS
5000
4000
3000
2000
1000
0

SUN
RA = 8ʰ 8ᵐ 3ˢ.8
DEC = 20° 10' 53".0
SD = 15' 44".5
HP = 0° 0' 8".7

MOON
RA = 8ʰ 7ᵐ 16ˢ.7
DEC = 19° 36' 14".0
SD = 16' 24".3
HP = 1° 0' 12".3

SAROS 146 (28/76)   JD = 2461974.623   ΔT = 89.9 SEC

GREATEST   LAT = 15° 36'.0S   ALT = 52°.7   DURATION = 5: 9.7
ECLIPSE :   LONG = 126° 46'.2E   AZ = 16°.7   WIDTH = 230.2 KM

Figure 123
235

PARTIAL SOLAR ECLIPSE   —   14 JAN 2029

GREATEST = 17:12:14.8 UT
CONJUNCTION = 17:46:34.0 UT
GAMMA = 1.05534
MAG = 0.87097

P1 = 15: 1:36.5 UT
P4 = 19:22:44.5 UT

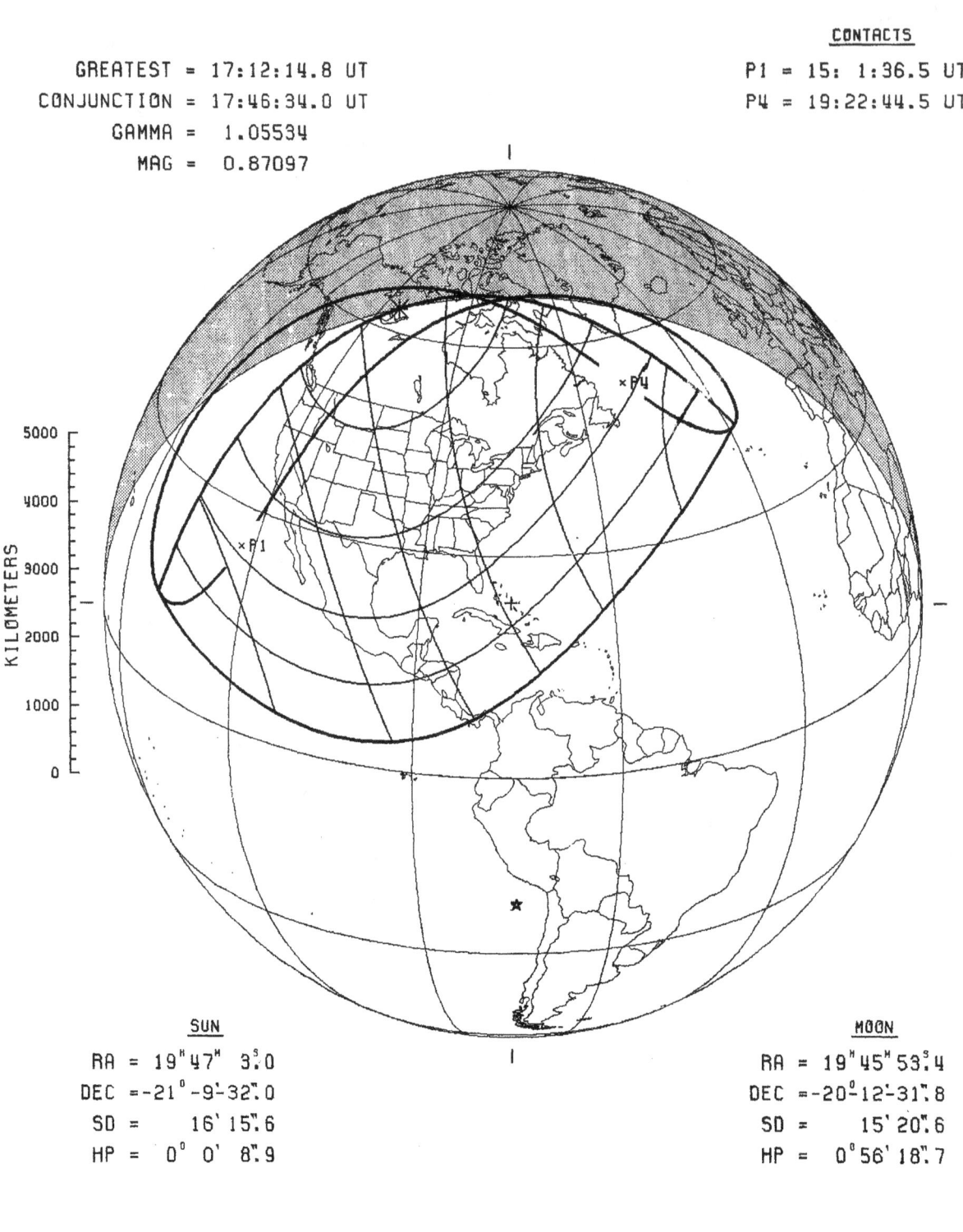

KILOMETERS

5000
4000
3000
2000
1000
0

SUN

RA = $19^h 47^m 3^s.0$
DEC = $-21° -9' -32''.0$
SD = $16' 15''.6$
HP = $0° 0' 8''.9$

MOON

RA = $19^h 45^m 53^s.4$
DEC = $-20° 12' -31''.8$
SD = $15' 20''.6$
HP = $0° 56' 18''.7$

SAROS 151 (15/72)   JD = 2462151.218   ΔT = 90.4 SEC

Figure 124

236

PARTIAL SOLAR ECLIPSE - 12 JUN 2029

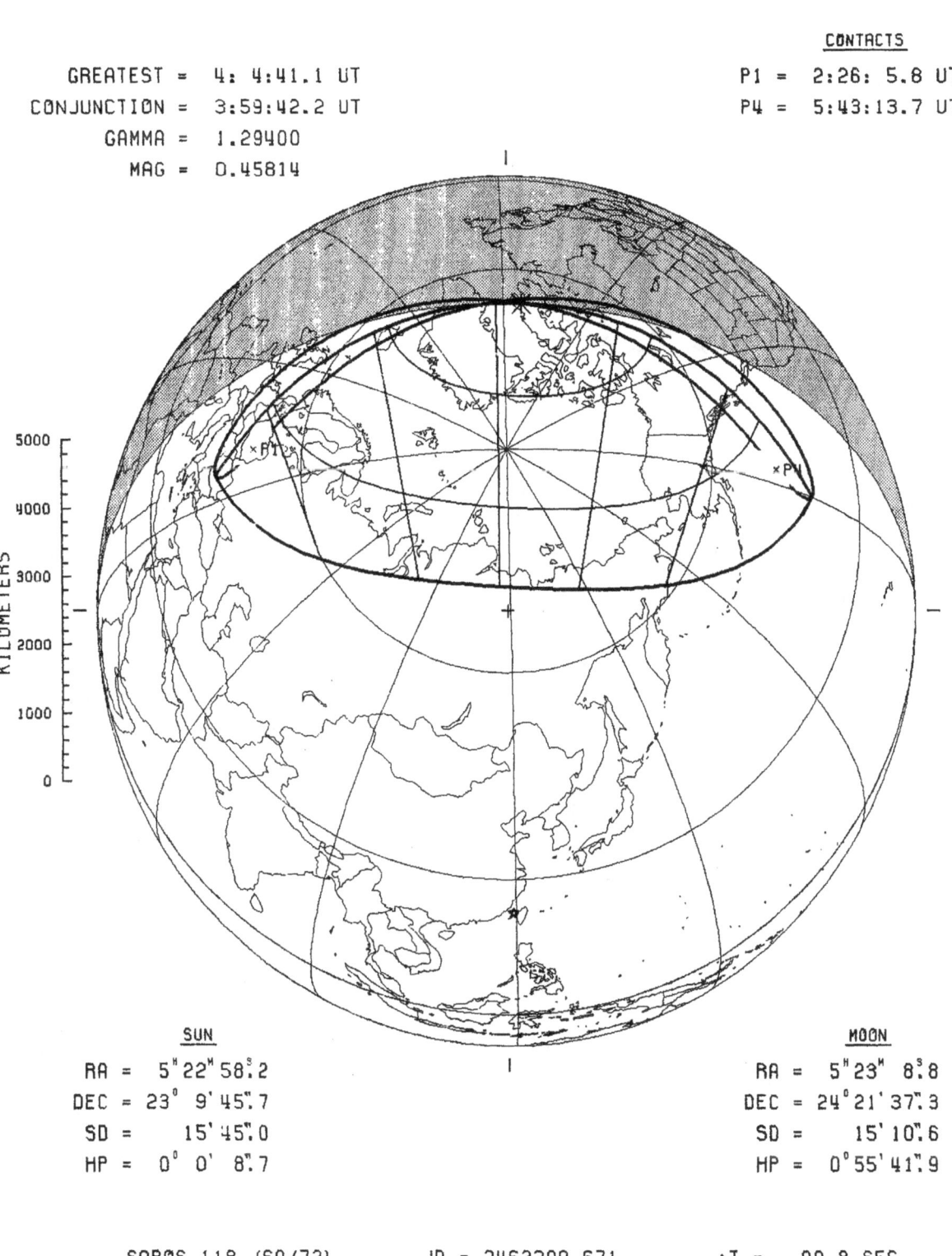

GREATEST = 4: 4:41.1 UT
CONJUNCTION = 3:59:42.2 UT
GAMMA = 1.29400
MAG = 0.45814

<u>CONTACTS</u>
P1 = 2:26: 5.8 UT
P4 = 5:43:13.7 UT

KILOMETERS

5000
4000
3000
2000
1000
0

SUN
RA = 5ʰ22ᴹ58ˢ2
DEC = 23° 9'45".7
SD = 15'45".0
HP = 0° 0' 8".7

MOON
RA = 5ʰ23ᴹ 8ˢ8
DEC = 24°21'37".3
SD = 15'10".6
HP = 0°55'41".9

SAROS 118 (69/72)   JD = 2462299.671   ΔT = 90.8 SEC

Figure 125

237

PARTIAL SOLAR ECLIPSE — 11 JUL 2029

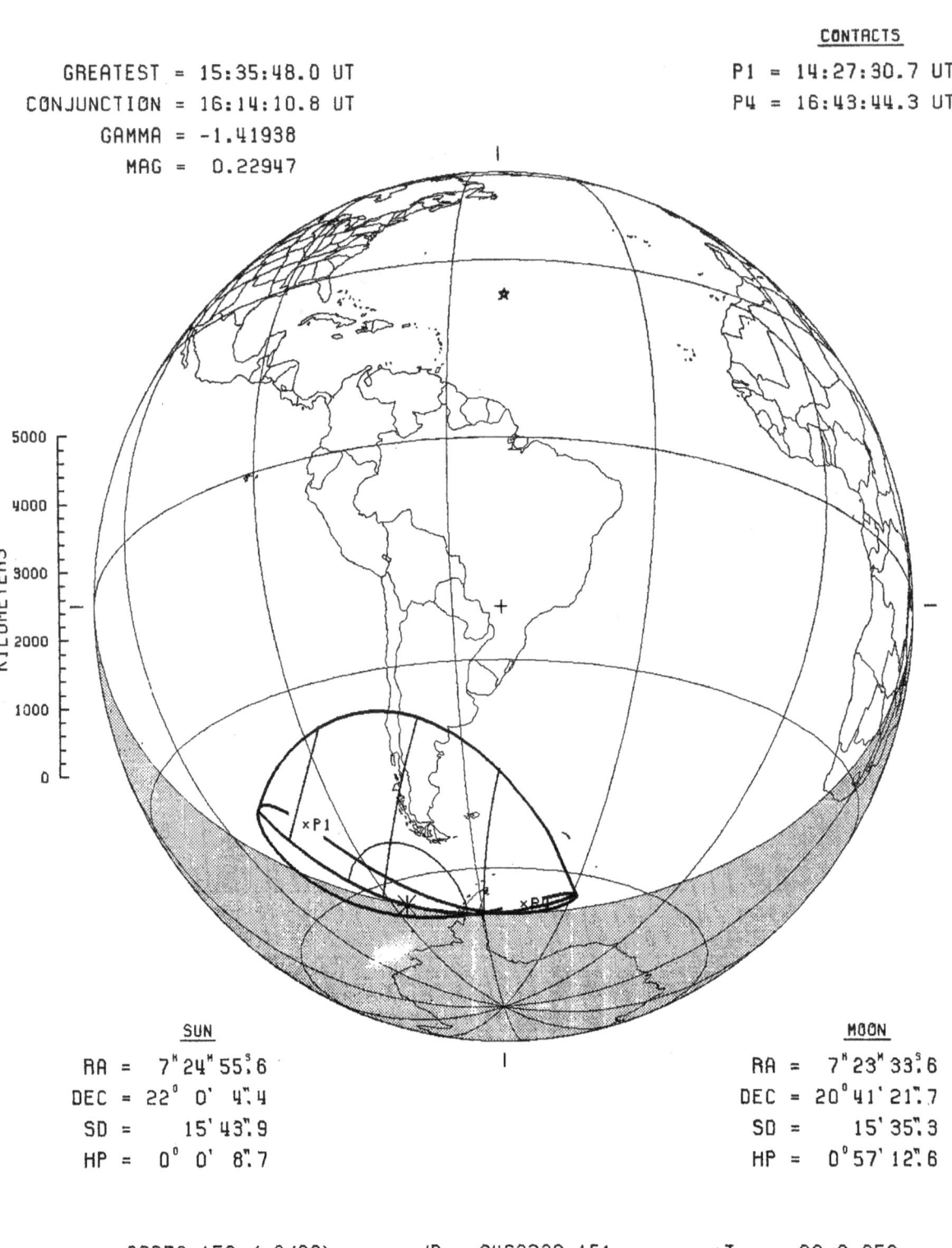

Figure 126

238

PARTIAL SOLAR ECLIPSE — 5 DEC 2029

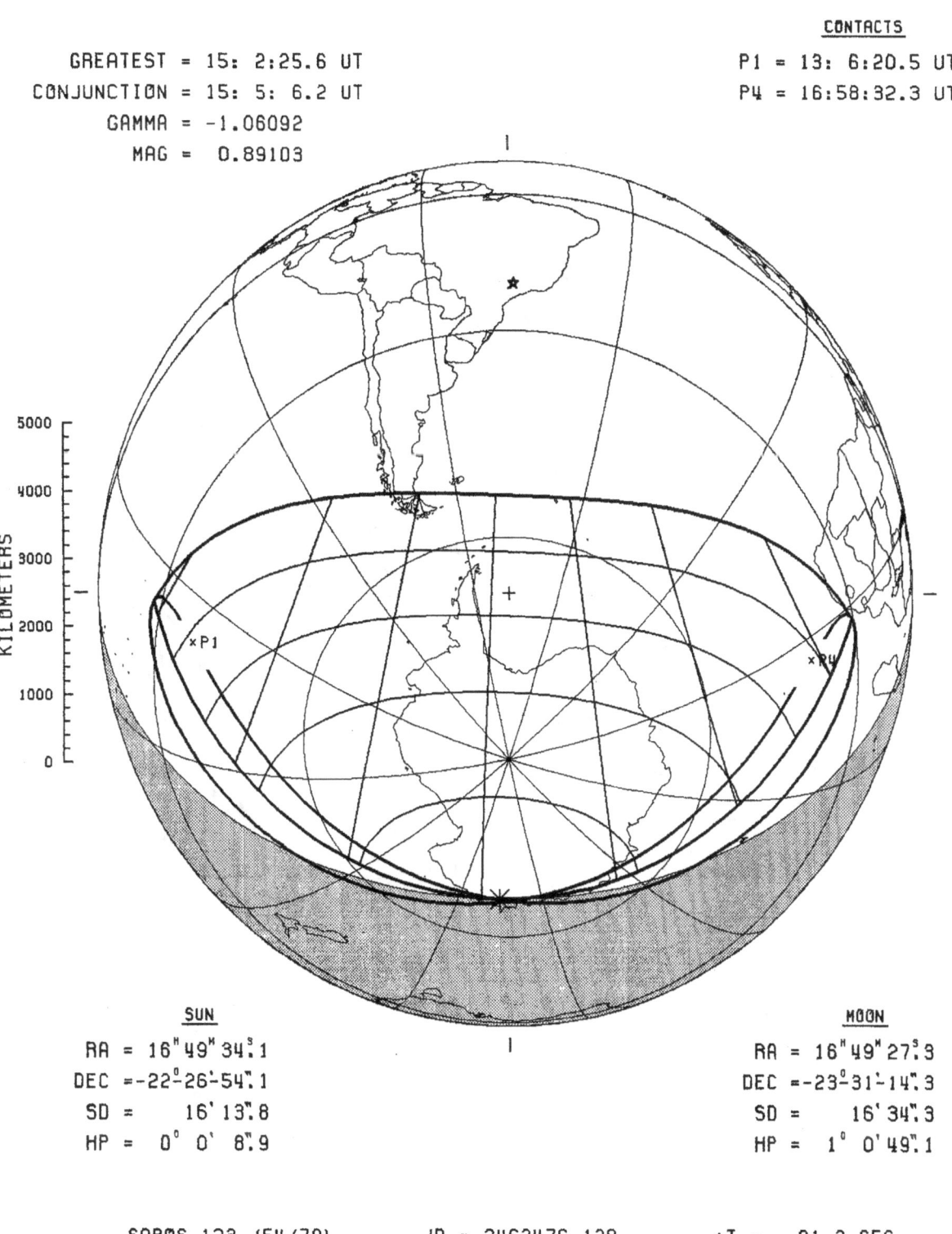

GREATEST = 15: 2:25.6 UT
CONJUNCTION = 15: 5: 6.2 UT
GAMMA = -1.06092
MAG = 0.89103

<u>CONTACTS</u>

P1 = 13: 6:20.5 UT
P4 = 16:58:32.3 UT

×P1

+

×P4

KILOMETERS

5000
4000
3000
2000
1000
0

<u>SUN</u>

RA = 16ʰ49ᵐ34ˢ.1
DEC =-22°26'54".1
SD = 16'13".8
HP = 0° 0' 8".9

<u>MOON</u>

RA = 16ʰ49ᵐ27ˢ.3
DEC =-23°31'14".3
SD = 16'34".3
HP = 1° 0'49".1

SAROS 123 (54/70)   JD = 2462476.128   ΔT =   91.3 SEC

Figure 127

239

ANNULAR SOLAR ECLIPSE — 1 JUN 2030

GREATEST = 6:27:41.5 UT
CONJUNCTION = 6:30:26.5 UT
GAMMA = 0.56233
RATIO = 0.94425

CONTACTS

P1 = 3:34:21.4 UT
U1 = 4:46:53.0 UT
U4 = 8: 8:27.3 UT
P4 = 9:20:58.9 UT

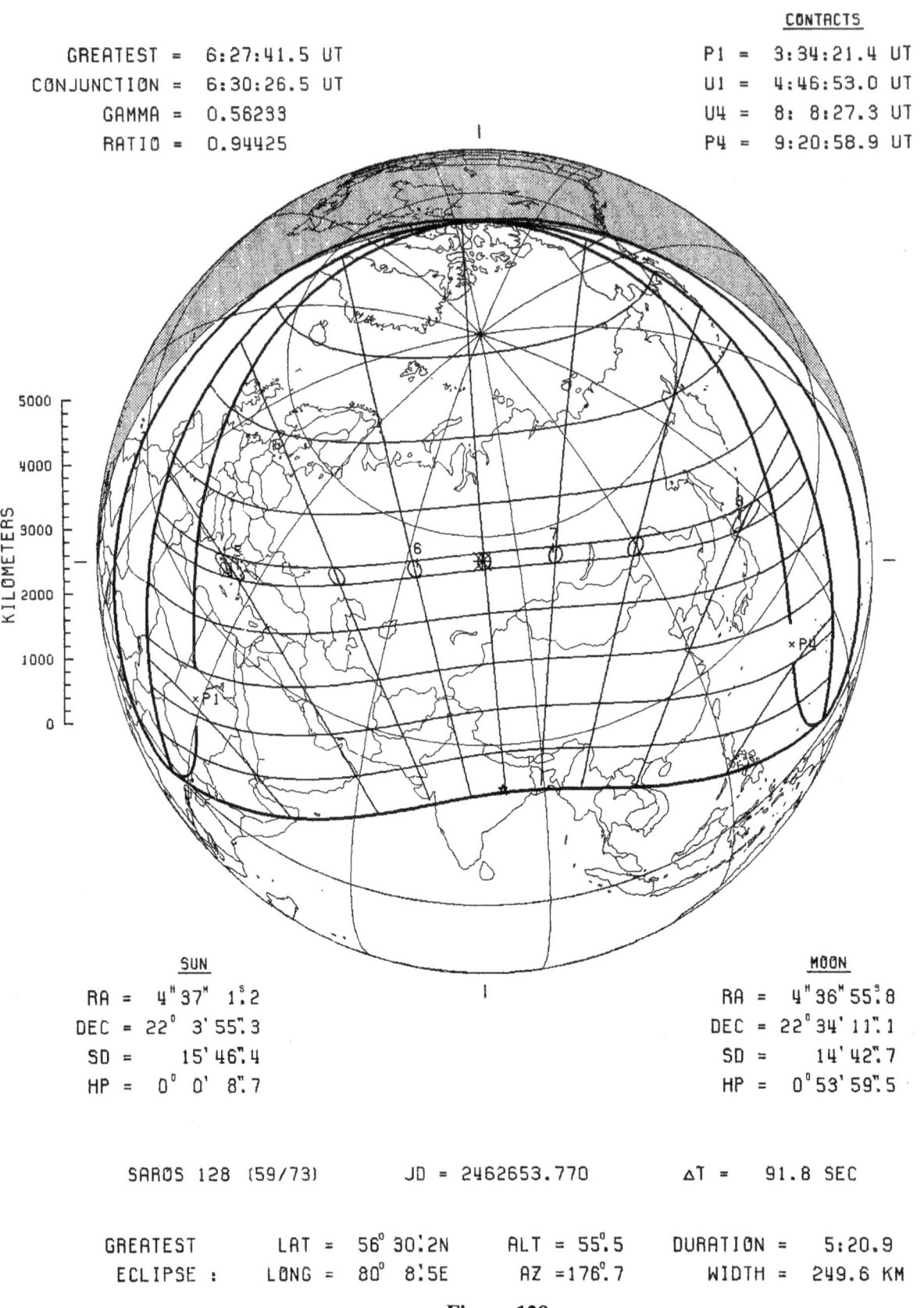

SUN

RA = 4ʰ 37ᴹ 1ˢ.2
DEC = 22° 3' 55".3
SD = 15' 46".4
HP = 0° 0' 8".7

MOON

RA = 4ʰ 36ᴹ 55ˢ.8
DEC = 22° 34' 11".1
SD = 14' 42".7
HP = 0° 53' 59".5

SAROS 128 (59/73)   JD = 2462653.770   ΔT =   91.8 SEC

GREATEST   LAT = 56° 30'.2N   ALT = 55°.5   DURATION =   5:20.9
ECLIPSE :   LONG = 80° 8'.5E   AZ =176°.7   WIDTH = 249.6 KM

Figure 128

TOTAL SOLAR ECLIPSE -   25 NOV 2030

GREATEST = 6:50: 3.9 UT
CONJUNCTION = 6:53:52.5 UT
GAMMA = -0.38679
RATIO = 1.04685

CONTACTS

P1 = 4:16:22.8 UT
U1 = 5:14: 1.4 UT
U4 = 8:26: 4.4 UT
P4 = 9:23:41.8 UT

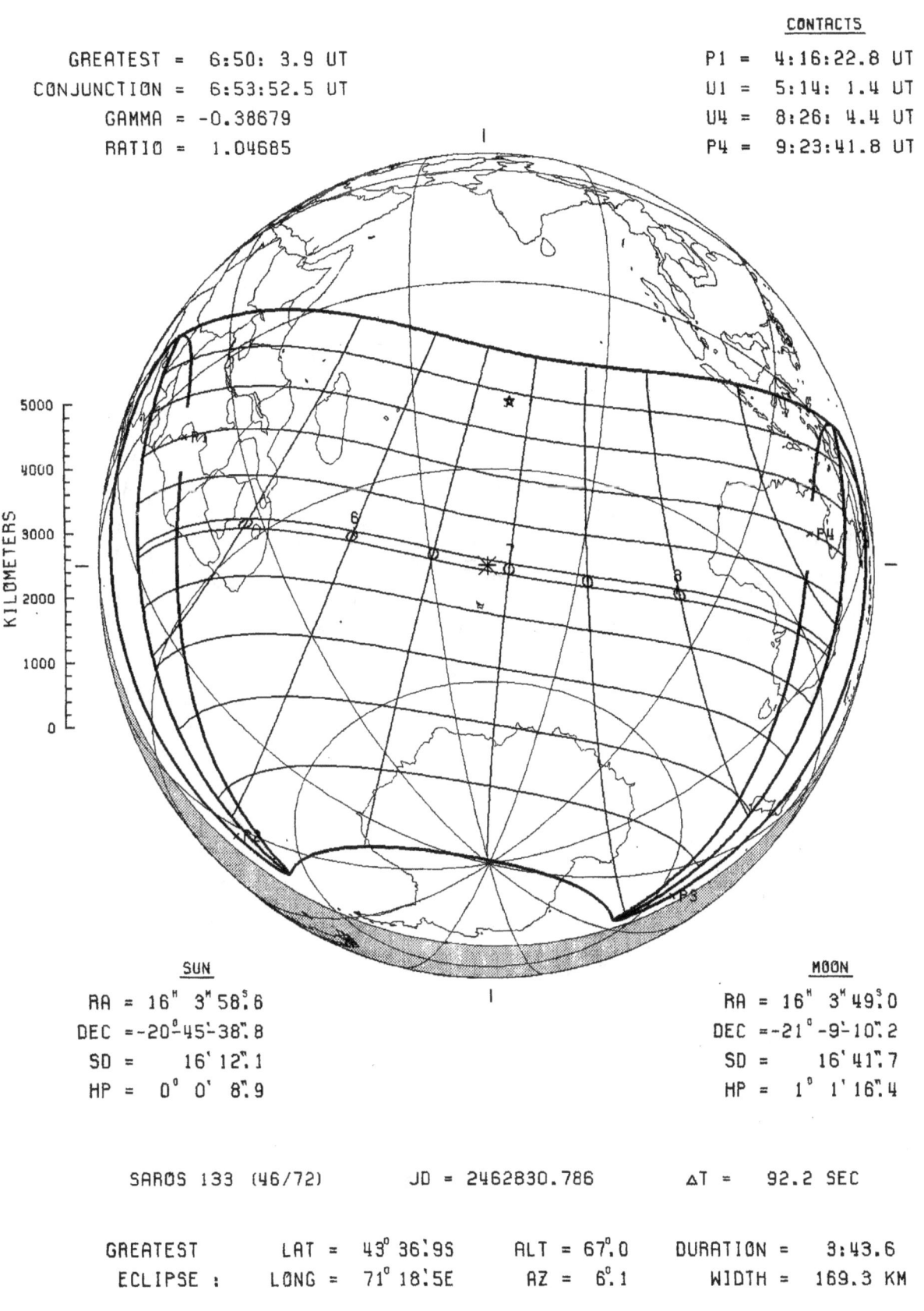

SUN

RA = 16^h 3^{m}58^{s}6
DEC = -20°45,38,,8
SD = 16,12,,1
HP = 0° 0, 8,,9

MOON

RA = 16^h 3^{m}49^{s}0
DEC = -21°-9,10,,2
SD = 16,41,,7
HP = 1° 1,16,,4

SAROS 133 (46/72)   JD = 2462830.786   ΔT =   92.2 SEC

GREATEST   LAT = 43° 36,9S   ALT = 67°0   DURATION =   3:43.6
ECLIPSE :   LONG = 71° 18,5E   AZ = 6°1   WIDTH = 169.3 KM

Figure 129
241

ANNULAR SOLAR ECLIPSE — 21 MAY 2031

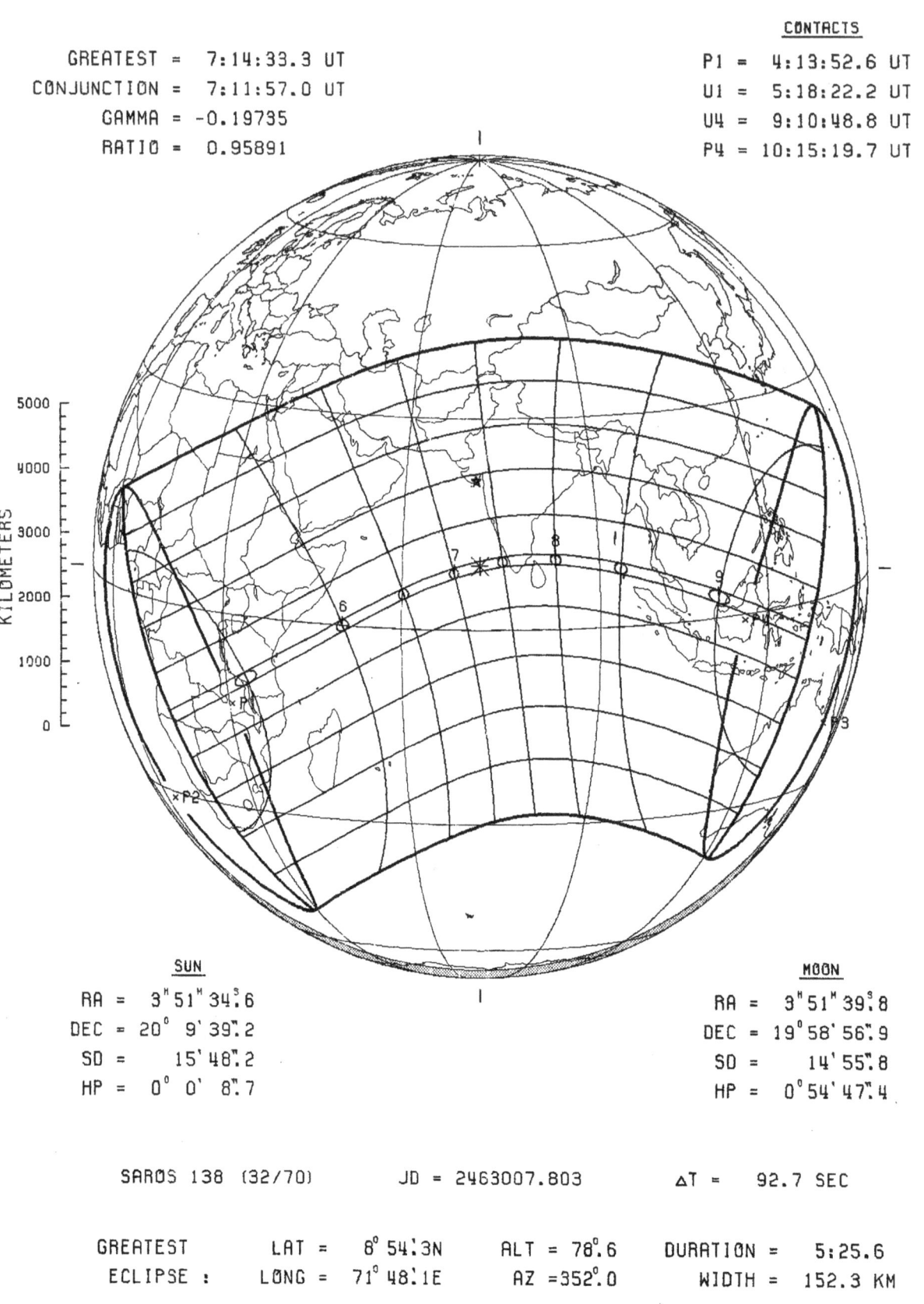

GREATEST = 7:14:33.3 UT
CONJUNCTION = 7:11:57.0 UT
GAMMA = -0.19735
RATIO = 0.95891

<u>CONTACTS</u>

P1 = 4:13:52.6 UT
U1 = 5:18:22.2 UT
U4 = 9:10:48.8 UT
P4 = 10:15:19.7 UT

KILOMETERS
5000
4000
3000
2000
1000
0

<u>SUN</u>

RA = 3ʰ51ᴹ34ˢ6
DEC = 20° 9' 39".2
SD = 15' 48".2
HP = 0° 0' 8".7

<u>MOON</u>

RA = 3ʰ51ᴹ39ˢ8
DEC = 19°58' 56".9
SD = 14' 55".8
HP = 0°54' 47".4

SAROS 138 (32/70)   JD = 2463007.803   ΔT = 92.7 SEC

GREATEST   LAT = 8° 54".3N   ALT = 78°.6   DURATION = 5:25.6
ECLIPSE :   LONG = 71° 48".1E   AZ =352°.0   WIDTH = 152.3 KM

Figure 130
242

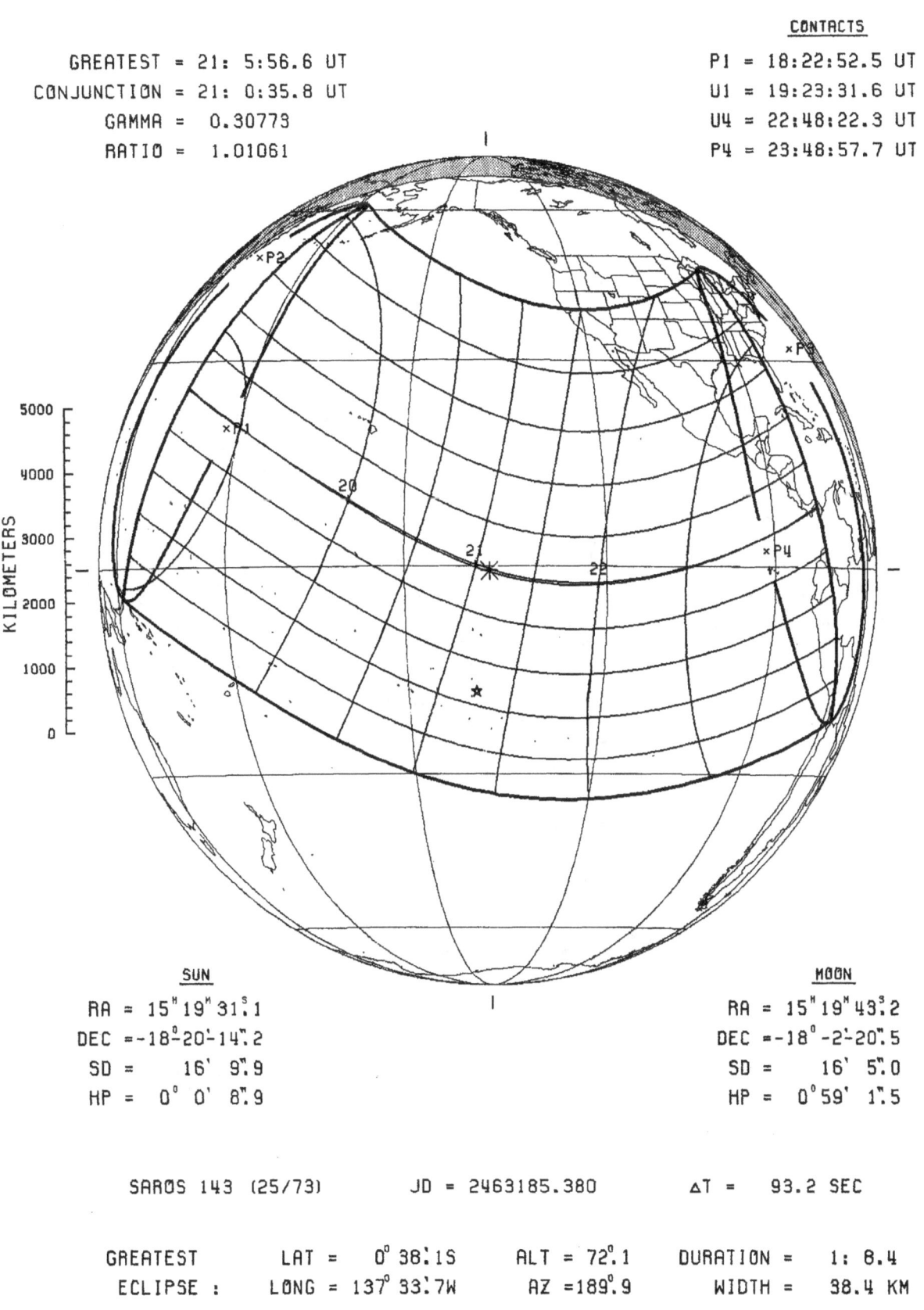

ANN/TOT SOLAR ECLIPSE - 14 NOV 2031

GREATEST = 21: 5:56.6 UT
CONJUNCTION = 21: 0:35.8 UT
GAMMA = 0.30773
RATIO = 1.01061

CONTACTS

P1 = 18:22:52.5 UT
U1 = 19:23:31.6 UT
U4 = 22:48:22.3 UT
P4 = 23:48:57.7 UT

KILOMETERS

5000
4000
3000
2000
1000
0

SUN

RA = $15^h 19^m 31^s.1$
DEC = $-18°20'14".2$
SD = $16' 9".9$
HP = $0° 0' 8".9$

MOON

RA = $15^h 19^m 49^s.2$
DEC = $-18° 2'20".5$
SD = $16' 5".0$
HP = $0°59' 1".5$

SAROS 143 (25/73)   JD = 2463185.380   ΔT = 93.2 SEC

GREATEST   LAT = $0° 38'.1S$ ALT = $72°.1$ DURATION = 1: 8.4
ECLIPSE :   LONG = $137° 33'.7W$ AZ = $189°.9$ WIDTH = 38.4 KM

Figure 131
243

ANNULAR SOLAR ECLIPSE — 9 MAY 2032

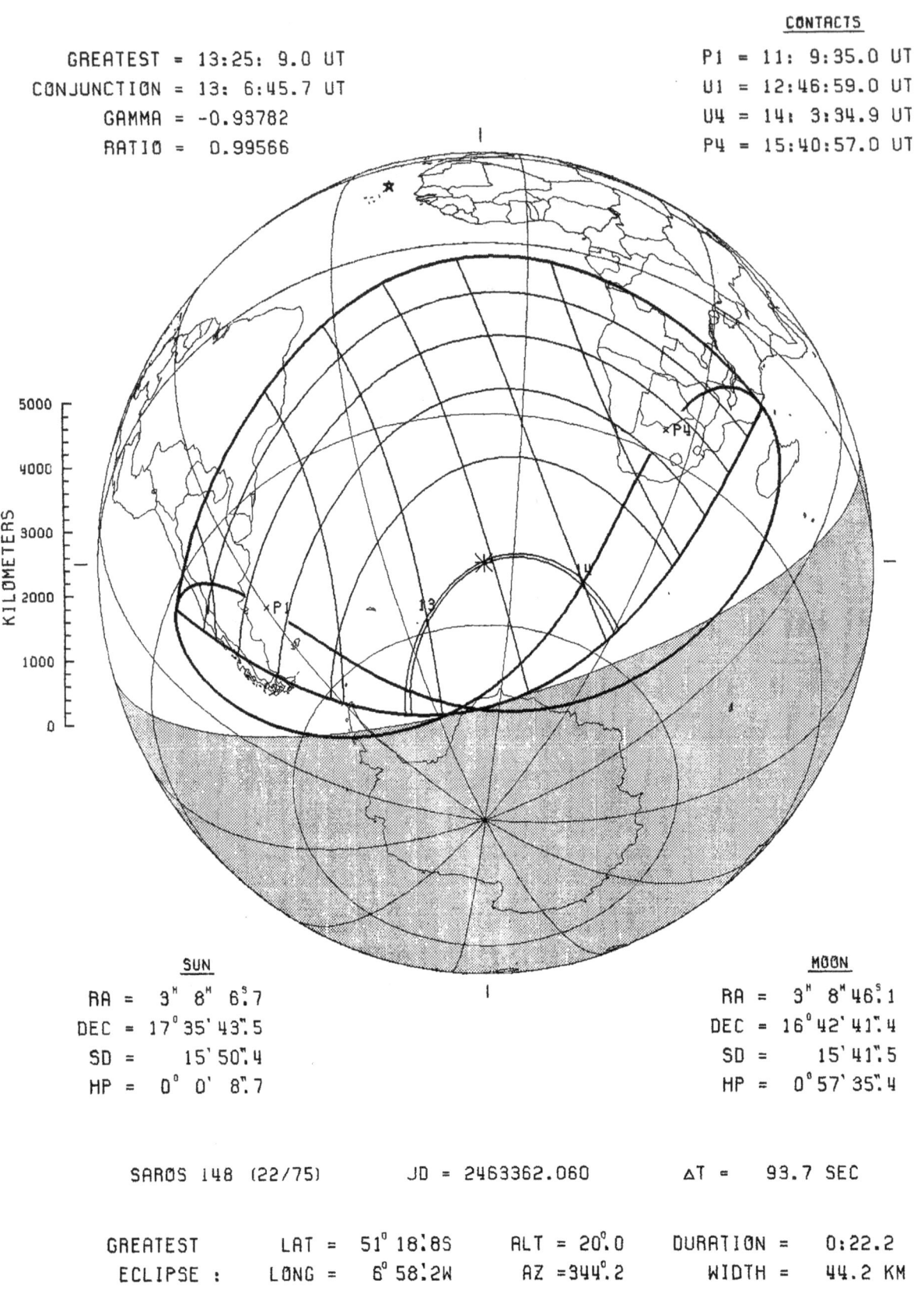

GREATEST = 13:25: 9.0 UT
CONJUNCTION = 13: 6:45.7 UT
GAMMA = -0.93782
RATIO = 0.99566

CONTACTS

P1 = 11: 9:35.0 UT
U1 = 12:46:59.0 UT
U4 = 14: 3:34.9 UT
P4 = 15:40:57.0 UT

KILOMETERS

5000
4000
3000
2000
1000
0

SUN

RA = 3ʰ 8ᴹ 6ˢ.7
DEC = 17° 35' 43".5
SD = 15' 50".4
HP = 0° 0' 8".7

MOON

RA = 3ʰ 8ᴹ 46ˢ.1
DEC = 16° 42' 41".4
SD = 15' 41".5
HP = 0° 57' 35".4

SAROS 148 (22/75)   JD = 2463362.060   ΔT = 93.7 SEC

GREATEST   LAT = 51° 18'.8S   ALT = 20°.0   DURATION = 0:22.2
ECLIPSE :   LONG = 6° 58'.2W   AZ = 344°.2   WIDTH = 44.2 KM

Figure 132

244

PARTIAL SOLAR ECLIPSE – 3 NOV 2032

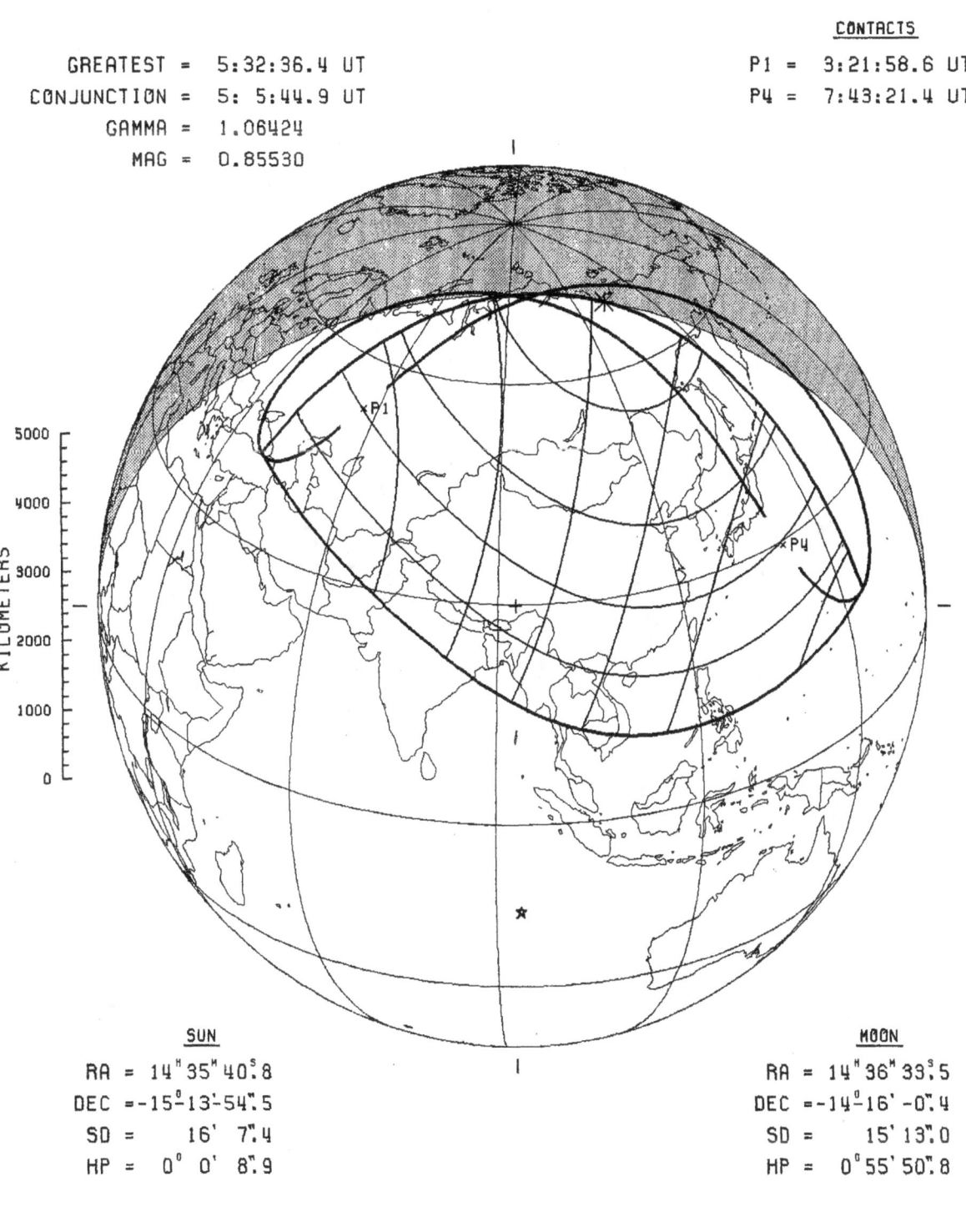

GREATEST = 5:32:36.4 UT

CONJUNCTION = 5: 5:44.9 UT

GAMMA = 1.06424

MAG = 0.85530

<u>CONTACTS</u>

P1 = 3:21:58.6 UT

P4 = 7:43:21.4 UT

KILOMETERS

5000

4000

3000

2000

1000

0

<u>SUN</u>

RA = $14^h 35^m 40^s.8$

DEC = $-15^{\circ} 13' 54''.5$

SD = $16' 7''.4$

HP = $0^{\circ} 0' 8''.9$

<u>MOON</u>

RA = $14^h 36^m 33^s.5$

DEC = $-14^{\circ} 16' -0''.4$

SD = $15' 13''.0$

HP = $0^{\circ} 55' 50''.8$

SAROS 153 (10/70)   JD = 2463539.732   ∆T =   94.2 SEC

Figure 133

TOTAL SOLAR ECLIPSE   -   30 MAR 2033

GREATEST = 18: 1: 0.7 UT
CONJUNCTION = 18:32:51.3 UT
GAMMA =   0.97760
RATIO =   1.04616

CONTACTS

P1 = 15:59:10.1 UT
U1 = 17:35:23.4 UT
U4 = 18:26:14.5 UT
P4 = 20: 2:37.1 UT

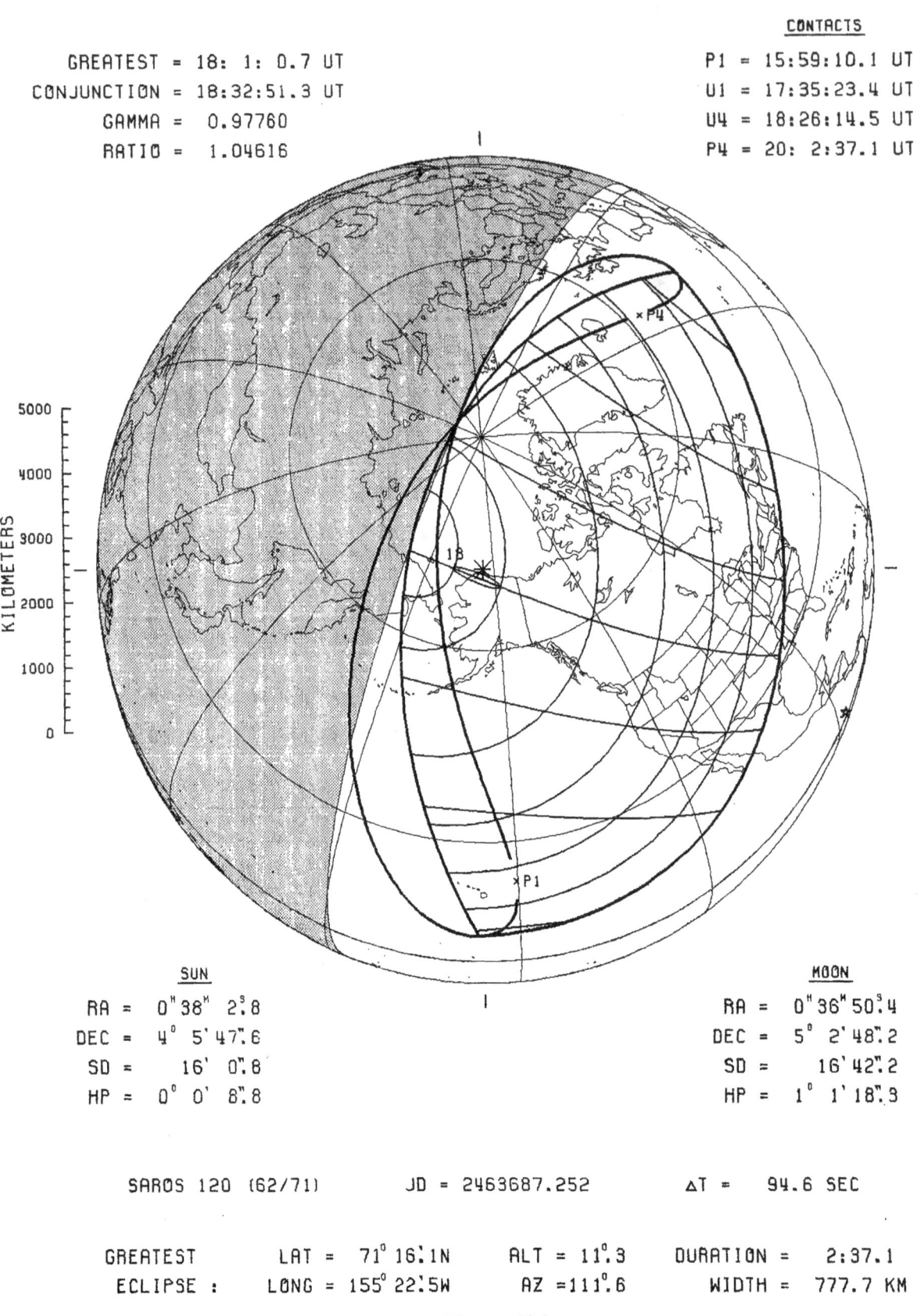

SUN

RA = 0ʰ38ᴹ 2ˢ.8
DEC = 4° 5' 47".6
SD =   16' 0".8
HP = 0° 0' 8".8

MOON

RA = 0ʰ36ᴹ 50ˢ.4
DEC = 5° 2' 48".2
SD =   16' 42".2
HP = 1° 1' 18".9

SAROS 120 (62/71)   JD = 2463687.252   ΔT =   94.6 SEC

GREATEST   LAT = 71° 16'.1N   ALT = 11°.3   DURATION =   2:37.1
ECLIPSE :   LONG = 155° 22'.5W   AZ =111°.6   WIDTH =   777.7 KM

Figure 134

246

PARTIAL SOLAR ECLIPSE — 23 SEP 2033

CONTACTS

GREATEST = 13:52:53.9 UT

CONJUNCTION = 14:37: 0.3 UT

GAMMA = -1.15839

MAG = 0.68829

P1 = 11:47:30.2 UT

P4 = 15:57:58.7 UT

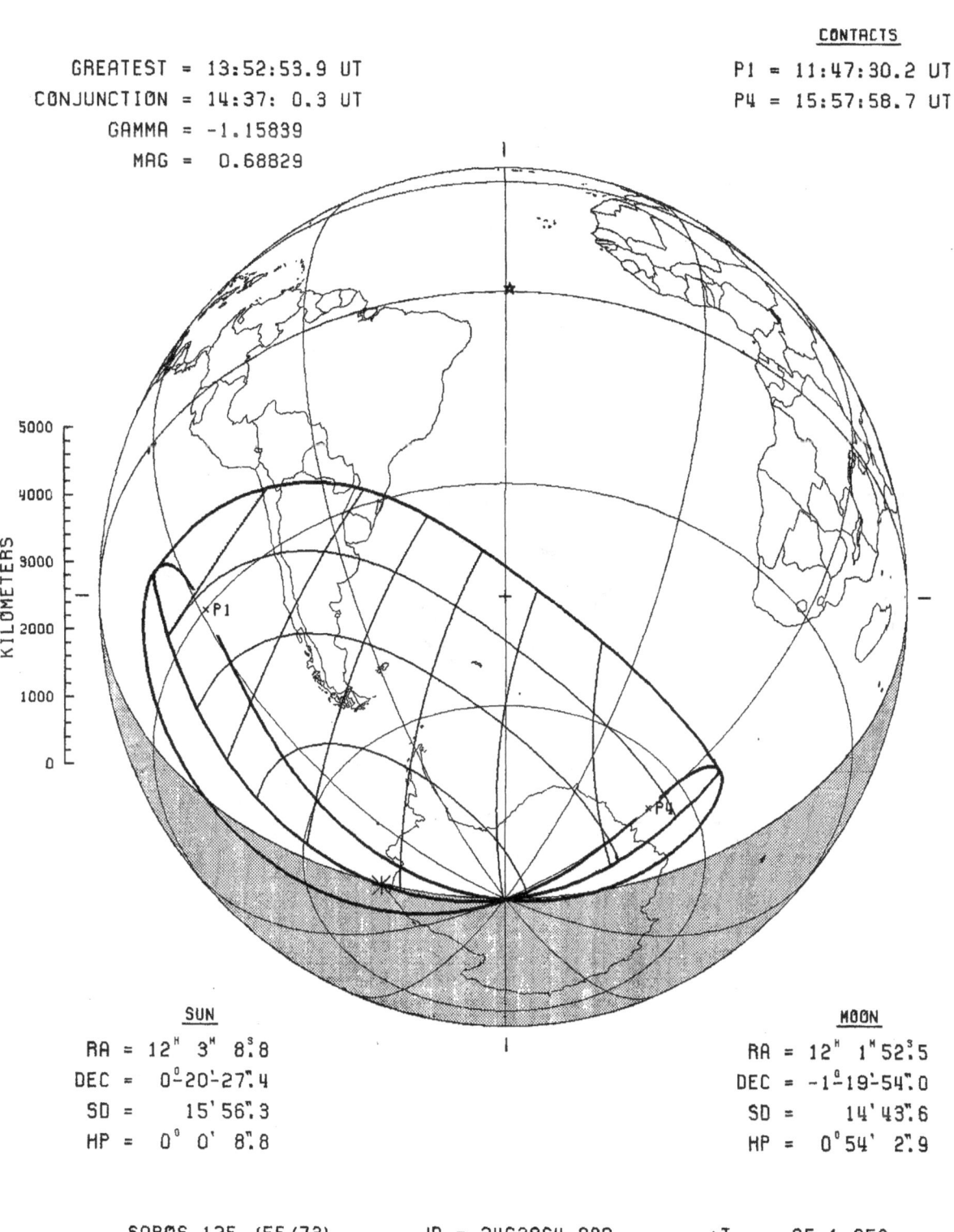

SUN

RA = 12^H 3^M 8^{S}8

DEC = 0^{0}20'27".4

SD = 15'56".3

HP = 0^0 0' 8".8

MOON

RA = 12^H 1^{M}52^{S}5

DEC = -1^{0}19'54".0

SD = 14'43".6

HP = 0^{0}54' 2".9

SAROS 125 (55/73)   JD = 2463864.080   ΔT = 95.1 SEC

Figure 135

247

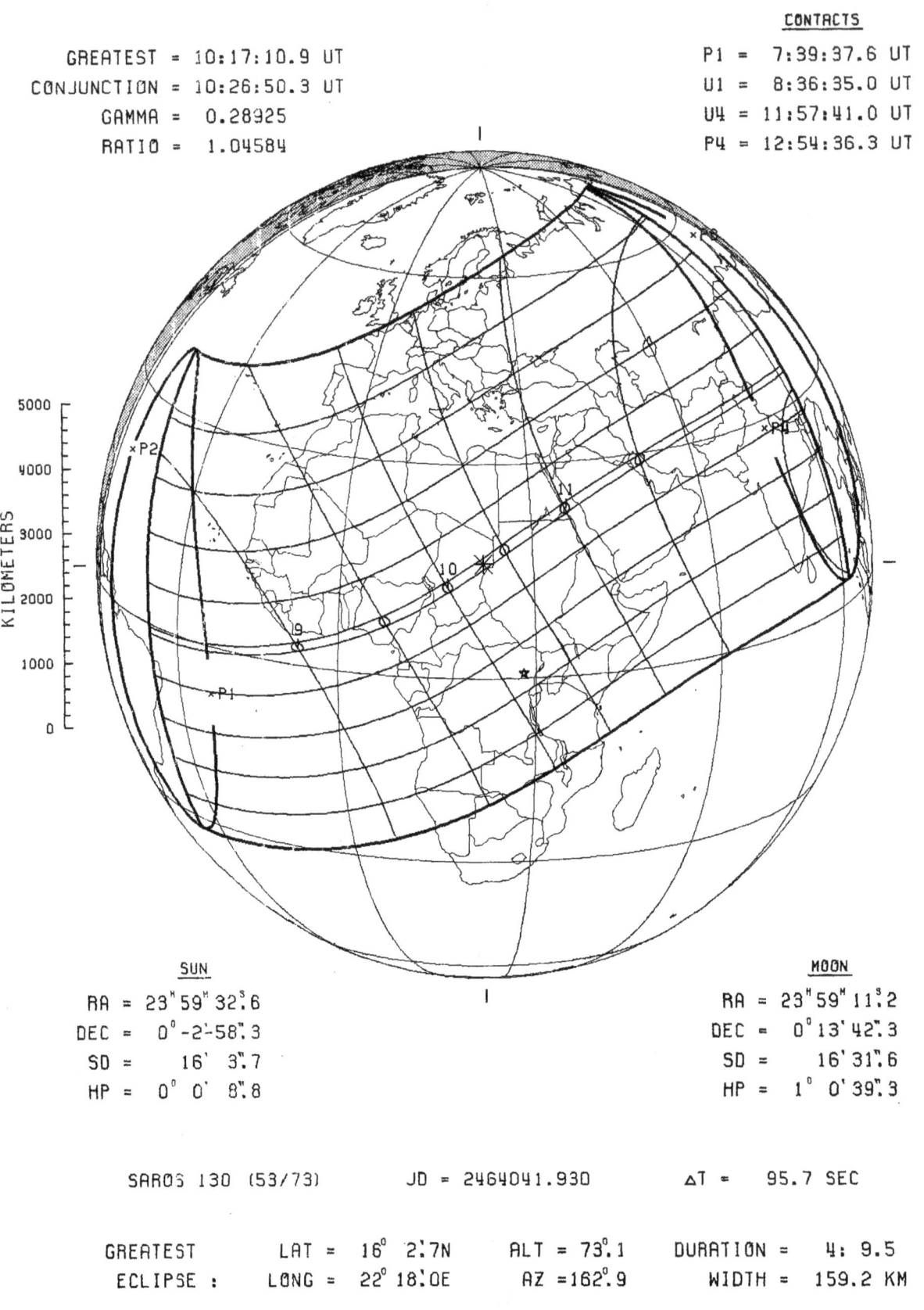

TOTAL SOLAR ECLIPSE – 20 MAR 2034

GREATEST = 10:17:10.9 UT
CONJUNCTION = 10:26:50.3 UT
GAMMA = 0.28925
RATIO = 1.04584

<u>CONTACTS</u>

P1 = 7:39:37.6 UT
U1 = 8:36:35.0 UT
U4 = 11:57:41.0 UT
P4 = 12:54:36.3 UT

<u>SUN</u>

RA = $23^h 59^m 32^s.6$
DEC = $0° -2'-58".3$
SD = $16' 3".7$
HP = $0° 0' 8".8$

<u>MOON</u>

RA = $23^h 59^m 11^s.2$
DEC = $0° 13' 42".3$
SD = $16' 31".6$
HP = $1° 0' 39".3$

SAROS 130 (53/73)   JD = 2464041.930   ΔT = 95.7 SEC

GREATEST   LAT = $16° 2'.7N$ ALT = $73°.1$ DURATION = 4: 9.5
ECLIPSE :   LONG = $22° 18'.0E$ AZ = $162°.9$ WIDTH = 159.2 KM

Figure 136
248

ANNULAR SOLAR ECLIPSE – 12 SEP 2034

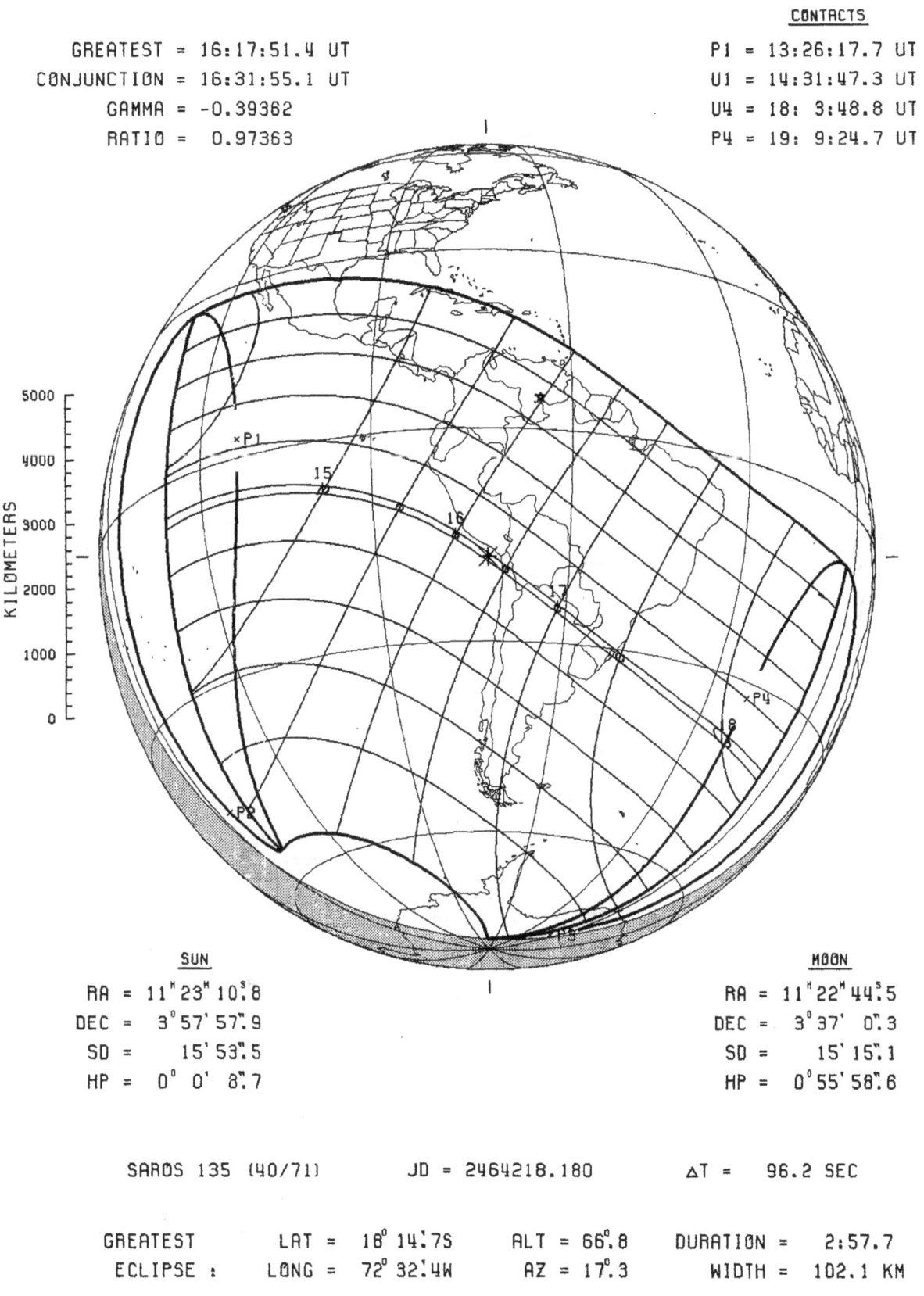

GREATEST = 16:17:51.4 UT
CONJUNCTION = 16:31:55.1 UT
GAMMA = -0.39362
RATIO = 0.97363

CONTACTS
P1 = 13:26:17.7 UT
U1 = 14:31:47.3 UT
U4 = 18: 3:48.8 UT
P4 = 19: 9:24.7 UT

SUN
RA = $11^h 23^m 10\overset{s}{.}8$
DEC = $3° 57' 57\overset{"}{.}9$
SD = $15' 53\overset{"}{.}5$
HP = $0° 0' 8\overset{"}{.}7$

MOON
RA = $11^h 22^m 44\overset{s}{.}5$
DEC = $3° 37' 0\overset{"}{.}3$
SD = $15' 15\overset{"}{.}1$
HP = $0° 55' 58\overset{"}{.}6$

SAROS 135 (40/71)   JD = 2464218.180   ΔT = 96.2 SEC

GREATEST   LAT = $18° 14\overset{.}{.}7S$ ALT = $66°.8$ DURATION = 2:57.7
ECLIPSE :   LONG = $72° 32\overset{.}{.}4W$ AZ = $17°.3$ WIDTH = 102.1 KM

Figure 137
249

ANNULAR SOLAR ECLIPSE - 9 MAR 2035

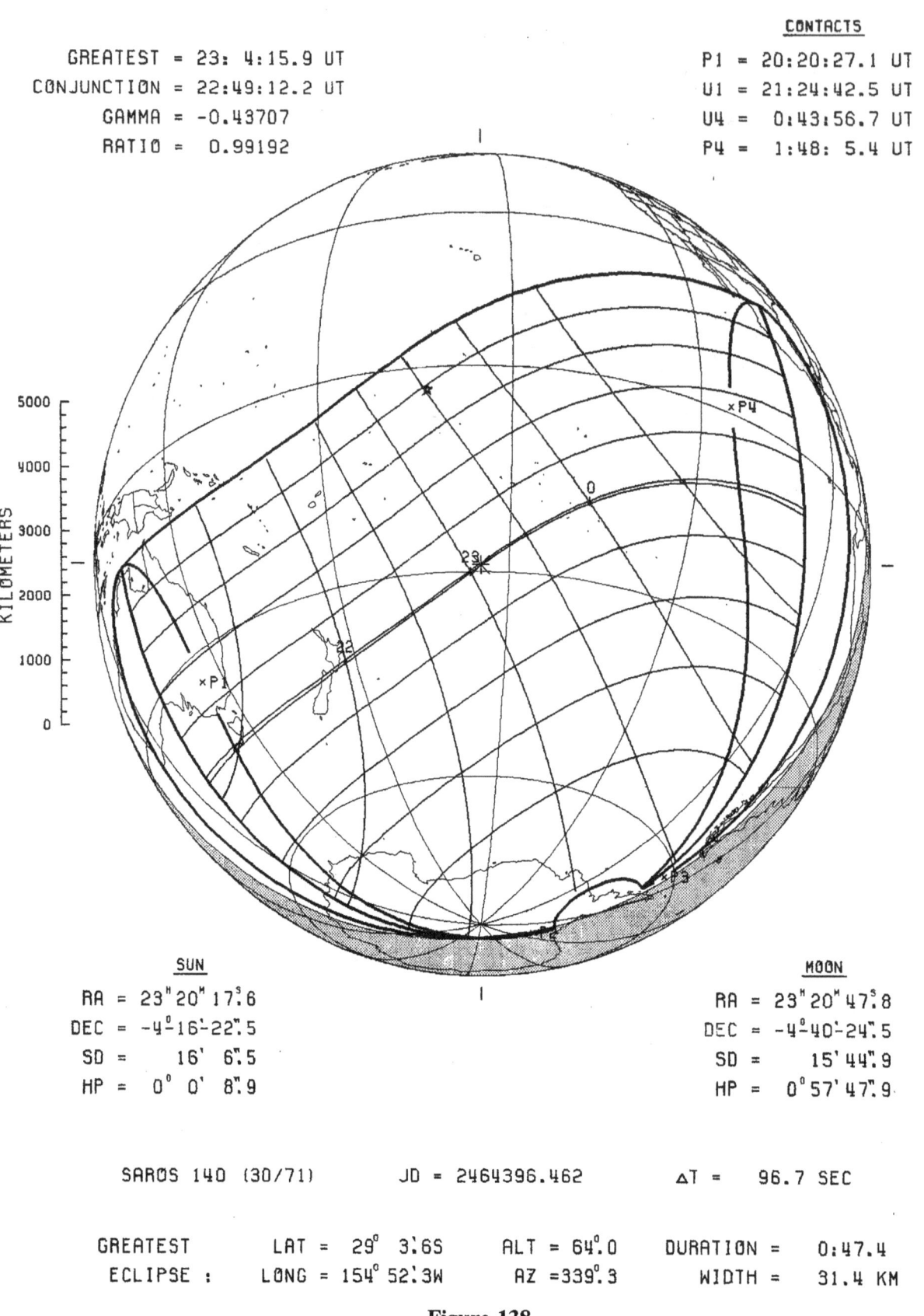

GREATEST = 23: 4:15.9 UT
CONJUNCTION = 22:49:12.2 UT
GAMMA = -0.43707
RATIO = 0.99192

P1 = 20:20:27.1 UT
U1 = 21:24:42.5 UT
U4 = 0:43:56.7 UT
P4 = 1:48: 5.4 UT

KILOMETERS

5000
4000
3000
2000
1000
0

SUN

RA = $23^h 20^m 17^s.6$
DEC = $-4^o 16' 22".5$
SD = $16' 6".5$
HP = $0^o 0' 8".9$

MOON

RA = $23^h 20^m 47^s.8$
DEC = $-4^o 40' 24".5$
SD = $15' 44".9$
HP = $0^o 57' 47".9$

SAROS 140 (30/71)   JD = 2464396.462   ∆T = 96.7 SEC

GREATEST   LAT = $29^o 3'.6$S   ALT = $64^o.0$ DURATION = 0:47.4
ECLIPSE :   LONG = $154^o 52'.3$W   AZ = $339^o.3$ WIDTH = 31.4 KM

Figure 138
250

TOTAL SOLAR ECLIPSE –   2 SEP 2035

GREATEST = 1:55: 8.8 UT
CONJUNCTION = 1:43:23.5 UT
GAMMA = 0.37278
RATIO = 1.03202

P1 = 23:15: 8.4 UT
U1 = 0:15:28.1 UT
U4 = 3:34:56.1 UT
P4 = 4:35:20.2 UT

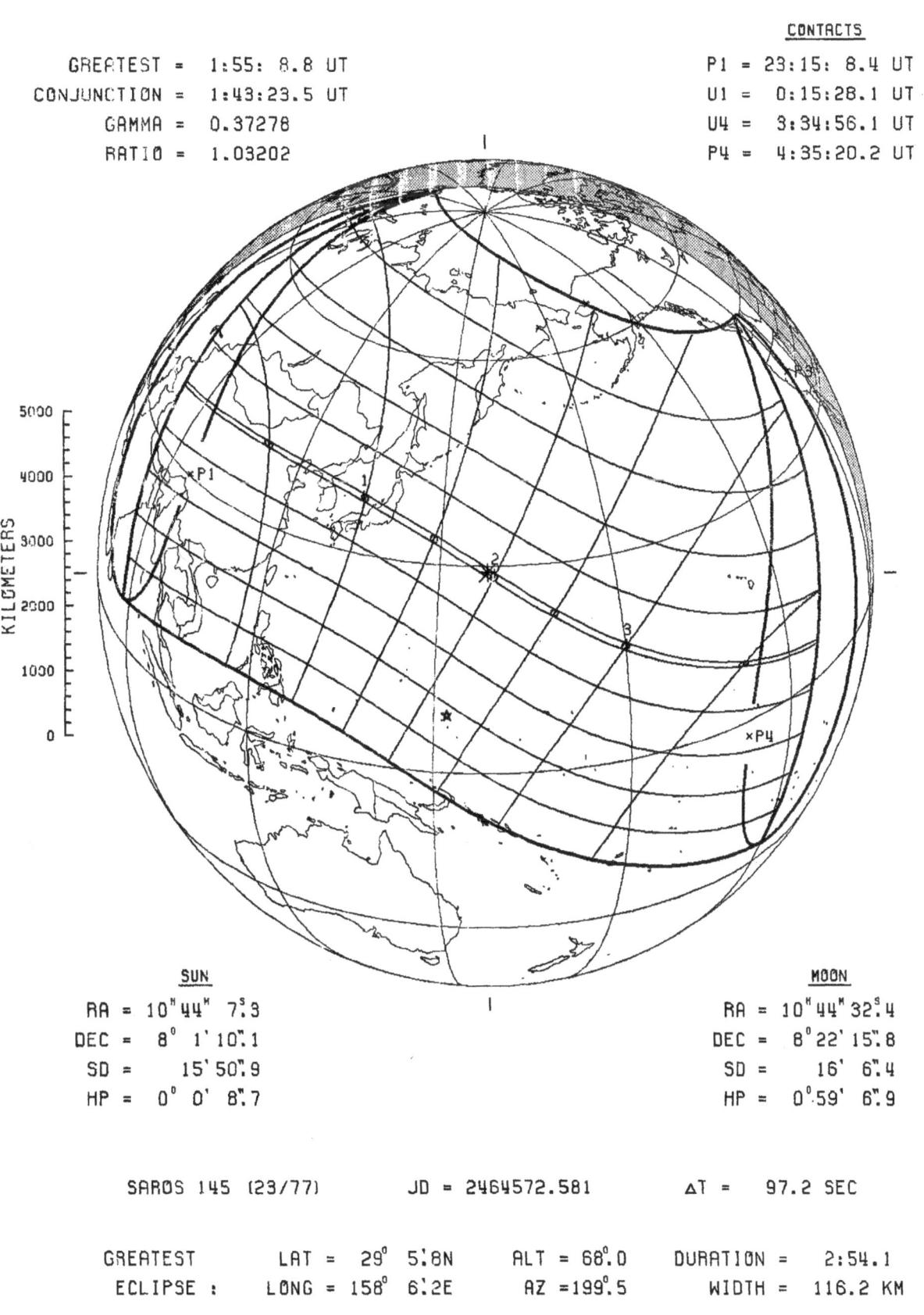

KILOMETERS

5000
4000
3000
2000
1000
0

SUN

RA = 10^{h}44^m 7^{s}3
DEC = 8^0 1' 10".1
SD =   15' 50".9
HP = 0^0 0' 8".7

MOON

RA = 10^{h}44^{m}32^{s}4
DEC = 8^{0}22' 15".8
SD =   16' 6".4
HP = 0^{0}59' 6".9

SAROS 145 (23/77)   JD = 2464572.581   ΔT =   97.2 SEC

GREATEST   LAT = 29^0 5.8N   ALT = 68^0.0   DURATION =   2:54.1
ECLIPSE :   LONG = 158^0 6.2E   AZ =199^0.5   WIDTH = 116.2 KM

Figure 139
251

FIFTY YEAR CANON OF SOLAR ECLIPSES : 1986 - 2035

APPENDIX A - <u>SOLAR ECLIPSES</u>

GEOMETRY OF SOLAR ECLIPSES

One of the most remarkable coincidences found in nature is the fact that the Moon and Sun both appear the same size as seen from the Earth. The Moon, a small, cold, dark body, is only 3500 km in diameter while the Sun, a self luminous, gaseous giant, is 1,400,000 km across. The coincidence arises from the fact that although the Sun is 400 times larger than the Moon, it's also 400 times farther from the Earth. A direct consequence of this fortuitous geometry is that during a total solar eclipse, the Moon occults the Sun with a nearly perfect fit.

The fundamental basis of the solar eclipse is the alignment of the Sun, Moon and the Earth such that some region of the Earth passes through the Moon's shadow. This shadow is composed of two parts: the outer or penumbral shadow and the inner or umbral shadow. From within the penumbra, only part of the Sun is obscured. In contrast, the dark, central umbra is the shadow of complete or total eclipse. During a total eclipse, the umbra sweeps across the Earth from west to east and the course it travels is called the path of totality. Anyone standing within this zone will see the Sun completely obscured by the Moon for as long as 7 1/2 minutes. At this time, the solar corona is visible as a halo about the Moon and the landscape takes on the appearance of an eerie twilight. Outside the path of totality but still within the penumbra, a partial eclipse is seen. The path of the umbra is rarely more than 300 km wide while that of the penumbra is about 7000 km wide. Sometimes the umbral shadow misses the Earth entirely and only a partial eclipse occurs.

Eclipse geometry is complicated by the fact that the Earth's orbit around the Sun is elliptical. As a result, the Sun's apparent semi-diameter varies from 944 arc-seconds at aphelion to 976 arc-seconds at perihelion. This 3% range in apparent size is, of course, quite indistinguishable to the naked eye.

However, the orbit of the Moon about the Earth is also elliptical. The Moon is 406,700 km from the Earth's center at apogee and 356,400 km at perigee. This 12% range in distance causes the Moon's apparent semi-diameter to vary between 882 and 1006 arc-seconds. Thus, during a total solar eclipse, the Moon's apparent diameter can exceed the Sun's by as much as 7% or 2 arc-minutes. Conversely, it's also possible for the Moon to appear almost 10% or 3 arc-minutes smaller than the Sun.

Such a geometry results in the case of an annular solar eclipse. When the Moon is near apogee, it's umbral shadow falls short of the

Earth and the Moon appears smaller than the Sun. An observer stationed in the path of totality would see the Moon completely silhouetted against the Sun's bright photosphere. This type of eclipse takes its name from the ring or annulus of sunlight which surrounds the Moon at maximum eclipse. Unfortunately, the blindingly bright photosphere hides the corona from view and precludes any measurements or photographs of the Sun's outer atmosphere.

The term "central eclipse" is reserved for any eclipse in which the axis of the Moon's shadow intersects the Earth. From the previous discussion, it's obvious that a central eclipse can be either total or annular in nature. When the Moon is at perigee and the Earth is at aphelion, the shadow extends 23,500 km beyond the Earth's center. However, when the Moon is at apogee and the Earth is at perihelion, the shadow falls 39,400 km short of the geocenter. These represent the extremes of the Moon's umbra (Figure 1). In the first case, the umbral cone can have a maximum diameter of 273 kilometers at the Earth. If the surface of the shadow cone in the second case is extended, it generates a negative or anti-umbra. As it strikes the Earth, the anti-umbra can have a maximum diameter of 313 km and an observer positioned there will see an annular eclipse.

A third type of central eclipse is possible which forms the transition between the annular and total eclipse. It occurs when the umbral shadow is just long enough to reach part of the Earth's surface, but not the Earth's center. In this case, the eclipse will be annular at either end of the eclipse path while it will be total along the middle section. The eclipses of 3 October 1986 and 30 March 1987 are two examples of the annular/total eclipse.

ECLIPSE FREQUENCY AND RECURRANCE

Having established the preliminary geometry for solar eclipses, a question immediately arises. Why doesn't a solar eclipse occur at every new Moon? Since the Moon cycles through its phases every 29 1/2 days or one synodic month, one would expect an eclipse to occur during each conjunction with the Sun. If the Moon's orbit around the Earth were in the same plane as the Earth's around the Sun, this is precisely what would happen. However, the Moon's orbit is inclined 5° to the Earth's. Our planet's natural satellite passes through the ecliptic only twice a month at a pair of points called the nodes (Figure 2). The rest of the time, the Moon is either above or below the plane of the Earth's orbit. Since an eclipse can only occur when the Sun, Moon and Earth lie in

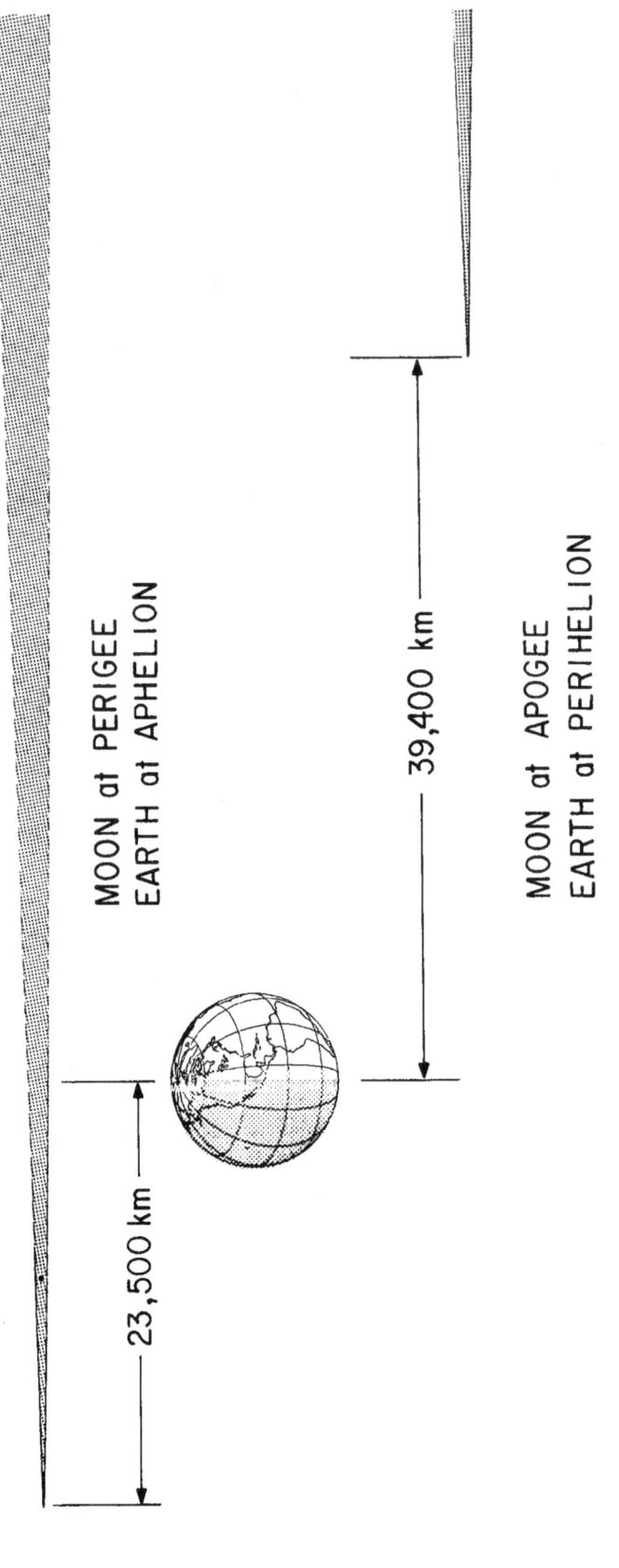

LUNAR SHADOW EXTREMES

MOON at PERIGEE
EARTH at APHELION

23,500 km

MOON at APOGEE
EARTH at PERIHELION

39,400 km

Figure 1

256

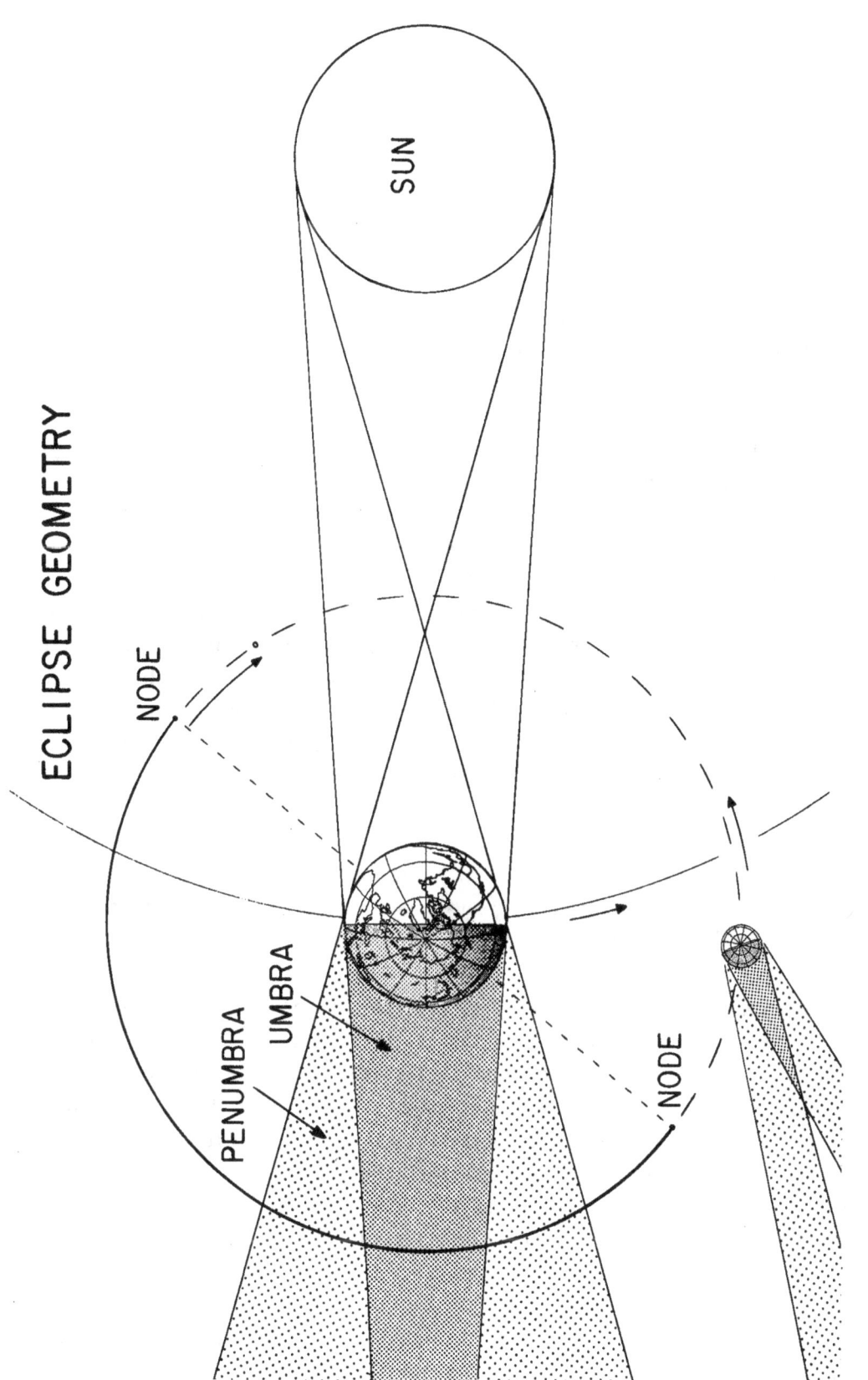

ECLIPSE GEOMETRY

SUN

NODE

PENUMBRA

UMBRA

NODE

Figure 2

257

the same plane, these conditions are met when new Moon takes place at one of the nodes.

Contrary to popular belief, solar eclipses are not at all rare. In fact, they're more common than lunar eclipses. An examination of the exterior tangents which delineate the Earth's umbra will substantiate this claim. An eclipse is possible only when the Moon is within that section of its orbit inscribed by the exterior tangents. However, the sunward arc of the Moon's orbit is clearly longer than the anti-sunward arc which passes through the shadow. The number of solar and lunar eclipses that occur are proportional to the lengths of these two arcs. Thus, solar eclipses out number lunar eclipses by almost 5 to 3. In this argument, the Earth's penumbral shadow has been ignored since penumbral lunar eclipses are essentially unobservable.

In any one calendar year, there are at least two and as many as five solar eclipses. On the other hand, there can be no more than three lunar eclipses per year and it's quite possible to have none at all. Combining both solar and lunar eclipses, it's possible for one calendar year to contain a maximum of seven eclipses. However, they can only occur in the combinations of five solar and two lunar or four solar and three lunar. In either case, the solar eclipses must all be partial. As a point of interest, 1982 happened to be one of the rare years containing seven eclipses. What made it even more remarkable was the fact that all three lunar eclipses were total. This will not happen again until the year 2485 AD.

The previous discussion contradicts common experience because lunar eclipses are observed more frequently than solar eclipses. The conflict is resolved since solar eclipses are only visible from isolated regions of the Earth while lunar eclipses are visible from the entire night time hemisphere of our planet.

An examination of the geometry of the nodes yields further clues on the subject of eclipse recurrence. Since the Sun and Moon both subtend significant angles, neither one has to be exactly at the nodes for an eclipse to occur. In addition, an observer's position on the surface of the Earth introduces a sizable parallax of 2° in ecliptic latitude. These factors make a solar eclipse possible whenever the Sun is within 18.5° of a node. The Sun travels along the ecliptic at about 1° per day and requires about 37 days to cross through the eclipse zone centered on each node. New Moon occurs every 29 1/2 days and thus guarantees at least one eclipse during each of the Sun's node crossings.

The period during which the Sun is near a node is called an eclipse season and there are two eclipse seasons each year. If the line of nodes were fixed in space, then eclipse seasons would occur six months apart and at the same time each year. Actually, the line of nodes slowly drifts westward at the rate of 19 degrees per year. As a result, eclipse seasons occur every 173.3 days. Two eclipse seasons constitute an eclipse year of 346.6 days. This is 18.6 days short of a solar year and is equal to the time required by the Sun to cross the same node twice.

In order to find a periodicity in the mechanics of solar eclipses, we must search for a commensurability between the synodic month and the eclipse year. Fortunately, 19 eclipse years are almost exactly equal to 223 synodic months; they differ by only 11 hours. The coincidence is all the more remarkable when compared to a period known as the anomalistic month. This is the time required for the Moon to pass from perigee to perigee and is approximately 27 1/2 days. The anomalistic month is important because the Moon's geocentric distance is the primary factor determining the annular or total nature of a solar eclipse. As unlikely as it may seem, 239 anomalistic months are also equal to 223 synodic months to within 6 hours.

This is the origin of the famous Saros cycle of 6585 1/3 days or 18 years, 11 days and 8 hours. Any two eclipses separated by one Saros cycle share very similar mechanical characteristics. They occur at the same node with the Moon at the same distance from Earth and at the same time of year. Because the Saros does not contain an integral number of days, its biggest drawback is that subsequent eclipses are visible from different parts of the globe. Although the 1/3 day displacement shifts the eclipse path 120° westward with each cycle, the series returns to the same geographic region every 3 Saroses or 56 years and 34 days.

A Saros series doesn't last indefinitely because the various periods are not perfectly commensurate with one another. In particular, 19 eclipse years are 1/2 day longer than the Saros. As a result, the node shifts eastward by about 0.5° with each cycle.

A typical Saros series begins when new Moon occurs about 18 degrees east of a node. If the first eclipse occurs at the Moon's decending node, the Moon's umbral shadow will pass 3500 km below the Earth and a partial eclipse will be visible from the south polar region. On the following return, the umbra will pass about 300 km closer to the Earth and a partial eclipse of slightly larger magnitude will result. After ten or eleven Saros cycles (about 200 years), the first central eclipse will occur near the south pole of the Earth. Over the course of the next 950 years, a central eclipse will occur at each Saros but will be displaced northward by an average of 300 km. Halfway through this period, eclipses of long duration will occur near the equator. The last central eclipse of the series will occur near the north pole. The next ten eclipses will be partial with successively smaller magnitudes. Finally, the Saros series will end some 13 centuries after it began at the opposite pole. A typical series may be comprised of 70 to 80 eclipses, about 50 of which are central. If a Saros series begins near the ascending node, the first eclipse will be partial from the northern polar region and the previous sequence of events is reversed.

Since at least two solar eclipses occur every year, there are obviously many different Saros series in progress simultaneously. For instance, during the later half of the twentieth century, there are 41 individual series and 26 of them are producing central eclipses. As old series terminate, new ones are always beginning and take their places.

To illustrate, the total solar eclipses of 1925, 1943, 1961, 1979, 1997, 2015 and 2033 are all members of Saros 120 (Figure 3). The series began with a partial eclipse at the south pole in 915 AD. The 2033 event is the last central eclipse of the series. Note that the paths of the last four eclipses grow progressively broader as the umbral shadow cone passes closer to the limb of the Earth. The next eclipse in the series will be a partial eclipse in 2051. Saros 120 will end with a partial eclipse near the north pole in 2195.

MODERN ECLIPSE PREDICTION

In order to predict the general characteristics as well as the local circumstances for a solar eclipse, high accuracy ephemerides for both the Sun and the Moon are required. Conventional ephemerides tabulate the positions and distances of these bodies with respect to the Earth's center. However, the eclipse calculator is primarily interested in the position, dimensions and velocity of the Moon's shadow on the Earth's surface.

In 1824, the Prussian astronomer and mathematician Friedrich Bessel introduced a new theory for the prediction of eclipses. It was so successful that it remains today as the most powerful technique, even with the application of the digital computer. What Bessel did was to express the ephemerides of the Sun and Moon in terms of the Moon's shadow. This change in the frame of reference greatly simplifies the mathematics and geometry without any sacrifice in accuracy.

To define the Besselian elements of an eclipse, a plane is passed through the center of the Earth which is fixed perpendicular to axis of the lunar shadow. This is called the fundamental plane and on it is

SOLAR ECLIPSES OF SAROS 120

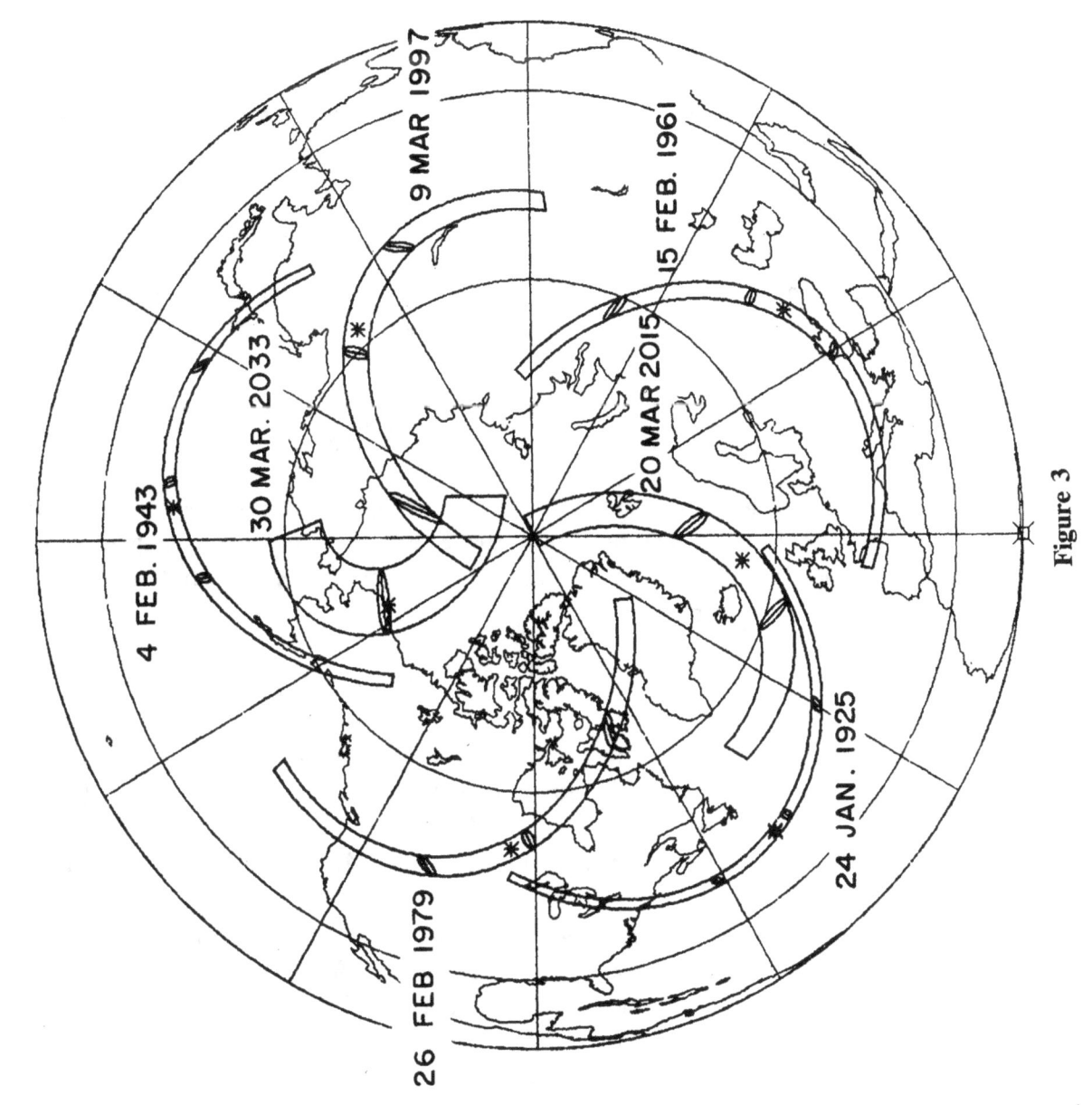

9 MAR 1997

15 FEB. 1961

20 MAR 2015

30 MAR. 2033

4 FEB. 1943

24 JAN. 1925

26 FEB 1979

Figure 3

constructed an X-Y rectangular coordinate system with its origin at the geocenter. The axes of this system are oriented with north in the positive Y direction and east in the positive X direction. The Z axis is perpendicular to the fundamental plane and parallel to the shadow axis.

The X-Y coordinates of the shadow axis can now be expressed in units of the equatorial radius of the Earth. The radii of the penumbral and umbral shadows on the fundamental plane are also tabulated as L_1 and L_2, respectively. The direction of the shadow axis on the celestial sphere is defined by its declination 'd' and ephemeris hour angle 'μ'.

Finally, the angles which the penumbral and umbral shadow cones make with the shadow axis are expressed as f_1 and f_2, respectively. These eight parameters, tabulated at hourly intervals serve as the only input needed to characterize an eclipse. The details of actual eclipse calculations can be found in Chauvenet [1891] or the Explanatory Supplement [1974].

GEOMETRY OF THE UMBRAL SHADOW

The length of the Moon's cone-shaped umbral shadow is a function of the Moon's distance from the Sun. For any given eclipse, it can be determined by first calculating the half-angle 'f_2' which the umbral cone makes with its axis:

$$\sin f_2 = (\sin s_0 - k \sin \pi_0) / D_{sm}$$

where: s_0 = semi-diameter of Sun at mean distance (= 959.63")
 π_0 = horizontal parallax of Sun at mean distance (= 8.794")
 k = mean radius of Moon in Earth radii (= 0.272281)
 D_{sm} = distance between the Moon and Sun in Earth radii

The distance between the Moon and the Sun 'D_{sm}' is readily calculated as the difference in distance between the Moon and Earth 'D_m' and the distance between the Sun and the Earth 'D_s' as:

$$D_{sm} = D_s - D_m$$

$$D_m = 1 / \tan \pi_m$$

and $$D_s = 1 / \tan \pi_s$$

where: D_m = distance of Moon from Earth in Earth radii
 D_s = distance of Sun from Earth in Earth radii

π_m = horizontal parallax of Moon (from Section 4)
π_s = horizontal parallax of Sun (from Section 4)

The length of the umbral shadow 'D_u' (in Earth radii) then follows as:

$$D_u = k \ / \ \tan f_2$$

An observer somewhere along the eclipse path can calculate their topocentric distance (in Earth radii) from the Moon 'D_t' as (approximately):

$$D_t = k \ / \ \tan (q \ s_s)$$

where: q = ratio of apparent diameters of the Moon and Sun
as seen by the observer (from Section 3)
s_s = geocentric semi-diameter of Sun (from Section 4)

Then, the radius of the umbral shadow 'r_u' seen by an observer at topocentric distance 'D_t' is just:

$$r_u = k \ (D_u\text{-}D_t) \ / \ D_u$$

The local geometry of the umbral shadow at any one point along the eclipse path is rather complicated and changes rapidly. Ignoring the curvature of the Earth's surface, the shadow is elliptical with its major axis aligned along the azimuth of the Sun; the minor axis is equal to the actual diameter of the umbral cone at that point 'r_u'. The eccentricity 'e' of the shadow is related to the Sun's altitude 'A' by:

$$e = \sqrt{1 - \sin^2 A}$$

The major and minor semi-axes (in kilometers) of the umbral shadow ('a' and 'b', respectively) can also be used to define the eccentricity as:

$$e = \sqrt{a^2 - b^2} \ / \ a$$

where: $b = r_e \ r_u$

$$a = b \ / \ \sin A$$

263

and r_e = equatorial radius of Earth (= 6378.388 km)

Unfortunately, the azimuth of the shadow's path is independent of the parameters describing its shape. As a result, the width of the path of totality (or annularity) can take on any value between the major and minor axes' dimensions. In order to determine the relationship between the distance from the center line and the duration of totality, it is necessary to adopt a different frame of reference which will simplify this geometry.

Figure 4 represents the appearance of the umbral shadow, seen from the Moon as it sweeps over the Earth's surface. Neglecting the tiny irregularities in the Moon's limb profile, the shadow is quite circular. If the shadow is moving from the left to the right, the center line runs along diameter 'D' and totality lasts 'T' seconds. A hypothetical observer at point 'P' is clearly located perpendicular distances 'd' from the center line and 'h' from the path edge. Such information is readily obtained from a large scale map on which the path of totality has been plotted. As the shadow passes over the observer, he will lie on chord 'C'. The key question is : how long will totality last at point 'P' ?

From plane geometry, the length of chord 'C' is related to a circle's diameter 'D' and perpendicular distance 'd' by:

$$C = \sqrt{D^2 - 4d^2}$$

But the length of chord 'C' is directly proportional to the duration of totality at any point on the chord. Therefore, the duration of totality 't' at point 'P' is just:

$$t = T \cdot C / D = T \cdot \sqrt{D^2 - 4d^2} / D \text{ seconds}$$

where 'D' is the width of the path, 'd' is the distance of the observer from the center line and 'T' is the duration on the center line.

Expressing the duration of totality as a function of distance from the path edge, the relation for chord 'C' becomes:

$$C = \sqrt{4h(D-h)}$$

and the duration of totality 't' is:

$$t = T \cdot C / D = T \cdot \sqrt{4h(D-h)} / D \text{ seconds}$$

264

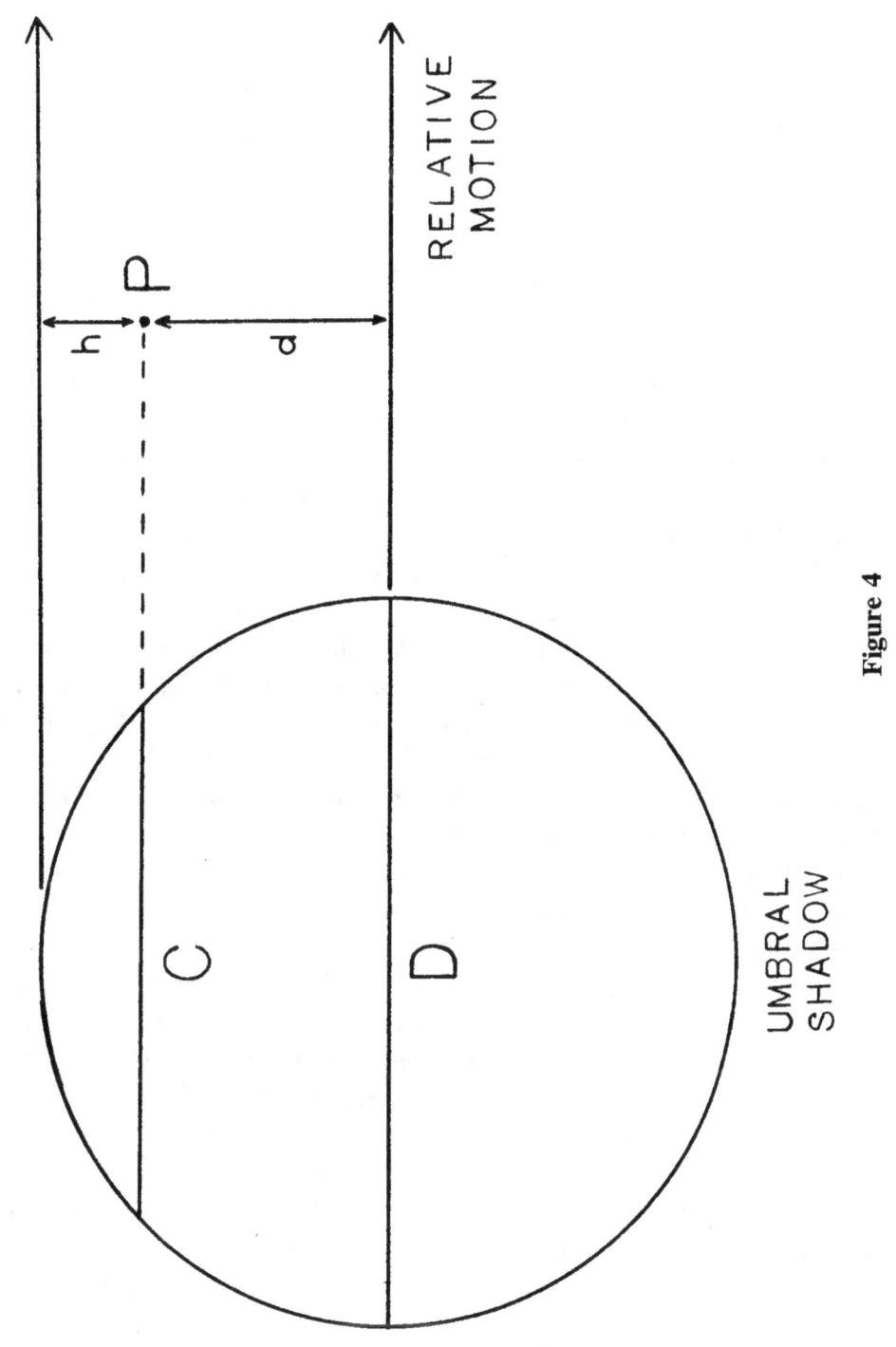

RELATIVE MOTION

P

h

d

C

D

UMBRAL SHADOW

Figure 4

265

It should be pointed out that observers within 5 to 10 kilometers of the path edge must consider the effects of the lunar limb profile in determining the duration of totality.

TIME DETERMINATION

The measurement of time is of fundamental importance to all branches of science, but to none more so than astronomy. In fact, astronomy was born through man's first attempts to measure the passage of time by observing the motions of the Sun and the Moon. It should come as no surprise then, that time reckoning remains intricately entwined with astronomy even today. However, the Sun's apparent motion no longer plays the pivotal role as the ultimate temporal yardstick. It's been known for thousands of years that the length of the solar day is not constant but varies with an annual cycle. What was not known before Kepler's time was that the Earth's elliptical orbit about the Sun, coupled with the inclination in the planet's axis were responsible for the periodic variations.

Mean Solar Time can be conceptualized as time kept by a fictitious or mean Sun which moves eastward along the celestial equator at the average rate of the true Sun. Greenwich Mean Time (GMT) or Universal Time (UT) is simply Mean Solar Time as measured from Greenwich, England and was used in navigation and surveying for hundreds of years. Unfortunately, this too has fallen by the wayside because the Earth does not turn on its axis at a uniform and constant rate. As the Earth spins, a tidal friction is imposed on it through the gravitational interaction with the Moon and, to a lesser extent, the Sun. This secular acceleration gradually transfers angular momentum from the Earth to the Moon. As the Earth loses energy and slows down, the Moon gains this energy and its distance from the Earth increases. In fact, the Moon's distance will eventually increase to the point where its angular diameter is always smaller than the Sun's and total solar eclipses will no longer be possible. Although still in its infancy, the technique of lunar laser ranging has shown that the Moon's average distance from the Earth is increasing by about four centimeters per year. At this rate, total solar eclipses will continue to occur for the next 750 million years.

It should be pointed out that the secular acceleration of the Moon is very poorly known and may not be constant. Careful records for its derivation only go back as far as 100 years or so. Before then, spurious and often incomplete eclipse observations from medieval and ancient manuscripts comprise the data base. In any case, the current value implies an increase in the length of the day by about 0.001 seconds per century. Such a trivially small amount may seem insignificant, but it has very measurable cummulative effects. In one century, the Earth loses 45 seconds, while in one millennium, the planet is one and a quarter hours "behind schedule".

The Earth's rotation on its axis is also subject to short term fluctuations for periods of up to several decades. It is believed that these fluctuations may be due to fluid motions in the Earth's core which interact with and disturb the rotation of the mantle. However, climatological changes and variations in sea-level may also play a significant role since they should alter the Earth's moment of inertia. Whatever the mechanism is, it is clear that its effects cannot be predicted with the current state of knowledge.

A better standard than diurnal rotation for the absolute measurement of time is the use of solar system dynamics. The orbital motions of the planets and of the Moon are predictable to very high accuracy and are directly verifiable through observations. The resulting time is referred to as Ephemeris Time (ET).

In 1957, the International Astronomical Union adopted Ephemeris Time as the standard and defined the ephemeris second as 1/31,556,925.9747 of the tropical year 1900 at January 0 at 12 hours Universal Time. The difference between Ephemeris Time and Universal Time (delta T or ΔT) is obtained through observations of the Moon. The Moon's position is predicted in terms of Ephemeris Time but it's observed with respect to Universal Time. Between 1900 and 1980, the slowing of the Earth's rotation on its axis had caused Universal Time to lag 50.54 seconds (= delta T) behind Ephemeris Time.

Ephemeris Time remained the basis of all time measurements until 1984. With the technological development of the atomic clock, a method of time measurement became available which has a permanence and stability unmatched by even celestial mechanics. The atomic or SI (for Systeme International) second is defined as 9,192,631,770 periods of the radiation corresponding to the transition between two hyperfine levels of the ground state of the Cesium 133 atom. The SI second was carefully chosen to agree as closely as possible to the ephemeris second. In 1984, the SI second was adopted as the newest time standard and Terrestrial Dynamical Time (TDT) replaced Ephemeris Time. For consistency, the time scale for Terrestrial Dynamical Time was chosen to agree with 1984 Ephemeris Time.

Solar eclipse predictions are now based on Terrestrial Dynamical Time but the position of the central eclipse path still depends on

Universal Time. Unfortunately, it's impossible to predict how delta T (i.e.: TDT-UT) will vary in the future. At best, the current trends can be extrapolated but the resulting values of delta T will inevitably diverge from actual observations. As such observations become available, corrections to the eclipse path longitudes can be calculated as follows:

$$\text{Shift (in degrees)} = 0.00417807 * (\Delta T1 - \Delta T2)$$

where: $\Delta T1$ = table value of delta T (in seconds)
$\Delta T2$ = true or observed delta T (in seconds).

This shift is added to the tabulated path longitudes to calculate the correct longitudes of the central path. Changes in the value of delta T have no effect on the tabulated latitudes.

FIFTY YEAR CANON OF SOLAR ECLIPSES : 1986 - 2035

APPENDIX B - **Program SUNECL**

```
                 1         2         3         4         5         6         7
        1234567890123456789012345678901234567890123456789012345678901234567890123456789012

001     C****PROGRAM : SUNECL
002     C****
003     C****PROGRAM SUNECL SEARCHES FOR ALL SOLAR ECLIPSES
004     C****OCCURRING WITHIN A GIVEN DATE INTERVAL.
005     C****THE GENERAL CHARACTERISTICS AND TIMES FOR EACH ECLIPSE ARE
006     C****THEN CALCULATED.
007     C****THE PREDICTED ECLIPSE CHARACTERISTICS ARE STORED IN
008     C****COMMON/ZERO/ WHERE :
009     C****   MONTH,IDAY,IYEAR - CALENDAR DATE OF ECLIPSE.
010     C****             ITYPE - TYPE OF ECLIPSE WHERE :
011     C****                 =0  -  NO ECLIPSE OCCURS.
012     C****                 =1  -  TOTAL SOLAR ECLIPSE.
013     C****                 =2  -  ANNULAR SOLAR ECLIPSE.
014     C****                 =3  -  PARTIAL SOLAR ECLIPSE.
015     C****                 =4  -  ANNULAR/TOTAL SOLAR ECLIPSE.
016     C****                 =5  -  PARTIAL ECLIPSE IS POSSIBLE.
017     C****               FJD - JULIAN DATE OF INSTANT OF GREATEST ECLIPSE.
018     C****             FTIME - TIME (TDT) OF GREATEST ECLIPSE.
019     C****             GAMMA - DISTANCE OF LUNAR SHADOW AXIS FROM EARTH'S
020     C****                     CENTER (IN UNITS OF EARTH RADII).
021     C****              ZMAG - MAGNITUDE OF ECLIPSE (FRACTION SUN OBSCURED).
022     C****                     (FOR TOTALS : DIAMETER RATIO OF SUN/MOON).
023     C****             T1,T2 - SEMI-DURATION OF PENUMBRAL AND UMBRAL PHASES.
024     C****WRITTEN BY F. ESPENAK - OCTOBER 1983.
025     C****LAST MODIFIED - NOV 1986.
026           IMPLICIT REAL*8(A-H,O-Z)
027           COMMON/ZERO/MONTH,IDAY,IYEAR,ITYPE,FJD,FTIME,DELTA,GAMMA,ZMAG,
028          1            NSAR,NCN,LNS,T1,T2
029           DATA SYNOD/29.530589D0/,K/0/
030     C****READ DATE INTERVALS OF ECLIPSE SEARCH.
031     C****    IM1,ID1,IY1 - MONTH, DAY AND YEAR OF START OF SEARCH INTERVAL.
032     C****    IM2,ID2,IY2 - MONTH, DAY AND YEAR OF END OF SEARCH INTERVAL.
033           READ(20,100) IM1,ID1,IY1,IM2,ID2,IY2
034       100 FORMAT(2I3,I5,2I3,I5)
035           IF(IY1.EQ.0.OR.IY2.EQ.0) GO TO 99
036     C****CONVERT GREGORIAN (CALENDAR) DATES TO JULIAN DATES.
037           CALL JULDAT(DJ1,IW1,ID1,IM1,IY1,0,0,0.0)
038           CALL JULDAT(DJ2,IW2,ID2,IM2,IY2,0,0,0.0)
039           IDAY=ID1
040           MONTH=IM1

        1234567890123456789012345678901234567890123456789012345678901234567890123456789012
                 1         2         3         4         5         6         7
```

```
                 1         2         3         4         5         6         7
        12345678901234567890123456789012345678901234567890123456789012345678901 2

041          IYEAR=IY1
042          FJD=DJ1-SYNOD
043          WRITE(6,200) IM1,ID1,IY1,IM2,ID2,IY2
044      200 FORMAT(/5X,'***** SOLAR ECLIPSE SEARCH FROM ',
045         1 I3,'/',I2,'/',I5,' TO ',I2,'/',I2,'/',I5/)
046  C****CALCULATE THE INSTANT OF NEW MOON SYZYGY AND DETERMINE WHETHER
047  C****AN ECLIPSE IS POSSIBLE.
048        1 XJD=FJD+SYNOD
049          IF(XJD.GT.DJ2) GO TO 99
050          CALL PRESEC(XJD)
051          IF(ITYPE.EQ.O.AND.DABS(GAMMA).GT.1.25) GO TO 1
052          IF(FJD.GT.DJ2) GO TO 99
053  C****IF ECLIPSE IS PREDICTED, PRINT ITS CHARACTERISTICS.
054          K=K+1
055          WRITE(6,210) MONTH,IDAY,IYEAR,FTIME,GAMMA,ZMAG,ITYPE,K
056      210 FORMAT(5X,'***** SOLAR ECLIPSE ON',I3,'/',I2,'/',I5,
057         1 4X,'TIME =',F6.2,' TDT',4X,'GAMMA =',F6.3,
058         2 4X,'MAG =',F6.3,4X,'ITYPE =',I2,4X,'K =',I2)
059          GO TO 1
060  C****EXIT PROGRAM ECLIPSE.
061       99 WRITE(6,220) K
062      220 FORMAT(/5X,'***** A TOTAL OF',I4,' ECLIPSES WERE PREDICTED FOR ',
063         1 'THIS DATE INTERVAL.'/)
064          STOP
065          END

        12345678901234567890123456789012345678901234567890123456789012345678901 2
                 1         2         3         4         5         6         7
```

272

```
                1           2           3           4           5           6           7
      12345678901234567890123456789012345678901234567890123456789012345678901234567890123456789012

001          SUBROUTINE PRESEC(EJD)
002   C****SUBROUTINE PRESEC PREDICTS THE INSTANT OF NEW MOON SYZYGY
003   C****NEAREST TO THE INPUT JULIAN DATE 'EJD'.
004   C****THIS CORRESPONDS TO THE INSTANT WHEN THE MOON'S SHADOW AXIS
005   C****PASSES NEAREST TO THE EARTH'S CENTER.
006   C****SUBROUTINE PRESEC THEN TESTS THE ALIGNMENT TO DETERMINE WHETHER
007   C****A SOLAR ECLIPSE WILL OCCUR AND CALCULATES ITS CHARACTERISTICS.
008   C****BASED ON ALGORITHMS FROM
009   C****   "ASTRONOMICAL FORMULAE FOR CALCULATORS", MEEUS, PAGES 153-160.
010   C****WRITTEN BY F. ESPENAK - MAR 1982, LAST MODIFIED - NOV 1986.
011          IMPLICIT REAL*8(A-H,O-Z)
012          COMMON/ZERO/MONTH,IDAY,IYEAR,ITYPE,FJD,FTIME,DELTA,GAMMA,ZMAG,
013         1            NSAR,NCN,LNS,IXX,T1,T2
014          DATA SYNOD/29.53058868D0/,PI/3.1415926535D0/
015          DATA DTR,RTD/0.017453292519943D0,57.2957795131D0/
016          KNT=0
017   C****CALCULATE TIME ELLAPSED IN LUNAR MONTHS SINCE FIRST NEW MOON
018   C****OF 1900. (I.E. - 1/1/1900 13:34:05ET)
019        1 KNT=KNT+1
020          Z=(EJD-2415021.065D0)/SYNOD
021          KP=IDINT(Z+0.5)
022          IF(Z.LT.0.0) KP=IDINT(Z-0.5)
023          P=DFLOAT(KP)
024          Q=SYNOD*P/36525.D0
025   C****CALCULATE JULIAN DATE OF MEAN PHASE.
026          PJD=2415020.75933D0+29.53058868D0*P+1.178D-04*Q*Q-1.55D-07*Q*Q*Q
027         1    +3.3D-04*DSIN(DTR*(166.56+132.87*Q-9.173D-03*Q*Q))
028   C****CALCULATE THE MEAN ANOMALIES OF THE SUN AND MOON.
029          ZM=359.2242D0+29.10535608D0*P-3.33D-05*Q*Q-3.47D-06*Q*Q*Q
030          XM=306.0253D0+385.81691806D0*P+1.07306D-02*Q*Q+1.236D-05*Q*Q*Q
031          ZM=DMOD(ZM,360.D0)
032          XM=DMOD(XM,360.D0)
033   C****CALCULATE THE MOON'S ARGUEMENT OF LATITUDE.
034          XF=21.2964D0+390.67050646*P-1.6528D-03*Q*Q-2.39D-06*Q*Q*Q
035          XF=DMOD(XF,360.D0)
036   C****CALCULATE DATE CORRECTION FOR ECLIPSE TEST.
037          EPC=+(0.1734-3.93D-04*Q)*DSIN(DTR*ZM)+0.0021*DSIN(DTR*(ZM+ZM))
038         1    -0.4068*DSIN(DTR*XM)+0.0161*DSIN(DTR*(XM+XM))
039         2    -0.0051*DSIN(DTR*(ZM+XM))-0.0074*DSIN(DTR*(ZM-XM))
040         3    -0.0104*DSIN(DTR*(XF+XF))

      12345678901234567890123456789012345678901234567890123456789012345678901234567890123456789012
                1           2           3           4           5           6           7
```

```
                1          2          3          4          5          6          7
       1234567890123456789012345678901234567890123456789012345678901234567890123456789012

041  C****CALCULATE INSTANT OF MAXIMUM ECLIPSE.
042        EJD=PJD+EPC
043        IF(KNT.LT.3) GO TO 1
044        FJD=EJD
045        CALL CALDAT(FJD,IW,NDAY,IDAY,MONTH,IYEAR,
046       1             IHR,MIN,ISEC,FTIME,FMIN,SEC)
047  C****CALCULATE SHADOW AXIS DISTANCE AT INSTANT OF MAXIMUM ECLIPSE.
048        S=+5.19595-0.0048*DCOS(DTR*ZM)+0.0020*DCOS(DTR*2.*ZM)
049       1  -0.3283*DCOS(DTR*XM)-0.0060*DCOS(DTR*(ZM+XM))
050       2  +0.0041*DCOS(DTR*(ZM-XM))
051        C=+0.2070*DSIN(DTR*ZM)+0.0024*DSIN(DTR*2.*ZM)-0.0390*DSIN(DTR*XM)
052       1  +0.0115*DSIN(DTR*2.*XM)-0.0073*DSIN(DTR*(ZM+XM))
053       2  -0.0067*DSIN(DTR*(ZM-XM))+0.0117*DSIN(DTR*2.*XF)
054        GAMMA=S*DSIN(DTR*XF)+C*DCOS(DTR*XF)
055  C****CALCULATE THE RADII OF THE UMBRAL AND PENUMBRAL SHADOWS.
056        ZL2=+0.0059+0.0046*DCOS(DTR*ZM)-0.0182*DCOS(DTR*XM)
057       1   +0.0004*DCOS(DTR*2.*XM)-0.0005*DCOS(DTR*(ZM+XM))
058        ZL1=ZL2+0.5460
059  C****DETERMINE TYPE OF SOLAR ECLIPSE AND ECLIPSE MAGNITUDE.
060  C****     ITYPE=0  -  NO ECLIPSE OCCURS.
061  C****     ITYPE=1  -  TOTAL SOLAR ECLIPSE.
062  C****     ITYPE=2  -  ANNULAR SOLAR ECLIPSE.
063  C****     ITYPE=3  -  PARTIAL SOLAR ECLIPSE.
064  C****     ITYPE=4  -  ANNULAR/TOTAL SOLAR ECLIPSE.
065  C****     ITYPE=5  -  PARTIAL ECLIPSE IS POSSIBLE.
066        ITYPE=0
067        NCN=1
068        LNS=-1
069        IF(GAMMA.GT.0.0) LNS=+1
070        T1=0.0
071        T2=0.0
072        ZMAG=0.0
073        AGAM=DABS(GAMMA)
074        IF(AGAM.GT.(1.5432+ZL2+0.2)) GO TO 999
075        ITYPE=5
076        IF(AGAM.LE.(1.5432+ZL2)) ITYPE=3
077        ZMAG=(1.5432+ZL2-AGAM)/(0.5460+2.*ZL2)
078        IF(AGAM.GT.(0.9972+DABS(ZL2))) GO TO 5
079        IF(AGAM.GT.(1.5432+ZL2)) GO TO 5
080  C****CALCULATE THE SOLAR RADIUS VECTOR (RAU) AND SEMI-DIAMETER (SDS).

       1234567890123456789012345678901234567890123456789012345678901234567890123456789012
                1          2          3          4          5          6          7
```

```
                    1         2         3         4         5         6         7
          12345678901234567890123456789012345678901234567890123456789012345678901234567890

081             T=(FJD-2415020.0D0)/36525.0D0
082             X=296.104608D0+477198.849109D0*T+9.1917D-03*T*T+1.439D-05*T*T*T
083             D=350.737486D0+445267.114217D0*T-1.4361D-03*T*T+1.889D-06*T*T*T
084             Z=358.47583D0+35999.04975D0*T-1.50D-04*T*T-3.3D-06*T*T*T
085             X=DTR*DMOD(X,360.D0)
086             D=DTR*DMOD(D,360.D0)
087             Z=DTR*DMOD(Z,360.D0)
088             ENOM=DATAN2(DSIN(Z),(DCOS(Z)-E))
089             RAU=1.0000002*(1.0-E*DCOS(ENOM))
090             SDS=959.63/RAU
091       C****CALCULATE THE MOON'S EQUATORIAL HORIZONTAL PARALLAX.
092       C****(FROM "ASTRONOMICAL FORMULAE FOR CALCULATORS", MEEUS, PAGE 141)
093             PIM=0.950724+0.051818*DCOS(X)
094            1     +0.009531*DCOS(D+D-X)+0.007843*DCOS(D+D)
095            2     +0.002824*DCOS(X+X)+0.000857*DCOS(D+D+X)
096             DZ=0.0
097             IF(DABS(GAMMA).LT.1.0) DZ=DSQRT(1.0-GAMMA**2)
098             PIM=2.*RTD*DATAN(1.0/((1.0/DTAN(0.5*DTR*PIM))-DZ))
099             SDM=0.272476D0*3600.0*PIM
100       C****CALCULATE MAGNITUDE FOR CENTRAL ECLIPSES.
101             NCN=0
102             ZMAG=SDM/SDS
103             IF(ZL2.LE.0.0) ITYPE=1
104             IF(ZL2.GT.0.0) ITYPE=2
105             IF(ZL2.GT.0.0) ZMAG=-ZMAG
106             W=DSIGN(0.5*PI,GAMMA)
107             IF(DABS(GAMMA).LT.1.0) W=DASIN(GAMMA)
108             OMEGA=0.00464*DCOS(W)
109             IF(ZL2.GT.0.0.AND.ZL2.LT.0.0047.AND.OMEGA.GT.ZL2) ITYPE=4
110             IF(AGAM.GT.0.9972) NCN=+1
111             IF(AGAM.LE.0.9972) LNS=0
112       C****CALCULATE THE SOLAR ECLIPSE SEMI-DURATIONS.
113           5 S1=1.00+ZL1
114             S2=1.00+ZL2
115             ZN=0.5458+0.0400*DCOS(DTR*XM)
116             IF(DABS(S1).GT.DABS(GAMMA)) T1=DSQRT(S1*S1-GAMMA*GAMMA)/ZN
117             IF(DABS(S2).GT.DABS(GAMMA)) T2=DSQRT(S2*S2-GAMMA*GAMMA)/ZN
118       C****EXIT SUBROUTINE PRESEC.
119         999 RETURN
120             END

          12345678901234567890123456789012345678901234567890123456789012345678901234567890
                    1         2         3         4         5         6         7
```

```
                1           2           3           4           5           6           7
       1234567890123456789012345678901234567890123456789012345678901234567890123456789012

001            SUBROUTINE JULDAT(DJ,IW,ID,IM,IY,IHOUR,IMIN,SEC)
002    C****SUBROUTINE JULDAT COMPUTES THE JULIAN DECIMAL DATE (DJ) FROM
003    C****THE GREGORIAN (OR JULIAN) CALENDAR DATE.
004    C****THE GREGORIAN CALENDAR REFORM OCCURRED ON 1582 OCT 15.
005    C****THIS IS 1582 OCT 5 BY THE JULIAN CALENDAR.
006    C****INPUT :          ID,IM,IY - DAY,MONTH,YEAR.
007    C****            IHR,IMIN,SEC - HOUR,MINUTE,SECOND.
008    C****OUTPUT :          DJ - JULIAN DECIMAL DATE
009    C****                     (= 0 FOR B.C. 4713 JAN 1, 12 GMT).
010    C****                  IW - DAY OF WEEK (1=SUNDAY).
011    C****REFERENCE : "ASTRONOMICAL FORMULAE FOR CALCULATORS", MEEUS, P.23.
012    C****WRITTEN BY F. ESPENAK - APRIL 1982.
013    C****LAST MODIFIED - APRIL 1982.
014            REAL*8 DJ,SEC,FRAC,GYR
015    C****CALCULATE DECIMAL DAY FRACTION.
016            FRAC=DFLOAT(IHOUR)/24.+DFLOAT(IMIN)/1440.+SEC/86400.
017    C****CONVERT DATE TO FORMAT YYYY.MMDDdd
018            GYR=DFLOAT(IY)+0.01*DFLOAT(IM)+0.0001*DFLOAT(ID)+0.0001*FRAC
019          1   +1.0D-09
020    C****CALCULATE CONVERSION FACTORS.
021            IYO=IY
022            IMO=IM
023            IF(IM.LE.2) IYO=IY-1
024            IF(IM.LE.2) IMO=IM+12
025            IA=IYO/100
026            IB=2-IA+IA/4
027    C****CALCULATE JULIAN DATE.
028            JD=IDINT(365.25D0*IYO)+IDINT(30.6001D0*(IMO+1))+ID+1720994
029            IF(IY.LT.0) JD=IDINT(365.25D0*IYO-0.75)+IDINT(30.6001D0*(IMO+1))
030          1                +ID+1720994
031            IF(GYR.GE.1582.1015D0) JD=JD+IB
032            DJ=DFLOAT(JD)+FRAC+0.5D0
033    C****CALCULATE DAY OF WEEK.
034            JD=IDINT(DJ+0.5)
035            IW=JMOD((JD+1),7)+1
036            RETURN
037            END

       1234567890123456789012345678901234567890123456789012345678901234567890123456789012
                1           2           3           4           5           6           7
```

```
                1         2         3         4         5         6         7
       12345678901234567890123456789012345678901234567890123456789012345678901234567890
```
```
001          SUBROUTINE CALDAT(DJ,IW,ND,ID,IM,IY,IHR,IMIN,ISEC,AHR,AMIN,ASEC)
002    C****SUBROUTINE CALDAT CALCULATES THE DAY OF THE WEEK, THE DAY OF
003    C****THE YEAR, THE GREGORIAN (OR JULIAN) CALENDAR DATE AND
004    C****THE UNIVERSAL TIME FROM THE JULIAN DECIMAL DATE.
005    C****THE GREGORIAN CALENDAR REFORM OCCURRED ON 1582 OCT 15.
006    C****THIS IS 1582 OCT 5 BY THE JULIAN CALENDAR.
007    C****INPUT :             DJ - JULIAN DECIMAL DATE
008    C****                         (= 0 FOR B.C. 4713 JAN 1, 12 GMT).
009    C****OUTPUT :            IW - DAY OF THE WEEK (1=SUNDAY).
010    C****                    ND - DAY OF THE YEAR (1 JAN = 1).
011    C****            ID,IM,IY - CALENDAR DAY,MONTH,YEAR.
012    C****     IHR,IMIN,ISEC - INTEGER HOUR,MINUTE,SECOND.
013    C****      AHR,AMIN,ASEC - DECIMAL HOUR,MINUTE,SECOND.
014    C****REFERENCE : "ASTRONOMICAL FORMULAE FOR CALCULATORS", MEEUS, P.23.
015    C****WRITTEN BY F. ESPENAK - APRIL 1982.
016    C****LAST MODIFIED - 22 JULY 1986.
017          REAL*8 DJ,FRAC,AHR,AMIN,ASEC
018    C****CALCULATE INTERGER JULIAN DATE.
019          JD=IDINT(DJ+0.5)
020    C****CALCULATE DAY FRACTION.
021          FRAC=DJ+0.5-DFLOAT(JD)+1.0D-10
022    C****CALCULATE CONVERSION FACTORS.
023          KA=JD
024          IF(JD.LT.2299161) GO TO 10
025          IALP=IDINT((JD-1867216.25D0)/36524.25D0)
026          KA=JD+1+IALP-IALP/4
027       10 KB=KA+1524
028          KC=IDINT((KB-122.1)/365.25D0)
029          KD=IDINT(365.25D0*KC)
030          KE=IDINT((KB-KD)/30.6001D0)
031    C****CALCULATE THE CALENDAR DAY, MONTH AND YEAR.
032          ID=KB-KD-IDINT(30.6001D0*KE)
033          IM=KE-1
034          IF(KE.GT.13) IM=KE-13
035          IF(IM.EQ.2.AND.ID.GT.28) ID=29
036          IY=KC-4715
037          IF(IM.GT.2) IY=KC-4716
038          IF(IM.EQ.2.AND.ID.EQ.29.AND.KE.EQ.3) IY=KC-4716
039    C****CALCULATE THE UNIVERSAL TIME FROM THE FRACTIONAL DAY.
040          AHR=FRAC*24.
```
```
       12345678901234567890123456789012345678901234567890123456789012345678901234567890
                1         2         3         4         5         6         7
```

```
                    1         2         3         4         5         6         7
          1234567890123456789012345678901234567890123456789012345678901234567890123456789012

041               IHR=AHR
042               AMIN=(AHR-IHR)*60.
043               IMIN=AMIN
044               ASEC=(AMIN-IMIN)*60.
045               ISEC=ASEC
046       C****CALCULATE THE DAY OF THE WEEK.
047               IW=JMOD((JD+1),7)+1
048       C****CALCULATE THE DAY OF THE YEAR.
049               LYR=4*(IY/4)
050               ND=(275*IM)/9-2*((IM+9)/12)+ID-30
051               IF(IY.EQ.LYR) ND=(275*IM)/9-((IM+9)/12)+ID-30
052       C       WRITE(6,200) IM,ID,IY,JD,IALP,KA,KB,KC,KD,KE
053         200 FORMAT(I4,'/',I2,'/',I5,2X,I8,I4,
054             1 2X,'KA-E =',5I8)
055               RETURN
056               END

          1234567890123456789012345678901234567890123456789012345678901234567890123456789012
                    1         2         3         4         5         6         7
```